Proceedings of the
Fourteenth Lake Louise Winter Institute

ELECTROWEAK

PHYSICS

Proceedings of the
Fourteenth Lake Louise Winter Institute

ELECTROWEAK

PHYSICS

Lake Louise, Alberta, Canada; 14–20 February 1999

Editors

A Astbury
B A Campbell
F C Khanna
J Pinfold
M G Vincter

 **World Scientific**
Singapore • New Jersey • London • Hong Kong

Published by

World Scientific Publishing Co. Pte. Ltd.

P O Box 128, Farrer Road, Singapore 912805

USA office: Suite 1B, 1060 Main Street, River Edge, NJ 07661

UK office: 57 Shelton Street, Covent Garden, London WC2H 9HE

British Library Cataloguing-in-Publication Data
A catalogue record for this book is available from the British Library.

ELECTROWEAK PHYSICS

ISBN 981-02-4068-6

Printed in Singapore by Regal Press (S) Pte. Ltd.

PREFACE

The fourteenth annual Lake Louise Winter Institute, entitled "Electroweak Physics", was held from February 14–20, 1999 at the Chateau Lake Louise located in the scenic Canadian Rocky Mountains. The format of the Winter Institute was three days of pedagogical lectures followed by a topical workshop of short contributed talks. As usual, the sessions were scheduled in the mornings and in the evenings leaving the afternoons free to enjoy the many aspects of the Rocky Mountains covered in winter white.

This Institute primarily focused on the electroweak interactions of matter from both an experimental and theoretical point of view while covering a wide spectrum of energies. A detailed discussion of low energy tests of electroweak interactions was presented to the participants. Recent experimental results from data taken at the LEP accelerator at CERN demonstrated the incredible precision to which the Standard Model electroweak theory is known. A series of lectures cleanly laid out what could be expected from physics beyond the Standard Model. A detailed survey of neutrino physics was presented along with some very recent and exciting results on the neutrino mass. CP-violation and rare decays of kaons and B-mesons were well presented in the context of the existing electroweak theory and the most recent experimental measurements in this field. Contributed talks on a variety of topics supplemented the pedagogical talks.

We wish to express our most sincere gratitude to Audrey Schaapman and Lee Grimard for their efforts, patience, and incredible organizational skills in bringing this Winter Institute to fruition from the very first email to the publication of these proceedings. They were well aided by David Shaw and Jeff de Jong in the logistics and transportation of the participants to the Institute. We are indebted to them.

Finally, we wish to acknowledge the generous financial support of the University of Alberta, the Institute of Particle Physics, and TRIUMF.

organizing committee: A. Astbury
B.A. Campbell
F.C. Khanna
J.L. Pinfold
M.G. Vincter

CONTENTS

CP VIOLATION AND RARE DECAYS OF K AND B MESONS

ANDREJ J. BURAS

Technische Universität München
Physik Department
D-85748 Garching, Germany

These lectures describe CP violation and rare decays of K and B mesons and consist of ten chapters: i) Grand view of the field including CKM matrix and the unitarity triangle, ii) General aspects of the theoretical framework, iii) Particle-antiparticle mixing and CP violation, iv) Standard analysis of the unitarity triangle, v) The ratio ε'/ε including most recent developments, vi) Rare decays $K^+ \to \pi^+\nu\bar\nu$ and $K_L \to \pi^0\nu\bar\nu$, vii) Express review of other rare decays, viii) Express review of CP violation in B decays, ix) A brief look beyond the Standard Model including connections between ε'/ε and CP violating rare K decays, and x) Final messages.

1 Grand View

1.1 Preface

CP violation and rare decays of K and B mesons play an important role in the tests of the Standard Model and of its extensions. The prime examples, already observed experimentally, are $K^0 - \bar K^0$ and $B_d^0 - \bar B_d^0$ mixings, CP violation in $K_L \to \pi\pi$ and the rare decays $B \to X\gamma$, $K_L \to \mu\bar\mu$ and $K^+ \to \pi^+\nu\bar\nu$. In the coming years CP violation in B decays, $B_s^0 - \bar B_s^0$ mixing and rare decays $K_L \to \pi^0\nu\bar\nu$, $K_L \to \pi^0 e^+e^-$, $B \to X_{s,d}l^+l^-$, $B_{d,s} \to l^+l^-$ and $B \to X_{s,d}\nu\bar\nu$ will hopefully be included in this list.

These lectures provide a non-technical description of this fascinating field. There is unavoidably an overlap with my Les Houches lectures [1] and with the reviews [2,3]. On the other hand new developments are included and all numerical results updated.

1.2 Some Facts about the Standard Model

Throughout these lectures we will dominantly work in the context of the Standard Model with three generations of quarks and leptons and the interactions described by the gauge group $SU(3)_C \otimes SU(2)_L \otimes U(1)_Y$ spontaneously broken to $SU(3)_C \otimes U(1)_Q$. There are excellent text books on the dynamics of the Standard Model. Let us therfore collect here only those ingredients of this model which are fundamental for the subject of weak decays.

- The strong interactions are mediated by eight gluons G_a, and the electroweak interactions by $W^\pm$, Z^0 and γ.

- Concerning *Electroweak Interactions*, the left-handed leptons and quarks are put into $SU(2)_L$ doublets:

$$\begin{pmatrix} \nu_e \\ e^- \end{pmatrix}_L \quad \begin{pmatrix} \nu_\mu \\ \mu^- \end{pmatrix}_L \quad \begin{pmatrix} \nu_\tau \\ \tau^- \end{pmatrix}_L \qquad (1.1)$$

$$\begin{pmatrix} u \\ d' \end{pmatrix}_L \quad \begin{pmatrix} c \\ s' \end{pmatrix}_L \quad \begin{pmatrix} t \\ b' \end{pmatrix}_L \qquad (1.2)$$

  with the corresponding right-handed fields transforming as singlets under $SU(2)_L$. The primes in (1.2) will be discussed in a moment.

- The charged current processes mediated by $W^\pm$ are flavour violating with the strength of violation given by the gauge coupling g_2 and effectively at low energies by the Fermi constant

$$\frac{G_F}{\sqrt{2}} = \frac{g_2^2}{8M_W^2} \qquad (1.3)$$

  and a *unitary* 3×3 CKM matrix.

- The CKM matrix [4,5] connects the *weak eigenstates* (d', s', b') and the corresponding *mass eigenstates* d, s, b through

$$\begin{pmatrix} d' \\ s' \\ b' \end{pmatrix} = \begin{pmatrix} V_{ud} & V_{us} & V_{ub} \\ V_{cd} & V_{cs} & V_{cb} \\ V_{td} & V_{ts} & V_{tb} \end{pmatrix} \begin{pmatrix} d \\ s \\ b \end{pmatrix} = \hat{V}_{CKM} \begin{pmatrix} d \\ s \\ b \end{pmatrix}. \qquad (1.4)$$

  In the leptonic sector the analogous mixing matrix is a unit matrix due to the masslessness of neutrinos in the Standard Model.

- The unitarity of the CKM matrix assures the absence of flavour changing neutral current transitions at the tree level. This means that the elementary vertices involving neutral gauge bosons (G_a, Z^0, γ) are flavour conserving. This property is known under the name - GIM mechanism [6].

- The fact that the V_{ij}'s can a priori be complex numbers allows CP violation in the Standard Model [5].

1.3 CKM Matrix

General Remarks

We know from the text books that the CKM matrix can be parametrized by three angles and a single complex phase. This phase leading to an imaginary part of the CKM matrix is a necessary ingredient to describe CP violation within the framework of the Standard Model.

Many parametrizations of the CKM matrix have been proposed in the literature. We will use two parametrizations in these lectures: the standard parametrization[7] recommended by the Particle Data Group[8] and the Wolfenstein parametrization[9].

Standard Parametrization

With $c_{ij} = \cos\theta_{ij}$ and $s_{ij} = \sin\theta_{ij}$ $(i, j = 1, 2, 3)$, the standard parametrization is given by:

$$\hat{V}_{CKM} = \begin{pmatrix} c_{12}c_{13} & s_{12}c_{13} & s_{13}e^{-i\delta} \\ -s_{12}c_{23} - c_{12}s_{23}s_{13}e^{i\delta} & c_{12}c_{23} - s_{12}s_{23}s_{13}e^{i\delta} & s_{23}c_{13} \\ s_{12}s_{23} - c_{12}c_{23}s_{13}e^{i\delta} & -s_{23}c_{12} - s_{12}c_{23}s_{13}e^{i\delta} & c_{23}c_{13} \end{pmatrix},$$
(1.5)

where δ is the phase necessary for CP violation. c_{ij} and s_{ij} can all be chosen to be positive and δ may vary in the range $0 \le \delta \le 2\pi$. However, the measurements of CP violation in K decays force δ to be in the range $0 < \delta < \pi$.

From phenomenological applications we know that s_{13} and s_{23} are small numbers: $\mathcal{O}(10^{-3})$ and $\mathcal{O}(10^{-2})$, respectively. Consequently to an excellent accuracy $c_{13} = c_{23} = 1$ and the four independent parameters are given as

$$s_{12} = |V_{us}|, \quad s_{13} = |V_{ub}|, \quad s_{23} = |V_{cb}|, \quad \delta. \tag{1.6}$$

The first three can be extracted from tree level decays mediated by the transitions $s \to u$, $b \to u$ and $b \to c$ respectively. The phase δ can be extracted from CP violating transitions or loop processes sensitive to $|V_{td}|$. The latter fact is based on the observation that for $0 \le \delta \le \pi$, as required by the analysis of CP violation in the K system, there is a one–to–one correspondence between δ and $|V_{td}|$ given by

$$|V_{td}| = \sqrt{a^2 + b^2 - 2ab\cos\delta}, \quad a = |V_{cd}V_{cb}|, \quad b = |V_{ud}V_{ub}|. \tag{1.7}$$

The main phenomenological advantages of (1.5) over other parametrizations proposed in the literature are basically these two:

- s_{12}, s_{13} and s_{23} being related in a very simple way to $|V_{us}|$, $|V_{ub}|$ and $|V_{cb}|$ respectively, can be measured independently in three decays.

- The CP violating phase is always multiplied by the very small s_{13}. This shows clearly the suppression of CP violation independently of the actual size of δ.

4

For numerical evaluations the use of the standard parametrization is quite strongly recommended. However once the four parameters in (1.6) have been determined it is often useful to make a change of basic parameters in order to see the structure of the result more transparently. This brings us to the Wolfenstein parametrization [9] and its generalization given in ref. [10].

Wolfenstein Parameterization

The Wolfenstein parametrization is an approximate parametrization of the CKM matrix in which each element is expanded as a power series in the small parameter $\lambda = |V_{us}| = 0.22$,

$$\hat{V} = \begin{pmatrix} 1 - \frac{\lambda^2}{2} & \lambda & A\lambda^3(\varrho - i\eta) \\ -\lambda & 1 - \frac{\lambda^2}{2} & A\lambda^2 \\ A\lambda^3(1 - \varrho - i\eta) & -A\lambda^2 & 1 \end{pmatrix} + \mathcal{O}(\lambda^4), \qquad (1.8)$$

and the set (1.6) is replaced by

$$\lambda, \qquad A, \qquad \varrho, \qquad \eta. \qquad (1.9)$$

Because of the smallness of λ and the fact that for each element the expansion parameter is actually λ^2, it is sufficient to keep only the first few terms in this expansion.

The Wolfenstein parametrization is certainly more transparent than the standard parametrization. However, if one requires sufficient level of accuracy, the higher order terms in λ have to be included in phenomenological applications. This can be done in many ways. The point is that since (1.8) is only an approximation. The *exact* definition of the parameters in (1.9) is not unique by terms of the neglected order $\mathcal{O}(\lambda^4)$. This situation is familiar from any perturbative expansion, where different definitions of expansion parameters (coupling constants) are possible. This is also the reason why in different papers in the literature different $\mathcal{O}(\lambda^4)$ terms in (1.8) can be found. They simply correspond to different definitions of the parameters in (1.9). Since the physics does not depend on a particular definition, it is useful to make a choice for which the transparency of the original Wolfenstein parametrization is not lost. Here we present one way of achieving this.

Wolfenstein Parametrization beyond LO

An efficient and systematic way of finding higher order terms in λ is to go back to the standard parametrization (1.5) and to *define* the parameters $(\lambda, A, \varrho, \eta)$

through [10,11]

$$s_{12} = \lambda, \qquad s_{23} = A\lambda^2, \qquad s_{13}e^{-i\delta} = A\lambda^3(\varrho - i\eta) \qquad (1.10)$$

to *all orders* in λ. It follows that

$$\varrho = \frac{s_{13}}{s_{12}s_{23}}\cos\delta, \qquad \eta = \frac{s_{13}}{s_{12}s_{23}}\sin\delta. \qquad (1.11)$$

Eq's. (1.10) and (1.11) represent simply the change of variables from (1.6) to (1.9). Making this change of variables in the standard parametrization (1.5) we find the CKM matrix as a function of $(\lambda, A, \varrho, \eta)$ which satisfies unitarity exactly. Expanding next each element in powers of λ we recover the matrix in (1.8) and in addition find explicit corrections of $\mathcal{O}(\lambda^4)$ and higher order terms:

$$V_{ud} = 1 - \frac{1}{2}\lambda^2 - \frac{1}{8}\lambda^4 + \mathcal{O}(\lambda^6) \qquad (1.12)$$

$$V_{us} = \lambda + \mathcal{O}(\lambda^7), \qquad V_{ub} = A\lambda^3(\varrho - i\eta) \qquad (1.13)$$

$$V_{cd} = -\lambda + \frac{1}{2}A^2\lambda^5[1 - 2(\varrho + i\eta)] + \mathcal{O}(\lambda^7) \qquad (1.14)$$

$$V_{cs} = 1 - \frac{1}{2}\lambda^2 - \frac{1}{8}\lambda^4(1 + 4A^2) + \mathcal{O}(\lambda^6) \qquad (1.15)$$

$$V_{cb} = A\lambda^2 + \mathcal{O}(\lambda^8), \qquad V_{tb} = 1 - \frac{1}{2}A^2\lambda^4 + \mathcal{O}(\lambda^6) \qquad (1.16)$$

$$V_{td} = A\lambda^3\left[1 - (\varrho + i\eta)(1 - \frac{1}{2}\lambda^2)\right] + \mathcal{O}(\lambda^7) \qquad (1.17)$$

$$V_{ts} = -A\lambda^2 + \frac{1}{2}A(1 - 2\varrho)\lambda^4 - i\eta A\lambda^4 + \mathcal{O}(\lambda^6). \qquad (1.18)$$

We note that by definition V_{ub} remains unchanged and the corrections to V_{us} and V_{cb} appear only at $\mathcal{O}(\lambda^7)$ and $\mathcal{O}(\lambda^8)$, respectively. Consequently to an an excellent accuracy we have:

$$V_{us} = \lambda, \qquad V_{cb} = A\lambda^2, \qquad (1.19)$$

$$V_{ub} = A\lambda^3(\varrho - i\eta), \qquad V_{td} = A\lambda^3(1 - \bar{\varrho} - i\bar{\eta}) \qquad (1.20)$$

with

$$\bar{\varrho} = \varrho(1 - \frac{\lambda^2}{2}), \qquad \bar{\eta} = \eta(1 - \frac{\lambda^2}{2}). \qquad (1.21)$$

The advantage of this generalization of the Wolfenstein parametrization over other generalizations found in the literature is the absence of relevant corrections to V_{us}, V_{cb} and V_{ub} and an elegant change in V_{td} which allows a simple generalization of the so-called unitarity triangle beyond LO.

Finally let us collect useful approximate analytic expressions for $\lambda_i = V_{id}V_{is}^*$ with $i = c, t$:

$$\text{Im}\lambda_t = -\text{Im}\lambda_c = \eta A^2 \lambda^5 = \mid V_{ub} \parallel V_{cb} \mid \sin\delta \qquad (1.22)$$

$$\text{Re}\lambda_c = -\lambda(1 - \frac{\lambda^2}{2}) \qquad (1.23)$$

$$\text{Re}\lambda_t = -(1 - \frac{\lambda^2}{2})A^2\lambda^5(1 - \bar{\varrho}). \qquad (1.24)$$

Expressions (1.22) and (1.23) represent to an accuracy of 0.2% the exact formulae obtained using (1.5). The expression (1.24) deviates by at most 2% from the exact formula in the full range of parameters considered. For ϱ close to zero this deviation is below 1%. After inserting the expressions (1.22)–(1.24) in the exact formulae for quantities of interest, a further expansion in λ should not be made.

Unitarity Triangle

The unitarity of the CKM-matrix implies various relations between its elements. In particular, we have

$$V_{ud}V_{ub}^* + V_{cd}V_{cb}^* + V_{td}V_{tb}^* = 0. \qquad (1.25)$$

Phenomenologically this relation is very interesting as it involves simultaneously the elements V_{ub}, V_{cb} and V_{td} which are under extensive discussion at present.

The relation (1.25) can be represented as a "unitarity" triangle in the complex $(\bar{\varrho}, \bar{\eta})$ plane. The invariance of (1.25) under any phase-transformations implies that the corresponding triangle is rotated in the $(\bar{\varrho}, \bar{\eta})$ plane under such transformations. Since the angles and the sides (given by the moduli of the elements of the mixing matrix) in these triangles remain unchanged, they are phase convention independent and are physical observables. Consequently they can be measured directly in suitable experiments. The area of the unitarity triangle is related to the measure of CP violation J_{CP} [12,13]:

$$\mid J_{\text{CP}} \mid = 2 \cdot A_\Delta, \qquad (1.26)$$

where A_Δ denotes the area of the unitarity triangle.

The construction of the unitarity triangle proceeds as follows:

- We note first that
$$V_{cd}V_{cb}^* = -A\lambda^3 + \mathcal{O}(\lambda^7).\tag{1.27}$$
Thus to an excellent accuracy $V_{cd}V_{cb}^*$ is real with $|V_{cd}V_{cb}^*| = A\lambda^3$.

- Keeping $\mathcal{O}(\lambda^5)$ corrections and rescaling all terms in (1.25) by $A\lambda^3$ we find

$$\frac{1}{A\lambda^3}V_{ud}V_{ub}^* = \bar{\varrho} + i\bar{\eta}, \qquad \frac{1}{A\lambda^3}V_{td}V_{tb}^* = 1 - (\bar{\varrho} + i\bar{\eta})\tag{1.28}$$

with $\bar{\varrho}$ and $\bar{\eta}$ defined in (1.21).

- Thus we can represent (1.25) as the unitarity triangle in the complex $(\bar{\varrho}, \bar{\eta})$ plane as shown in Fig. 1.

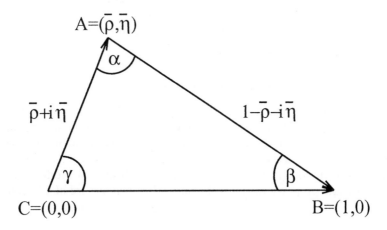

Figure 1: Unitarity Triangle.

Let us collect useful formulae related to this triangle:

- Using simple trigonometry one can express $\sin(2\phi_i)$, $\phi_i = \alpha, \beta, \gamma$, in terms of $(\bar{\varrho}, \bar{\eta})$ as follows:

$$\sin(2\alpha) = \frac{2\bar{\eta}(\bar{\eta}^2 + \bar{\varrho}^2 - \bar{\varrho})}{(\bar{\varrho}^2 + \bar{\eta}^2)((1 - \bar{\varrho})^2 + \bar{\eta}^2)}\tag{1.29}$$

$$\sin(2\beta) = \frac{2\bar{\eta}(1 - \bar{\varrho})}{(1 - \bar{\varrho})^2 + \bar{\eta}^2}\tag{1.30}$$

$$\sin(2\gamma) = \frac{2\bar{\varrho}\bar{\eta}}{\bar{\varrho}^2 + \bar{\eta}^2} = \frac{2\varrho\eta}{\varrho^2 + \eta^2}. \tag{1.31}$$

- The lengths CA and BA in the rescaled triangle to be denoted by R_b and R_t, respectively, are given by

$$R_b \equiv \frac{|V_{ud}V_{ub}^*|}{|V_{cd}V_{cb}^*|} = \sqrt{\bar{\varrho}^2 + \bar{\eta}^2} = (1 - \frac{\lambda^2}{2})\frac{1}{\lambda}\left|\frac{V_{ub}}{V_{cb}}\right| \tag{1.32}$$

$$R_t \equiv \frac{|V_{td}V_{tb}^*|}{|V_{cd}V_{cb}^*|} = \sqrt{(1 - \bar{\varrho})^2 + \bar{\eta}^2} = \frac{1}{\lambda}\left|\frac{V_{td}}{V_{cb}}\right|. \tag{1.33}$$

- The angles β and γ of the unitarity triangle are related directly to the complex phases of the CKM-elements V_{td} and V_{ub}, respectively, through

$$V_{td} = |V_{td}|e^{-i\beta}, \quad V_{ub} = |V_{ub}|e^{-i\gamma}. \tag{1.34}$$

- The angle α can be obtained through the relation

$$\alpha + \beta + \gamma = 180° \tag{1.35}$$

expressing the unitarity of the CKM-matrix.

The triangle depicted in Fig. 1 together with $|V_{us}|$ and $|V_{cb}|$ gives a full description of the CKM matrix. Looking at the expressions for R_b and R_t, we observe that within the Standard Model the measurements of four CP *conserving* decays sensitive to $|V_{us}|$, $|V_{ub}|$, $|V_{cb}|$ and $|V_{td}|$ can tell us whether CP violation ($\eta \neq 0$) is predicted in the Standard Model. This is a very remarkable property of the Kobayashi-Maskawa picture of CP violation: quark mixing and CP violation are closely related to each other.

1.4 Grand Picture

What do we know about the CKM matrix and the unitarity triangle on the basis of *tree level* decays? A detailed answer to this question can be found in the reports of the Particle Data Group [8] as well as other reviews [14,15], where references to the relevant experiments and related theoretical work can be found. In particular we have

$$|V_{us}| = \lambda = 0.2205 \pm 0.0018, \quad |V_{cb}| = 0.040 \pm 0.002, \tag{1.36}$$

$$\frac{|V_{ub}|}{|V_{cb}|} = 0.089 \pm 0.016, \quad |V_{ub}| = (3.56 \pm 0.56) \cdot 10^{-3}. \tag{1.37}$$

Using (1.19) and (1.32) we find then

$$A = 0.826 \pm 0.041, \qquad R_b = 0.39 \pm 0.07 . \qquad (1.38)$$

This tells us only that the apex A of the unitarity triangle lies in the band shown in Fig. 2. In order to answer the question where the apex A lies on this "unitarity clock" we have to look at different decays. Most promising in this respect are the so-called "loop induced" decays and transitions which are the subject of several sections in these lectures and CP asymmetries in B-decays which will be briefly discussed in Sec. 8. These two different routes

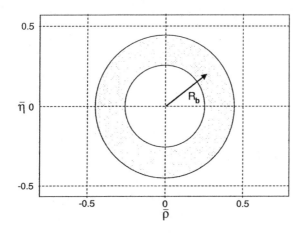

Figure 2: "Unitarity Clock".

for explorations of the CKM matrix and of the related unitarity triangle may answer the important question: whether the Kobayashi-Maskawa picture of CP violation is correct and more generally whether the Standard Model offers a correct description of weak decays of hadrons. Indeed, in order to answer these important questions it is essential to calculate as many branching ratios as possible, measure them experimentally and check if they all can be described by the same set of the parameters $(\lambda, A, \varrho, \eta)$. In the language of the unitarity triangle this means that the various curves in the $(\bar{\varrho}, \bar{\eta})$ plane extracted from different decays should cross each other at a single point as shown in Fig. 3.

Moreover the angles (α, β, γ) in the resulting triangle should agree with those extracted one day from CP-asymmetries in B-decays. For artistic reasons the value of $\bar{\eta}$ in Fig. 3 has been chosen to be higher than the fitted central value $\bar{\eta} \approx 0.35$.

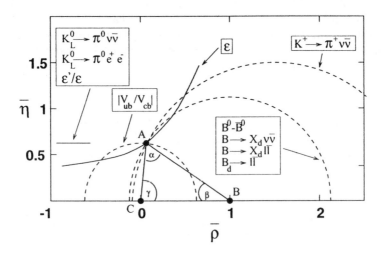

Figure 3: The ideal Unitarity Triangle.

On the other hand if new physics contributes to weak decays, the different curves based on the Standard Model expressions will not cross each other at a single point and the angles (α, β, γ) extracted one day from CP-asymmetries in B-decays will disagree with the ones determined from rare K and B decays. Clearly the plot in Fig. 3 is highly idealized because in order to extract such nice curves from various decays one needs perfect experiments and perfect theory. One of the goals of these lectures is to identify those decays for which at least the theory is under control. For such decays, if they can be measured with a sufficient precision, the curves in Fig. 3 are not fully unrealistic. Let us then briefly discuss the theoretical framework for weak decays.

2   Theoretical Framework

2.1   OPE and Renormalization Group

The basis for any serious phenomenology of weak decays of hadrons is the *Operator Product Expansion* (OPE) [16,17], which allows to write the effective weak Hamiltonian simply as follows

$$\mathcal{H}_{eff} = \frac{G_F}{\sqrt{2}} \sum_i V^i_{\text{CKM}} C_i(\mu) Q_i . \tag{2.1}$$

Here G_F is the Fermi constant and Q_i are the relevant local operators which govern the decays in question. They are built out of quark and lepton fields. The Cabibbo-Kobayashi-Maskawa factors V^i_{CKM}[4,5] and the Wilson coefficients C_i[16] describe the strength with which a given operator enters the Hamiltonian. An amplitude for a decay of a given meson $M = K, B, ..$ into a final state $F = \pi\nu\bar{\nu}$, $\pi\pi$, DK is then simply given by

$$A(M \to F) = \langle F|\mathcal{H}_{eff}|M\rangle = \frac{G_F}{\sqrt{2}} \sum_i V^i_{CKM} C_i(\mu)\langle F|Q_i(\mu)|M\rangle, \tag{2.2}$$

where $\langle F|Q_i(\mu)|M\rangle$ are the hadronic matrix elements of Q_i between M and F.

The essential virtue of OPE is this one. It allows to separate the problem of calculating the amplitude $A(M \to F)$ into two distinct parts: the *short distance* (perturbative) calculation of the coefficients $C_i(\mu)$ and the *long-distance* (generally non-perturbative) calculation of the matrix elements $\langle Q_i(\mu)\rangle$. The scale μ separates the physics contributions into short distance contributions contained in $C_i(\mu)$ and the long distance contributions contained in $\langle Q_i(\mu)\rangle$. Thus C_i include the top quark contributions and contributions from other heavy particles such as W, Z-bosons and charged Higgs particles or supersymmetric particles in the supersymmetric extensions of the Standard Model. Consequently $C_i(\mu)$ depend generally on m_t and also on the masses of new particles if extensions of the Standard Model are considered. This dependence can be found by evaluating so-called *box* and *penguin* diagrams with full W- Z-, top- and new particles exchanges and *properly* including short distance QCD effects. The latter govern the μ-dependence of $C_i(\mu)$.

The value of μ can be chosen arbitrarily but the final result must be μ-independent. Therefore the μ-dependence of $C_i(\mu)$ has to cancel the μ-dependence of $\langle Q_i(\mu)\rangle$. In other words it is a matter of choice what exactly belongs to $C_i(\mu)$ and what to $\langle Q_i(\mu)\rangle$. This cancellation of μ-dependence involves generally several terms in the expansion in (2.2). The coefficients $C_i(\mu)$ depend also on the renormalization scheme. This scheme dependence must also be cancelled by the one of $\langle Q_i(\mu)\rangle$ so that the physical amplitudes are renormalization scheme independent. Again, as in the case of the μ-dependence, the cancellation of the renormalization scheme dependence involves generally several terms in the expansion (2.2).

Although μ is in principle arbitrary, it is customary to choose μ to be of the order of the mass of the decaying hadron. This is $\mathcal{O}(m_b)$ and $\mathcal{O}(m_c)$

for B-decays and D-decays respectively. In the case of K-decays the typical choice is $\mu = \mathcal{O}(1 - 2\ GeV)$ instead of $\mathcal{O}(m_K)$, which is much too low for any perturbative calculation of the couplings C_i. Now due to the fact that $\mu \ll M_{W,Z}$, m_t, large logarithms $\ln M_W/\mu$ compensate in the evaluation of $C_i(\mu)$. The smallness of the QCD coupling constant α_s and terms $\alpha_s^n(\ln M_W/\mu)^n$, $\alpha_s^n(\ln M_W/\mu)^{n-1}$ etc ... have to be resummed to all orders in α_s before a reliable result for C_i can be obtained. This can be done very efficiently by means of the renormalization group methods. The resulting *renormalization group improved* perturbative expansion for $C_i(\mu)$ in terms of the effective coupling constant $\alpha_s(\mu)$ does not involve large logarithms and is more reliable.

All this looks rather formal but in fact should be familiar. Indeed, in the simplest case of the β-decay, $\mathcal{H}_{eff}$ takes the familiar form

$$\mathcal{H}_{eff}^{(\beta)} = \frac{G_F}{\sqrt{2}} \cos\theta_c [\bar{u}\gamma_\mu(1 - \gamma_5)d \otimes \bar{e}\gamma^\mu(1 - \gamma_5)\nu_e]\,, \qquad (2.3)$$

where V_{ud} has been expressed in terms of the Cabibbo angle. In this particular case the Wilson coefficient is equal to unity and the local operator, the object between the square brackets, is given by a product of two $V - A$ currents. Eq. (2.3) represents the Fermi theory for β-decays as formulated by Sudarshan and Marshak[18] and Feynman and Gell-Mann[19] forty years ago, except that in (2.3) the quark language has been used and following Cabibbo a small departure of V_{ud} from unity has been incorporated. In this context the basic formula (2.1) can be regarded as a generalization of the Fermi Theory to include all known quarks and leptons as well as their strong and electroweak interactions as summarized by the Standard Model.

Due to the interplay of electroweak and strong interactions the structure of the local operators is much richer than in the case of the β-decay. They can be classified with respect to the Dirac structure, colour structure and the type of quarks and leptons relevant for a given decay. Of particular interest are the operators involving quarks only. In the case of the $\Delta S = 1$ transitions the relevant set of operators is given as follows:

Current–Current :

$$Q_1 = (\bar{s}_\alpha u_\beta)_{V-A}\,(\bar{u}_\beta d_\alpha)_{V-A}\,, \quad Q_2 = (\bar{s}u)_{V-A}\,(\bar{u}d)_{V-A} \qquad (2.4)$$

QCD–Penguins :

$$Q_3 = (\bar{s}d)_{V-A} \sum_{q=u,d,s} (\bar{q}q)_{V-A}\,, \quad Q_4 = (\bar{s}_\alpha d_\beta)_{V-A} \sum_{q=u,d,s} (\bar{q}_\beta q_\alpha)_{V-A} \qquad (2.5)$$

$$Q_5 = (\bar{s}d)_{V-A} \sum_{q=u,d,s} (\bar{q}q)_{V+A}, \ Q_6 = (\bar{s}_\alpha d_\beta)_{V-A} \sum_{q=u,d,s} (\bar{q}_\beta q_\alpha)_{V+A} \qquad (2.6)$$

Electroweak–Penguins :

$$Q_7 = \frac{3}{2}(\bar{s}d)_{V-A} \sum_{q=u,d,s} e_q (\bar{q}q)_{V+A}, \ Q_8 = \frac{3}{2}(\bar{s}_\alpha d_\beta)_{V-A} \sum_{q=u,d,s} e_q(\bar{q}_\beta q_\alpha)_{V+A}$$

$$(2.7)$$

$$Q_9 = \frac{3}{2}(\bar{s}d)_{V-A} \sum_{q=u,d,s} e_q(\bar{q}q)_{V-A}, \ Q_{10} = \frac{3}{2}(\bar{s}_\alpha d_\beta)_{V-A} \sum_{q=u,d,s} e_q (\bar{q}_\beta q_\alpha)_{V-A}.$$

$$(2.8)$$

Here, e_q denotes the electric quark charges reflecting the electroweak origin of $Q_7, \ldots, Q_{10}$.

Clearly, in order to calculate the amplitude $A(M \to F)$, the matrix elements $\langle Q_i(\mu) \rangle$ have to be evaluated. Since they involve long distance contributions one is forced in this case to use non-perturbative methods such as lattice calculations, the $1/N$ expansion (N is the number of colours), QCD sum rules, hadronic sum rules, chiral perturbation theory and so on. In the case of certain B-meson decays, the *Heavy Quark Effective Theory* (HQET) also turns out to be a useful tool. Needless to say, all these non-perturbative methods have some limitations. Consequently the dominant theoretical uncertainties in the decay amplitudes reside in the matrix elements $\langle Q_i(\mu) \rangle$.

The fact that in most cases the matrix elements $\langle Q_i(\mu) \rangle$ cannot be reliably calculated at present is very unfortunate. One of the main goals of the experimental studies of weak decays is the determination of the CKM factors V_{CKM} and the search for the physics beyond the Standard Model. Without a reliable estimate of $\langle Q_i(\mu) \rangle$ this goal cannot be achieved unless these matrix elements can be determined experimentally or removed from the final measurable quantities by taking the ratios or suitable combinations of amplitudes or branching ratios. However, this can be achieved only in a handful of decays and generally one has to face directly the calculation of $\langle Q_i(\mu) \rangle$. We will discuss these issues later on.

2.2 Inclusive Decays

So far I have discussed only *exclusive* decays. It turns out that in the case of *inclusive* decays of heavy mesons, like B-mesons, things turn out to be easier. In an inclusive decay one sums over all (or over a special class) of accessible final states so that the amplitude for an inclusive decay takes the form:

$$A(B \to X) = \frac{G_F}{\sqrt{2}} \sum_{f \in X} V_{\text{CKM}}^i C_i(\mu) \langle f|Q_i(\mu)|B \rangle . \qquad (2.9)$$

At first sight things look as complicated as in the case of exclusive decays. It turns out, however, that the resulting branching ratio can be calculated in the expansion in inverse powers of m_b with the leading term described by the spectator model in which the B-meson decay is modelled by the decay of the b-quark:

$$\text{Br}(B \to X) = \text{Br}(b \to q) + \mathcal{O}(\frac{1}{m_b^2}) \ . \tag{2.10}$$

This formula is known under the name of Heavy Quark Expansion (HQE) [20]. Since the leading term in this expansion represents the decay of the quark, it can be calculated in perturbation theory or more correctly in the renormalization group improved perturbation theory. It should be realized that also here the basic starting point is the effective Hamiltonian (2.1) and that the knowledge of $C_i(\mu)$ is essential for the evaluation of the leading term in (2.10). But there is an important difference relative to the exclusive case: the matrix elements of the operators Q_i can be "effectively" evaluated in perturbation theory. This means, in particular, that their μ and renormalization scheme dependences can be evaluated and the cancellation of these dependences by those present in $C_i(\mu)$ can be investigated.

Clearly in order to complete the evaluation of $Br(B \to X)$ also the remaining terms in (2.10) have to be considered. These terms are of a non-perturbative origin, but fortunately they are suppressed by at least two powers of m_b. They have been studied by several authors in the literature with the result that they affect various branching ratios by less then 10% and often by only a few percent. Consequently the inclusive decays give generally more precise theoretical predictions at present than the exclusive decays. On the other hand their measurements are harder. There are of course some important theoretical issues related to the validity of HQE in (2.10) which appear in the literature under the name of quark-hadron duality. Since these matters are rather involved I will not discuss them here.

2.3   Status of NLO Calculations

In order to achieve sufficient precision for the theoretical predictions it is desirable to have accurate values of $C_i(\mu)$. Indeed it has been realized at the end of the eighties that the leading term (LO) in the renormalization group improved perturbation theory, in which the terms $\alpha_s^n (\ln M_W/\mu)^n$ are summed, is generally insufficient and the inclusion of next-to-leading corrections (NLO) which correspond to summing the terms $\alpha_s^n (\ln M_W/\mu)^{n-1}$ is necessary. In particular, unphysical left-over μ-dependences in the decay amplitudes and branching ratios resulting from the truncation of the perturbative series are considerably reduced by including NLO corrections. These corrections are known by now for the most important and interesting decays and will be taken into account in these lectures. The review of all existing NLO calculations can be found in ref's [2,21].

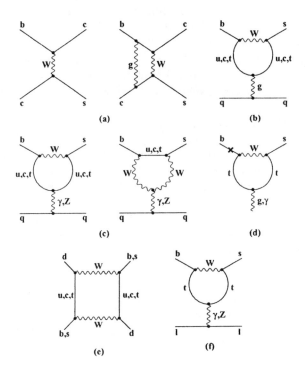

Figure 4: Typical Penguin and Box Diagrams.

2.4   Penguin–Box Expansion

The rare and CP violating decays of K and B mesons are governed by various penguin and box diagrams with internal top quark and charm quark exchanges. Some examples are shown in Fig. 4. Evaluating these diagrams one finds a set of basic universal (process independent) m_t-dependent functions $F_r(x_t)$ [22] where $x_t = m_t^2/M_W^2$. Explicit expressions for these functions will be given below.

It is useful to express the OPE formula (2.2) directly in terms of the functions $F_r(x_t)$ [23]:

$$A(M \to F) = P_0(M \to F) + \sum_r P_r(M \to F)\, F_r(x_t), \qquad (2.11)$$

where the sum runs over all possible functions contributing to a given amplitude. P_0 summarizes contributions stemming from internal quarks other than the top, in particular the charm quark. The coefficients P_0 and P_r are process dependent and include QCD corrections. They depend also on hadronic matrix elements of local operators and the relevant CKM factors. I would like to call (2.11) *Penguin-Box Expansion* (PBE). We will encounter many examples of PBE in the course of these lectures.

Originally PBE was designed to expose the m_t-dependence of FCNC processes [23]. After the top quark mass has been measured precisely this role of PBE is less important. On the other hand, PBE is very well suited for the study of the extentions of the Standard Model in which new particles are exchanged in the loops. If there are no new local operators the mere change is to modify the functions $F_r(x_t)$ which now acquire the dependence on the masses of new particles such as charged Higgs particles and supersymmetric particles. The process dependent coefficients P_0 and P_r remain unchanged. The effects of new physics can be then transparently seen. However, if new effective operators with different Dirac and colour structures are present the values of P_0 and P_r are modified.

Let us denote by B_0, C_0 and D_0 the functions $F_r(x_t)$ resulting from $\Delta F = 1$ (F stands for flavour) box diagram, Z^0-penguin and γ-penguin diagram respectively. These diagrams are gauge dependent and it is useful to introduce gauge independent combinations [23]

$$X_0 = C_0 - 4B_0, \qquad Y_0 = C_0 - B_0, \qquad Z_0 = C_0 + \frac{1}{4}D_0. \qquad (2.12)$$

Then the set of gauge independent basic functions which govern the FCNC processes in the Standard Model is given to a very good approximation as follows:

$$S_0(x_t) = 2.46 \left(\frac{m_t}{170\,\text{GeV}}\right)^{1.52}, \qquad S_0(x_c) = x_c \qquad (2.13)$$

$$S_0(x_c, x_t) = x_c \left[\ln\frac{x_t}{x_c} - \frac{3x_t}{4(1 - x_t)} - \frac{3x_t^2 \ln x_t}{4(1 - x_t)^2}\right]. \qquad (2.14)$$

$$X_0(x_t) = 1.57 \left(\frac{m_t}{170\,\text{GeV}}\right)^{1.15}, \qquad Y_0(x_t) = 1.02 \left(\frac{m_t}{170\,\text{GeV}}\right)^{1.56}, \qquad (2.15)$$

$$Z_0(x_t) = 0.71 \left(\frac{m_t}{170\,\text{GeV}}\right)^{1.86}, \qquad E_0(x_t) = 0.26 \left(\frac{m_t}{170\,\text{GeV}}\right)^{-1.02}, \qquad (2.16)$$

$$D_0'(x_t) = 0.38 \left(\frac{m_t}{170\,\text{GeV}}\right)^{0.60}, \qquad E_0'(x_t) = 0.19 \left(\frac{m_t}{170\,\text{GeV}}\right)^{0.38}. \quad (2.17)$$

The first three functions correspond to $\Delta F = 2$ box diagrams with (t, t), (c, c) and (t, c) exchanges. E_0 results from QCD penguin diagram with off-shell gluon, D_0' and E_0' from γ and QCD penguins with on-shell photons and gluons respectively. The subscript "0" indicates that these functions do not include QCD corrections to the relevant penguin and box diagrams.

In the range $150\,\text{GeV} \leq m_t \leq 200\,\text{GeV}$ these approximations reproduce the exact expressions to an accuracy of better than 1%. These formulae will allow us to exhibit elegantly the m_t dependence of various branching ratios in the phenomenological sections of these lectures. Exact expressions for all functions can be found in ref. [1].

Generally, several basic functions contribute to a given decay, although decays exist which depend only on a single function. We have the following correspondence between the most interesting FCNC processes and the basic functions:

$K^0 - \bar{K}^0$-mixing	$S_0(x_t)$, $S_0(x_c, x_t)$, $S_0(x_c)$
$B^0 - \bar{B}^0$-mixing	$S_0(x_t)$
$K \to \pi\nu\bar{\nu}$, $B \to X_{d,s}\nu\bar{\nu}$	$X_0(x_t)$
$K_\text{L} \to \mu\bar{\mu}$, $B \to l\bar{l}$	$Y_0(x_t)$
$K_\text{L} \to \pi^0 e^+ e^-$	$Y_0(x_t)$, $Z_0(x_t)$, $E_0(x_t)$
ε'	$X_0(x_t)$, $Y_0(x_t)$, $Z_0(x_t)$, $E_0(x_t)$
$B \to X_s\gamma$	$D_0'(x_t)$, $E_0'(x_t)$
$B \to X_s\mu^+\mu^-$	$Y_0(x_t)$, $Z_0(x_t)$, $E_0(x_t)$, $D_0'(x_t)$, $E_0'(x_t)$.

3   Particle-Antiparticle Mixing and CP Violation

3.1   Preliminaries

Let us next discuss particle–antiparticle mixing which in the past has been of fundamental importance in testing the Standard Model and often has proven to be an undefeatable challenge for suggested extensions of this model. Let us just recall that from the calculation of the $K_\text{L} - K_\text{S}$ mass difference, Gaillard and Lee [24] were able to estimate the value of the charm quark mass before charm discovery. On the other hand $B_d^0 - \bar{B}_d^0$ mixing [25] gave the first indication of a large top quark mass. Finally, particle–antiparticle mixing in the $K^0 - \bar{K}^0$ system offers within the Standard Model a plausible description of CP violation in $K_L \to \pi\pi$ discovered in 1964 [26].

In this section we will predominantly discuss the parameter ε describing the *indirect* CP violation in the K system and the mass differences $\Delta M_{d,s}$

which describe the size of $B^0_{d,s} - \bar{B}^0_{d,s}$ mixings. In the Standard Model these phenomena appear first at the one–loop level and as such they are sensitive measures of the top quark couplings $V_{ti}(i = d, s, b)$ and and in particular of the phase $\delta = \gamma$. They allow then to construct the unitarity triangle.

Let us next enter some details. The following subsection borrows a lot from ref's [27,28]. A nice review of CP violation can also be found in ref. [29].

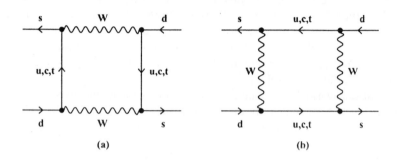

Figure 5: Box diagrams contributing to $K^0 - \bar{K}^0$ mixing in the Standard Model.

3.2 Express Review of $K^0 - \bar{K}^0$ Mixing

$K^0 = (\bar{s}d)$ and $\bar{K}^0 = (s\bar{d})$ are flavour eigenstates which in the Standard Model may mix via weak interactions through the box diagrams in Fig. 5. We will choose the phase conventions so that

$$CP|K^0\rangle = -|\bar{K}^0\rangle, \qquad CP|\bar{K}^0\rangle = -|K^0\rangle. \tag{3.1}$$

In the absence of mixing the time evolution of $|K^0(t)\rangle$ is given by

$$|K^0(t)\rangle = |K^0(0)\rangle \exp(-iHt) , \qquad H = M - i\frac{\Gamma}{2} , \tag{3.2}$$

where M is the mass and Γ the width of K^0. Similar formula for $\bar{K}^0$ exists.

On the other hand, in the presence of flavour mixing the time evolution of the $K^0 - \bar{K}^0$ system is described by

$$i\frac{d\psi(t)}{dt} = \hat{H}\psi(t), \qquad \psi(t) = \begin{pmatrix} |K^0(t)\rangle \\ |\bar{K}^0(t)\rangle \end{pmatrix} \tag{3.3}$$

where

$$\hat{H} = \hat{M} - i\frac{\hat{\Gamma}}{2} = \begin{pmatrix} M_{11} - i\frac{\Gamma_{11}}{2} & M_{12} - i\frac{\Gamma_{12}}{2} \\ M_{21} - i\frac{\Gamma_{21}}{2} & M_{22} - i\frac{\Gamma_{22}}{2} \end{pmatrix} \tag{3.4}$$

with $\hat{M}$ and $\hat{\Gamma}$ being hermitian matrices having positive (real) eigenvalues in analogy with M and Γ. M_{ij} and Γ_{ij} are the transition matrix elements from virtual and physical intermediate states respectively. Using

$$M_{21} = M_{12}^* , \qquad \Gamma_{21} = \Gamma_{12}^* , \qquad \text{(hermiticity)} \qquad (3.5)$$

$$M_{11} = M_{22} \equiv M , \qquad \Gamma_{11} = \Gamma_{22} \equiv \Gamma , \qquad \text{(CPT)} \qquad (3.6)$$

we have

$$\hat{H} = \begin{pmatrix} M - i\frac{\Gamma}{2} & M_{12} - i\frac{\Gamma_{12}}{2} \\ M_{12}^* - i\frac{\Gamma_{12}^*}{2} & M - i\frac{\Gamma}{2} \end{pmatrix} . \qquad (3.7)$$

We can next diagonalize the system to find:

Eigenstates:

$$K_{L,S} = \frac{(1 + \bar{\varepsilon})K^0 \pm (1 - \bar{\varepsilon})\bar{K}^0}{\sqrt{2(1 + |\bar{\varepsilon}|^2)}} \qquad (3.8)$$

where $\bar{\varepsilon}$ is a small complex parameter given by

$$\frac{1 - \bar{\varepsilon}}{1 + \bar{\varepsilon}} = \sqrt{\frac{M_{12}^* - i\frac{1}{2}\Gamma_{12}^*}{M_{12} - i\frac{1}{2}\Gamma_{12}}} = \frac{\Delta M - i\frac{1}{2}\Delta\Gamma}{2M_{12} - i\Gamma_{12}} \equiv r \exp(i\kappa) . \qquad (3.9)$$

with $\Delta\Gamma$ and ΔM given below.

Eigenvalues:

$$M_{L,S} = M \pm \mathrm{Re}Q \qquad \Gamma_{L,S} = \Gamma \mp 2\mathrm{Im}Q \qquad (3.10)$$

where

$$Q = \sqrt{(M_{12} - i\frac{1}{2}\Gamma_{12})(M_{12}^* - i\frac{1}{2}\Gamma_{12}^*)}. \qquad (3.11)$$

Consequently we have

$$\Delta M = M_L - M_S = 2\mathrm{Re}Q \qquad \Delta\Gamma = \Gamma_L - \Gamma_S = -4\mathrm{Im}Q. \qquad (3.12)$$

It should be noted that the mass eigenstates K_S and K_L differ from CP eigenstates

$$K_1 = \frac{1}{\sqrt{2}}(K^0 - \bar{K}^0), \qquad CP|K_1\rangle = |K_1\rangle , \qquad (3.13)$$

$$K_2 = \frac{1}{\sqrt{2}}(K^0 + \bar{K}^0), \qquad CP|K_2\rangle = -|K_2\rangle , \qquad (3.14)$$

by a small admixture of the other CP eigenstate:

$$K_S = \frac{K_1 + \bar{\varepsilon} K_2}{\sqrt{1 + |\bar{\varepsilon}|^2}}, \qquad K_L = \frac{K_2 + \bar{\varepsilon} K_1}{\sqrt{1 + |\bar{\varepsilon}|^2}}. \tag{3.15}$$

It should be stressed that the small parameter $\bar{\varepsilon}$ depends on the phase convention chosen for K^0 and $\bar{K}^0$. Therefore it may not be taken as a physical measure of CP violation. On the other hand $\mathrm{Re}\bar{\varepsilon}$ and r are independent of phase conventions. In particular the departure of r from 1 measures CP violation in the $K^0 - \bar{K}^0$ mixing:

$$r = 1 + \frac{2|\Gamma_{12}|^2}{4|M_{12}|^2 + |\Gamma_{12}|^2} \mathrm{Im}\left(\frac{M_{12}}{\Gamma_{12}}\right). \tag{3.16}$$

Since $\bar{\varepsilon}$ is $\mathcal{O}(10^{-3})$, one has to a very good approximation:

$$\Delta M_K = 2\mathrm{Re}M_{12}, \qquad \Delta\Gamma_K = 2\mathrm{Re}\Gamma_{12}, \tag{3.17}$$

where we have introduced the subscript K to stress that these formulae apply only to the $K^0 - \bar{K}^0$ system.

The $K_L - K_S$ mass difference is experimentally measured to be [8]

$$\Delta M_K = M(K_L) - M(K_S) = (3.489 \pm 0.009) \cdot 10^{-15}\,\mathrm{GeV}. \tag{3.18}$$

In the Standard Model roughly 70% of the measured ΔM_K is described by the real parts of the box diagrams with charm quark and top quark exchanges, whereby the contribution of the charm exchanges is by far dominant. This is related to the smallness of the real parts of the CKM top quark couplings compared with the corresponding charm quark couplings. Some non-negligible contribution comes from the box diagrams with simultaneous charm and top exchanges. The remaining 20% of the measured ΔM_K is attributed to long distance contributions which are difficult to estimate [30]. Further information with the relevant references can be found in ref. [31].

The situation with $\Delta\Gamma_K$ is rather different. It is fully dominated by long distance effects. Experimentally one has $\Delta\Gamma_K \approx -2\Delta M_K$.

3.3 The First Look at ε and ε'

Since a two pion final state is CP even while a three pion final state is CP odd, K_S and K_L preferably decay to 2π and 3π, respectively via the following CP conserving decay modes:

$$K_L \to 3\pi \;\; (\text{via } K_2), \qquad K_S \to 2\pi \;\; (\text{via } K_1). \tag{3.19}$$

This difference is responsible for the large disparity in their life-times; a factor of 579. However, K_L and K_S are not CP eigenstates and may decay with small branching fractions as follows:

$$K_L \rightarrow 2\pi \quad (\text{via } K_1), \qquad K_S \rightarrow 3\pi \quad (\text{via } K_2). \tag{3.20}$$

This violation of CP is called *indirect* as it proceeds not via explicit breaking of the CP symmetry in the decay itself but via the admixture of the CP state with opposite CP parity to the dominant one. The measure for this indirect CP violation is defined as

$$\varepsilon = \frac{A(K_L \rightarrow (\pi\pi)_{I=0})}{A(K_S \rightarrow (\pi\pi)_{I=0})}. \tag{3.21}$$

Following the derivation in ref. [27] one finds

$$\varepsilon = \frac{\exp(i\pi/4)}{\sqrt{2}\Delta M_K} \left(\text{Im} M_{12} + 2\xi \text{Re} M_{12} \right), \qquad \xi = \frac{\text{Im} A_0}{\text{Re} A_0} \tag{3.22}$$

where the term involving $\text{Im} M_{12}$ represents $\bar{\varepsilon}$ defined in (3.9). The phase convention dependence of the term involving ξ cancels the convention dependence of $\bar{\varepsilon}$ so that ε is free from this dependence.

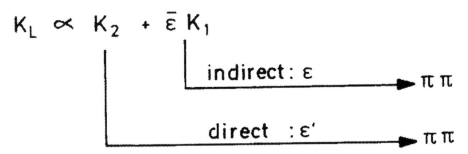

Figure 6: Indirect versus direct CP violation in $K_L \rightarrow \pi\pi$.

While *indirect* CP violation reflects the fact that the mass eigenstates are not CP eigenstates, so-called *direct* CP violation is realized via a direct transition of a CP odd to a CP even state or vice versa (see Fig. 6). A measure of such a direct CP violation in $K_L \rightarrow \pi\pi$ is characterized by a complex parameter ε' defined as

$$\varepsilon' = \frac{1}{\sqrt{2}} \text{Im} \left(\frac{A_2}{A_0} \right) \exp(i\Phi_{\varepsilon'}), \qquad \Phi_{\varepsilon'} = \frac{\pi}{2} + \delta_2 - \delta_0, \tag{3.23}$$

where the isospin amplitudes A_I in $K \to \pi\pi$ decays are introduced through

$$A(K^+ \to \pi^+\pi^0) = \sqrt{\frac{3}{2}} A_2 e^{i\delta_2} \tag{3.24}$$

$$A(K^0 \to \pi^+\pi^-) = \sqrt{\frac{2}{3}} A_0 e^{i\delta_0} + \sqrt{\frac{1}{3}} A_2 e^{i\delta_2} \tag{3.25}$$

$$A(K^0 \to \pi^0\pi^0) = \sqrt{\frac{2}{3}} A_0 e^{i\delta_0} - 2\sqrt{\frac{1}{3}} A_2 e^{i\delta_2} \, . \tag{3.26}$$

Here the subscript $I = 0, 2$ denotes states with isospin $0, 2$ equivalent to $\Delta I = 1/2$ and $\Delta I = 3/2$ transitions, respectively, and $\delta_{0,2}$ are the corresponding strong phases. The weak CKM phases are contained in A_0 and A_2. The isospin amplitudes A_I are complex quantities which depend on phase conventions. On the other hand, ε' measures the difference between the phases of A_2 and A_0 and is a physical quantity. The strong phases $\delta_{0,2}$ can be extracted from $\pi\pi$ scattering resulting in $\Phi_{\varepsilon'} \approx \pi/4$.

Experimentally ε and ε' can be found by measuring the ratios

$$\eta_{00} = \frac{A(K_L \to \pi^0\pi^0)}{A(K_S \to \pi^0\pi^0)}, \qquad \eta_{+-} = \frac{A(K_L \to \pi^+\pi^-)}{A(K_S \to \pi^+\pi^-)}. \tag{3.27}$$

Indeed, assuming ε and ε' to be small numbers one finds

$$\eta_{00} = \varepsilon - \frac{2\varepsilon'}{1 - \sqrt{\omega}} \simeq \varepsilon - 2\varepsilon', \qquad \eta_{+-} = \varepsilon + \frac{\varepsilon'}{1 + \omega/\sqrt{2}} \simeq \varepsilon + \varepsilon' \tag{3.28}$$

where experimentally $\omega = \mathrm{Re}A_2/\mathrm{Re}A_0 = 0.045$.

In the absence of direct CP violation $\eta_{00} = \eta_{+-}$. The ratio ε'/ε can then be measured through

$$\left| \frac{\eta_{00}}{\eta_{+-}} \right|^2 \simeq 1 - 6\,\mathrm{Re}\left(\frac{\varepsilon'}{\varepsilon}\right). \tag{3.29}$$

3.4 Basic Formula for ε

With all this information at hand let us derive a formula for ε which can be efficiently used in pheneomenological applications. The off-diagonal element M_{12} in the neutral K-meson mass matrix representing K^0-$\bar{K}^0$ mixing is given by

$$2m_K M_{12}^* = \langle \bar{K}^0 | \mathcal{H}_{\mathrm{eff}}(\Delta S = 2) | K^0 \rangle \, , \tag{3.30}$$

where $\mathcal{H}_{\text{eff}}(\Delta S = 2)$ is the effective Hamiltonian for the $\Delta S = 2$ transitions. That M_{12}^* and not M_{12} stands on the l.h.s of this formula, is evident from (3.7). The factor $2m_K$ reflects our normalization of external states.

To lowest order in electroweak interactions $\Delta S = 2$ transitions are induced through the box diagrams of Fig. 5. Including QCD corrections one has

$$
\mathcal{H}_{\text{eff}}^{\Delta S = 2} = \frac{G_F^2}{16\pi^2} M_W^2 \left[\lambda_c^2 \eta_1 S_0(x_c) + \lambda_t^2 \eta_2 S_0(x_t) + 2\lambda_c \lambda_t \eta_3 S_0(x_c, x_t) \right] \times
$$
$$
\times \left[\alpha_s^{(3)}(\mu) \right]^{-2/9} \left[1 + \frac{\alpha_s^{(3)}(\mu)}{4\pi} J_3 \right] Q(\Delta S = 2) + h.c. \qquad (3.31)
$$

where $\lambda_i = V_{is}^* V_{id}$, $\mu < \mu_c = \mathcal{O}(m_c)$ and $\alpha_s^{(3)}$ is the strong coupling constant in an effective three flavour theory. In (3.31), the relevant operator

$$
Q(\Delta S = 2) = (\bar{s}d)_{V-A}(\bar{s}d)_{V-A}, \qquad (3.32)
$$

is multiplied by the corresponding coefficient function. This function is decomposed into a charm-, a top- and a mixed charm-top contribution. The functions S_0 are given in (2.13) and (2.14).

Short-distance QCD effects are described through the correction factors η_1, η_2, η_3 and the explicitly α_s-dependent terms in (3.31). The NLO values of η_i are given as follows [31,32,33]:

$$
\eta_1 = 1.38 \pm 0.20, \qquad \eta_2 = 0.57 \pm 0.01, \qquad \eta_3 = 0.47 \pm 0.04 . \qquad (3.33)
$$

The quoted errors reflect the remaining theoretical uncertainties due to leftover μ-dependences at $\mathcal{O}(\alpha_s^2)$ and $\Lambda_{\overline{MS}}$, the scale in the QCD running coupling.

Defining the renormalization group invariant parameter $\hat{B}_K$ by

$$
\hat{B}_K = B_K(\mu) \left[\alpha_s^{(3)}(\mu) \right]^{-2/9} \left[1 + \frac{\alpha_s^{(3)}(\mu)}{4\pi} J_3 \right] \qquad (3.34)
$$

$$
\langle \bar{K}^0 | (\bar{s}d)_{V-A}(\bar{s}d)_{V-A} | K^0 \rangle \equiv \frac{8}{3} B_K(\mu) F_K^2 m_K^2 \qquad (3.35)
$$

and using (3.31) one finds

$$
M_{12} = \frac{G_F^2}{12\pi^2} F_K^2 \hat{B}_K m_K M_W^2 \left[\lambda_c^{*2} \eta_1 S_0(x_c) + \lambda_t^{*2} \eta_2 S_0(x_t) + 2\lambda_c^* \lambda_t^* \eta_3 S_0(x_c, x_t) \right],
$$
$$
\qquad (3.36)
$$

where F_K is the K-meson decay constant and m_K the K-meson mass.

To proceed further we neglect the last term in (3.22) as it constitutes at most a 2 % correction to ε. This is justified in view of other uncertainties, in particular those connected with $\hat{B}_K$. Inserting (3.36) into (3.22) we find

$$\varepsilon = C_\varepsilon \hat{B}_K \text{Im}\lambda_t \left\{ \text{Re}\lambda_c \left[\eta_1 S_0(x_c) - \eta_3 S_0(x_c, x_t) \right] - \text{Re}\lambda_t \eta_2 S_0(x_t) \right\} \exp(i\pi/4) , \tag{3.37}$$

where we have used the unitarity relation $\text{Im}\lambda_c^* = \text{Im}\lambda_t$ and have neglected $\text{Re}\lambda_t/\text{Re}\lambda_c = \mathcal{O}(\lambda^4)$ in evaluating $\text{Im}(\lambda_c^*\lambda_t^*)$. The numerical constant C_ε is given by

$$C_\varepsilon = \frac{G_F^2 F_K^2 m_K M_W^2}{6\sqrt{2}\pi^2 \Delta M_K} = 3.84 \cdot 10^4 . \tag{3.38}$$

To this end we have used the experimental value of ΔM_K in (3.18).

Using the standard parametrization of (1.5) to evaluate $\text{Im}\lambda_i$ and $\text{Re}\lambda_i$, setting the values for s_{12}, s_{13}, s_{23} and m_t in accordance with experiment and taking a value for $\hat{B}_K$ (see below), one can determine the phase δ by comparing (3.37) with the experimental value for ε

$$\varepsilon_{exp} = (2.280 \pm 0.013) \cdot 10^{-3} \exp i\Phi_\varepsilon, \qquad \Phi_\varepsilon = \frac{\pi}{4}. \tag{3.39}$$

Once δ has been determined in this manner one can find the apex $(\bar{\varrho}, \bar{\eta})$ of the unitarity triangle in Fig. 1 by using

$$\varrho = \frac{s_{13}}{s_{12}s_{23}} \cos\delta, \qquad \eta = \frac{s_{13}}{s_{12}s_{23}} \sin\delta \tag{3.40}$$

and

$$\bar{\varrho} = \varrho(1 - \frac{\lambda^2}{2}), \qquad \bar{\eta} = \eta(1 - \frac{\lambda^2}{2}). \tag{3.41}$$

For a given set $(s_{12}, s_{13}, s_{23}, m_t, \hat{B}_K)$ there are two solutions for δ and consequently two solutions for $(\bar{\varrho}, \bar{\eta})$. This will be evident from the analysis of the unitarity triangle discussed in detail below.

Finally we have to say a few words about the non-perturbative parameter $\hat{B}_K$, the main uncertainty in this analysis. Reviews are given in ref's[34,35]. Here we only collect in Table 1 values for $\hat{B}_K$ obtained in various non-perturbative approaches. In our numerical analysis presented below we will use

$$\hat{B}_K = 0.80 \pm 0.15 \tag{3.42}$$

which is in the ball park of various lattice and large-N estimates.

Table 1: $\hat{B}_K$ obtained using various methods. WA stands for recent world avarage.

Method	$\hat{B}_K$	Reference
Chiral QM	1.1 ± 0.2	36
Lattice (APE)	0.93 ± 0.16	37
Lattice (JLQCD)	0.86 ± 0.06	38
Lattice (GKS)	0.85 ± 0.05	39
Lattice (WA)	0.89 ± 0.13	40
Large-N	0.70 ± 0.10	41,42
Large-N	$0.4 - 0.7$	43
QCDS	$0.5 - 0.6$	44
CHPTH	0.42 ± 0.06	45
QCD HD	0.39 ± 0.10	46
SU(3)+PCAC	0.33	47

3.5 Basic Formula for B^0-$\bar{B}^0$ Mixing

The strength of the $B^0_{d,s} - \bar{B}^0_{d,s}$ mixings is described by the mass differences

$$\Delta M_{d,s} = M_H^{d,s} - M_L^{d,s} \qquad (3.43)$$

with "H" and "L" denoting *Heavy* and *Light* respectively. In contrast to ΔM_K, in this case the long distance contributions are estimated to be very small and $\Delta M_{d,s}$ is very well approximated by the relevant box diagrams. Moreover, due $m_{u,c} \ll m_t$ only the top sector can contribute significantly to $B^0_{d,s} - \bar{B}^0_{d,s}$ mixings. The charm sector and the mixed top-charm contributions are entirely negligible.

$\Delta M_{d,s}$ can be expressed in terms of the off-diagonal element in the neutral B-meson mass matrix by using the formulae developed previously for the K-meson system. One finds

$$\Delta M_q = 2|M_{12}^{(q)}|, \qquad q = d, s. \qquad (3.44)$$

This formula differs from $\Delta M_K = 2\mathrm{Re}M_{12}$ because in the B-system $\Gamma_{12} \ll M_{12}$.

The off-diagonal term M_{12} in the neutral B-meson mass matrix is then given by a formula analogous to (3.30)

$$2m_{B_q}|M_{12}^{(q)}| = |\langle \bar{B}^0_q|\mathcal{H}_{\mathrm{eff}}(\Delta B = 2)|B^0_q\rangle|, \qquad (3.45)$$

where in the case of $B_d^0 - \bar{B}_d^0$ mixing

$$\mathcal{H}_{\text{eff}}^{\Delta B=2} = \frac{G_F^2}{16\pi^2} M_W^2 \left(V_{tb}^* V_{td}\right)^2 \eta_B S_0(x_t) \times$$

$$\times \left[\alpha_s^{(5)}(\mu_b)\right]^{-6/23} \left[1 + \frac{\alpha_s^{(5)}(\mu_b)}{4\pi} J_5\right] Q(\Delta B = 2) + h.c. \quad (3.46)$$

Here $\mu_b = \mathcal{O}(m_b)$,

$$Q(\Delta B = 2) = (\bar{b}d)_{V-A}(\bar{b}d)_{V-A} \quad (3.47)$$

and [32]

$$\eta_B = 0.55 \pm 0.01. \quad (3.48)$$

In the case of $B_s^0 - \bar{B}_s^0$ mixing one should simply replace $d \to s$ in (3.46) and (3.47) with all other quantities unchanged.

Defining the renormalization group invariant parameters $\hat{B}_q$ in analogy to (3.34) and (3.35) one finds using (3.46)

$$\Delta M_q = \frac{G_F^2}{6\pi^2} \eta_B m_{B_q} (\hat{B}_{B_q} F_{B_q}^2) M_W^2 S_0(x_t) |V_{tq}|^2, \quad (3.49)$$

where F_{B_q} is the B_q-meson decay constant. This implies two useful formulae

$$\Delta M_d = 0.50/\text{ps} \cdot \left[\frac{\sqrt{\hat{B}_{B_d}} F_{B_d}}{200\,\text{MeV}}\right]^2 \left[\frac{\overline{m}_t(m_t)}{170\,\text{GeV}}\right]^{1.52} \left[\frac{|V_{td}|}{8.8 \cdot 10^{-3}}\right]^2 \left[\frac{\eta_B}{0.55}\right] \quad (3.50)$$

and

$$\Delta M_s = 15.1/\text{ps} \cdot \left[\frac{\sqrt{\hat{B}_{B_s}} F_{B_s}}{240\,\text{MeV}}\right]^2 \left[\frac{\overline{m}_t(m_t)}{170\,\text{GeV}}\right]^{1.52} \left[\frac{|V_{ts}|}{0.040}\right]^2 \left[\frac{\eta_B}{0.55}\right]. \quad (3.51)$$

There is a vast literature on the calculations of F_{B_d} and $\hat{B}_d$. The most recent lattice results are summarized in ref. [48]. They are compatible with the results obtained with the help of QCD sum rules [49]. In our numerical analysis we will use the value for $F_{B_d}\sqrt{\hat{B}_{B_d}}$ given in Table 2. The experimental situation on $\Delta M_{d,s}$ is also given there.

4 Standard Analysis of the Unitarity Triangle

4.1 Basic Procedure

With all these formulae at hand we can now summarize the standard analysis of the unitarity triangle in Fig. 1. It proceeds in five steps.

Step 1:
From $b \to c$ transition in inclusive and exclusive leading B meson decays one finds $|V_{cb}|$ and consequently the scale of the unitarity triangle:

$$|V_{cb}| \implies \lambda |V_{cb}| = \lambda^3 A \qquad (4.52)$$

Step 2:
From $b \to u$ transition in inclusive and exclusive B meson decays one finds $|V_{ub}/V_{cb}|$ and consequently the side $CA = R_b$ of the unitarity triangle:

$$\left| \frac{V_{ub}}{V_{cb}} \right| \implies R_b = \sqrt{\bar{\varrho}^2 + \bar{\eta}^2} = 4.44 \cdot \left| \frac{V_{ub}}{V_{cb}} \right| \qquad (4.53)$$

Step 3:
From the experimental value of ε (3.39) and the formula (3.37) one derives, using the approximations (1.22)–(1.24), the constraint

$$\bar{\eta} \left[(1 - \bar{\varrho}) A^2 \eta_2 S_0(x_t) + P_0(\varepsilon) \right] A^2 \hat{B}_K = 0.224, \qquad (4.54)$$

where

$$P_0(\varepsilon) = [\eta_3 S_0(x_c, x_t) - \eta_1 x_c] \frac{1}{\lambda^4}, \qquad x_t = \frac{m_t^2}{M_W^2}. \qquad (4.55)$$

$P_0(\varepsilon) = 0.31 \pm 0.05$ summarizes the contributions of box diagrams with two charm quark exchanges and the mixed charm-top exchanges. The main uncertainties in the constraint (4.54) reside in $\hat{B}_K$ and to some extent in A^4 which multiplies the leading term. Equation (4.54) specifies a hyperbola in the $(\bar{\varrho}, \bar{\eta})$ plane. This hyperbola intersects the circle found in step 2 in two points which correspond to the two solutions for δ mentioned earlier. This is illustrated in Fig. 7. The position of the hyperbola (4.54) in the $(\bar{\varrho}, \bar{\eta})$ plane depends on m_t, $|V_{cb}| = A\lambda^2$ and $\hat{B}_K$. With decreasing m_t, $|V_{cb}|$ and $\hat{B}_K$ the ε-hyperbola moves away from the origin of the $(\bar{\varrho}, \bar{\eta})$ plane.

Step 4: From the observed $B_d^0 - \bar{B}_d^0$ mixing parametrized by ΔM_d the side $BA = R_t$ of the unitarity triangle can be determined:

$$R_t = \frac{1}{\lambda} \frac{|V_{td}|}{|V_{cb}|} = 1.0 \cdot \left[\frac{|V_{td}|}{8.8 \cdot 10^{-3}} \right] \left[\frac{0.040}{|V_{cb}|} \right] \qquad (4.56)$$

with

$$|V_{td}| = 8.8 \cdot 10^{-3} \left[\frac{200 \, \text{MeV}}{\sqrt{B_{B_d}} F_{B_d}} \right] \left[\frac{170 \, GeV}{\overline{m}_t(m_t)} \right]^{0.76} \left[\frac{\Delta M_d}{0.50/\text{ps}} \right]^{0.5} \sqrt{\frac{0.55}{\eta_B}}. \qquad (4.57)$$

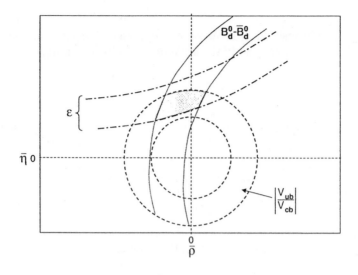

Figure 7: Schematic determination of the Unitarity Triangle.

Since m_t, ΔM_d and η_B are already rather precisely known, the main uncertainty in the determination of $|V_{td}|$ from $B_d^0 - \bar{B}_d^0$ mixing comes from $F_{B_d}\sqrt{B_{B_d}}$. Note that R_t suffers from additional uncertainty in $|V_{cb}|$, which is absent in the determination of $|V_{td}|$ this way. The constraint in the $(\bar{\varrho}, \bar{\eta})$ plane coming from this step is illustrated in Fig. 7.

Step 5:

The measurement of $B_s^0 - \bar{B}_s^0$ mixing parametrized by ΔM_s together with ΔM_d allows to determine R_t in a different way. Using (3.49) one finds

$$\frac{|V_{td}|}{|V_{ts}|} = \xi \sqrt{\frac{m_{B_s}}{m_{B_d}}} \sqrt{\frac{\Delta M_d}{\Delta M_s}}, \qquad \xi = \frac{F_{B_s}\sqrt{B_{B_s}}}{F_{B_d}\sqrt{B_{B_d}}}. \qquad (4.58)$$

Using next $\Delta M_d^{\max} = 0.487/ps$ and $|V_{ts}/V_{cb}|^{\max} = 0.991$ one finds a useful approximate formula

$$(R_t)_{\max} = 1.0 \cdot \xi \sqrt{\frac{10.2/ps}{(\Delta M_s)_{\min}}}. \qquad (4.59)$$

One should note that m_t and $|V_{cb}|$ dependences have been eliminated this way

Table 2: Collection of input parameters.

Quantity	Central	Error	Reference
$\lvert V_{cb} \rvert$	0.040	± 0.002	8
$\lvert V_{ub} \rvert$	$3.56 \cdot 10^{-3}$	$\pm 0.56 \cdot 10^{-3}$	15
$\hat{B}_K$	0.80	± 0.15	See Text
$\sqrt{B_d} F_{B_d}$	200 MeV	± 40 MeV	48
m_t	165 GeV	± 5 GeV	50
ΔM_d	0.471 ps^{-1}	± 0.016 ps^{-1}	51
ΔM_s	> 12.4 ps^{-1}	95%C.L.	51
ξ	1.14	± 0.08	48,52

and that ξ should in principle contain much smaller theoretical uncertainties than the hadronic matrix elements in ΔM_d and ΔM_s separately. The most recent values relevant for (4.59) are summarized in Table 2.

4.2 Numerical Results

Input Parameters

The input parameters needed to perform the standard analysis of the unitarity triangle are given in Table 2, where m_t refers to the running current top quark mass defined at $\mu = m_t^{Pole}$. It corresponds to $m_t^{Pole} = 174.3 \pm 5.1$ GeV measured by CDF and D0 [50].

Output of the Standard Analysis

In what follows we will present two types of numerical analyses [35,53]:

- Method 1: The experimentally measured numbers are used with Gaussian errors and for the theoretical input parameters we take a flat distribution in the ranges given in Table 2.

- Method 2: Both the experimentally measured numbers and the theoretical input parameters are scanned independently within the ranges given in Table 2.

The results are shown in Table 3. The allowed region for $(\bar{\varrho}, \bar{\eta})$ is presented in Fig. 8. It is the shaded area on the right hand side of the solid circle which represents the upper bound for $(\Delta M)_d/(\Delta M)_s$. The hyperbolas give

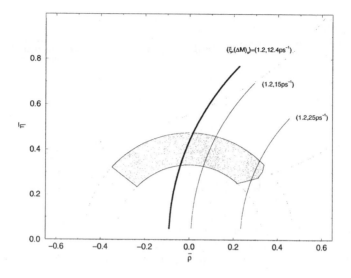

Figure 8: Unitarity Triangle 1999.

the constraint from ε and the two circles centered at $(0,0)$ the constraint from $|V_{ub}/V_{cb}|$. The white areas between the lower ε-hyperbola and the shaded region are excluded by $B_d^0 - \bar{B}_d^0$ mixing. We observe that the region $\bar{\varrho} < 0$ is practically excluded. The results in Fig. 8 correspond to a simple independent scanning of all parameters within one standard deviation. We find that whereas the angle β is rather constrained, the uncertainties in α and γ are substantially larger:

$$66° \leq \alpha \leq 113° , \quad 17° \leq \beta \leq 29° , \quad 44° \leq \gamma \leq 97° . \qquad (4.60)$$

The result for $\sin 2\beta$ is consistent with the recent measurement of CP asymmetry in $B \to \psi K_S$ by CDF [54], although the large experimental error precludes any definite conclusion.

Other studies of the unitarity triangle can be found in ref's [14,15,55,56].

4.3 Final Remarks

In this section we have completed the determination of the CKM matrix. It is given by the values of $|V_{us}|$, $|V_{cb}|$ and $|V_{ub}|$ in (1.36) and (1.37), the results in Table 3 and the unitarity triangle shown in Fig. 8. Clearly the accuracy of this determination is not yet impressive. We should stress, however, that in a few years from now the standard analysis may give much more accurate results.

Table 3: Output of the Standard Analysis. $\lambda_t = V_{ts}^* V_{td}$.

Quantity	Scanning	Gaussian
$\mid V_{td} \mid /10^{-3}$	$7.0 - 9.8$	8.1 ± 0.6
$\mid V_{ts}/V_{cb} \mid$	$0.975 - 0.991$	0.984 ± 0.004
$\mid V_{td}/V_{ts} \mid$	$0.17 - 0.24$	0.201 ± 0.017
$\sin(2\beta)$	$0.57 - 0.84$	0.73 ± 0.09
$\sin(2\alpha)$	$-0.72 - 1.0$	-0.03 ± 0.36
$\sin(\gamma)$	$0.69 - 1.0$	0.89 ± 0.08
$\text{Im}\lambda_t/10^{-4}$	$1.04 - 1.63$	1.33 ± 0.14

In particular a single precise measurement of ΔM_s will have a very important impact on the allowed area in the $(\bar{\varrho}, \bar{\eta})$ plane. Such a measurement should come from SLD and later from LHC.

Having the values of CKM parameters at hand, we can use them to predict various branching ratios for rare and CP-violating decays. This we will do in the subsequent sections.

5 ε'/ε in the Standard Model

5.1   Preliminaries

Direct CP violation remains one of the important targets of contemporary particle physics. In the case of $K \to \pi\pi$, a non-vanishing value of the ratio $\text{Re}(\varepsilon'/\varepsilon)$ defined in (3.23) would give the first signal for direct CP violation ruling out superweak models [57]. Until recently the experimental situation on ε'/ε was rather unclear:

$$\text{Re}(\varepsilon'/\varepsilon) = \begin{cases} (23 \pm 7) \cdot 10^{-4} & \text{(NA31)} \, [58] \\ (7.4 \pm 5.9) \cdot 10^{-4} \text{(E731)}. \, [59] \end{cases}$$

(5.1)

While the result of the NA31 collaboration at CERN [58] clearly indicated direct CP violation, the value of E731 at Fermilab [59] was compatible with superweak theories [57] in which $\varepsilon'/\varepsilon = 0$. This controversy is now settled with the very recent measurement by KTeV at Fermilab [60]

$$\text{Re}(\varepsilon'/\varepsilon) = (28.0 \pm 4.1) \cdot 10^{-4} \quad \text{(KTeV)}$$

(5.2)

which together with the NA31 result confidently establishes direct CP violation in nature. The grand average including NA31, E731 and KTeV results read[60]

$$\mathrm{Re}(\varepsilon'/\varepsilon) = (21.8 \pm 3.0) \cdot 10^{-4} \qquad (5.3)$$

very close to the NA31 result but with a smaller error. The error should be further reduced once the first data from NA48 collaboration at CERN are available and complete data from both collaborations have been analyzed. It is also of great interest to see what value for ε'/ε will be measured by KLOE at Frascati, which uses a different experimental technique than KTeV and NA48.

There is a long history of calculations of ε'/ε in the Standard Model. The first calculation of ε'/ε for $m_t \ll M_W$ without the inclusion of renormalization group effects can be found in ref. [61]. Renormalization group effects in the leading logarithmic approximation have been first presented in ref. [62]. For $m_t \ll M_W$ only QCD penguins play a substantial role. First extensive phenomenological analyses in this approximation can be found in ref. [63]. Over the eighties these calculations were refined through the inclusion of QED penguin effects for $m_t \ll M_W$ [64,65,66], the inclusion of isospin breaking in the quark masses [65,66,67], and through improved estimates of hadronic matrix elements in the framework of the $1/N$ approach [68]. This era of ε'/ε ref's [69,70], where QCD penguins, electroweak penguins (γ and Z^0 penguins) and the relevant box diagrams were included for arbitrary top quark masses. The strong cancellation between QCD penguins and electroweak penguins for $m_t > 150$ GeV found in these papers was confirmed by other authors [71].

During the nineties considerable progress has been made by calculating complete NLO corrections to ε' [72–76]. Together with the NLO corrections to ε and $B^0 - \bar{B}^0$ mixing [31,32,33], this allowed a complete NLO analysis of ε'/ε including constraints from the observed indirect CP violation (ε) and $B_{d,s}^0 - \bar{B}_{d,s}^0$ mixings ($\Delta M_{d,s}$). The improved determination of the V_{ub} and V_{cb} elements of the CKM matrix, the improved estimates of hadronic matrix elements using the lattice approach as well as other non-perturbative approaches and in particular the determination of the top quark mass m_t had of course also an important impact on ε'/ε.

Now, ε'/ε is given by (5.5) where in a crude approximation (not to be used for any serious analysis)

$$F_{\varepsilon'} \approx 13 \cdot \left[\frac{110\,\text{MeV}}{m_s(2\ \text{GeV})} \right]^2 \left[B_6^{(1/2)} (1 - \Omega_{\eta+\eta'}) - \right.$$

$$\left. 0.4 \cdot B_8^{(3/2)} \left(\frac{m_t}{165\,\text{GeV}} \right)^{2.5} \left(\frac{\Lambda_{\overline{\text{MS}}}^{(4)}}{340\ \text{MeV}} \right) \right].$$

$$(5.4)$$

Here $\Omega_{\eta+\eta'} \approx 0.25$ represents isospin breaking corrections and B_i are hadronic parameters which we will define later on. This formula exhibits very clearly the dominant uncertainties in $F_{\varepsilon'}$ which reside in the values of m_s, $B_6^{(1/2)}$, $B_8^{(3/2)}$, $\Lambda_{\overline{\text{MS}}}^{(4)}$ and $\Omega_{\eta+\eta'}$. Moreover, the partial cancellation between QCD penguin ($B_6^{(1/2)}$) and electroweak penguin ($B_8^{(3/2)}$) contributions requires accurate values of $B_6^{(1/2)}$ and $B_8^{(3/2)}$ for an acceptable estimate of ε'/ε. Because of the accurate value $m_t(m_t) = 165 \pm 5$ GeV, the uncertainty in ε'/ε due to the top quark mass amounts only to a few percent. A more accurate formula for $F_{\varepsilon'}$ will be given below.

Now, it has been known for some time that for central values of the input parameters the size of ε'/ε in the Standard Model is well below the NA31 value of $(23.0 \pm 6.5) \cdot 10^{-4}$. Indeed, extensive NLO analyses with lattice and large-N estimates of $B_6^{(1/2)} \approx 1$ and $B_8^{(3/2)} \approx 1$ performed first in ref's [74,75] and after the top discovery in ref's [77,78,79] have found ε'/ε in the ball park of $(3-7) \cdot 10^{-4}$ for $m_s(2\ \text{GeV}) \approx 130$ MeV. On the other hand it has been stressed repeatedly in ref's [1,78] that for extreme values of $B_6^{(1/2)}$, $B_8^{(3/2)}$ and m_s still consistent with lattice, QCD sum rules and large-N estimates as well as sufficiently high values of $\text{Im}\lambda_t$ and $\Lambda_{\overline{\text{MS}}}^{(4)}$, a ratio ε'/ε as high as $(2-3) \cdot 10^{-3}$ could be obtained within the Standard Model. Yet, it has also been admitted that such simultaneously extreme values of all input parameters and consequently values of ε'/ε close to the NA31 result are rather improbable in the Standard Model. Different conclusions have been reached in ref. [80], where values $(1-2) \cdot 10^{-3}$ for ε'/ε can be found. Also the Trieste group [81], which calculated the parameters $B_6^{(1/2)}$ and $B_8^{(3/2)}$ in the chiral quark model, found $\varepsilon'/\varepsilon = (1.7 \pm 1.4) \cdot 10^{-3}$. On the other hand using an effective chiral lagrangian approach, the authors in ref. [83] found ε'/ε consistent with zero.

After these general remarks let us discuss ε'/ε in explicit terms. Other reviews of ε'/ε can be found in ref's [82,81].

5.2 Basic Formulae

The parameter ε' is given in terms of the isospin amplitudes A_I in (3.23). Applying OPE to these amplitudes one finds

$$\frac{\varepsilon'}{\varepsilon} = \mathrm{Im}\lambda_t \cdot F_{\varepsilon'}, \tag{5.5}$$

where

$$F_{\varepsilon'} = \left[P^{(1/2)} - P^{(3/2)} \right] \exp(i\Phi), \tag{5.6}$$

with

$$P^{(1/2)} = r \sum y_i \langle Q_i \rangle_0 (1 - \Omega_{\eta+\eta'}) , \tag{5.7}$$

$$P^{(3/2)} = \frac{r}{\omega} \sum y_i \langle Q_i \rangle_2 . \tag{5.8}$$

Here

$$r = \frac{G_F \omega}{2|\varepsilon|\mathrm{Re}A_0} , \qquad \langle Q_i \rangle_I \equiv \langle (\pi\pi)_I | Q_i | K \rangle , \qquad \omega = \frac{\mathrm{Re}A_2}{\mathrm{Re}A_0}. \tag{5.9}$$

Since

$$\Phi = \Phi_{\varepsilon'} - \Phi_\varepsilon \approx 0, \tag{5.10}$$

$F_{\varepsilon'}$ and ε'/ε are real to an excellent approximation. The operators Q_i have been given already in (2.4)-(2.8). The Wilson coefficient functions $y_i(\mu)$ were calculated including the complete next-to-leading order (NLO) corrections in [72-76]. The details of these calculations can be found there and in review [2]. Their numerical values for $\Lambda_{\overline{MS}}^{(4)}$ corresponding to $\alpha_{\overline{MS}}^{(5)}(M_Z) = 0.119 \pm 0.003$ and two renormalization schemes (NDR and HV) are given in Table 4 [35].

It is customary in phenomenological applications to take $\mathrm{Re}A_0$ and ω from experiment, i.e.

$$\mathrm{Re}A_0 = 3.33 \cdot 10^{-7} \,\mathrm{GeV}, \qquad \omega = 0.045, \tag{5.11}$$

where the last relation reflects the so-called $\Delta I = 1/2$ rule. This strategy avoids to a large extent the hadronic uncertainties in the real parts of the isospin amplitudes A_I.

The sum in (5.7) and (5.8) runs over all contributing operators. $P^{(3/2)}$ is fully dominated by electroweak penguin contributions. $P^{(1/2)}$ on the other hand is governed by QCD penguin contributions which are suppressed by isospin breaking in the quark masses ($m_u \neq m_d$). The latter effect is described by

Table 4: $\Delta S = 1$ Wilson coefficients at $\mu = m_c = 1.3\,\text{GeV}$ for $m_t = 165\,\text{GeV}$ and $f = 3$ effective flavours. $y_1 = y_2 \equiv 0$.

Scheme	$\Lambda_{\overline{\text{MS}}}^{(4)} = 290\,\text{MeV}$		$\Lambda_{\overline{\text{MS}}}^{(4)} = 340\,\text{MeV}$		$\Lambda_{\overline{\text{MS}}}^{(4)} = 390\,\text{MeV}$	
	NDR	HV	NDR	HV	NDR	HV
y_3	0.027	0.030	0.030	0.034	0.033	0.038
y_4	−0.054	−0.056	−0.059	−0.061	−0.064	−0.067
y_5	0.006	0.015	0.005	0.016	0.003	0.017
y_6	−0.082	−0.074	−0.092	−0.083	−0.105	−0.093
y_7/α	−0.038	−0.037	−0.037	−0.036	−0.037	−0.034
y_8/α	0.118	0.127	0.134	0.143	0.152	0.161
y_9/α	−1.410	−1.410	−1.437	−1.437	−1.466	−1.466
y_{10}/α	0.496	0.502	0.539	0.546	0.585	0.593

$$\Omega_{\eta+\eta'} = \frac{1}{\omega} \frac{(\text{Im}\,A_2)_{\text{I.B.}}}{\text{Im}\,A_0}. \tag{5.12}$$

For $\Omega_{\eta+\eta'}$ we will first set

$$\Omega_{\eta+\eta'} = 0.25, \tag{5.13}$$

which is in the ball park of the values obtained in the $1/N$ approach [66] and in chiral perturbation theory [65,67]. $\Omega_{\eta+\eta'}$ is independent of m_t. We will investigate the sensitivity of ε'/ε to $\Omega_{\eta+\eta'}$ later on.

5.3 Hadronic Matrix Elements

The main source of uncertainty in the calculation of ε'/ε are the hadronic matrix elements $\langle Q_i \rangle_I$. They generally depend on the renormalization scale μ and on the scheme used to renormalize the operators Q_i. These two dependences are canceled by those present in the Wilson coefficients $y_i(\mu)$ so that the resulting physical ε'/ε does not (in principle) depend on μ and on the renormalization scheme of the operators. Unfortunately, the accuracy of the present non-perturbative methods used to evalutate $\langle Q_i \rangle_I$ is not sufficient to have the μ and scheme dependences of $\langle Q_i \rangle_I$ fully under control. We believe that this situation will change once the lattice calculations and QCD sum rule calculations improve. A brief review of the existing methods including most recent developments will be given below.

In view of this situation it has been suggested in ref. [74] to determine as many matrix elements $\langle Q_i \rangle_I$ as possible from the leading CP conserving $K \to \pi\pi$ decays, for which the experimental data is summarized in (5.11). To

36

this end it turned out to be very convenient to determine $\langle Q_i \rangle_I$ in the three-flavour effective theory at a scale $\mu \approx m_c$. The details of this approach will not be discussed here. It suffices to say that this method allows to determine only the matrix elements of the $(V - A) \otimes (V - A)$ operators. For the central value of $\text{Im}\lambda_t$ these operators give a negative contribution to ε'/ε of about $-2.5 \cdot 10^{-4}$. This shows that these operators are only relevant if ε'/ε is below $1 \cdot 10^{-3}$. Unfortunately the matrix elements of the dominant $(V - A) \otimes (V + A)$ operators cannot be determined by the CP conserving data and one has to use non-perturbative methods to estimate them.

Concerning the $(V - A) \otimes (V + A)$ operators $Q_5 - Q_8$, it is customary to express their matrix elements $\langle Q_i \rangle_I$ in terms of non-perturbative parameters $B_i^{(1/2)}$ and $B_i^{(3/2)}$ as follows:

$$\langle Q_i \rangle_0 \equiv B_i^{(1/2)} \langle Q_i \rangle_0^{(\text{vac})}, \qquad \langle Q_i \rangle_2 \equiv B_i^{(3/2)} \langle Q_i \rangle_2^{(\text{vac})}. \qquad (5.14)$$

The label "vac" stands for the vacuum insertion estimate of the hadronic matrix elements in question for which $B_i^{(1/2)} = B_i^{(3/2)} = 1$.

As the numerical analysis in ref. [74] shows ε'/ε is only weakly sensitive to the values of the parameters $B_3^{(1/2)}$, $B_5^{(1/2)}$, $B_7^{(1/2)}$, $B_8^{(1/2)}$ and $B_7^{(3/2)}$ as long as their absolute values are not substantially larger than 1. As in ref. [74] our strategy is to set

$$B_{3,7,8}^{(1/2)}(m_c) = 1, \qquad B_5^{(1/2)}(m_c) = B_6^{(1/2)}(m_c), \qquad B_7^{(3/2)}(m_c) = B_8^{(3/2)}(m_c) \qquad (5.15)$$

and to treat $B_6^{(1/2)}(m_c)$ and $B_8^{(3/2)}(m_c)$ as free parameters.

The approach in ref. [74] allows then in a good approximation to express ε'/ε or equivalently $F_{\varepsilon'}$ in terms of $\Lambda_{\overline{\text{MS}}}^{(4)}$, m_t, m_s and the two non-perturbative parameters $B_6^{(1/2)} \equiv B_6^{(1/2)}(m_c)$ and $B_8^{(3/2)} \equiv B_8^{(3/2)}(m_c)$ which cannot be fixed by the CP conserving data.

5.4 An Analytic Formula for ε'/ε

As shown in ref. [84], it is possible to cast the formal expressions for ε'/ε in (5.5)–(5.8) into an analytic formula which exhibits the m_t dependence together with the dependence on m_s, $\Lambda_{\overline{\text{MS}}}^{(4)}$, $B_6^{(1/2)}$ and $B_8^{(3/2)}$. To this end the approach for hadronic matrix elements presented above is used and $\Omega_{\eta+\eta'}$ is set to 0.25. The analytic formula given below, while being rather accurate, exhibits various features which are not transparent in a pure numerical analysis. It can be used in phenomenological applications if one is satisfied with a few percent accuracy.

Needless to say, in the numerical analysis[35] presented below we have used exact expressions.

In this formulation the function $F_{\epsilon'}$ is given simply as follows ($x_t = m_t^2/M_W^2$):

$$F_{\epsilon'} = P_0 + P_X\, X_0(x_t) + P_Y\, Y_0(x_t) + P_Z\, Z_0(x_t) + P_E\, E_0(x_t) \qquad (5.16)$$

with the m_t-dependent functions given in subsection 2.4.

The coefficients P_i are given in terms of $B_6^{(1/2)} \equiv B_6^{(1/2)}(m_c)$, $B_8^{(3/2)} \equiv B_8^{(3/2)}(m_c)$ and $m_s(m_c)$ as follows:

$$P_i = r_i^{(0)} + r_i^{(6)} R_6 + r_i^{(8)} R_8 \qquad (5.17)$$

where

$$R_6 \equiv B_6^{(1/2)}\left[\frac{137\,\mathrm{MeV}}{m_s(m_c) + m_d(m_c)}\right]^2 , \qquad R_8 \equiv B_8^{(3/2)}\left[\frac{137\,\mathrm{MeV}}{m_s(m_c) + m_d(m_c)}\right]^2 . \qquad (5.18)$$

The P_i are renormalization scale and scheme independent. They depend, however, on $\Lambda_{\overline{\mathrm{MS}}}^{(4)}$. In Table 5 we give the numerical values of $r_i^{(0)}$, $r_i^{(6)}$ and $r_i^{(8)}$ for different values of $\Lambda_{\overline{\mathrm{MS}}}^{(4)}$ at $\mu = m_c$ in the NDR renormalization scheme[35]. Actually at NLO only r_0 coefficients are renormalization scheme dependent. The last row gives them in the HV scheme. The inspection of Table 5 shows that the terms involving $r_0^{(6)}$ and $r_Z^{(8)}$ dominate the ratio ε'/ε. Moreover, the function $Z_0(x_t)$ representing a gauge invariant combination of Z^0- and γ-penguins grows rapidly with m_t and due to $r_Z^{(8)} < 0$ these contributions suppress ε'/ε strongly for large m_t [69,70].

5.5 The Status of m_s, $B_6^{(1/2)}$, $B_8^{(3/2)}$, $\Omega_{\eta+\eta'}$ and $\Lambda_{\overline{\mathrm{MS}}}^{(4)}$

The present status of these parameters has been recently reviewed in details in ref. [35]. Therefore our presentation will be very brief.

m_s

The present values for $m_s(2\,\mathrm{GeV})$ extracted from lattice calculations and QCD sum rules are

$$m_s(2\,\mathrm{GeV}) = \begin{cases} (110 \pm 20)\ \mathrm{MeV}\ \ (\mathrm{Lattice})\ ^{34,85} \\ (124 \pm 22)\ \mathrm{MeV}\,(\mathrm{QCDS}).\ ^{86} \end{cases} \qquad (5.19)$$

Table 5: Coefficients in the formula (5.17) for various $\Lambda_{\overline{MS}}^{(4)}$ in the NDR scheme. The last row gives the r_0 coefficients in the HV scheme.

i	$\Lambda_{\overline{MS}}^{(4)} = 290\,\text{MeV}$			$\Lambda_{\overline{MS}}^{(4)} = 340\,\text{MeV}$		
	$r_i^{(0)}$	$r_i^{(6)}$	$r_i^{(8)}$	$r_i^{(0)}$	$r_i^{(6)}$	$r_i^{(8)}$
0	−2.771	9.779	1.429	−2.811	11.127	1.267
X_0	0.532	0.017	0	0.518	0.021	0
Y_0	0.396	0.072	0	0.381	0.079	0
Z_0	0.354	−0.013	−9.404	0.409	−0.015	−10.230
E_0	0.182	−1.144	0.411	0.167	−1.254	0.461
0	−2.749	8.596	1.050	−2.788	9.638	0.871

i	$\Lambda_{\overline{MS}}^{(4)} = 390\,\text{MeV}$		
	$r_i^{(0)}$	$r_i^{(6)}$	$r_i^{(8)}$
0	−2.849	12.691	1.081
X_0	0.506	0.024	0
Y_0	0.367	0.087	0
Z_0	0.367	0.087	0
Z_0	0.470	−0.017	−11.164
E_0	0.153	−1.375	0.517
0	−2.825	10.813	0.669

The value for QCD sum rules is an average over the results given in ref. [86]. QCD sum rules also allow to derive lower bounds on the strange quark mass. It was found that generally $m_s(2\,\text{GeV}) \gtrsim 100\,\text{MeV}$ [87]. If these bounds hold, they would rule out the very low strange mass values found in unquenched lattice QCD simulations given above.

Finally, one should also mention the very recent determination of the strange mass from the hadronic τ-spectral function [88,89]:

$$m_s(2\,\text{GeV}) = (170^{+44}_{-55})\,\text{MeV}.$$

We observe that the central value is much larger than the corresponding results given above although the error is still large. In the future, however, improved experimental statistics and a better understanding of perturbative QCD corrections should make the determination of m_s from the τ-spectral function competitive to the other methods. On the other hand a very recent estimate using new τ-like ϕ-meson sum rules gives $m_s(2\,\text{GeV}) = (136 \pm 16)\,\text{MeV}$ [90].

We conclude that the error on m_s is still rather large. In our numerical analysis of ε'/ε, where m_s is evaluated at the scale m_c, we will set

$$m_s(m_c) = (130 \pm 25) \text{ MeV}, \tag{5.20}$$

roughly corresponding to $m_s(2 \text{ GeV})$ obtained in the lattice approach.

$B_6^{(1/2)}$ and $B_8^{(3/2)}$

The values for $B_6^{(1/2)}$ and $B_8^{(3/2)}$ obtained in various approaches are collected in table 6. The lattice results have been obtained at $\mu = 2\,\text{GeV}$. The results in the large–N approach and the chiral quark model correspond to scales below $1\,\text{GeV}$. However, as a detailed numerical analysis in ref. [74] showed, $B_6^{(1/2)}$ and $B_8^{(3/2)}$ are only weakly dependent on μ. Consequently the comparison of these parameters obtained in different approaches at different μ is meaningful.

Next, the values coming from lattice and chiral quark model are given in the NDR renormalization scheme. The corresponding values in the HV scheme can be found using approximate relations [35]

$$(B_6^{(1/2)})_{\text{HV}} \approx 1.2 (B_6^{(1/2)})_{\text{NDR}}, \qquad (B_8^{(3/2)})_{\text{HV}} \approx 1.2 (B_8^{(3/2)})_{\text{NDR}}. \tag{5.21}$$

The results in the large-N approach are unfortunately not sensitive to the renormalization scheme.

Concerning the lattice results for $B_6^{(1/2)}$, the old results read $B_{5,6}^{(1/2)}(2 \text{ GeV})$ $= 1.0 \pm 0.2$ [93,94]. More accurate estimates for $B_6^{(1/2)}$ are given in ref. [95]: $B_6^{(1/2)}(2 \text{ GeV}) = 0.67 \pm 0.04 \pm 0.05$ (quenched) and $B_6^{(1/2)}(2 \text{ GeV}) = 0.76 \pm 0.03 \pm 0.05$ ($f = 2$). However, a recent work in ref. [96] shows that lattice calculations of $B_6^{(1/2)}$ are very uncertain and one has to conclude that there are no solid predictions for $B_6^{(1/2)}$ from the lattice at present.

Table 6: Results for $B_6^{(1/2)}$ and $B_8^{(3/2)}$ obtained in various approaches.

Method	$B_6^{(1/2)}$	$B_8^{(3/2)}$
Lattice [39,91,37]	–	$0.69 - 1.06$
Large–N [92,43]	$0.72 - 1.10$	$0.42 - 0.64$
ChQM [81]	$1.07 - 1.58$	$0.75 - 0.79$

We observe that most non-perturbative approaches discussed above found $B_8^{(3/2)}$ below unity. The suppression of $B_8^{(3/2)}$ below unity is rather modest

40

(at most 20%) in the lattice approaches and in the chiral quark model. In the $1/N$ approach $B_8^{(3/2)}$ is rather strongly suppressed and can be as low as 0.5.

Concerning $B_6^{(1/2)}$ the situation is worse. As we stated above there is no solid prediction for this parameter in the lattice approach. On the other hand while the average value of $B_6^{(1/2)}$ in the $1/N$ approach is close to 1.0, the chiral quark model gives in the NDR scheme the value for $B_6^{(1/2)}$ as high as 1.33 ± 0.25. Interestingly both approaches give the ratio $B_6^{(1/2)}/B_8^{(3/2)}$ in the ball park of 1.7.

Guided by the results presented above and biased to some extent by the results from the large-N approach and lattice calculations, we will use in our numerical analysis below $B_6^{(1/2)}$ and $B_8^{(3/2)}$ in the ranges:

$$B_6^{(1/2)} = 1.0 \pm 0.3, \qquad B_8^{(3/2)} = 0.8 \pm 0.2 \qquad (5.22)$$

keeping always $B_6^{(1/2)} \geq B_8^{(3/2)}$.

$\Omega_{\eta+\eta'}$ and $\Lambda_{\overline{\rm MS}}^{(4)}$

The dependence of ε'/ε on $\Omega_{\eta+\eta'}$ can be studied numerically by using the formula (5.7) or incorporated approximately into the analytic formula (5.16) by simply replacing $B_6^{(1/2)}$ with an effective parameter

$$(B_6^{(1/2)})_{\rm eff} = B_6^{(1/2)} \frac{(1 - 0.9\,\Omega_{\eta+\eta'})}{0.775} . \qquad (5.23)$$

A numerical analysis shows that using $(1 - \Omega_{\eta+\eta'})$ overestimates the role of $\Omega_{\eta+\eta'}$. In our numerical analysis we have incorporated the uncertainty in $\Omega_{\eta+\eta'}$ by increasing the error in $B_6^{(1/2)}$ from ± 0.2 to ± 0.3.

The last estimates of $\Omega_{\eta+\eta'}$ were done more than ten years ago [65-67] and it is desirable to update these analyses which can be summarized by

$$\Omega_{\eta+\eta'} = 0.25 \pm 0.08 . \qquad (5.24)$$

In Table 7 we summarize the input parameters used in the numerical analysis of ε'/ε below. The range for $\Lambda_{\overline{\rm MS}}^{(4)}$ in Table 7 corresponds roughly to $\alpha_s(M_Z) = 0.119 \pm 0.003$.

5.6 Numerical Results for ε'/ε

In order to make predictions for ε'/ε we need the value of ${\rm Im}\lambda_t$. This can be obtained from the standard analysis of the unitarity triangle as discussed in Sec. 4.

Table 7: Collection of input parameters. We impose $B_6^{(1/2)} \geq B_8^{(3/2)}$.

Quantity	Central	Error	Reference
$\Lambda_{\overline{MS}}^{(4)}$	340 MeV	±50 MeV	8,97
$m_s(m_c)$	130 MeV	±25 MeV	See Text
$B_6^{(1/2)}$	1.0	±0.3	See Text
$B_8^{(3/2)}$	0.8	±0.2	See Text

In what follows we will present two types of numerical analyses of ε'/ε which use the methods 1 and 2 discussed already in Sec. 4. This analysis is based on ref. [35].

Using the first method we find the probability density distributions for ε'/ε in Fig. 9. From this distribution we deduce the following results:

$$\varepsilon'/\varepsilon = \begin{cases} (7.7 \, ^{+6.0}_{-3.5}) \cdot 10^{-4} & \text{(NDR)} \\ (5.2 \, ^{+4.6}_{-2.7}) \cdot 10^{-4} & \text{(HV)}. \end{cases} \qquad (5.25)$$

The difference between these two results indicates the left over renormalization scheme dependence. Since the resulting probability density distributions for ϵ'/ϵ are very asymmetric with very long tails towards large values we quote the medians and the 68%(95%) confidence level intervals. This means that 68%(95%) of our data can be found inside the corresponding error interval and that 50% of our data has smaller ϵ'/ϵ than our median.

We observe that negative values of ϵ'/ϵ can be excluded at 95% C.L. For completeness we quote the mean and the standard deviation for ϵ'/ϵ:

$$\varepsilon'/\varepsilon = \begin{cases} (9.1 \pm 6.2) \cdot 10^{-4} & \text{(NDR)} \\ (6.3 \pm 4.8) \cdot 10^{-4} & \text{(HV)}. \end{cases} \qquad (5.26)$$

Using the second method and the parameters in Table 2 we find :

$$1.05 \cdot 10^{-4} \leq \varepsilon'/\varepsilon \leq 28.8 \cdot 10^{-4} \qquad \text{(NDR)} \qquad (5.27)$$

and

$$0.26 \cdot 10^{-4} \leq \varepsilon'/\varepsilon \leq 22.0 \cdot 10^{-4} \qquad \text{(HV)}. \qquad (5.28)$$

We observe that ε'/ε is generally lower in the HV scheme if the same values for $B_6^{(1/2)}$ and $B_8^{(3/2)}$ are used in both schemes. Since the present non-perturbative methods do not have renormalization scheme dependence fully under control we think that such treatment of $B_6^{(1/2)}$ and $B_8^{(3/2)}$ is the proper way of estimating scheme dependences at present. Assuming, on the other hand, that the values in (5.22) correspond to the NDR scheme and using the relation (5.21), we find for the HV scheme the range $0.58 \cdot 10^{-4} \leq \varepsilon'/\varepsilon \leq 26.9 \cdot 10^{-4}$ which is much closer to the NDR result in (5.27). This exercise shows that it is very desirable to have the scheme dependence under control.

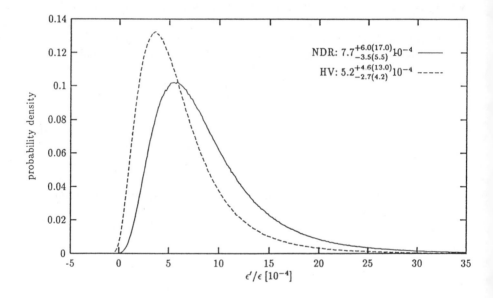

Figure 9: Probability density distributions for ε'/ε in NDR and HV schemes.

We observe that the most probable values for ε'/ε in the NDR scheme are in the ball park of $1 \cdot 10^{-3}$. They are lower by roughly 30% in the HV scheme if the same values for $(B_6^{(1/2)}, B_8^{(3/2)})$ are used. On the other hand the ranges in (5.27) and (5.28) show that for particular choices of the input parameters, values for ε'/ε as high as $(2-3) \cdot 10^{-3}$ cannot be excluded at present. Let us study this in more detail.

In Table 8, taken from ref. [35], we show the values of ε'/ε in units of 10^{-4} for specific values of $B_6^{(1/2)}$, $B_8^{(3/2)}$ and $m_s(m_c)$ as calculated in the NDR scheme. The corresponding values in the HV scheme are lower as discussed above. The fourth column shows the results for central values of all remaining parameters. The comparison of the the fourth and the fifth column demonstrates how ε'/ε is increased when $\Lambda_{\overline{MS}}^{(4)}$ is raised from 340 MeV to 390 MeV. As stated in (5.4) ε'/ε is roughly proportional to $\Lambda_{\overline{MS}}^{(4)}$. Finally, in the last column maximal values of ε'/ε are given. To this end we have scanned all parameters relevant for the analysis of $\text{Im}\lambda_t$ within one standard deviation and have chosen the highest value of $\Lambda_{\overline{MS}}^{(4)} = 390\,\text{MeV}$. Comparison of the last two columns demonstrates the impact of the increase of $\text{Im}\lambda_t$ from its central to its maximal value and of the variation of m_t.

Table 8 gives a good insight in the dependence of ε'/ε on various parameters which is roughly described by (5.4). We observe the following hierarchies:

- The largest uncertainties reside in m_s, $B_6^{(1/2)}$ and $B_8^{(3/2)}$. ε'/ε increases universally by roughly a factor of 2.3 when $m_s(m_c)$ is changed from 155 MeV to 105 MeV. The increase of $B_6^{(1/2)}$ from 1.0 to 1.3 increases ε'/ε by $(55 \pm 10)\%$, depending on m_s and $B_8^{(3/2)}$. The corresponding changes due to $B_8^{(3/2)}$ are approximately $(40 \pm 15)\%$.

- The combined uncertainty due to $\text{Im}\lambda_t$ and m_t, present both in $\text{Im}\lambda_t$ and $F_{\varepsilon'}$, is approximately $\pm25\%$. The uncertainty due to m_t alone is only $\pm5\%$.

- The uncertainty due to $\Lambda_{\overline{MS}}^{(4)}$ is approximately $\pm16\%$.

- The uncertainty due to $\Omega_{\eta+\eta'}$ is approximately $\pm12\%$.

The large sensitivity of ε'/ε to m_s has been known since the analyses in the eighties. In the context of the KTeV result this issue has been analyzed in ref. [98]. It has been found that provided $2B_6^{(1/2)} - B_8^{(3/2)} \leq 2$ the consistency of the Standard Model with the KTeV result requires the 2σ bound $m_s(2\,\text{GeV}) \leq 110\,\text{MeV}$. Our analysis is compatible with these findings.

5.7 Summary

As we have seen, the estimates of ε'/ε in the Standard Model are typically below the experimental data. However, as our scanning analysis shows, for suitably chosen parameters, ε'/ε in the Standard Model can be made consistent with data. However, this happens only if all relevant parameters are

Table 8: Values of ε'/ε in units of 10^{-4} for specific values of $B_6^{(1/2)}$, $B_8^{(3/2)}$ and $m_s(m_c)$ and other parameters as explained in the text.

$B_6^{(1/2)}$	$B_8^{(3/2)}$	$m_s(m_c)[\,\mathrm{MeV}]$	Central	$\Lambda_{\overline{\mathrm{MS}}}^{(4)} = 390\,\mathrm{MeV}$	Maximal
1.3	0.6	105	20.2	23.3	28.8
		130	12.8	14.8	18.3
		155	8.5	9.9	12.3
1.3	0.8	105	18.1	20.8	26.0
		130	11.3	13.1	16.4
		155	7.5	8.7	10.9
1.3	1.0	105	15.9	18.3	23.2
		130	9.9	11.5	14.5
		155	6.5	7.6	9.6
1.0	0.6	105	13.7	15.8	19.7
		130	8.4	9.8	12.2
		155	5.4	6.4	7.9
1.0	0.8	105	11.5	13.3	16.9
		130	7.0	8.1	10.4
		155	4.4	5.2	6.6
1.0	1.0	105	9.4	10.9	14.1
		130	5.5	6.5	8.5
		155	3.3	4.0	5.2

simultaneously close to their extreme values. This is clearly seen in Table 8. Moreover, the probability density distributions for ε'/ε in Fig. 9 indicates that values of ε'/ε in the ball park of NA31 and KTeV results are rather improbable.

Unfortunately, in view of very large hadronic and substantial parametric uncertainties, it is impossible to conclude at present whether new physics contributions are indeed required to fit the data. Similarly it is difficult to conclude what is precisely the impact of the ε'/ε-data on the CKM matrix. However, as analyzed in ref. [35] there are indications that the lower limit on $\mathrm{Im}\lambda_t$ is improved. The same applies to the lower limits for the branching ratios for $K_L \to \pi^0 \nu\bar\nu$ and $K_L \to \pi^0 e^+ e^-$ decays discussed in the following sections.

It is also clear that the ε'/ε data puts models in which there are new positive contributions to ε and negative contibutions to ε' in serious difficulties. In particular as analyzed in ref. [35] the two Higgs Doublet Model II[99] can either be ruled out with improved hadronic matrix elements or a powerful lower bound on $\tan\beta$ can be obtained from ε'/ε. In the Minimal Supersymmetric Standard Model, in addition to charged Higgs exchanges in loop diagrams, also charginos contribute. For suitable choice of supersymmetric parameters, the chargino contribution can enhance ε'/ε with respect to Standard Model expectations[100]. Yet, generally the most conspicuous effect of minimal supersymmetry is a depletion of ε'/ε. The situation can be different in more general models in which there are more parameters than in the two Higgs doublet model II and in the MSSM, in particular new CP violating phases. As an example, in more general supersymmetric models ε'/ε can be made consistent with experimental findings [101,102]. Unfortunately, in view of the large number of free parameters such models are not very predictive.

The future of ε'/ε in the Standard Model and in its extensions depends on the progress in the reduction of parametric and hadronic uncertainties. In any case ε'/ε already played a decisive role in establishing direct CP violation in nature and its rather large value gives additional strong motivation for searching for this phenomenon in cleaner K decays like $K_L \to \pi^0 \nu\bar\nu$ and $K_L \to \pi^0 e^+ e^-$, in B decays, in D decays and elsewhere. We now turn to discuss some of these topics.

6 The Decays $K^+ \to \pi^+\nu\bar\nu$ and $K_L \to \pi^0\nu\bar\nu$

6.1 General Remarks

We will now move on to discuss the semileptonic rare FCNC transitions $K^+ \to \pi^+\nu\bar\nu$ and $K_L \to \pi^0\nu\bar\nu$. Within the Standard Model these decays are loop-induced semileptonic FCNC processes determined only by Z^0-penguin and box diagrams and are governed by the single function $X_0(x_t)$ given in (2.15).

A particular and very important virtue of $K \to \pi\nu\bar\nu$ is their clean theoretical character. This is related to the fact that the low energy hadronic matrix elements required are just the matrix elements of quark currents between hadron states, which can be extracted from the leading (non-rare) semileptonic decays. Other long-distance contributions are negligibly small [103,104]. As a consequence of these features, the scale ambiguities, inherent to perturbative QCD, essentially constitute the only theoretical uncertainties present in the analysis of these decays. These theoretical uncertainties have been considerably reduced through the inclusion of next-to-leading QCD corrections in ref's [105–109].

The investigation of these low energy rare decay processes in conjunction with their theoretical cleanliness, allows to probe, albeit indirectly, high energy scales of the theory and in particular to measure V_{td} and $\text{Im}\lambda_t = \text{Im}V_{ts}^* V_{td}$ from $K^+ \to \pi^+\nu\bar\nu$ and $K_L \to \pi^0\nu\bar\nu$ respectively. However, the very fact that these processes are based on higher order electroweak effects implies that their branching ratios are expected to be very small and not easy to access experimentally.

6.2   The Decay $K^+ \to \pi^+ \nu \bar{\nu}$

The effective Hamiltonian

The effective Hamiltonian for $K^+ \to \pi^+ \nu \bar{\nu}$ can be written as

$$\mathcal{H}_{\text{eff}} = \frac{G_F}{\sqrt{2}} \frac{\alpha}{2\pi \sin^2 \Theta_W} \sum_{l=e,\mu,\tau} \left(V_{cs}^* V_{cd} X_{NL}^l + V_{ts}^* V_{td} X(x_t) \right) (\bar{s}d)_{V-A} (\bar{\nu}_l \nu_l)_{V-A} . \tag{6.1}$$

The index $l=e, \mu, \tau$ denotes the lepton flavour. The dependence on the charged lepton mass resulting from the box-graph is negligible for the top contribution. In the charm sector this is the case only for the electron and the muon but not for the τ-lepton.

The function $X(x_t)$ relevant for the top part is given by

$$X(x_t) = X_0(x_t) + \frac{\alpha_s}{4\pi} X_1(x_t) = \eta_X \cdot X_0(x_t), \qquad \eta_X = 0.994, \tag{6.2}$$

with the QCD correction [106,108,109]

$$X_1(x_t) = \tilde{X}_1(x_t) + 8x_t \frac{\partial X_0(x_t)}{\partial x_t} \ln x_\mu . \tag{6.3}$$

Here $x_\mu = \mu_t^2 / M_W^2$ with $\mu_t = \mathcal{O}(m_t)$ and $\tilde{X}_1(x_t)$ is a complicated function given in ref's [106,108,109]. The μ_t-dependence of the last term in (6.3) cancels to the considered order the μ_t-dependence of the leading term $X_0(x_t(\mu))$. The leftover μ_t-dependence in $X(x_t)$ is below 1%. The factor η_X summarizes the NLO corrections represented by the second term in (6.2). With $m_t \equiv \overline{m}_t(m_t)$ the QCD factor η_X is practically independent of m_t and $\Lambda_{\overline{MS}}$ and is very close to unity.

The expression corresponding to $X(x_t)$ in the charm sector is the function X_{NL}^l. It results from the NLO calculation [107] and is given explicitly in ref. [109]. The inclusion of NLO corrections reduced considerably the large μ_c dependence (with $\mu_c = \mathcal{O}(m_c)$) present in the leading order expressions for the charm contribution [110]. Varying μ_c in the range $1\,\text{GeV} \leq \mu_c \leq 3\,\text{GeV}$ changes X_{NL} by roughly 24% after the inclusion of NLO corrections to be compared with 56% in the leading order. Further details can be found in ref's [107,2]. The impact of the μ_c uncertainties on the resulting branching ratio $Br(K^+ \to \pi^+ \nu \bar{\nu})$ is discussed below.

The numerical values for X_{NL}^l for $\mu = m_c$ and several values of $\Lambda_{\overline{MS}}^{(4)}$ and $m_c(m_c)$ can be found in ref. [109]. The net effect of QCD corrections is to suppress the charm contribution by roughly 30%. For our purposes we need only

$$P_0(X) = \frac{1}{\lambda^4} \left[\frac{2}{3} X_{NL}^e + \frac{1}{3} X_{NL}^\tau \right] = 0.42 \pm 0.06 \qquad (6.4)$$

where the error results from the variation of $\Lambda_{\overline{MS}}^{(4)}$ and $m_c(m_c)$.

Deriving the Branching Ratio

The relevant hadronic matrix element of the weak current $(\bar{s}d)_{V-A}$ in (6.1) can be extracted with the help of isospin symmetry from the leading decay $K^+ \to \pi^0 e^+ \nu$. Consequently the resulting theoretical expression for the branching fraction $Br(K^+ \to \pi^+ \nu\bar{\nu})$ can be related to the experimentally well known quantity $Br(K^+ \to \pi^0 e^+ \nu)$. Let us demonstrate this.

The effective Hamiltonian for the tree level decay $K^+ \to \pi^0 e^+ \nu$ is given by

$$\mathcal{H}_{eff}(K^+ \to \pi^0 e^+ \nu) = \frac{G_F}{\sqrt{2}} V_{us}^* (\bar{s}u)_{V-A} (\bar{\nu}_e e)_{V-A} . \qquad (6.5)$$

Using isospin symmetry we have

$$\langle \pi^+ | (\bar{s}d)_{V-A} | K^+ \rangle = \sqrt{2} \langle \pi^0 | (\bar{s}u)_{V-A} | K^+ \rangle . \qquad (6.6)$$

Consequently neglecting differences in the phase space of these two decays due to $m_{\pi^+} \neq m_{\pi^0}$ and $m_e \neq 0$ we find

$$\frac{Br(K^+ \to \pi^+ \nu\bar{\nu})}{Br(K^+ \to \pi^0 e^+ \nu)} = \frac{\alpha^2}{|V_{us}|^2 2\pi^2 \sin^4 \Theta_W} \sum_{l=e,\mu,\tau} \left| V_{cs}^* V_{cd} X_{NL}^l + V_{ts}^* V_{td} X(x_t) \right|^2 . \qquad (6.7)$$

Basic Phenomenology

Using (6.7) and including isospin breaking corrections one finds

$$Br(K^+ \to \pi^+ \nu\bar{\nu}) = \kappa_+ \cdot \left[\left(\frac{Im\lambda_t}{\lambda^5} X(x_t) \right)^2 + \left(\frac{Re\lambda_c}{\lambda} P_0(X) + \frac{Re\lambda_t}{\lambda^5} X(x_t) \right)^2 \right] , \qquad (6.8)$$

$$\kappa_+ = r_{K+} \frac{3\alpha^2 Br(K^+ \to \pi^0 e^+ \nu)}{2\pi^2 \sin^4 \Theta_W} \lambda^8 = 4.11 \cdot 10^{-11} , \qquad (6.9)$$

48

where we have used

$$\alpha = \frac{1}{129}, \qquad \sin^2 \Theta_W = 0.23, \qquad Br(K^+ \to \pi^0 e^+ \nu) = 4.82 \cdot 10^{-2}. \quad (6.10)$$

Here $\lambda_i = V_{is}^* V_{id}$ with λ_c being real to a very high accuracy. $r_{K+} = 0.901$ summarizes isospin breaking corrections in relating $K^+ \to \pi^+ \nu \bar{\nu}$ to $K^+ \to \pi^0 e^+ \nu$. These isospin breaking corrections are due to quark mass effects and electroweak radiative corrections and have been calculated in ref. [111]. Finally $P_0(X)$ is given in (6.4).

Using the improved Wolfenstein parametrization and the approximate formulae (1.22) – (1.24) we can next put (6.8) into a more transparent form [10]:

$$Br(K^+ \to \pi^+ \nu \bar{\nu}) = 4.11 \cdot 10^{-11} A^4 X^2(x_t) \frac{1}{\sigma} \left[(\sigma \bar{\eta})^2 + (\varrho_0 - \bar{\varrho})^2 \right], \quad (6.11)$$

where

$$\sigma = \left(\frac{1}{1 - \frac{\lambda^2}{2}} \right)^2. \quad (6.12)$$

The measured value of $Br(K^+ \to \pi^+ \nu \bar{\nu})$ then determines an ellipse in the $(\bar{\varrho}, \bar{\eta})$ plane centered at $(\varrho_0, 0)$ with

$$\varrho_0 = 1 + \frac{P_0(X)}{A^2 X(x_t)} \quad (6.13)$$

and having the squared axes

$$\bar{\varrho}_1^2 = r_0^2, \qquad \bar{\eta}_1^2 = \left(\frac{r_0}{\sigma} \right)^2 \quad (6.14)$$

where

$$r_0^2 = \frac{1}{A^4 X^2(x_t)} \left[\frac{\sigma \cdot Br(K^+ \to \pi^+ \nu \bar{\nu})}{4.11 \cdot 10^{-11}} \right]. \quad (6.15)$$

Note that r_0 depends only on the top contribution. The departure of ϱ_0 from unity measures the relative importance of the internal charm contributions.

The ellipse defined by r_0, ϱ_0 and σ given above intersects with the circle (1.32). This allows to determine $\bar{\varrho}$ and $\bar{\eta}$ with

$$\bar{\varrho} = \frac{1}{1 - \sigma^2} \left(\varrho_0 - \sqrt{\sigma^2 \varrho_0^2 + (1 - \sigma^2)(r_0^2 - \sigma^2 R_b^2)} \right), \qquad \bar{\eta} = \sqrt{R_b^2 - \bar{\varrho}^2}$$

$$(6.16)$$

and consequently

$$R_t^2 = 1 + R_b^2 - 2\bar{\varrho}, \quad (6.17)$$

where $\bar{\eta}$ is assumed to be positive. Given $\bar{\varrho}$ and $\bar{\eta}$ one can determine V_{td}:

$$V_{td} = A\lambda^3(1 - \bar{\varrho} - i\bar{\eta}), \qquad |V_{td}| = A\lambda^3 R_t. \qquad (6.18)$$

The determination of $|V_{td}|$ and of the unitarity triangle requires the knowledge of V_{cb} (or A) and of $|V_{ub}/V_{cb}|$. Both values are subject to theoretical uncertainties present in the existing analyses of tree level decays. Whereas the dependence on $|V_{ub}/V_{cb}|$ is rather weak, the very strong dependence of $Br(K^+ \to \pi^+\nu\bar{\nu})$ on A or V_{cb} makes a precise prediction for this branching ratio difficult at present. We will return to this below. The dependence of $Br(K^+ \to \pi^+\nu\bar{\nu})$ on m_t is also strong. However m_t is known already within $\pm 4\%$ and consequently the related uncertainty in $Br(K^+ \to \pi^+\nu\bar{\nu})$ is substantialy smaller than the corresponding uncertainty due to V_{cb}.

Numerical Analysis of $K^+ \to \pi^+\nu\bar{\nu}$

The uncertainties in the prediction for $Br(K^+ \to \pi^+\nu\bar{\nu})$ and in the determination of $|V_{td}|$ related to the choice of the renormalization scales μ_t and μ_c in the top part and the charm part, respectively have been inestigated in ref. [2]. To this end the scales μ_c and μ_t entering $m_c(\mu_c)$ and $m_t(\mu_t)$, respectively, have been varied in the ranges $1\,\text{GeV} \le \mu_c \le 3\,\text{GeV}$ and $100\,\text{GeV} \le \mu_t \le 300\,\text{GeV}$. It has been found that including the full next-to-leading corrections reduces the uncertainty in the determination of $|V_{td}|$ from $\pm 14\%$ (LO) to $\pm 4.6\%$ (NLO). The main bulk of this theoretical error stems from the charm sector. In the case of $Br(K^+ \to \pi^+\nu\bar{\nu})$, the theoretical uncertainty due to $\mu_{c,t}$ is reduced from $\pm 22\%$ (LO) to $\pm 7\%$ (NLO).

Scanning the input parameters of Table 2 we find

$$Br(K^+ \to \pi^+\nu\bar{\nu}) = (7.9 \pm 3.1) \cdot 10^{-11} \qquad (6.19)$$

where the error comes dominantly from the uncertainties in the CKM parameters.

It is possible to derive an upper bound on $Br(K^+ \to \pi^+\nu\bar{\nu})$ [109]:

$$Br(K^+ \to \pi^+\nu\bar{\nu})_{\max} = \frac{\kappa_+}{\sigma} \left[P_0(X) + A^2 X(x_t) \frac{r_{sd}}{\lambda} \sqrt{\frac{\Delta M_d}{\Delta M_s}} \right]^2 \qquad (6.20)$$

where $r_{ds} = \xi\sqrt{m_{B_s}/m_{B_d}}$. This equation translates a lower bound on ΔM_s into an upper bound on $Br(K^+ \to \pi^+\nu\bar{\nu})$. This bound is very clean and does not involve theoretical hadronic uncertainties except for r_{sd}. Using

$$\sqrt{\frac{\Delta M_d}{\Delta M_s}} < 0.2 , \quad A < 0.87 , \quad P_0(X) < 0.48 , \quad X(x_t) < 1.56 , \quad r_{sd} < 1.2 \qquad (6.21)$$

we find

$$Br(K^+ \rightarrow \pi^+ \nu\bar{\nu})_{\max} = 12.2 \cdot 10^{-11} \ . \tag{6.22}$$

This limit could be further strengthened with improved input. However, this bound is strong enough to indicate a clear conflict with the Standard Model if $Br(K^+ \rightarrow \pi^+ \nu\bar{\nu})$ should be measured at $2 \cdot 10^{-10}$.

$|V_{td}|$ from $K^+ \rightarrow \pi^+ \nu\bar{\nu}$

Once $Br(K^+ \rightarrow \pi^+ \nu\bar{\nu}) \equiv Br(K^+)$ is measured, $|V_{td}|$ can be extracted subject to various uncertainties:

$$\frac{\sigma(|V_{td}|)}{|V_{td}|} = \pm 0.04_{scale} \pm \frac{\sigma(|V_{cb}|)}{|V_{cb}|} \pm 0.7 \frac{\sigma(\bar{m}_c)}{\bar{m}_c} \pm 0.65 \frac{\sigma(Br(K^+))}{Br(K^+)} \ . \tag{6.23}$$

Taking $\sigma(|V_{cb}|) = 0.002$, $\sigma(\bar{m}_c) = 100\,\mathrm{MeV}$ and $\sigma(Br(K^+)) = 10\%$ and adding the errors in quadrature we find that $|V_{td}|$ can be determined with an accuracy of $\pm 10\%$. This number is increased to $\pm 11\%$ once the uncertainties due to m_t, α_s and $|V_{ub}|/|V_{cb}|$ are taken into account. Clearly this determination can be improved although a determination of $|V_{td}|$ with an accuracy better than $\pm 5\%$ seems rather unrealistic.

Summary and Outlook

The accuracy of the Standard Model prediction for $Br(K^+ \rightarrow \pi^+ \nu\bar{\nu})$ has improved considerably during the last five years. This progress can be traced back to the improved values of m_t and $|V_{cb}|$ and to the inclusion of NLO QCD corrections which considerably reduced the scale uncertainties in the charm sector.

Now, what about the experimental status of this decay? One of the highlights of 97 was the observation by BNL787 collaboration at Brookhaven[112] of one event consistent with the signature expected for this decay. The branching ratio:

$$Br(K^+ \rightarrow \pi^+ \nu\bar{\nu}) = (4.2^{+9.7}_{-3.5}) \cdot 10^{-10} \tag{6.24}$$

has the central value by a factor of 5 above the Standard Model expectation but in view of large errors the result is compatible with the Standard Model. The analysis of additional data on $K^+ \rightarrow \pi^+ \nu\bar{\nu}$ present on tape at BNL787 should narrow this range in the near future considerably. In view of the clean character of this decay a measurement of its branching ratio at the level of $2 \cdot 10^{-10}$ would signal the presence of physics beyond the Standard Model. The Standard Model sensitivity is expected to be reached at AGS around the year 2000[113]. Also Fermilab with the Main Injector could measure this decay (see ref.[114]).

6.3 The Decay $K_{\rm L} \to \pi^0 \nu \bar\nu$

The effective Hamiltonian

The effective Hamiltonian for $K_{\rm L} \to \pi^0 \nu \bar\nu$ is given as follows:

$$\mathcal{H}_{\rm eff} = \frac{G_{\rm F}}{\sqrt{2}} \frac{\alpha}{2\pi \sin^2 \Theta_{\rm W}} V_{ts}^* V_{td} X(x_t) (\bar s d)_{V-A} (\bar\nu\nu)_{V-A} + h.c. , \qquad (6.25)$$

where the function $X(x_t)$, present already in $K^+ \to \pi^+ \nu \bar\nu$, includes NLO corrections and is given in (6.2).

As we will demonstrate shortly, $K_{\rm L} \to \pi^0 \nu \bar\nu$ proceeds in the Standard Model almost entirely through direct CP violation [115]. It is completely dominated by short-distance loop diagrams with top quark exchanges. The charm contribution can be fully neglected and the theoretical uncertainties present in $K^+ \to \pi^+ \nu \bar\nu$ due to m_c, μ_c and $\Lambda_{\overline{MS}}$ are absent here. Consequently the rare decay $K_{\rm L} \to \pi^0 \nu \bar\nu$ is even cleaner than $K^+ \to \pi^+ \nu \bar\nu$ and is very well suited for the determination of the Wolfenstein parameter η and in particular $\mathrm{Im}\lambda_t$.

It is usually stated in the literature that the decay $K_{\rm L} \to \pi^0 \nu \bar\nu$ is dominated by *direct* CP violation. Now the standard definition of the direct CP violation requires the presence of strong phases which are completely negligible in $K_{\rm L} \to \pi^0 \nu \bar\nu$. Consequently the violation of CP symmetry in $K_{\rm L} \to \pi^0 \nu \bar\nu$ arises through the interference between $K^0 - \bar K^0$ mixing and the decay amplitude. This type of CP violation is often called *mixing-induced* CP violation. However, as already pointed out by Littenberg [115] and demonstrated explictly in a moment, the contribution of CP violation to $K_{\rm L} \to \pi^0 \nu \bar\nu$ via $K^0 - \bar K^0$ mixing alone is tiny. It gives $Br(K_{\rm L} \to \pi^0 \nu \bar\nu) \approx 2 \cdot 10^{-15}$. Consequently, in this sence, CP violation in $K_{\rm L} \to \pi^0 \nu \bar\nu$ with $Br(K_{\rm L} \to \pi^0 \nu \bar\nu) = \mathcal{O}(10^{-11})$ is a manifestation of CP violation in the decay and as such deserves the name of *direct* CP violation. In other words the difference in the magnitude of CP violation in $K_{\rm L} \to \pi\pi$ (ε) and $K_{\rm L} \to \pi^0 \nu \bar\nu$ is a signal of direct CP violation and measuring $K_{\rm L} \to \pi^0 \nu \bar\nu$ at the expected level would be another signal of this phenomenon. More details on this issue can be found in ref's [116,117,118].

Deriving the Branching Ratio

Let us derive the basic formula for $Br(K_{\rm L} \to \pi^0 \nu \bar\nu)$ in a manner analogous to the one for $Br(K^+ \to \pi^+ \nu \bar\nu)$. To this end we consider one neutrino flavour and define the complex function:

$$F = \frac{G_{\rm F}}{\sqrt{2}} \frac{\alpha}{2\pi \sin^2 \Theta_{\rm W}} V_{ts}^* V_{td} X(x_t). \qquad (6.26)$$

Then the effective Hamiltonian in (6.25) can be written as

$$\mathcal{H}_{\text{eff}} = F(\bar{s}d)_{V-A}(\bar{\nu}\nu)_{V-A} + F^*(\bar{d}s)_{V-A}(\bar{\nu}\nu)_{V-A} . \tag{6.27}$$

Now, from (3.8) we have

$$K_L = \frac{1}{\sqrt{2}}[(1+\bar{\varepsilon})K^0 + (1-\bar{\varepsilon})\bar{K}^0] \tag{6.28}$$

where we have neglected $|\bar{\varepsilon}|^2 \ll 1$. Thus the amplitude for $K_L \to \pi^0 \nu \bar{\nu}$ is given by

$$\begin{aligned} A(K_L \to \pi^0 \nu \bar{\nu}) &= \frac{1}{\sqrt{2}} \left[F(1+\bar{\varepsilon})\langle \pi^0|(\bar{s}d)_{V-A}|K^0\rangle \right. \\ &+ \left. F^*(1-\bar{\varepsilon})\langle \pi^0|(\bar{d}s)_{V-A}|\bar{K}^0\rangle(\bar{\nu}\nu)_{V-A}. \end{aligned}$$

$$\tag{6.30}$$

Recalling

$$CP|K^0\rangle = -|\bar{K}^0\rangle, \qquad C|K^0\rangle = |\bar{K}^0\rangle \tag{6.31}$$

we have

$$\langle \pi^0|(\bar{d}s)_{V-A}|\bar{K}^0\rangle = -\langle \pi^0|(\bar{s}d)_{V-A}|K^0\rangle, \tag{6.32}$$

where the minus sign is crucial for the subsequent steps.

Thus we can write

$$A(K_L \to \pi^0 \nu \bar{\nu}) = \frac{1}{\sqrt{2}} \left[F(1+\bar{\varepsilon}) - F^*(1-\bar{\varepsilon}) \right] \langle \pi^0|(\bar{s}d)_{V-A}|K^0\rangle(\bar{\nu}\nu)_{V-A}.$$

$$\tag{6.33}$$

Now the terms $\bar{\varepsilon}$ can be safely neglected in comparision with unity, which implies that the indirect CP violation (CP violation in the $K^0 - \bar{K}^0$ mixing) is negligible in this decay. We have then

$$F(1+\bar{\varepsilon}) - F^*(1-\bar{\varepsilon}) = \frac{G_F}{\sqrt{2}} \frac{\alpha}{\pi \sin^2 \Theta_W} \text{Im}(V_{ts}^* V_{td}) \cdot X(x_t). \tag{6.34}$$

Consequently using isospin relation

$$\langle \pi^0|(\bar{d}s)_{V-A}|\bar{K}^0\rangle = \langle \pi^0|(\bar{s}u)_{V-A}|K^+\rangle \tag{6.35}$$

together with (6.5) and taking into account the difference in the lifetimes of K_L and K^+ we have after summation over three neutrino flavours

$$\frac{Br(K_L \to \pi^0 \nu \bar{\nu})}{Br(K^+ \to \pi^0 e^+ \nu)} = 3 \frac{\tau(K_L)}{\tau(K^+)} \frac{\alpha^2}{|V_{us}|^2 2\pi^2 \sin^4 \Theta_W} \left[\text{Im}\lambda_t \cdot X(x_t)\right]^2 \tag{6.36}$$

where $\lambda_t = V_{ts}^* V_{td}$.

Master Formulae for $Br(K_L \to \pi^0 \nu \bar{\nu})$

Using (6.36) we can write $Br(K_L \to \pi^0 \nu \bar{\nu})$ simply as follows

$$Br(K_L \to \pi^0 \nu \bar{\nu}) = \kappa_L \cdot \left(\frac{\text{Im}\lambda_t}{\lambda^5} X(x_t) \right)^2, \qquad (6.37)$$

$$\kappa_L = \frac{r_{K_L}}{r_{K^+}} \frac{\tau(K_L)}{\tau(K^+)} \kappa_+ = 1.80 \cdot 10^{-10} \qquad (6.38)$$

with κ_+ given in (6.9) and $r_{K_L} = 0.944$ summarizing isospin breaking corrections in relating $K_L \to \pi^0 \nu \bar{\nu}$ to $K^+ \to \pi^0 e^+ \nu$ [111].

Using the Wolfenstein parametrization and (6.2) we can rewrite (6.37) as

$$Br(K_L \to \pi^0 \nu \bar{\nu}) = 3.0 \cdot 10^{-11} \left[\frac{\eta}{0.39} \right]^2 \left[\frac{\overline{m}_t(m_t)}{170 \text{ GeV}} \right]^{2.3} \left[\frac{|V_{cb}|}{0.040} \right]^4. \qquad (6.39)$$

The determination of η using $Br(K_L \to \pi^0 \nu \bar{\nu})$ requires the knowledge of V_{cb} and m_t. The very strong dependence on V_{cb} or A makes a precise prediction for this branching ratio difficult at present.

On the other hand inverting (6.37) and using (6.2) one finds [118]:

$$\text{Im}\lambda_t = 1.36 \cdot 10^{-4} \left[\frac{170 \text{ GeV}}{\overline{m}_t(m_t)} \right]^{1.15} \left[\frac{Br(K_L \to \pi^0 \nu \bar{\nu})}{3 \cdot 10^{-11}} \right]^{1/2} \qquad (6.40)$$

without any uncertainty in $|V_{cb}|$. (6.40) offers the cleanest method to measure $\text{Im}\lambda_t$; even better than the CP asymmetries in B decays discussed briefly in Sec. 8.

Numerical Analysis of $K_L \to \pi^0 \nu \bar{\nu}$

The μ_t-uncertainties present in the function $X(x_t)$ have already been discussed in connection with $K^+ \to \pi^+ \nu \bar{\nu}$. After the inclusion of NLO corrections they are so small that they can be neglected for all practical purposes. Scanning the input parameters of Table 2 we find

$$Br(K_L \to \pi^0 \nu \bar{\nu}) = (2.8 \pm 1.1) \cdot 10^{-11} \qquad (6.41)$$

where the error comes dominantly from the uncertainties in the CKM parameters.

54

Summary and Outlook

The accuracy of the Standard Model prediction for $Br(K_L \to \pi^0 \nu\bar{\nu})$ has improved considerably during the last five years. This progress can be traced back mainly to the improved values of m_t and $|V_{cb}|$ and to some extent to the inclusion of NLO QCD corrections.

The present upper bound on $Br(K_L \to \pi^0 \nu\bar{\nu})$ from FNAL experiment E799[119] is

$$Br(K_L \to \pi^0 \nu\bar{\nu}) < 1.6 \cdot 10^{-6} . \tag{6.42}$$

This is about five orders of magnitude above the Standard Model expectation (6.41). Moreover this bound is substantially weaker than the *model independent* bound[116] from isospin symmetry:

$$Br(K_L \to \pi^0 \nu\bar{\nu}) < 4.4 \cdot Br(K^+ \to \pi^+ \nu\bar{\nu}) \tag{6.43}$$

which through (6.24) gives

$$Br(K_L \to \pi^0 \nu\bar{\nu}) < 6.1 \cdot 10^{-9} \tag{6.44}$$

Now FNAL-E799 expects to reach the accuracy $\mathcal{O}(10^{-8})$ and a very interesting new experiment at Brookhaven (BNL E926)[113] expects to reach the single event sensitivity $2 \cdot 10^{-12}$ allowing a 10% measurement of the expected branching ratio. There are furthermore plans to measure this gold-plated decay with comparable sensitivity at Fermilab[120] and KEK[121].

6.4   Unitarity Triangle and $\sin 2\beta$ from $K \to \pi\nu\bar{\nu}$

The measurement of $Br(K^+ \to \pi^+ \nu\bar{\nu})$ and $Br(K_L \to \pi^0 \nu\bar{\nu})$ can determine the unitarity triangle completely, (see Fig. 10), provided m_t and V_{cb} are known[122]. Using these two branching ratios simultaneously allows to eliminate $|V_{ub}/V_{cb}|$ from the analysis which removes a considerable uncertainty. Indeed it is evident from (6.8) and (6.37) that, given $Br(K^+ \to \pi^+ \nu\bar{\nu})$ and $Br(K_L \to \pi^0 \nu\bar{\nu})$, one can extract both $\text{Im}\lambda_t$ and $\text{Re}\lambda_t$. One finds[122,2]

$$\text{Im}\lambda_t = \lambda^5 \frac{\sqrt{B_2}}{X(x_t)}, \qquad \text{Re}\lambda_t = -\lambda^5 \frac{\frac{\text{Re}\lambda_c}{\lambda} P_0(X) + \sqrt{B_1 - B_2}}{X(x_t)}, \tag{6.45}$$

where we have defined the "reduced" branching ratios

$$B_1 = \frac{Br(K^+ \to \pi^+ \nu\bar{\nu})}{4.11 \cdot 10^{-11}}, \qquad B_2 = \frac{Br(K_L \to \pi^0 \nu\bar{\nu})}{1.80 \cdot 10^{-10}} . \tag{6.46}$$

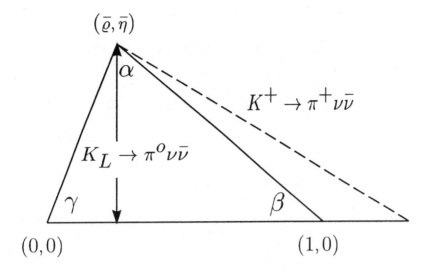

Figure 10: Unitarity triangle from $K \to \pi \nu \bar{\nu}$.

Using next the expressions for $\mathrm{Im}\lambda_t$, $\mathrm{Re}\lambda_t$ and $\mathrm{Re}\lambda_c$ given in (1.22)–(1.24) we find

$$\bar{\varrho} = 1 + \frac{P_0(X) - \sqrt{\sigma(B_1 - B_2)}}{A^2 X(x_t)}, \qquad \bar{\eta} = \frac{\sqrt{B_2}}{\sqrt{\sigma} A^2 X(x_t)} \qquad (6.47)$$

with σ defined in (6.12). An exact treatment of the CKM matrix shows that the formulae (6.47) are rather precise [122].

Using (6.47) one finds subsequently [122]

$$r_s = r_s(B_1, B_2) \equiv \frac{1 - \bar{\varrho}}{\bar{\eta}} = \cot\beta, \qquad \sin 2\beta = \frac{2r_s}{1 + r_s^2} \qquad (6.48)$$

with

$$r_s(B_1, B_2) = \sqrt{\sigma}\frac{\sqrt{\sigma(B_1 - B_2)} - P_0(X)}{\sqrt{B_2}}. \qquad (6.49)$$

Thus within the approximation of (6.47) $\sin 2\beta$ is independent of V_{cb} (or A) and m_t.

It should be stressed that $\sin 2\beta$ determined this way depends only on two measurable branching ratios and on the function $P_0(X)$ which is completely calculable in perturbation theory. Consequently this determination is free from

56

Table 9: Illustrative example of the determination of CKM parameters from $K \to \pi\nu\bar{\nu}$.

	$\sigma(|V_{cb}|) = \pm0.002$	$\sigma(|V_{cb}|) = \pm0.001$
$\sigma(|V_{td}|)$	$\pm10\%$	$\pm9\%$
$\sigma(\bar{\varrho})$	±0.16	±0.12
$\sigma(\bar{\eta})$	±0.04	±0.03
$\sigma(\sin 2\beta)$	±0.05	±0.05
$\sigma(\mathrm{Im}\lambda_t)$	$\pm5\%$	$\pm5\%$

any hadronic uncertainties and its accuracy can be estimated with a high degree of confidence.

An extensive numerical analysis of the formulae above has been presented in ref's [122,118]. Assuming that the branching ratios are known to within $\pm10\%$ and m_t within ±3 GeV one finds the results in Table 9 [118]. We observe that respectable determinations of all considered quantities except for $\bar{\varrho}$ can be obtained. Of particular interest are the accurate determinations of $\sin 2\beta$ and of $\mathrm{Im}\lambda_t$. The latter quantity as seen in (6.40) can be obtained from $K_L \to \pi^0\nu\bar{\nu}$ alone and does not require knowledge of V_{cb}. The importance of measuring accurately $\mathrm{Im}\lambda_t$ is evident. It plays a central role in the phenomenology of CP violation in K decays and is furthermore equivalent to the Jarlskog parameter J_{CP} [12], the invariant measure of CP violation in the Standard Model, $J_{\mathrm{CP}} = \lambda(1 - \lambda^2/2)\mathrm{Im}\lambda_t$.

The accuracy to which $\sin 2\beta$ can be obtained from $K \to \pi\nu\bar{\nu}$ is, in the example discussed above, comparable to the one expected in determining $\sin 2\beta$ from CP asymmetries in B decays prior to LHC experiments. In this case $\sin 2\beta$ is determined best by measuring CP violation in $B_d \to J/\psi K_S$. Using the formula for the corresponding time-integrated CP asymmetry one finds an interesting connection between rare K decays and B physics [122]

$$\frac{2r_s(B_1, B_2)}{1 + r_s^2(B_1, B_2)} = -a_{\mathrm{CP}}(B_d \to J/\psi K_S)\frac{1 + x_d^2}{x_d} \qquad (6.50)$$

which must be satisfied in the Standard Model. Here x_d is a $B_d^0 - \bar{B}_d^0$ parameter. We stress that except for $P_0(X)$ all quantities in (6.50) can be directly measured in experiment and that this relationship is essentially independent of m_t and V_{cb}. Due to very small theoretical uncertainties in (6.50), this relation is particularly suited for tests of CP violation in the Standard Model and offers a powerful tool to probe the physics beyond it.

7 Express Review of Rare K and B Decays

7.1 The Decays $B \to X_{s,d}\nu\bar{\nu}$

The decays $B \to X_{s,d}\nu\bar{\nu}$ are the theoretically cleanest decays in the field of rare B-decays. They are dominated by the same Z^0-penguin and box diagrams involving top quark exchanges which we encountered already in the case of $K^+ \to \pi^+\nu\bar{\nu}$ and $K_L \to \pi^0\nu\bar{\nu}$ except for the appropriate change of the external quark flavours. Since the change of external quark flavours has no impact on the m_t dependence, the latter is fully described by the function $X(x_t)$ in (6.2) which includes the NLO corrections. The charm contribution is fully negligible here and the resulting effective Hamiltonian is very similar to the one for $K_L \to \pi^0\nu\bar{\nu}$ given in (6.25). For the decay $B \to X_s\nu\bar{\nu}$ it reads

$$\mathcal{H}_{\text{eff}} = \frac{G_F}{\sqrt{2}} \frac{\alpha}{2\pi \sin^2\Theta_W} V_{tb}^* V_{ts} X(x_t)(\bar{b}s)_{V-A}(\bar{\nu}\nu)_{V-A} + h.c. \qquad (7.1)$$

with s replaced by d in the case of $B \to X_d\nu\bar{\nu}$.

The theoretical uncertainties related to the renormalization scale dependence are as in $K_L \to \pi^0\nu\bar{\nu}$ and can be essentially neglected. The same applies to long distance contributions considered in ref. [123]. The calculation of the branching fractions for $B \to X_{s,d}\nu\bar{\nu}$ can be done in the spectator model corrected for short distance QCD effects. Normalizing to $Br(B \to X_c e\bar{\nu})$ and summing over three neutrino flavours one finds

$$\frac{Br(B \to X_s\nu\bar{\nu})}{Br(B \to X_c e\bar{\nu})} = \frac{3\alpha^2}{4\pi^2 \sin^4\Theta_W} \frac{|V_{ts}|^2}{|V_{cb}|^2} \frac{X^2(x_t)}{f(z)} \frac{\kappa(0)}{\kappa(z)}. \qquad (7.2)$$

Here $f(z)$ is the phase-space factor for $B \to X_c e\bar{\nu}$ with $z = m_c^2/m_b^2$ and $\kappa(z) = 0.88$ is the corresponding QCD correction[124,125]. The factor $\kappa(0) = 0.83$ represents the QCD correction to the matrix element of the $b \to s\nu\bar{\nu}$ transition due to virtual and bremsstrahlung contributions. In the case of $B \to X_d\nu\bar{\nu}$ one has to replace V_{ts} by V_{td} which results in a decrease of the branching ratio by roughly an order of magnitude.

Setting $Br(B \to X_c e\bar{\nu}) = 10.4\%$, $f(z) = 0.54$, $\kappa(z) = 0.88$ and using the values in (6.10) we have

$$Br(B \to X_s\nu\bar{\nu}) = 3.7 \cdot 10^{-5} \frac{|V_{ts}|^2}{|V_{cb}|^2} \left[\frac{\overline{m}_t(m_t)}{170\,\text{GeV}}\right]^{2.30}. \qquad (7.3)$$

Taking next, $f(z) = 0.54 \pm 0.04$ and $Br(B \to X_c e\bar{\nu}) = (10.4 \pm 0.4)\%$ and scanning the input parameters of Table 2 we find

$$Br(B \to X_s\nu\bar{\nu}) = (3.5 \pm 0.7) \cdot 10^{-5} \qquad (7.4)$$

to be compared with the experimental upper bound:

$$Br(B \to X_s \nu \bar{\nu}) < 7.7 \cdot 10^{-4} \quad (90\% \text{ C.L.}) \tag{7.5}$$

obtained for the first time by ALEPH[126]. This is only a factor of 20 above the Standard Model expectation. Even if the actual measurement of this decay is extremely difficult, all efforts should be made to measure it. One should also make attempts to measure $Br(B \to X_d \nu \bar{\nu})$. Indeed

$$\frac{Br(B \to X_d \nu \bar{\nu})}{Br(B \to X_s \nu \bar{\nu})} = \frac{|V_{td}|^2}{|V_{ts}|^2} \tag{7.6}$$

offers the cleanest direct determination of $|V_{td}|/|V_{ts}|$ as all uncertainties related to m_t, $f(z)$ and $Br(B \to X_c e \bar{\nu})$ cancel out.

7.2 The Decays $B_{s,d} \to l^+ l^-$

The decays $B_{s,d} \to l^+ l^-$ are after $B \to X_{s,d} \nu \bar{\nu}$ the theoretically cleanest decays in the field of rare B-decays. They are dominated by the Z^0-penguin and box diagrams involving top quark exchanges which we encountered already in the case of $B \to X_{s,d} \nu \bar{\nu}$ except that due to charged leptons in the final state the charge flow in the internal lepton line present in the box diagram is reversed. This results in a different m_t dependence summarized by the function $Y(x_t)$, the NLO generalization[106,108,109] of the function $Y_0(x_t)$ given in (2.15). The charm contributions are fully negligible here and the resulting effective Hamiltonian is given for $B_s \to l^+ l^-$ as follows:

$$\mathcal{H}_{\text{eff}} = -\frac{G_F}{\sqrt{2}} \frac{\alpha}{2\pi \sin^2 \Theta_W} V_{tb}^* V_{ts} Y(x_t) (\bar{b}s)_{V-A} (\bar{l}l)_{V-A} + h.c. \tag{7.7}$$

with s replaced by d in the case of $B_d \to l^+ l^-$.

The function $Y(x)$ is given by

$$Y(x_t) = Y_0(x_t) + \frac{\alpha_s}{4\pi} Y_1(x_t) \equiv \eta_Y Y_0(x_t), \qquad \eta_Y = 1.012 \tag{7.8}$$

where $Y_1(x_t)$ can be found in ref's [106,108,109]. The leftover μ_t-dependence in $Y(x_t)$ is tiny and amounts to an uncertainty of $\pm 1\%$ at the level of the branching ratio. With $m_t \equiv \bar{m}_t(m_t)$ the QCD factor η_Y depends only very weakly on m_t. The dependence on $\Lambda_{\overline{MS}}$ can be neglected.

The branching ratio for $B_s \to l^+ l^-$ is given by ref.[106]

$$Br(B_s \to l^+l^-) = \tau(B_s)\frac{G_F^2}{\pi}\left(\frac{\alpha}{4\pi \sin^2 \Theta_W}\right)^2 F_{B_s}^2 m_l^2 m_{B_s}$$

$$\sqrt{1 - 4\frac{m_l^2}{m_{B_s}^2}}|V_{tb}^* V_{ts}|^2 Y^2(x_t) \qquad (7.9)$$

where B_s denotes the flavour eigenstate $(\bar{b}s)$ and F_{B_s} is the corresponding decay constant. Using (6.10) and (7.8) we find in the case of $B_s \to \mu^+\mu^-$

$$Br(B_s \to \mu^+\mu^-) = 3.5 \cdot 10^{-9} \left[\frac{\tau(B_s)}{1.6\text{ps}}\right]\left[\frac{F_{B_s}}{210\,\text{MeV}}\right]^2\left[\frac{|V_{ts}|}{0.040}\right]^2\left[\frac{\overline{m}_t(m_t)}{170\,\text{GeV}}\right]^{3.12} .$$
$$(7.10)$$

The main uncertainty in this branching ratio results from the uncertainty in F_{B_s}. Scanning the input parameters of Table 2 together with $\tau(B_s) = 1.6$ ps and $F_{B_s} = (210 \pm 30)\,\text{MeV}$ we find

$$Br(B_s \to \mu^+\mu^-) = (3.2 \pm 1.5) \cdot 10^{-9} . \qquad (7.11)$$

For $B_d \to \mu^+\mu^-$ a similar formula holds with obvious replacements of labels $(s \to d)$. Provided the decay constants F_{B_s} and F_{B_d} will have been calculated reliably by non-perturbative methods or measured in leading leptonic decays one day, the rare processes $B_s \to \mu^+\mu^-$ and $B_d \to \mu^+\mu^-$ should offer clean determinations of $|V_{ts}|$ and $|V_{td}|$. In particular the ratio

$$\frac{Br(B_d \to \mu^+\mu^-)}{Br(B_s \to \mu^+\mu^-)} = \frac{\tau(B_d)}{\tau(B_s)}\frac{m_{B_d}}{m_{B_s}}\frac{F_{B_d}^2}{F_{B_s}^2}\frac{|V_{td}|^2}{|V_{ts}|^2} \qquad (7.12)$$

having smaller theoretical uncertainties than the separate branching ratios should offer a useful measurement of $|V_{td}|/|V_{ts}|$. Since $Br(B_d \to \mu^+\mu^-) = \mathcal{O}(10^{-10})$ this is, however, a very difficult task. For $B_s \to \tau^+\tau^-$ and $B_s \to e^+e^-$ one expects branching ratios $\mathcal{O}(10^{-6})$ and $\mathcal{O}(10^{-13})$, respectively, with the corresponding branching ratios for B_d-decays by one order of magnitude smaller.

The bounds on $B_{s,d} \to l\bar{l}$ are still many orders of magnitude away from Standard Model expectations. The best bounds come from CDF[127]. One has:

$$Br(B_s \to \mu^+\mu^-) \leq 2.6 \cdot 10^{-6} \qquad (95\%C.L.) \qquad (7.13)$$

and $Br(B_d \to \mu^+\mu^-) \leq 8.6 \cdot 10^{-7}$. CDF should reach in Run II the sensitivity of $1 \cdot 10^{-8}$ and $4 \cdot 10^{-8}$ for $B_d \to \mu\bar{\mu}$ and $B_s \to \mu\bar{\mu}$, respectively. It is hoped that these decays will be observed at LHC-B. The experimental status of $B \to \tau^+\tau^-$ and its usefulness in tests of the physics beyond the Standard Model is discussed in ref. [128].

7.3 $B \to X_s \gamma$ and $B \to X_s l^+ l^-$

In view of space limitations I will be very brief on these two decays.

A lot of effort has been put into predicting the branching ratio for the inclusive radiative decay $B \to X_s \gamma$ including NLO QCD corrections and higher order electroweak corrections. The relevant references can be found in ref's[1,21], where also theoretical details are given. The final result of these efforts can be summarized by

$$Br(B \to X_s \gamma)_{th} = (3.30 \pm 0.15(\text{scale}) \pm 0.26(\text{par})) \cdot 10^{-4} \qquad (7.14)$$

where the first error represents residual scale dependences and the second error is due to uncertainties in input parameters. The main achievement is the reduction of the scale dependence through NLO calculations, in particular those given in ref's[129,130]. In the leading order the corresponding error would be roughly ± 0.6.

The theoretical result in (7.14) should be compared with experimental data:

$$Br(B \to X_s \gamma)_{exp} = \begin{cases} (3.15 \pm 0.35 \pm 0.41) \cdot 10^{-4} \,, & \text{CLEO} \\ (3.11 \pm 0.80 \pm 0.72) \cdot 10^{-4} \,, & \text{ALEPH,} \end{cases} \qquad (7.15)$$

which implies the combined branching ratio:

$$Br(B \to X_s \gamma)_{exp} = (3.14 \pm 0.48) \cdot 10^{-4} \,. \qquad (7.16)$$

Clearly, the Standard Model result agrees well with the data. In order to see whether any new physics can be seen in this decay, the theoretical and in particular experimental errors should be reduced. This is certainly a very difficult task.

The rare decays $B \to X_{s,d} l^+ l^-$ have been the subject of many theoretical studies. It is clear that once these decays have been observed, they will offer useful tests of the Standard Model and of its extentions. Most recent reviews can be found in ref's[131,132].

7.4 $K_L \to \pi^0 e^+ e^-$

There are three contributions to this decay: CP conserving, indirectly CP violating and directly CP violating. Unfortunately out of these three contributions only the directly CP violating can be calculated reliably. Including NLO corrections[133] and scanning the input parameters of Table 2 we find

$$Br(K_L \to \pi^0 e^+ e^-)_{dir} = (4.6 \pm 1.8) \cdot 10^{-12} \,, \qquad (7.17)$$

where the errors come dominantly from the uncertainties in the CKM parameters. The remaining two contributions to this decay are plagued by theoretical uncertainties [134]. They are expected to be $\mathcal{O}(10^{-12})$ but generally smaller than $Br(K_L \rightarrow \pi^0 e^+ e^-)_{\text{dir}}$. This implies that within the Standard Model $Br(K_L \rightarrow \pi^0 e^+ e^-)$ is expected to be at most 10^{-11}.

Experimentally we have the bound [135]

$$Br(K_L \rightarrow \pi^0 e^+ e^-) < 4.3 \cdot 10^{-9} \qquad (7.18)$$

and considerable improvements are expected in the coming years.

7.5 $K_L \rightarrow \mu^+ \mu^-$

The $K_L \rightarrow \mu^+ \mu^-$ branching ratio can be decomposed generally as follows:

$$BR(K_L \rightarrow \mu^+ \mu^-) = |\text{Re}A|^2 + |\text{Im}A|^2, \qquad (7.19)$$

where $\text{Re}A$ denotes the dispersive contribution and $\text{Im}A$ the absorptive one. The latter contribution can be determined in a model independent way from the $K_L \rightarrow \gamma\gamma$ branching ratio. The resulting $|\text{Im}A|^2$ is very close to the experimental branching ratio $Br(K_L \rightarrow \mu^+ \mu^-) = (7.2 \pm 0.5) \cdot 10^{-9}$ [136] so that $|\text{Re}A|^2$ is substantially smaller and extracted to be [136]

$$|\text{Re}A_{\text{exp}}|^2 < 5.6 \cdot 10^{-10} \qquad (90\% \text{ C.L.}). \qquad (7.20)$$

Now $\text{Re}A$ can be decomposed as

$$\text{Re}A = \text{Re}A_{\text{LD}} + \text{Re}A_{\text{SD}}, \qquad (7.21)$$

with

$$|\text{Re}A_{\text{SD}}|^2 \equiv Br(K_L \rightarrow \mu^+ \mu^-)_{\text{SD}} \qquad (7.22)$$

representing the short-distance contribution which can be calculated reliably. An improved estimate of the long-distance contribution $\text{Re}A_{\text{LD}}$ has been recently presented in ref. [137]

$$|\text{Re}A_{LD}| < 2.9 \cdot 10^{-5} \qquad (90\% \text{ C.L.}). \qquad (7.23)$$

Together with (7.20) this gives

$$Br(K_L \rightarrow \mu^+ \mu^-)_{\text{SD}} < 2.8 \cdot 10^{-9}. \qquad (7.24)$$

This result is very close to one presented by Gomez Dumm and Pich [138]. More pesimistic view on the extraction of the short distance part from $Br(K_L \rightarrow \mu^+ \mu^-)$ can be found in ref. [139].

The bound in (7.24) should be compared with the short distance contribution within the Standard Model for which we find

$$Br(K_{\rm L} \to \mu^+\mu^-)_{\rm SD} = (8.7 \pm 3.6) \cdot 10^{-10}. \qquad (7.25)$$

This implies there is considerable room for new physics contributions. We return to this point in Sec. 9. Reviews of rare K decays are listed in ref. [140].

8 Express Review of CP Violation in B Decays

8.1 CP-Asymmetries in B-Decays: General Picture

CP violation in B-decays is certainly one of the most important targets of B-factories and of dedicated B-experiments at hadron facilities. It is well known that CP violating effects are expected to occur in a large number of channels at a level attainable at forthcoming experiments. Moreover there exist channels which offer the determination of CKM phases essentially without any hadronic uncertainties. Since extensive reviews on CP violation in B decays can be found in the literature [141,142,3] and I am running out of space, let me concentrate only on a few points beginning with a quick review of classic methods for the determination of the angles α, β and γ in the unitarity triangle.

The classic determination of α by means of the time dependent CP asymmetry in the decay $B_d^0 \to \pi^+\pi^-$ is affected by the "QCD penguin pollution" which has to be taken care of in order to extract α. The recent CLEO results for penguin dominated decays indicate that this pollution could be substantial as stressed in particular in ref. [143]. The most popular strategy to deal with this "penguin problem" is the isospin analysis of Gronau and London [144]. It requires however the measurement of $Br(B^0 \to \pi^0\pi^0)$ which is expected to be below 10^{-6}: a very difficult experimental task. For this reason several rather involved strategies [145] have been proposed which avoid the use of $B_d \to \pi^0\pi^0$ in conjunction with $a_{CP}(\pi^+\pi^-, t)$. They are reviewed in ref. [3]. It is to be seen which of these methods will eventually allow us to measure α with a respectable precision. It is however clear that the determination of this angle is a real challenge for both theorists and experimentalists.

The CP-asymmetry in the decay $B_d \to \psi K_S$ allows in the Standard Model a direct measurement of the angle β in the unitarity triangle without any theoretical uncertainties [146]. Of considerable interest [142,147] is also the pure penguin decay $B_d \to \phi K_S$, which is expected to be sensitive to physics beyond the Standard Model. Comparision of β extracted from $B_d \to \phi K_S$ with the one from $B_d \to \psi K_S$ should be important in this respect. An analogue of $B_d \to \psi K_S$ in B_s-decays is $B_s \to \psi\phi$. The CP asymmetry measures here η[148]

in the Wolfenstein parametrization. It is very small, however, and this fact makes it a good place to look for the physics beyond the Standard Model. In particular the CP violation in $B_s^0 - \bar{B}_s^0$ mixing from new sources beyond the Standard Model should be probed in this decay.

The two theoretically cleanest methods for the determination of γ are: i) the full time dependent analysis of $B_s \rightarrow D_s^+ K^-$ and $\bar{B}_s \rightarrow D_s^- K^+$ [149] and ii) the well known triangle construction due to Gronau and Wyler [150] which uses six decay rates $B^\pm \rightarrow D_{CP}^0 K^\pm$, $B^+ \rightarrow D^0 K^+$, $\bar{D}^0 K^+$ and $B^- \rightarrow D^0 K^-$, $\bar{D}^0 K^-$. Both methods are unaffected by penguin contributions. The first method is experimentally very challenging because of the expected large $B_s^0 - \bar{B}_s^0$ mixing. The second method is problematic because of the small branching ratios of the colour supressed channel $B^+ \rightarrow D^0 K^+$ and its charge conjugate, giving a rather squashed triangle and thereby making the extraction of γ very difficult. Variants of the latter method which could be more promising have been proposed in ref's [151,152]. It appears that these methods will give useful results at later stages of CP-B investigations. In particular the first method will be feasible only at LHC-B. Other recent strategies for γ will be mentioned below.

8.2 B^0-Decays to CP Eigenstates

Let us demonstrate some of the statements made above explicitly.

A time dependent asymmetry in the decay $B^0 \rightarrow f$ with f being a CP eigenstate is given by

$$a_{CP}(t, f) = \mathcal{A}_{CP}^{dir}(B \rightarrow f) \cos(\Delta M t) + \mathcal{A}_{CP}^{mix-ind}(B \rightarrow f) \sin(\Delta M t) \quad (8.1)$$

where we have separated the *direct* CP-violating contributions from those describing *mixing-induced* CP violation:

$$\mathcal{A}_{CP}^{dir}(B \rightarrow f) \equiv \frac{1 - |\xi_f|^2}{1 + |\xi_f|^2}, \qquad \mathcal{A}_{CP}^{mix-ind}(B \rightarrow f) \equiv \frac{2\mathrm{Im}\xi_f}{1 + |\xi_f|^2}. \quad (8.2)$$

In (8.1), ΔM denotes the mass splitting of the physical B^0–$\bar{B}^0$–mixing eigenstates. The quantity ξ_f containing essentially all the information needed to evaluate the asymmetries (8.2) is given by

$$\xi_f = \exp(i2\phi_M)\frac{A(\bar{B} \rightarrow f)}{A(B \rightarrow f)} \quad (8.3)$$

with ϕ_M denoting the weak phase in the $B - \bar{B}$ mixing and $A(B \rightarrow f)$ the decay amplitude.

Generally several decay mechanisms with different weak and strong phases can contribute to $A(B \to f)$. These are tree diagram (current-current) contributions, QCD penguin contributions and electroweak penguin contributions. If they contribute with similar strength to a given decay amplitude the resulting CP asymmetries suffer from hadronic uncertainies related to matrix elements of the relevant operators Q_i.

An interesting case arises when a single mechanism dominates the decay amplitude or the contributing mechanisms have the same weak phases. Then

$$\xi_f = \exp(i2\phi_M)\exp(-i2\phi_D), \qquad |\xi_f|^2 = 1 \qquad (8.4)$$

where ϕ_D is the weak phase in the decay amplitude. In this particular case the hadronic matrix elements drop out, the direct CP violating contribution vanishes and the mixing-induced CP asymmetry is given entirely in terms of the weak phases ϕ_M and ϕ_D. In particular the time integrated asymmetry is given by

$$a_{CP}(f) = \pm \sin(2\phi_D - 2\phi_M)\frac{x_{d,s}}{1 + x_{d,s}^2} \qquad (8.5)$$

where $\pm$ refers to f being a $CP = \pm$ eigenstate and $x_{d,s}$ are the $B_{d,s}^0 - \bar{B}_{d,s}^0$ mixing parameters.

If a single tree diagram dominates, the factor $\sin(2\phi_D - 2\phi_M)$ can be calculated by using

$$\phi_D = \begin{cases} \gamma & b \to u \\ 0 & b \to c \end{cases} \qquad \phi_M = \begin{cases} -\beta & B_d^0 \\ 0 & B_s^0 \end{cases} \qquad (8.6)$$

where we have indicated the basic transition of the b-quark into a lighter quark. On the other hand if the penguin diagram with internal top exchange dominates one has

$$\phi_D = \begin{cases} -\beta & b \to d \\ 0 & b \to s \end{cases} \cdot \qquad \phi_M = \begin{cases} -\beta & B_d^0 \\ 0 & B_s^0 \end{cases} \cdot \qquad (8.7)$$

These rules have been obtained using the Wolfenstein parametrization in the leading order. Let us practice with these formulae. Assuming that $B_d \to \psi K_S$ and $B_d \to \pi^+\pi^-$ are dominated by tree diagrams with $b \to c$ and $b \to u$ transitions respectively we readily find

$$a_{CP}(\psi K_S) = -\sin(2\beta)\frac{x_d}{1 + x_d^2}, \qquad (8.8)$$

$$a_{CP}(\pi^+\pi^-) = -\sin(2\alpha)\frac{x_d}{1 + x_d^2}. \qquad (8.9)$$

Now in the case of $B_d \rightarrow \psi K_S$ the penguin diagrams have to a very good approximation the same phase ($\phi_D = 0$) as the tree contribution and moreover are Zweig suppressed. Consequently (8.8) is very accurate. This is not the case for $B_d \rightarrow \pi^+\pi^-$ where the penguin contribution could be substantial. Having weak phase $\phi_D = -\beta$, which differs from the tree phase $\phi_D = \gamma$, this penguin contribution changes effectively (8.9) to

$$a_{CP}(\pi^+\pi^-) = -\sin(2\alpha + \theta_P)\frac{x_d}{1 + x_d^2} \qquad (8.10)$$

where θ_P is a function of β and hadronic parameters. The isospin analysis [144] mentioned before is supposed to determine θ_P so that α can be extracted from $a_{CP}(\pi^+\pi^-)$.

Similarly the pure penguin dominated decay $B_d \rightarrow \phi K_S$ is governed by the $b \rightarrow s$ penguin with internal top exchange which implies that in this decay the angle β is measured. The accuracy of this measurement is a bit lower than using $B_d \rightarrow \psi K_S$ as penguins with internal u and c exchanges may introduce a small pollution.

Finally we can consider the asymmetry in $B_s \rightarrow \psi\phi$, an analog of $B_d \rightarrow \psi K_s$. In the leading order of the Wolfenstein parametrization the asymmetry $a_{CP}(\psi\phi)$ vanishes. Including higher order terms in λ one finds [148]

$$a_{CP}(\psi\phi) = 2\lambda^2\eta\frac{x_s}{1 + x_s^2} \qquad (8.11)$$

where λ and η are the Wolfenstein parameters.

8.3 Recent Developments

All this has been known already for some time and is well documented in the literature. The most recent developments are related to the extraction of the angle γ from the decays $B \rightarrow PP$ (P=pseudoscalar) and their charge conjugates [153–156]. Some of these modes have been observed by the CLEO collaboration [157]. In the future they should allow us to obtain direct information on γ at B-factories (BaBar, BELLE, CLEO III) (for interesting feasibility studies, see ref's [154,155,131]). At present, there are only experimental results available for the combined branching ratios of these modes, i.e. averaged over decay and its charge conjugate, suffering from large hadronic uncertainties.

There has been large activity in this field during the last two years. The main issues here are the final state interactions, SU(3) symmetry breaking effects and the importance of electroweak penguin contributions. Several interesting ideas have been put forward to extract the angle γ in spite of large hadronic uncertainties in $B \to \pi K$ decays [153,154]. Also other $B \to PP$ decays have been investigated. As this field became rather technical, I decided not to include it in these lectures. A subset of relevant papers is listed in ref's [153,154,156,158,159,160], where further references can be found. In particular in ref's [156,159] general parametrizations for the study of the final state interactions, SU(3) symmetry breaking effects and the importance of electroweak penguin contributions have been presented. Moreover, upper bounds on the latter contributions following from SU(3) symmetry have been derived [160]. Recent reviews can be found in ref's [161,162]. New strategies for γ which include $B_s \to \psi K_S$ and $B_s \to K^+ K^-$ have been suggested very recently in ref. [163].

There is no doubt that these new ideas will be helpful in the future. They are, however, rather demanding for experimentalist as often several branching ratios have to be studied simultaneously and each has to be measured precisely in order to obtain an acceptable measurement of γ. On the other hand various suggested bounds on γ may either exclude the region around $90°$ [153] or give an improved lower bound on it [160,162,164] which would remove a large portion of the allowed range from the analysis of the unitarity triangle. In this context it has been pointed out in ref. [165] (see also ref. [164]) that generally charmless hadronic B decay results from CLEO seem to prefer negative values of $\cos \gamma$ which is not the case in the standard analysis of Sec. 4.

Finally I would like to mention a recent interesting paper of Lenz, Nierste and Ostermaier [166], where inclusive direct CP-asymmetries in charmless $B^{\pm}$-decays including QCD effects have been studied. These asymmetries should offer additional useful means to constrain the unitarity triangle.

8.4 CP-Asymmetries in B-Decays versus $K \to \pi \nu \bar{\nu}$

Let us next compare the potentials of the CP asymmetries in determining the parameters of the Standard Model with those of the cleanest rare K-decays: $K_L \to \pi^0 \nu \bar{\nu}$ and $K^+ \to \pi^+ \nu \bar{\nu}$.

Measuring $\sin 2\alpha$ and $\sin 2\beta$ from CP asymmetries in B decays allows, in principle, to fix the parameters $\bar{\eta}$ and $\bar{\varrho}$, which can be expressed as [167]

$$\bar{\eta} = \frac{r_-(\sin 2\alpha) + r_+(\sin 2\beta)}{1 + r_+^2(\sin 2\beta)}, \qquad \bar{\varrho} = 1 - \bar{\eta} r_+(\sin 2\beta), \qquad (8.12)$$

where $r_{\pm}(z) = (1 \pm \sqrt{1 - z^2})/z$. In general the calculation of $\bar{\varrho}$ and $\bar{\eta}$ from $\sin 2\alpha$ and $\sin 2\beta$ involves discrete ambiguities. As described in ref. [167] they can be resolved by using further information, e.g. bounds on $|V_{ub}/V_{cb}|$, so that eventually the solution (8.12) is singled out.

Table 10: Illustrative example of the determination of CKM parameters from $K \to \pi\nu\bar{\nu}$ and B-decays. We use $\sigma(|V_{cb}|) = \pm 0.002(0.001)$.

	$K \to \pi\nu\bar{\nu}$	Scenario I	Scenario II
$\sigma(|V_{td}|)$	$\pm 10\%(9\%)$	$\pm 5.5\%(3.5\%)$	$\pm 5.0\%(2.5\%)$
$\sigma(\bar{\varrho})$	$\pm 0.16(0.12)$	± 0.03	± 0.01
$\sigma(\bar{\eta})$	$\pm 0.04(0.03)$	± 0.04	± 0.01
$\sigma(\sin 2\beta)$	± 0.05	± 0.06	± 0.02
$\sigma(\mathrm{Im}\lambda_t)$	$\pm 5\%$	$\pm 14\%(11\%)$	$\pm 10\%(6\%)$

Let us then consider two scenarios for the measurements of CP asymmetries in $B_d \to \pi^+\pi^-$ and $B_d \to J/\psi K_S$, expressed in terms of $\sin 2\alpha$ and $\sin 2\beta$:

$$\sin 2\alpha = 0.40 \pm 0.10, \qquad \sin 2\beta = 0.70 \pm 0.06 \qquad \text{(scenario I)} \qquad (8.13)$$

$$\sin 2\alpha = 0.40 \pm 0.04, \qquad \sin 2\beta = 0.70 \pm 0.02 \qquad \text{(scenario II)}. \qquad (8.14)$$

Scenario I corresponds to the accuracy being aimed for at B-factories and HERA-B prior to the LHC era. An improved precision can be anticipated from LHC experiments, which we illustrate with the scenario II. We assume that the problems with the determination of α will be solved somehow.

In Table 10 this way of the determination of the Standard Model parameters is compared [118] with the analogous analysis using $K_L \to \pi^0\nu\bar{\nu}$ and $K^+ \to \pi^+\nu\bar{\nu}$ which has been presented in section 6. As can be seen in Table 10, the CKM determination using $K \to \pi\nu\bar{\nu}$ is competitive with the one based on CP violation in B decays in scenario I, except for $\bar{\varrho}$ which is less constrained by the rare kaon processes. On the other hand as advertised previously $\mathrm{Im}\lambda_t$ is better determined in $K \to \pi\nu\bar{\nu}$ even if scenario II is considered. The virtue of the comparision of the determinations of various parameters using CP-B asymmetries with the determinations in very clean decays $K \to \pi\nu\bar{\nu}$ is that any substantial deviations from these two determinations would signal new physics beyond the Standard Model. Formula (6.50) is an example of such a comparison. There are other strategies for determination of the unitarity triangle using combinations of CP asymmetries and rare decays. They are reviewed in ref. [1].

9 A Brief Look Beyond the Standard Model

9.1 General Remarks

We begin the discussion of the Physics beyond the Standard Model with a few general remarks. As the new particles in the extensions of the Standard Model are generally substantally heavier than $W^{\pm}$, the impact of new physics on charged current tree level decays should be marginal. On the other hand these new contributions could have in principle an important impact on loop induced decays. From these two observations we conclude:

- New physics should have only marginal impact on the determination of $|V_{us}|$, $|V_{cb}|$ and $|V_{ub}|$.

- There is no impact on the calculations of the low energy non-perturbative parameters B_i except that new physics can bring new local operators implying new parameters B_i.

- New physics could have substantial impact on rare and CP violating decays and consequently on the determination of the unitarity triangle.

9.2 Classification of New Physics

Let us then group the extensions of the Standard Model in three classes.
Class A

- There are no new complex phases and quark mixing is described by the CKM matrix.

- There are new contributions to rare and CP violating decays through diagrams involving new internal particles.

These new contributions will have impact on the determination of α, β, γ, $|V_{td}|$ and λ_t and will be signaled by

- Inconsistencies in the determination of $(\bar{\varrho}, \bar{\eta})$ through ε, $B^0_{s,d} - \bar{B}^0_{s,d}$ mixing and rare decays.

- Disagreement of $(\bar{\varrho}, \bar{\eta})$ extracted from loop induced decays with $(\bar{\varrho}, \bar{\eta})$ extracted using CP asymmetries.

Examples are two Higgs doublet model II and the constrained MSSM.

Class B

- Quark mixing is described by the CKM matrix.

- There are new phases in the new contributions to rare and CP violating decays.

This kind of new physics will also be signaled by inconsistencies in the $(\bar{\varrho}, \bar{\eta})$ plane. However, new complication arises. Because of new phases CP violating asymmetries measure generally different quantities than α, β and γ. For instance the CP asymmetry in $B \to \psi K_S$ will no longer measure β but $\beta + \theta_{NP}$ where θ_{NP} is a new phase. Strategies for dealling with such a situation have been developed. See for instance ref's [116,168] and references therein.

Examples are multi-Higgs models with complex phases in the Higgs sector, general SUSY models, models with spontaneous CP violation and left-right symmetric models.

Class C

- The unitarity of the three generation CKM matrix does not hold.

Examples are four generation models and models with tree level FCNC transitions. If this type of physics is present, the unitarity triangle will not close or some inconsistencies in the $(\bar{\varrho}, \bar{\eta})$ plane take place.

Clearly in order to sort out which type of new physics is responsible for deviations from the Standard Model expectations one has to study many loop induced decays and many CP asymmeteries. Some ideas in this direction can be found in ref's [168,116].

9.3 Upper Bounds on $K \to \pi \nu \bar{\nu}$ and $K_L \to \pi^0 e^+ e^-$ from ε'/ε and $K_L \to \mu^+ \mu^-$

We have seen in previous sections that the rare kaon decays $K_L \to \pi^0 \nu \bar{\nu}$, $K^+ \to \pi^+ \nu \bar{\nu}$ and $K_L \to \pi^0 e^+ e^-$ are governed by Z-penguin diagrams. Within the Standard Model the branching ratios for these decays have been found to be

$$Br(K_L \to \pi^0 \nu \bar{\nu}) = (2.8 \pm 1.1) \cdot 10^{-11}, \qquad (9.1)$$

$$Br(K^+ \to \pi^+ \nu \bar{\nu}) = (7.9 \pm 3.1) \cdot 10^{-11}, \qquad (9.2)$$

$$Br(K_L \to \pi^0 e^+ e^-)_{\text{dir}} = (4.6 \pm 1.8) \cdot 10^{-12}, \qquad (9.3)$$

where the errors come dominantly from the uncertainties in the CKM parameters. The branching ratio in (9.3) represents the so-called direct CP-violating contribution to $K_L \to \pi^0 e^+ e^-$. The remaining two contributions to this decay, the CP-conserving one and the indirect CP-violating one are plagued by theoretical uncertainties [134]. They are expected to be $\mathcal{O}(10^{-12})$ but generally smaller than $Br(K_{\mathrm{L}} \to \pi^0 e^+ e^-)_{\mathrm{dir}}$. This implies that within the Standard Model $Br(K_{\mathrm{L}} \to \pi^0 e^+ e^-)$ is expected to be at most 10^{-11}.

In this context a very interesting claim has been made by Colangelo and Isidori [169], who analyzing rare kaon decays in supersymmetric theories pointed out a possible large enhancement of the effective $\bar{s}dZ$ vertex leading to an enhancement of $Br(K^+ \to \pi^+ \nu\bar{\nu})$ by one order of magnitude and of $Br(K_{\mathrm{L}} \to \pi^0 \nu\bar{\nu})$ and $Br(K_{\mathrm{L}} \to \pi^0 e^+ e^-)$ by two orders of magnitude relative to the Standard Model expectations. Not surprisingly these results brought a lot of excitement among experimentalists.

Whether substantial enhancements of the branching ratios in question are indeed possible in supersymmetric theories is being investigated at present. On the other hand it can be shown [170] that in models in which the dominant new effect is an enhanced $\bar{s}dZ$ vertex, enhancements of $Br(K_{\mathrm{L}} \to \pi^0 \nu\bar{\nu})$ and $Br(K_{\mathrm{L}} \to \pi^0 e^+ e^-)$ as large as claimed in ref. [169] are already excluded by the existing data on ε'/ε in spite of large theoretical uncertainties. Similarly the large enhancement of $Br(K^+ \to \pi^+ \nu\bar{\nu})$ can be excluded by the data on ε'/ε and in particular by the present information on the short distance contribution to $K_{\mathrm{L}} \to \mu^+ \mu^-$. The latter can be bounded by analysing the data on $Br(K_{\mathrm{L}} \to \mu^+ \mu^-)$ in conjunction with improved estimates of long distance dispersive contributions [137,138]. In ref. [169] only constraints from $K_{\mathrm{L}} \to \mu^+ \mu^-$, the K_L–K_S mass difference ΔM_K and ε have been taken into account. As ε'/ε depends sensitively on the size of Z-penguin contributions and generally on the size of the effective $\bar{s}dZ$ vertex it is clear that the inclusion of the constraints from ε'/ε should have an important impact on the bounds for the rare decays in question. I will only describe the basic idea of ref. [170] and give numerical results. The relevant expressions can be found in this paper. Here we go.

In the Standard Model Z-penguins are represented by the function C_0 which enters the functions X_0, Y_0 and Z_0. In order to study the effect of an enhanced $\bar{s}dZ$ vertex one simply makes the following replacement in the formulae for ε'/ε, $K_{\mathrm{L}} \to \mu^+ \mu^-$ and rare decays in question:

$$\lambda_t C_0(x_t) \implies Z_{ds} \qquad (9.4)$$

where Z_{ds} denotes an effective $\bar{s}dZ$ vertex. The remaining contributions to ε'/ε, $K_{\mathrm{L}} \to \mu^+ \mu^-$ and rare K decays are evaluated in the Standard model as we assume that they are only marginally affected by new physics. We will, however, consider three scenarios for λ_t, which enters these remaining contributions.

Indeed there is the possibility that the value of λ_t is modified by new contributions to ε and $B_{d,s}^0 - \bar{B}_{d,s}^0$ mixings. We consider therefore three scenarios:

- **Scenario A:** λ_t is taken from the standard analysis of the unitarity triangle

- **Scenario B:** $\text{Im}\lambda_t = 0$ and $\text{Re}\lambda_t$ is varied in the full range consistent with the unitarity of the CKM matrix. In this scenario CP violation comes entirely from new physics contributions.

- **Scenario C:** λ_t is varied in the full range consistent with the unitarity of the CKM matrix. This means in particular that $\text{Im}\lambda_t$ can be negative.

Table 11: Upper bounds for the rare decays $K_L \to \pi^0 \nu\bar{\nu}$, $K_L \to \pi^0 e^+ e^-$ and $K^+ \to \pi^+ \nu\bar{\nu}$, obtained in various scenarios by imposing $\varepsilon'/\varepsilon \geq 2.5 \cdot 10^{-3}$, in the case $\text{Im}Z_{ds} > 0$.

Scenario	A	B	C	SM
$Br(K_L \to \pi^0 \nu\bar{\nu})[10^{-10}]$	0.5	–	0.7	0.4
$Br(K_L \to \pi^0 e^+ e^-)[10^{-11}]$	0.8	–	1.0	0.7
$Br(K^+ \to \pi^+ \nu\bar{\nu})[10^{-10}]$	1.8	–	2.2	1.1

Table 12: Upper bounds for the rare decays $K_L \to \pi^0 \nu\bar{\nu}$, $K_L \to \pi^0 e^+ e^-$ and $K^+ \to \pi^+ \nu\bar{\nu}$, obtained in various scenarios by imposing $\varepsilon'/\varepsilon \geq 1.5 \cdot 10^{-3}$, in the case $\text{Im}Z_{ds} > 0$.

Scenario	A	B	C	SM
$Br(K_L \to \pi^0 \nu\bar{\nu})[10^{-10}]$	1.1	–	1.2	0.4
$Br(K_L \to \pi^0 e^+ e^-)[10^{-11}]$	1.5	–	1.8	0.7
$Br(K^+ \to \pi^+ \nu\bar{\nu})[10^{-10}]$	1.9	–	2.3	1.1

Now Z_{ds} is a complex number. $\text{Im}Z_{ds}$ can be best bounded by ε'/ε. This implies bounds for $Br(K_L \to \pi^0 \nu\bar{\nu})$ and $Br(K_L \to \pi^0 e^+ e^-)$ which are sensitive functions of $\text{Im}Z_{ds}$. $\text{Re}Z_{ds}$ can be bounded by the present information on the short distance contribution to $K_L \to \mu^+ \mu^-$. This bound implies a bound on $Br(K^+ \to \pi^+ \nu\bar{\nu})$. Since $Br(K^+ \to \pi^+ \nu\bar{\nu})$ depends on both $\text{Re}Z_{ds}$ and $\text{Im}Z_{ds}$ also the bound on $\text{Im}Z_{ds}$ from ε'/ε matters in cases where $\text{Im}Z_{ds}$ is very enhanced over the Standard Model value.

The branching ratios $Br(K_L \to \pi^0 \nu\bar{\nu})$ and $Br(K_L \to \pi^0 e^+ e^-)$ are dominated by $(\text{Im}Z_{sd})^2$. Yet, the outcome of this analysis depends sensitively on the sign of $\text{Im}Z_{sd}$. Indeed, $\text{Im}Z_{sd} > 0$ results in the suppression of ε'/ε and as in the Standard Model the value for ε'/ε is generally below the data substantial enhancements of $\text{Im}Z_{sd}$ with $\text{Im}Z_{sd} > 0$ are not possible. The situation changes if new physics reverses the sign of $\text{Im}Z_{sd}$ so that it becomes negative. Then the upper bound on $\text{Im}Z_{sd}$ is governed by the upper bound on ε'/ε and with suitable choice of hadronic parameters and $\text{Im}\lambda_t$ (in particular in scenario C) large enhancements of $-\text{Im}Z_{sd}$ and of rare decay branching ratios are possible. The largest branching ratios are found when the neutral meson mixing is dominated by new physics contributions which force $\text{Im}\lambda_t$ to be as negative as possible within the unitarity of the CKM matrix. This possibility is quite remote. However, if this situation could be realized in some exotic model, then the branching ratios in question could be very high.

In Table 11 we show the upper bounds on rare decays for $\text{Im}Z_{sd} > 0$ for three scenarios in question and $\varepsilon'/\varepsilon \geq 2.5 \cdot 10^{-3}$. In Table 12 the corresponding bounds for $\varepsilon'/\varepsilon \geq 1.5 \cdot 10^{-3}$ are given. To this end all parameters relevant for ε'/ε have been scanned in the ranges used in Sec. 5. In Tables 13 and 14 the case $\text{Im}Z_{sd} < 0$ for $\varepsilon'/\varepsilon \leq 2.0 \cdot 10^{-3}$ and $\varepsilon'/\varepsilon \leq 3.0 \cdot 10^{-3}$ is considered respectively. In the last column we always give the upper bounds obtained in the Standard Model. Evidently for positive $\text{Im}Z_{sd}$ the enhancement of branching ratios are moderate but they can be very large when $\text{Im}Z_{sd} < 0$.

Table 13: Upper bounds for the rare decays $K_L \to \pi^0 \nu\bar{\nu}$, $K_L \to \pi^0 e^+ e^-$ and $K^+ \to \pi^+ \nu\bar{\nu}$, obtained in various scenarios by imposing $\varepsilon'/\varepsilon \leq 2.0 \cdot 10^{-3}$, in the case $\text{Im}Z_{ds} < 0$.

Scenario	A	B	C	SM
$BR(K_L \to \pi^0 \nu\bar{\nu})[10^{-10}]$	1.3	2.9	11.2	0.4
$BR(K_L \to \pi^0 e^+ e^-)[10^{-11}]$	2.9	5.1	18.2	0.7
$BR(K^+ \to \pi^+ \nu\bar{\nu})[10^{-10}]$	2.0	2.7	4.6	1.1

Table 14: Upper bounds for the rare decays $K_L \to \pi^0 \nu\bar{\nu}$, $K_L \to \pi^0 e^+ e^-$ and $K^+ \to \pi^+ \nu\bar{\nu}$, obtained in various scenarios by imposing $\varepsilon'/\varepsilon \leq 3.0 \cdot 10^{-3}$, in the case $\text{Im}Z_{ds} < 0$.

Scenario	A	B	C	SM
$BR(K_L \to \pi^0 \nu\bar{\nu})[10^{-10}]$	3.9	6.5	17.6	0.4
$BR(K_L \to \pi^0 e^+ e^-)[10^{-11}]$	7.9	11.5	28.0	0.7
$BR(K^+ \to \pi^+ \nu\bar{\nu})[10^{-10}]$	2.6	3.5	6.1	1.1

Other recent extensive analyses of supersymmetry effects in $K \to \pi\nu\bar{\nu}$ have been presented in ref's [116,171,172] where further references can be found. Model independent studies of these decays can be found in ref's [116,172]. The corresponding analyses in various no–supersymmetric extensions of the Standard Model are listed in ref. [173]. In particular, enhancement of $Br(K_L \to \pi^0 \nu\bar{\nu})$ by 1–2 orders of magnitude above the Standard Model expectations is according to ref. [174] still possible in four-generation models.

10 Summary and Outlook

I hope that I have convinced the students that the field of CP violation and rare decays plays an important role in the deeper understanding of the Standard Model and particle physics in general. Indeed the field of weak decays and of CP violation is one of the least understood sectors of the Standard Model. Even if the Standard Model is still consistent with the existing data for weak decay processes, the near future could change this picture dramatically through the advances in experiment and theory. In particular the experimental work done in the next ten years at BNL, CERN, CORNELL, DAΦNE, DESY, FNAL, KEK, SLAC and eventually LHC will certainly have considerable impact on this field.

Let us then make a list of things we could expect in the next ten years. This list is certainly very biased by my own interests but could be useful anyway. Here we go:

- The error on the CKM elements $|V_{cb}|$ and $|V_{ub}/V_{cb}|$ could be decreased below 0.002 and 0.01, respectively. This progress should come mainly from Cornell, B-factories and new theoretical efforts. It would have considerable impact on the unitarity triangle and would improve theoretical predictions for rare and CP-violating decays sensitive to these elements.

- The error on m_t should be decreased down to $\pm 3\,\mathrm{GeV}$ at Tevatron in the Main Injector era and to $\pm 1\,\mathrm{GeV}$ at LHC.

- The measurement of non-vanishing ratio of ε'/ε by NA31 and KTeV, excluding confidently the superweak models, has been an important achievement. The improved measurements of ε'/ε with the accuracy of $\pm(1-2)\cdot 10^{-4}$ from NA48, KTeV and KLOE should give some insight into the physics of direct CP violation inspite of large theoretical uncertainties. In this respect measurements of CP-violating asymmetries in charged B decays will also play an outstanding role. These experiments can be performed e.g. at CLEO since no time-dependences are needed. The situation concerning hadronic uncertainties is quite similar to ε'/ε. Therefore one should hope that some definite progress in calculating relevant hadronic matrix elements will also be made.

- More events for $K^+ \to \pi^+ \nu\bar{\nu}$ could in principle be reported from BNL already this year. In view of the theoretical cleanliness of this decay an observation of events at the $2\cdot 10^{-10}$ level would signal physics beyond the Standard Model. A detailed study of this very important decay requires, however, new experimental ideas and new efforts. The new efforts[113,114] in this direction allow to hope that a measurement of $Br(K^+ \to \pi^+\nu\bar{\nu})$ with an accuracy of $\pm 10\%$ should be possible before 2005. This would have a very important impact on the unitarity triangle and would constitute an important test of the Standard Model.

- The future improved inclusive $B \to X_{s,d}\gamma$ measurements confronted with improved Standard Model predictions could give the first signals of new physics. It appears that the errors on the input parameters could be lowered further and the theoretical error on $Br(B \to X_s\gamma)$ could be decreased confidently down to $\pm 8\%$ in the next years. The same accuracy in the experimental branching ratio will hopefully come from Cornell and later from KEK and SLAC. This may, however, be insufficient to disentangle new physics contributions although such an accuracy should put important constraints on the physics beyond the Standard Model. It would also be desirable to look for $B \to X_d\gamma$, but this is clearly a much harder task.

- Similar comments apply to transitions $B \to X_s l^+ l^-$ which appear to be even more sensitive to new physics contributions than $B \to X_{s,d}\gamma$. An observation of $B \to X_s\mu\bar{\mu}$ is expected from D0 and B-physics dedicated experiments at the beginning of the next decade. The distributions of various kind when measured should be very useful in the tests of the Standard Model and its extensions.

- The theoretical status of $K_{\rm L} \to \pi^0 e^+ e^-$ and of $K_{\rm L} \to \mu\bar{\mu}$, should be improved to confront future data. Experiments at DAΦNE should be very helpful in this respect. The first events of $K_{\rm L} \to \pi^0 e^+ e^-$ should come in the first years of the next decade from KAMI at FNAL. The experimental status of $K_{\rm L} \to \mu\bar{\mu}$, with the experimental error of $\pm 7\%$ to be decreased soon down to $\pm 1\%$, is truly impressive.

- The newly approved experiment at BNL to measure $Br(K_{\rm L} \to \pi^0\nu\bar{\nu})$ at the $\pm 10\%$ level before 2005 may make a decisive impact on the field of CP violation. In particular $K_{\rm L} \to \pi^0\nu\bar{\nu}$ seems to allow the cleanest determination of $Im\lambda_t$. Taken together with $K^+ \to \pi^+\nu\bar{\nu}$ a very clean determination of $\sin 2\beta$ can be obtained.

- The measurement of the $B_s^0 - \bar{B}_s^0$ mixing and in particular of $B \to X_{s,d}\nu\bar{\nu}$ and $B_{s,d} \to \mu\bar{\mu}$ will take most probably a longer time but as stressed in these lectures all efforts should be made to measure these transitions. Considerable progress on $B_s^0 - \bar{B}_s^0$ mixing should be expected from HERA-B, SLAC and TEVATRON in the first years of the next decade. LHC-B should measure it to a high precision. With the improved calculations of ξ in (4.58) this will have important impact on the determination of $|V_{td}|$ and on the unitarity triangle.

- Clearly future precise studies of CP violation at SLAC-B, KEK-B, HERA-B, CORNELL, FNAL and LHC-B providing first direct measurements of α, β and γ may totally revolutionize our field. In particular the first signals of new physics could be found in the $(\bar{\varrho}, \bar{\eta})$ plane. During the recent years several, in some cases quite sophisticated and involved strategies have been developed to extract these angles with small or even no hadronic uncertainties. Certainly the future will bring additional methods to determine α, β and γ. Obviously it is very desirable to have as many such strategies as possible available in order to overconstrain the unitarity triangle and to resolve certain discrete ambiguities which are a characteristic feature of these methods.

- The forbidden or strongly suppressed transitions such as $D^0 - \bar{D}^0$ mixing and $K_{\rm L} \to \mu e$ are also very important in this respect. Considerable progress in this area should come from the experiments at BNL, FNAL and KEK.

- On the theoretical side, one should hope that the non-perturbative methods will be considerably improved so that various B_i parameters will be calculated with sufficient precision. It is very important that simultaneously with advances in lattice QCD, further efforts are being made in finding efficient analytical tools for calculating QCD effects in the long distance regime. This is, in particular very important in the field of non-leptonic decays, where one should not expect too much from our lattice friends in the coming ten years unless somebody will get a brilliant idea which will revolutionize lattice calculations. The accumulation of data for non-leptonic B and D decays at Cornell, SLAC, KEK and FNAL should teach us more about the role of non-factorizable contributions and in particular about the final state interactions. In this context, in the case of K-decays, important lessons will come from DAΦNE which is an excellent machine for testing chiral perturbation theory and other non-perturbative methods.

In any case the field of weak decays and in particular of the FCNC transitions and of CP violation have a great future and one should expect that they could dominate particle physics in the first part of the next decade. Clearly the next ten years should be very exciting in this field.

76

Acknowledgements

I would like to thank Bruce Campbell, Faqir Khanna and Manuella Vincter for inviting me to such a wonderful Winter Institute and a great hospitality. I would also like to thank M. Gorbahn and L. Silvestrini for comments on the manuscript and the authors of ref. [35] for a most enjoyable collaboration.

References

1. A.J. Buras, hep-ph/9806471, to appear in *Probing the Standard Model of Particle Interactions*, eds. R. Gupta, A. Morel, E. de Rafael and F. David (Elsevier Science B.V., Amsterdam, 1998).
2. G. Buchalla, A.J. Buras and M. Lautenbacher, Rev. Mod. Phys. **68**, 1125, (1996).
3. A.J. Buras and R. Fleischer, hep-ph/9704376, in Heavy Flavours II, eds. A.J. Buras and M. Lindner, (World Scientific, 1998) page 65.
4. N. Cabibbo, Phys. Rev. Lett. **10**, 531, (1963).
5. M. Kobayashi and K. Maskawa, Prog. Theor. Phys. **49**, 652, (1973).
6. S.L. Glashow, J. Iliopoulos and L. Maiani, Phys. Rev. **D2**, 1285, (1970).
7. L.L. Chau and W.-Y. Keung, Phys. Rev. Lett. **53**, 1802, (1984).
8. Particle Data Group, Euro. Phys. J. **C3**, 1, (1998).
9. L. Wolfenstein, Phys. Rev. Lett. **51**, 1945, (1983).
10. A.J. Buras, M.E. Lautenbacher and G. Ostermaier, Phys. Rev. **D50**, 3433, (1994).
11. M. Schmidtler and K.R. Schubert, Z. Phys. **C53**, 347, (1992).
12. C. Jarlskog, Phys. Rev. Lett. **55**, 1039, (1985); Z. Phys. **C29**, 491, (1985).
13. C. Jarlskog and R. Stora, Phys. Lett. **B208**, 268, (1988).
14. A. Ali and D. London, hep-ph/9903535; F. Parodi, P. Roudeau, and A. Stocchi, hep-ex/9903063.
15. A. Stocchi, hep-ex/9902004.
16. K.G. Wilson, Phys. Rev. **179**, 1499, (1969); K.G. Wilson and W. Zimmermann, Comm. Math. Phys. **24**, 87, (1972).
17. W. Zimmermann, in Proc. 1970 Brandeis Summer Institute in Theor. Phys, (eds. S. Deser, M. Grisaru and H. Pendleton), MIT Press, 1971, p.396; Ann. Phys. **77**, 570, (1973).
18. E.C.G. Sudarshan and R.E. Marshak, Proc. Padua-Venice Conf. on Mesons and Recently Discovered Particles (1957).
19. R.P. Feynman and M. Gell-Mann, Phys. Rev. **109**, 193, (1958).
20. J. Chay, H. Georgi and B. Grinstein, Phys. Lett. **B247**, 399, (1990); I.I.

Bigi, N.G. Uraltsev and A.I. Vainshtein, Phys. Lett. **B293**, 430, (1992) [E: **B297**, 477, (1993)]; I.I. Bigi, M.A. Shifman, N.G. Uraltsev and A.I. Vainshtein, Phys. Rev. Lett. **71**, 496, (1993); B. Blok, L. Koyrakh, M.A. Shifman and A.I. Vainshtein, Phys. Rev. **D49**, 3356, (1994) [E: **D50**, 3572, (1994)]; A.V. Manohar and M.B. Wise, Phys. Rev. **D49**, 1310, (1994).

21. A.J. Buras, hep-ph/9901409.
22. T. Inami and C.S. Lim, Progr. Theor. Phys. **65**, 297, (1981).
23. G. Buchalla, A.J. Buras and M.K. Harlander, Nucl. Phys. **B349**, 1, (1991).
24. M.K. Gaillard and B.W. Lee, Phys. Rev. **D10**, 897, (1974).
25. H. Albrecht *et al.*, (ARGUS), Phys. Lett. **B192**, 245, (1987); M. Artuso *et al.*, (CLEO), Phys. Rev. Lett. **62**, 2233, (1989).
26. J.H. Christenson, J.W. Cronin, V.L. Fitch and R. Turlay, Phys. Rev. Lett. **13**, 128, (1964).
27. L.L. Chau, Physics Reports, **95**, 1, (1983).
28. A.J. Buras, W. Slominski and H. Steger, Nucl. Phys. **B245**, 369, (1984).
29. Y. Nir, SLAC-PUB-5874 (1992).
30. J. Bijnens, J.-M. Gérard and G. Klein, Phys. Lett. **B257**, 191, (1991).
31. S. Herrlich and U. Nierste, Nucl. Phys. **B419**, 292, (1994).
32. A.J. Buras, M. Jamin, and P.H. Weisz, Nucl. Phys. **B347**, 491, (1990); J. Urban, F. Krauss, U. Jentschura and G. Soff, Nucl. Phys. **B523**, 40, (1998).
33. S. Herrlich and U. Nierste, Phys. Rev. **D52**, 6505, (1995); Nucl. Phys. **B476**, 27, (1996).
34. R. Gupta, hep-ph/9801412.
35. S. Bosch, A.J. Buras, M. Gorbahn, S. Jäger, M. Jamin, M.E. Lautenbacher and L. Silvestrini, hep-ph/9904408.
36. S. Bertolini, J.O. Eeg, M. Fabbrichesi and E.I. Lashin, Nucl. Phys. **B514**, 63, (1998).
37. L. Conti, A. Donini, V. Gimenez, G. Martinelli, M. Talevi and A. Vladikas, Phys. Lett. **B421**, 273, (1998).
38. S. Aoki *et al.*, JLQCD collaboration, Phys. Rev. Lett. **80**, 5271, (1998); hep-lat/9901018.
39. G. Kilcup, R. Gupta and S.R. Sharpe, Phys. Rev. **D57**, 1654, (1998).
40. L. Lellouch, talk given at Recontres de Moriond, March 1999.
41. W.A. Bardeen, A.J. Buras and J.-M. Gérard, Phys. Lett. **B211**, 343, (1988); J-M. Gérard, Acta Physica Polonica **B21**, 257, (1990).
42. J. Bijnens and J. Prades, Nucl. Phys. **B444**, 523, (1995); hep-ph/9811472.

78

43. T. Hambye, G.O. Köhler and P.H. Soldan, hep-ph/9902334.
44. N. Bilic, C.A. Dominguez and B. Guberina, Z. Phys. **C39**, 351, (1988). R. Decker, Nucl. Phys. (Proc. Suppl.) **7A**, 180, (1989); S. Narison, Phys. Lett. **B351**, 369, (1995).
45. C. Bruno, Phys. Lett. **B320**, 135, (1994).
46. A. Pich and E. de Rafael, Phys. Lett. **B158**, 477, (1985); J. Prades *et al*, Z. Phys. **C51**, 287, (1991).
47. J.F. Donoghue, E. Golowich and B.R. Holstein, Phys. Lett. **B119**, 412, (1982).
48. T. Draper, hep-lat/9810065; S. Sharpe, hep-lat/9811006; C. Bernard *et al.*, Phys. Rev. Lett. **81**, 4812, (1998).
49. E. Bagan, P. Ball, V.M. Braun and H.G. Dosch, Phys. Lett. **B278**, 457, (1992); M. Neubert, Phys. Rev. **D45**, 2451, (1992) and references therein.
50. F. Abe *et al.*, (CDF Collaboration), Phys. Rev. Lett. **82**, 271, (1999); B. Abbott *et al.*, (D0 Collaboration), hep-ex/9808029.
51. The LEP B Oscillation Working Group, LEPBOSC 98/3.
52. S. Narison, Phys. Lett. **B322**, 247, (1994).
53. M. Gorbahn, unpublished.
54. CDF Collaboration, CDF/PUB/BOTTOM/CDF/4855, 1999.
55. Y. Grossman, Y. Nir, S. Plaszczynski and M. Schune, Nucl. Phys. **B511**, 69, (1998).
56. P. Paganini, F. Parodi, P. Roudeau and A. Stocchi, Phys. Scripta **58**, 556, (1998), hep-ph/9711261; F. Parodi, P. Roudeau and A. Stocchi, hep-ph/9802289.
57. L. Wolfenstein, Phys. Rev. Lett. **13**, 562, (1964).
58. G.D. Barr *et al.*, Phys. Lett. **B317**, 233, (1993).
59. L.K. Gibbons *et al.*, Phys. Rev. Lett. **70**, 1203, (1993).
60. Seminar presented by P. Shawhan for KTeV collaboration, Fermilab, Feb. 24 1999; http://fnphyx-www.fnal.gov/experiments/ktev/epsprime/epsprime.html.
61. J. Ellis, M.K. Gaillard and D.V. Nanopoulos, Nucl. Phys. **B109**, 213, (1976).
62. F.J. Gilman and M.B. Wise, Phys. Lett. **B83**, 83, (1979); B. Guberina and R.D. Peccei, Nucl. Phys. **B163**, 289, (1980).
63. F.J. Gilman and J.S. Hagelin, Phys. Lett. **B126**, 111, (1983); A.J. Buras, W. Slominski and H. Steger, Nucl. Phys. **B238**, 529, (1984); A.J. Buras and J.-M. Gérard, Phys. Lett. **B203**, 272, (1988).
64. J. Bijnens and M.B. Wise, Phys. Lett. **B137**, 245, (1984).

79

65. J.F. Donoghue, E. Golowich, B.R. Holstein and J. Trampetic, Phys. Lett. **B179**, 361, (1986).
66. A.J. Buras and J.-M. Gérard, Phys. Lett. **B192**, 156, (1987).
67. H.-Y. Cheng, Phys. Lett. **B201**, 155, (1988); M. Lusignoli, Nucl. Phys. **B325**, 33, (1989).
68. W.A. Bardeen, A.J. Buras and J.-M. Gérard, Phys. Lett. **B180**, 133, (1986); Nucl. Phys. **B293**, 787, (1987); Phys. Lett. **B192**, 138, (1987).
69. J.M. Flynn and L. Randall, Phys. Lett. **B224**, 221, (1989); erratum ibid. Phys. Lett. **B235**, 412, (1990).
70. G. Buchalla, A.J. Buras, and M.K. Harlander, Nucl. Phys. **B337**, 313, (1990).
71. E.A. Paschos and Y.L. Wu, Mod. Phys. Lett. **A6**, 93, (1991); M. Lusignoli, L. Maiani, G. Martinelli and L. Reina, Nucl. Phys. **B369**, 139, (1992).
72. A.J. Buras, M. Jamin, M.E. Lautenbacher and P.H. Weisz, Nucl. Phys. **B370**, 69, (1992); Nucl. Phys. **B400**, 37, (1993).
73. A.J. Buras, M. Jamin and M.E. Lautenbacher, Nucl. Phys. **B400**, 75, (1993).
74. A.J. Buras, M. Jamin and M.E. Lautenbacher, Nucl. Phys. **B408**, 209, (1993).
75. M. Ciuchini, E. Franco, G. Martinelli and L. Reina, Phys. Lett. **B301**, 263, (1993).
76. M. Ciuchini, E. Franco, G. Martinelli and L. Reina, Nucl. Phys. **B415**, 403, (1994).
77. M. Ciuchini, E. Franco, G. Martinelli, L. Reina and L. Silvestrini, Z. Phys. **C68**, 239, (1995).
78. A.J. Buras, M. Jamin, and M.E. Lautenbacher, Phys. Lett. **B389**, 749, (1996).
79. M. Ciuchini, Nucl. Phys. (Proc. Suppl.) **B59**, 149, (1997).
80. J. Heinrich, E.A. Paschos, J.-M. Schwarz, and Y.L. Wu, Phys. Lett. **B279**, 140, (1992); E.A. Paschos, review presented at the 27th Lepton-Photon Symposium, Beijing, China (August 1995).
81. S. Bertolini, M. Fabbrichesi and J.O. Eeg, hep-ph/9802405.
82. B. Winstein and L. Wolfenstein, Rev. Mod. Phys. **65**, 1113, (1993).
83. A.A. Belkov, G. Bohm, A.V. Lanyov and A.A. Moshkin, hep-ph/9704354.
84. A.J. Buras and M.E. Lautenbacher, Phys. Lett. **B318**, 212, (1993).
85. R.D. Kenway, Plenary talk at LATTICE 98, hep-ph/9810054.
86. M. Jamin and M. Münz, Z. Phys. **C66**, 633, (1995); S. Narison, Phys. Lett. **B358**, 113, (1995); K.G. Chetyrkin, D. Pirjol, and K. Schilcher, Phys. Lett. **B404**, 337, (1997); P. Colangelo, F. De Fazio, G. Nardulli,

and N. Paver, Phys. Lett. **B408**, 340, (1997); M. Jamin, Nucl. Phys. (Proc. Suppl.) **B64**, 250, (1998).

87. L. Lellouch, E. de Rafael, and J. Taron, Phys. Lett. **B414**, 195, (1997); F.J. Yndurain, Nucl. Phys. **B517**, 324, (1998). H.G. Dosch and S. Narison, Phys. Lett. **B417**, 173, (1998).
88. J. Prades and A. Pich, hep-ph/9811263, proceedings of QCD 98, Montpellier.
89. ALEPH collaboration, CERN-EP/99-026, hep-ex/9903015.
90. S. Narison, hep-ph/9905264.
91. R. Gupta, T. Bhattacharaya, and S.R. Sharpe, Phys. Rev. **D55**, 4036, (1997).
92. T. Hambye, G.O. Köhler, E.A. Paschos, P.H. Soldan and W.A. Bardeen, Phys. Rev. **D58**, 014017, (1998).
93. G.W. Kilcup, Nucl. Phys. (Proc. Suppl.) **B20**, 417, (1991).
94. S.R. Sharpe, Nucl. Phys. (Proc. Suppl.) **B20**, 429, (1991).
95. D. Pekurovsky and G. Kilcup, hep-lat/9709146.
96. D. Pekurovsky and G. Kilcup, hep-lat/9812019.
97. S. Bethke, hep-ex/9812026.
98. Y.-Y. Keum, U. Nierste and A.I. Sanda, hep-ph/9903230.
99. L.F. Abbot, P. Sikivie and M.B. Wise, Phys. Rev. **D21**, 1393, (1980).
100. E. Gabrielli and G.F. Giudice, Nucl. Phys. **B433**, 3, (1995).
101. E. Gabrielli, A. Masiero and L. Silvestrini, Phys. Lett. **B374**, 80, (1996); F. Gabbiani, E. Gabrielli, A. Masiero and L. Silvestrini, Nucl. Phys. **B477**, 321, (1996).
102. A. Masiero and H. Murayama, hep-ph/9903363.
103. D. Rein and L.M. Sehgal, Phys. Rev. **D39**, 3325, (1989); J.S. Hagelin and L.S. Littenberg, Prog. Part. Nucl. Phys. **23**, 1, (1989); M. Lu and M.B. Wise, Phys. Lett. **B324**, 461, (1994); S. Fajfer, [hep-ph/9602322]; C.Q. Geng, I.J. Hsu and Y.C. Lin, Phys. Rev. **D54**, 877, (1996).
104. G. Buchalla and G. Isidori, Phys. Lett. **B440**, 170, (1998).
105. G. Buchalla and A.J. Buras, Nucl. Phys. **B398**, 285, (1993).
106. G. Buchalla and A.J. Buras, Nucl. Phys. **B400**, 225, (1993).
107. G. Buchalla and A.J. Buras, Nucl. Phys. **B412**, 106, (1994).
108. M. Misiak and J. Urban, Phys. Lett. **B541**, 161, (1999).
109. G. Buchalla and A.J. Buras, hep-ph/9901288.
110. V.A. Novikov, A.I. Vainshtein, V.I. Zakharov and M.A. Shifman, Phys. Rev. **D16**, 223, (1977); J. Ellis and J.S. Hagelin, Nucl. Phys. **B217**, 189, (1983); C.O. Dib, I. Dunietz and F.J. Gilman, Mod. Phys. Lett. **A6**, 3573, (1991).
111. W. Marciano and Z. Parsa, Phys. Rev. **D53**, R1 (1996).

112. S. Adler *et al.*, Phys. Rev. Lett. **79**, 2204, (1997).
113. L. Littenberg and J. Sandweiss, eds., AGS2000, Experiments for the 21st Century, BNL 52512.
114. P. Cooper, M. Crisler, B. Tschirhart and J. Ritchie (CKM collaboration), EOI for measuring $Br(K^+ \to \pi^+\nu\bar{\nu})$ at the Main Injector, Fermilab EOI 14, 1996.
115. L. Littenberg,t Phys. Rev. **D39**, 3322, (1989).
116. Y. Grossman, Y. Nir and R. Rattazzi, hep-ph/9701231, in Heavy Flavours II, eds. A.J. Buras and M. Lindner, (World Scientific, 1998) page 755. Y. Nir, hep-ph/9904271.
117. G. Buchalla, hep-ph/9612307.
118. G. Buchalla and A.J. Buras, Phys. Rev. **D54**, 6782, (1996).
119. J. Adams *et al.*, Phys. Lett. **B447**, 240, (1999).
120. K. Arisaka *et al.*, KAMI conceptual design report, FNAL, June 1991.
121. T. Inagaki, T. Sato and T. Shinkawa, Experiment to search for the decay $K_L \to \pi^0\nu\bar{\nu}$ at KEK 12 GeV proton synchrotron, 30 Nov. 1991.
122. G. Buchalla and A.J. Buras, Phys. Lett. **B333**, 221, (1994).
123. G. Buchalla, G. Isidori and S.-J. Rey, Nucl. Phys. **B511**, 594, (1998).
124. N. Cabibbo and L. Maiani, Phys. Lett. **B79**, 109, (1978).
125. C.S. Kim and A.D. Martin, Phys. Lett **B225**, 186, (1989).
126. ALEPH Collaboration, Contribution (PA10-019) to the 28th International Conference on High Energy Physics, July 1996, Warsaw, Poland.
127. F. Abe *et al.*, (CDF), Phys. Rev. **D57**, R3811, (1998).
128. Y. Grossman, Z. Ligeti and E. Nardi, Phys. Rev. **D55**, 2768, (1997).
129. C. Greub, T. Hurth and D. Wyler, Phys. Lett. **B380**, 385, (1996); Phys. Rev. **D54**, 3350, (1996);
130. K.G. Chetyrkin, M. Misiak and M. Münz, Phys. Lett. **B400**, 206, (1997); Erratum-ibid. **B425**, 414, (1998).
131. The BaBar Physics Book, preprint SLAC-R-504.
132. A. Ali and G. Hiller, hep-ph/9812267.
133. A.J. Buras, M.E. Lautenbacher, M. Misiak and M. Münz, Nucl. Phys. **B423**, 349, (1994).
134. G. Ecker, A. Pich, and E. de Rafael, Nucl. Phys. **B291**, 692, (1987), Nucl. Phys. **B303**, 665, (1988); A.G. Cohen, G. Ecker, and A. Pich, Phys. Lett. **B304**, 347, (1993); P. Heiliger and L. Seghal, Phys. Rev. **D47**, 4920, (1993); C. Bruno and J. Prades, Z. Phys. **C57**, 585, (1993); J.F. Donoghue and F. Gabbiani, Phys. Rev. **D51**, 2187, (1995); A. Pich, hep-ph/9610243.
135. D.A. Harris *et al.*, Phys. Rev. Lett. **71**, 3918, (1993).
136. A.P. Heinson *et al.*, Phys. Rev. **D51**, 985, (1995); T. Akagi *et al.*, Phys.

Rev. **D51**, 2061, (1995).

137. G. D'Ambrosio, G. Isidori and J Portolés, Phys. Lett. **B423**, 385, (1998).

138. D. Gomez Dumm and A. Pich, hep-ph/9810523.

139. G. Valencia, hep-ph/9711377.

140. L. Littenberg and G. Valencia, Ann. Rev. Nucl. Part. Sci. **43**, 729, (1993); J.L. Ritchie and S.G. Wojcicki, Rev. Mod. Phys. **65**, 1149, (1993); A. Pich, hep-ph/9610243; G. D'Ambrosio and G. Isidori, hep-ph/9611284.

141. Y. Nir and H.R. Quinn Ann. Rev. Nucl. Part. Sci. **42**, 211, (1992) and in " B Decays ", ed S. Stone (World Scientific, 1994), p. 520; I. Dunietz, ibid p.550 and refs. therein.

142. R. Fleischer, Int. J. of Mod. Phys. **A12**, 2459, (1997).

143. M. Ciuchini, E. Franco, G. Martinelli, and L. Silvestrini, Nucl. Phys. **B501**, 271, (1997); M. Ciuchini, R. Contino, E. Franco, G. Martinelli, and L. Silvestrini, Nucl. Phys. **B512**, 3, (1998).

144. M. Gronau and D. London, Phys. Rev. Lett. **65**, 3381, (1990).

145. A. Snyder and H.R. Quinn, Phys. Rev. **D48**, 2139, (1993); A.J. Buras and R. Fleischer, Phys. Lett. **B360**, 138, (1995); J.P. Silva and L. Wolfenstein, Phys. Rev. **D49**, R1151, (1995); A.S. Dighe, M. Gronau and J. Rosner, Phys. Rev. **D54**, 3309, (1996); R. Fleischer and T. Mannel, Phys. Lett. **B397**, 269, (1997); C.S. Kim, D. London and T. Yoshikawa, Phys. Rev. **D57**, 4010, (1998).

146. I.I.Y. Bigi and A.I. Sanda, Nucl. Phys. **B193**, 85, (1981).

147. D. London and A. Soni, Phys. Lett. **B407**, 61, (1997); Y. Grossman and M.P. Worah, Phys. Lett. **B395**, 241, (1997); M. Ciuchini *et al.*, Phys. Rev. Lett. **79**, 978, (1997); R. Barbieri and A. Strumia, Nucl. Phys. **B508**, 3, (1997).

148. A.J. Buras, Nucl. Instr. Meth. **A368**, 1, (1995).

149. R. Aleksan, I. Dunietz and B. Kayser, Z. Phys. **C54**, 653, (1992); R. Fleischer and I. Dunietz, Phys. Lett. **B387**, 361, (1996).

150. M. Gronau and D. Wyler, Phys. Lett. **B265**, 172, (1991).

151. M. Gronau and D. London, Phys. Lett. **B253**, 483, (1991). I. Dunietz, Phys. Lett. **B270**, 75, (1991).

152. D. Atwood, I. Dunietz and A. Soni, Phys. Rev. Lett. **B78**, 3257, (1997).

153. R. Fleischer, Phys. Lett. **B365**, 399, (1996); R. Fleischer and T. Mannel, Phys. Rev. **D57**, 2752, (1998);

154. M. Gronau and J.L. Rosner, Phys. Rev. **D57**, 6843, (1998);

155. F. Würthwein and P. Gaidarev, hep-ph/9712531.

156. R. Fleischer, Eur. Phys. J. **C6**, 451, (1999); A.J. Buras and R. Fleischer, hep-ph/9810260.

157. R. Godang *et al.*, Phys. Rev. Lett. **80**, 3456, (1998); B.H. Behrens *et al.*, Phys. Rev. Lett. **80**, 3710, (1998); D.M. Asner *et al.*, Phys. Rev. **D53**, 1039, (1996);

158. L. Wolfenstein, Phys. Rev. **D52**, 537, (1995); J. Donoghue, E. Golowich, A. Petrov and J. Soares, Phys. Rev. Lett. **77**, 2178, (1996); B. Blok and I. Halperin, Phys. Lett. **B385**, 324, (1996); B. Blok, M. Gronau and J.L. Rosner, Phys. Rev. Lett. **78**, 3999, (1997); A.J. Buras, R. Fleischer and T. Mannel, Nucl. Phys. **B533**, 3, (1998); J.-M. Gérard and J. Weyers, hep-ph/9711469; M. Neubert, Phys. Lett. **B424**, 152, (1998); A.F. Falk, A.L. Kagan, Y. Nir and A.A. Petrov, Phys. Rev. **D57**, 4290, (1998); D. Atwood and A. Soni (1997), Phys. Rev. **D58**, 036005, (1998); R. Fleischer, Phys. Lett. **B435**, 221, (1998); M. Gronau and J.L. Rosner, Phys. Rev. **D58**, 113005, (1998);

159. M. Neubert, JHEP 9902, 014, (1998).

160. M. Neubert and J.L. Rosner, Phys. Lett. **B441**, 403, (1998); Phys. Rev. Lett. **81**, 5076, (1998);

161. R. Fleischer, hep-ph/9904313.

162. M. Neubert, hep-ph/9904321.

163. R. Fleischer, hep-ph/9903455, hep-ph/9903456.

164. Y. Gao and F. Würthwein, hep-ex/9904008.

165. X-G. He, W-S. Hou and K-Ch. Yang, hep-ph/9902256.

166. A. Lenz, U. Nierste and G. Ostermaier, Phys. Rev. **D59**, 034008, (1999); U. Nierste, hep-ph/9805388.

167. A.J. Buras, Phys. Lett. **B333**, 476, (1994).

168. M. Gronau and D. London, Phys. Rev. **D55**, 2845, (1997); Y. Grossman, Y. Nir and M.P. Worah, Phys. Lett. **B407**, 307, (1997).

169. G. Colangelo and G. Isidori, JHEP 09, 009, (1998).

170. A.J. Buras and L. Silvestrini, Nucl. Phys. **B546**, 299, (1999).

171. Y. Nir and M.P. Worah, Phys. Lett. **B423**, 319, (1998).

172. A.J. Buras, A. Romanino and L. Silvestrini, Nucl. Phys. **B520**, 3, (1998).

173. Y. Grossman and Y. Nir, Phys. Lett. **B398**, 163, (1997); C.E. Carlson, G.D. Dorada and M. Sher, Phys. Rev. **D54**, 4393, (1996); G. Burdman, Phys. Lett. **B409**, 443, (1997); A. Berera, T.W. Kephart and M. Sher, Phys. Rev. **D56**, 7457, (1997); Gi-Chol Cho, hep-ph/9804327.

174. T. Hattori, T. Hasuike and S. Wakaizumi, hep-ph/9804412.

NEUTRINO MASSES AND NEUTRINO OSCILLATIONS

L. DI LELLA

CERN, CH-1211 Geneva 23

These lectures review direct measurements of neutrino masses and the status of neutrino oscillation searches using both 'natural' neutrino sources (the Sun and cosmic rays interacting in the Earth atmosphere) and 'artificial' neutrinos (produced by nuclear reactors and accelerators). Finally, future experiments and plans are presented.

1 Introduction

At present, we know nothing about two basic neutrino properties: their mass, and whether neutrinos are their own antiparticles or whether ν and $\bar{\nu}$ differ.

In the Standard Model all neutrino masses are set equal to zero 'by hand'. Under this assumption the neutrino helicity (the spin component parallel to the momentum) is a good quantum number. If neutrinos are their own antiparticles (Majorana neutrinos), the two helicity states are just the two spin states of the same particle and lepton number is not conserved. If ν and $\bar{\nu}$ differ (Dirac neutrinos), left-handed neutrinos and right-handed antineutrinos are the only existing physical states; right-handed neutrinos and left-handed antineutrinos do not exist and lepton number is conserved.

However, if neutrinos are massive the helicity is not a good quantum number because it depends on the reference frame. Hence massive Dirac neutrinos and antineutrinos can exist in both helicity states but the interactions of right-handed neutrinos and left-handed antineutrinos with matter differ from those of left-handed neutrinos and right-handed antineutrinos.

Massive neutrinos could be an important component of hot dark matter in the Universe. According to Big Bang cosmology, the Universe is filled with a Fermi gas of neutrinos at a temperature of $\sim$ 1.9 K and with a density of $\sim$ $120/\text{cm}^3$ for each neutrino type.

The neutrino energy density, ρ_ν, normalized to the critical density of the Universe, ρ_c is given by

$$\Omega_\nu = \rho_\nu/\rho_c = \sum_\nu m_\nu/94 \, h_0^2 \qquad (1)$$

where $\rho_c = 3 \, H_0^2/8\pi G_N = 1.05 \times 10^4 \, h_0^2 \text{ eV/cm}^3$, the neutrino mass m_ν is expressed in eV, G_N is the Newton constant and $H_0 = 100 \, h_0$ km s^{-1} Mpc^{-1} is the Hubble constant. The normalized Hubble expansion rate h_0 is not precisely known but is believed to be in the interval $0.6 < h_0 < 0.8$ [1]. Thus a value of $\sum_\nu m_\nu$ between 34 and 60 eV would give $\Omega_\nu = 1$. Present cosmological models prefer Ω_ν values of the order of 0.2 [2], suggesting that the heaviest neutrino mass is in the eV range.

2 Direct measurements of neutrino masses

2.1 Electron-neutrino

The ν_e mass can be determined by measuring the electron energy spectrum from β-decay. The spectral shape has the form

$$\mathrm{d}n/\mathrm{d}E = CF(E)\, p\, W(E_0 - E)\sqrt{(E_0 - E)^2 - m_\nu^2} \qquad (2)$$

where C is a normalization constant, E, W and p are the electron kinetic energy, total energy and momentum, respectively, m_ν is the neutrino mass, E_0 is the electron kinetic energy at the end point of the spectrum (calculated for $m_\nu = 0$) and $F(E)$ is a calculable correction which takes into account the Coulomb interaction between the outgoing electron and the nucleus.

As shown by Eq. (2) the effect of m_ν on the spectrum is only visible near the end point. Tritium β-decay has been so far the best choice to determine the ν_e mass because of the very low E_0 value, $E_0 = 18.58$ KeV (the fraction of electrons emitted in the region of the spectrum where the effect of m_ν is measurable is proportional to E_0^{-3}).

Recent measurement of the Tritium β-decay spectrum are consistent with $m_\nu = 0$ and have provided upper limits to its value [3]. The most stringent limit, $m_\nu < 2.5$ eV (95% confidence) has been obtained by the Troitsk group [4].

It must be mentioned that the best fit to the data for all experiments, excluding the Troitsk one, gives a slightly negative value of m_ν^2, with a mean value $m_\nu^2 = -59 \pm 26$ eV^2 [5]. There is no clear explanation for this unphysical value which suggests the existence of a systematic effect. The Troitsk experiment [4], with an energy resolution of only 3.7 eV, has observed an excess of events concentrated in a narrow region of the spectrum a few eV below the end point. The origin of this excess is not understood but could be the origin of the unphysical m_ν^2 values found by previous experiments.

2.2   Muon-neutrino

The most stringent limit on the ν_μ mass has been obtained by measuring precisely the μ^+ momentum from $\pi^+ \to \mu^+ \nu_\mu$ decay at rest [6]:

$$m_\nu^2 = m_\pi^2 + m_\mu^2 - 2\, m_\pi \sqrt{p_\mu^2 + m_\mu^2} \;. \tag{3}$$

The muon momentum, p_μ, is measured to be [6]

$$p_\mu = 29.79207 \pm 0.00012 \text{ MeV} \;.$$

The charged pion mass, m_π, is obtained from a precision measurement of the $4f \to 3d$ transition energy of $\pi^-\ {}^{24}$Mg atoms [7] (at the large atomic radius corresponding to these quantum numbers strong interaction effects on the energy levels are negligible):

$$m_\pi = 139.56995 \pm 0.00035 \text{ MeV} \;.$$

The muon mass, m_μ, is known much more precisely from a combination of two measurements. Precision measurements of the muon spin precession in a magnetic field are used to determine the muon magnetic moment, $\mu_\mu = ge\hbar/2m_\mu$; and the g factor is measured independently in the CERN $g - 2$ experiment [8]. The result is

$$m_\mu = 105.658389 \pm 0.000034 \text{ MeV} \;.$$

When these values are used in Eq. (3) one obtains

$$m_\nu^2 = -0.022 \pm 0.023 \text{ MeV}^2$$

which gives the upper bound

$$m_\nu < 0.16 \text{ MeV}$$

at the 90% confidence level.

The contributions to the error on m_ν^2 are ± 0.009 MeV^2 from p_μ, ± 0.002 MeV^2 from m_μ and ± 0.021 MeV^2 from m_π. The last one is the dominant contribution, hence any improvement on the direct determination of m_ν from $\pi^+ \to \mu^+ \nu_\mu$ decay requires first a more precise knowledge of the π^+ mass.

2.3 Tau-neutrino

The most stringent upper bound on the ν_τ mass has been obtained by the ALEPH experiment at LEP [9] from a study of multi-prong hadronic decays of $\tau^\pm$ produced at the Z peak between 1991 and 95:

$$\tau^\pm \ \to \ \nu_\tau \ \pi^\pm \pi^+ \pi^- \ (2939 \text{ events});$$
$$\tau^\pm \ \to \ \nu_\tau \pi^\pm \pi^+ \pi^+ \pi^- \pi^- \ (52 \text{ events});$$
$$\tau^\pm \ \to \ \nu_\tau \pi^\pm \pi^+ \pi^+ \pi^- \pi^- \pi^0 \ (2 \text{ events}).$$

The ν_τ mass is determined using two variables, E_h and M_h, the total energy and invariant mass of the multi-pion system, respectively. Fig. 1 shows the event distribution in the plane M_h, E_h/E_{beam}, where E_{beam} is the LEP beam energy. Also shown in Fig. 1 are the boundaries of the allowed regions for $m_\nu = 0$ and 23 MeV. It is clear that the event distribution near the boundaries provides constraints to m_ν.

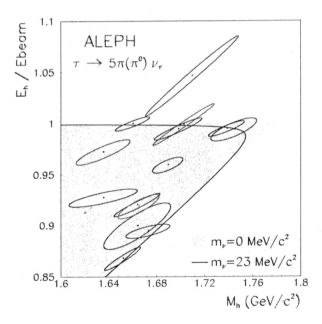

Figure 1: Distribution of $\tau^\pm \to \nu_\tau \pi^\pm \pi^+ \pi^+ \pi^- \pi^- (\pi^0)$ events. The grey area is the allowed region for $m_\nu = 0$. The boundary of the allowed region for $m_\mu = 23$ MeV is also shown. The ellipses represent the measurement errors.

88

A statistical analysis of this distribution gives the upper bound

$$m_\nu < 18.2 \text{ MeV}$$

at the 95% confidence level.

3 Neutrinoless double-β-decay

If neutrinos are massive and they are their own antiparticles, neutrinoless double-β-decay is expected to occur if energetically possible:

$$(A, Z) \rightarrow (A, Z+2) + e^- + e^- \tag{4}$$

where A and Z are the total number of nucleons and protons in the initial nucleus, respectively. This process violates lepton number conservation.

Reaction (4) is a second-order weak interaction process in which a neutrino is emitted by a neutron together with an e^- and is reabsorbed by another neutron in the same nucleus with the emission of another e^-. This process cannot occur for Dirac neutrinos ($\nu \neq \bar{\nu}$) because the emitted neutrino is in fact a $\bar{\nu}$ which cannot be absorbed by a neutron. In the case of Majorana neutrinos reaction (4) can only occur if the emitted neutrino undergoes helicity flip (otherwise it cannot be absorbed by a neutron). The helicity flip amplitude is proportional to m_ν/E_ν, where E_ν is the neutrino energy, thus reaction (4) can only occur if $m_\nu \neq 0$. For $m_\nu = 0$ neutrinoless double-β-decay is forbidden also for Majorana neutrinos.

Searches for neutrinoless double-β decay are based on the fact that the sum of the two electron energies from reaction (4) has a fixed value. The most sensitive search so far has been performed by the Heidelberg-Moscow experiment [10] in the Gran Sasso underground laboratory, using the reaction

$$^{76}\text{Ge}_{32} \rightarrow ^{76}\text{Se}_{34} + e^- + e^- \tag{5}$$

for which the sum of the two electron energies is 2038 keV. This experiment uses five enriched Germanium crystals which contain 86% of ^{76}Ge isotope for a total active mass of 19.96 kg (the fraction of ^{76}Ge isotope in natural Germanium is only 7.67%). Germanium crystals are solid-state detectors with an energy resolution of few keV at 2000 keV. The occurrence of reaction (5) is expected to result in an excess of events at 2038 keV above the flat background from other processes. The detector is surrounded by a shield of anticoincidence counters.

For an exposure of 24 kg · y the counting rate in the energy region where a signal from reaction (5) is expected is $\sim 0.2/(\text{kg} \cdot \text{y} \cdot \text{keV})$. No excess of events is observed in this region, providing a lower bound for the half-life of reaction (5):

$$T_{1/2} > 5.7 \times 10^{25} \ y$$

at the 90% confidence level [10]. This corresponds to the upper bound

$$m_\nu < 0.2 \ \text{eV} \tag{6}$$

for Majorana neutrinos. This bound is modified in the presence of neutrino mixing (see next Section).

It must be mentioned that ordinary double-β-decay

$$(A, Z) \rightarrow (A, Z + 2) + e^- + e^- + \bar{\nu}_e + \bar{\nu}_e$$

has been observed for a number of isotopes (e.g., ^{82}Se, ^{100}Mo, ^{150}Nd) with half-lives between 10^{19} and 10^{20} y [11].

4 Neutrino mixing and oscillations

4.1 Theory of neutrino oscillations in vacuum

As discussed in Sec. 2, direct measurements of neutrino masses are very far from being sensitive to ν_μ and ν_τ mass values of cosmological relevance. Searches for neutrinoless double-β-decay (see Sec. 3) are sensitive to ν_e masses lighter than 1 eV but only if neutrinos are Majorana particles.

At present, the most promising way to find evidence for $m_\nu > 0$ is to search for neutrino oscillations.

Neutrino oscillations are a consequence of the hypothesis of neutrino mixing first proposed by Pontecorvo [12] and independently by Maki *et al.* [13]. According to this hypothesis the three known neutrino flavours, ν_e, ν_μ and ν_τ, are not mass eigenstates but quantum-mechanical superpositions of three mass eigenstates, ν_1, ν_2 and ν_3, with mass eigenvalues m_1, m_2 and m_3, respectively:

$$\nu_\alpha = \sum_i U_{\alpha i} \ \nu_i . \tag{7}$$

In Eq. (7) $\alpha = e, \mu, \tau$ is the flavour index, $i = 1, 2, 3$ is the index of the mass eigenstates and U is a unitary 3×3 matrix. The relation

$$\nu_i = \sum_\alpha V_{i\alpha} \ \nu_\alpha$$

also holds, where $V = U^{-1}$ and $V_{i\alpha} = U^*_{\alpha i}$ because U is unitary.

From Eq. (7) it follows that the time evolution of a neutrino with momentum $\vec{p}$ produced in the state ν_α at time $t = 0$ is given by

$$\nu(t) = e^{i\vec{p}\cdot\vec{r}} \sum_i U_{\alpha i}\, e^{-iE_i t}\nu_i \tag{8}$$

where $E_i = \sqrt{p^2 + m_i^2}$. If the masses m_i are not all equal, the three terms of the sum in Eq. (8) get out of phase and the state $\nu(t)$ acquires components ν_β with $\beta \neq \alpha$.

The case of two-neutrino mixing is a particularly useful example. In this case the mixing matrix U is described by only one real parameter θ (the mixing angle), and Eqs. (7) and (8) become, respectively,

$$\begin{aligned} \nu_\alpha &= \cos\theta\, \nu_1 + \sin\theta\, \nu_2 \\ \nu_\beta &= -\sin\theta\, \nu_1 + \cos\theta\, \nu_2 \end{aligned}$$

and

$$\nu(t) = e^{i\vec{p}\cdot\vec{r}}(\cos\theta e^{-iE_1 t}\nu_1 + \sin\theta\, e^{-iE_2 t}\nu_2)\,.$$

The probability to detect a neutrino state ν_β at time t can then be easily calculated to be

$$P_{\alpha\beta}(t) = |\langle\nu_\beta|\nu(t)\rangle|^2 = \sin^2(2\theta)\,\sin^2\left(\frac{m_2^2 - m_1^2}{4p}\,t\right) \tag{9}$$

where we have used the approximation, valid for $m \ll p$,

$$E_i = \sqrt{p^2 + m_i^2} \approx p\left(1 + \frac{m_i^2}{2p^2}\right)\,.$$

It can be easily demonstrated that, for $\nu(0) = \nu_\beta$, $P_{\beta\alpha}(t)$ is also given by Eq. (9). Furthermore, we have

$$P_{\alpha\alpha}(t) = 1 - P_{\alpha\beta}(t)\,.$$

Eq. (9) is expressed in natural units. In more familiar units we can write

$$P_{\alpha\beta}(L) = \sin^2(2\theta)\sin^2\left(1.267\frac{\Delta m^2}{E}\,L\right) \tag{10}$$

where $L = ct$ is the distance from the source in metres, $\Delta m^2 = |m_2^2 - m_1^2|$ is measured in eV^2 and $E \approx p$ is the neutrino energy in MeV (the same equation holds if L is measured in km and E in GeV).

Eq. (10) describes an oscillation with amplitude equal to $\sin^2 (2\theta)$ and oscillation length λ given by

$$\lambda = 2.48 \; \frac{E}{\Delta m^2} \qquad (11)$$

where λ is expressed in metres (km), E in MeV (GeV) and Δm^2 in eV^2. We note that, if the oscillation length λ is much shorter than the size of the neutrino source or of the detector (or of both), the periodic term in Eq. (10) averages to $1/2$ and the oscillation probability becomes independent of L:

$$P_{\alpha\beta} = \frac{1}{2} \; \sin^2 (2\theta) \; .$$

4.2 Oscillation experiments

Experiments searching for neutrino oscillations can be subdivided into two categories:

4.2.1 Disappearance experiments

In these experiments the flux of a given neutrino flavour is measured at a certain distance L from the source. The presence of neutrino oscillations has the effect of reducing the flux with respect to the value expected in the absence of oscillations. The probability measured by these experiments is

$$P_{\alpha\alpha}(L) = 1 - \sum_{\beta \neq \alpha} P_{\alpha\beta}(L) \; .$$

The sensitivity of these experiments is limited by the systematic uncertainty on the knowledge of the neutrino flux from the source. To reduce this uncertainty a second detector close to the source is often used in order to measure directly the neutrino flux.

Disappearance experiments have been performed at nuclear reactors and at accelerators. The core of a nuclear reactor is an intense source of $\bar{\nu}_e$ with an average energy of ~ 3 MeV, which can be detected by observing the reaction $\bar{\nu}_e + p \rightarrow e^+ + n$. If a $\bar{\nu}_e$ turns into a $\bar{\nu}_\mu$ or a $\bar{\nu}_\tau$ it becomes invisible because μ^+ or τ^+ production is energetically forbidden.

Proton accelerators produce ν_μ's with energies between ~ 30 MeV and ~ 200 GeV. In disappearance experiments the ν_μ flux is measured by detecting the reaction $\nu_\mu + $ nucleon $\rightarrow \mu^- + $ hadrons. The energy threshold for the reaction $\nu_\mu + n \rightarrow \mu^- + p$ on a neutron at rest is 110.2 MeV.

Disappearance experiments are the only way to detect oscillations involving neutrinos with no coupling to the W and Z bosons ('sterile' neutrinos).

4.2.2 Appearance experiments

These experiments use beams containing predominantly one neutrino flavour and search for neutrinos of different flavour at a certain distance from the source.

The sensitivity of these experiments is often limited by the systematic uncertainty on the knowledge of the beam contamination by other neutrino flavours at the source. For example, in a typical ν_μ beam from a high-energy accelerator the ν_e contamination at the source is of the order of 1%.

Searches for ν_e and ν_τ appearance in a beam containing predominantly ν_μ have been performed at accelerators. In these experiments the presence of ν_e's (ν_τ) in the beam is detected by observing the reaction ν_e (ν_τ) + nucleon $\rightarrow e^-$ (τ^-) + hadrons.

4.2.3 Parameters of oscillation experiments

Typical parameters of oscillation experiments are listed in Table 1.

Table 1: Parameters of oscillation experiments

Neutrino source	Neutrino type	Baseline	Neutrino energy	Minimum accessible Δm^2
Sun	ν_e	$\sim 1.5 \times 10^{11}$ m	0.1–18 MeV	$\sim 10^{-11}$ eV^2
Cosmic ray $\pi^\pm$, $\mu^\pm$ decay	ν_μ ν_e $\bar{\nu}_\mu$ $\bar{\nu}_e$	~ 10 – 13000 km	0.2 GeV – several GeV	$\sim 10^{-4}$ eV^2
Nuclear power reactors	$\bar{\nu}_e$	~ 20 m –300 km	few MeV; $\langle E \rangle \sim 3$ MeV	$10^{-1} - 10^{-6}$ eV^2 depending on baseline
Accelerators	ν_μ ν_e $\bar{\nu}_\mu$ $\bar{\nu}_e$	~ 20 m – 732 km	~ 30 MeV – 100 GeV	$10^{-3} - 10$ eV^2 depending on baseline

4.3 Theory of neutrino oscillation in matter

It was first pointed out by Wolfenstein [14] that neutrino oscillations in dense matter differ from oscillations in vacuum if ν_e's are involved. This effect arises from coherent neutrino scattering at 0° which, in addition to the Z-boson exchange amplitude (the same for all three neutrino flavours), in the case of ν_e's has a contribution from W-boson exchange with the matter electrons (see the relevant Feynman graphs in Fig. 2).

Since scattering at 0° is a coherent process involving an extended target, the propagation of neutrinos in matter can be described by adding to the Hamiltonian a potential energy term which for the diagram of Fig. 2b is given by

$$V_W = \sqrt{2}\, G_F N_e \approx 7.63 \times 10^{-14}\, \frac{Z}{A}\rho\ \text{eV}$$

where G_F is the Fermi coupling constant, N_e is the number of electrons per unit volume, ρ is the matter density in g/cm^3 and the ratio Z/A is the number of electrons per nucleon.

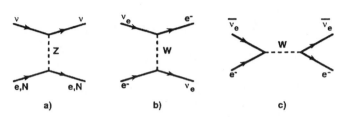

Figure 2: Feynman graphs for neutrino scattering in matter; a) neutrino-nucleon or neutrino-electron scattering by Z boson exchange; b) ν_e-electron and c) $\bar{\nu}_e$-electron scattering by W boson exchange.

We consider the case of two-neutrino mixing between ν_e and ν_μ:

$$\nu_e = \nu_1 \cos\theta_v + \nu_2 \sin\theta_v$$
$$\nu_\mu = -\nu_1 \sin\theta_v + \nu_2 \cos\theta_v$$

where θ_v is the mixing angle in vacuum. We assume that $\theta_v < 45°$ and $m_2 > m_1$, where $m_1(m_2)$ is the $\nu_1(\nu_2)$ mass value. The evolution equation is

$$i\frac{d\Psi}{dt} = H\Psi \tag{12}$$

where

$$\Psi = \begin{pmatrix} \nu_e \\ \nu_\mu \end{pmatrix}$$

is a two-component vector describing the neutrino state at time t and the Hamiltonian H is a 2×2 matrix:

$$H = \sqrt{p^2 + M^2} + V_Z \begin{vmatrix} 1 & 0 \\ 0 & 1 \end{vmatrix} + V_W \begin{vmatrix} 1 & 0 \\ 0 & 0 \end{vmatrix}$$

where M^2 is the square of the mass matrix and V_Z is the potential energy term resulting from Z boson exchange. By using the approximation

$$\sqrt{p^2 + M^2} \approx p + \frac{M^2}{2p} \approx E + \frac{M^2}{2E}$$

the Hamiltonian can be rewritten as

$$H = (E + V_z) \begin{vmatrix} 1 & 0 \\ 0 & 1 \end{vmatrix} + \frac{1}{2E} \begin{vmatrix} M_{ee}^2 + 2EV_W & M_{e\mu}^2 \\ M_{\mu e}^2 & M_{\mu\mu}^2 \end{vmatrix} \tag{13}$$

where

$$M_{ee}^2 = \frac{1}{2}(\mu^2 - \Delta m^2 \cos 2\theta_v)$$

$$M_{e\mu}^2 = M_{\mu e}^2 = \frac{1}{2}\Delta m^2 \sin 2\theta_v$$

$$M_{\mu\mu}^2 = \frac{1}{2}(\mu^2 + \Delta m^2 \cos 2\theta_v)$$

with $\mu^2 = m_1^2 + m_2^2$ and $\Delta m^2 = m_2^2 - m_1^2$. Obviously, the first term of the Hamiltonian (13) produces no mixing between ν_e and ν_μ.

The study of the ideal case of ν_e's produced in a medium of constant density is mathematically rather simple and is very useful to understand the physics of neutrino oscillations in matter. In this case the Hamiltonian is time-independent and the mass eigenstates can be found by diagonalising the second matrix in Eq. (13).

The two mass eigenvalues in matter are

$$m^2 = \frac{1}{2}(\mu^2 + \xi) \pm \frac{1}{2}\sqrt{(\Delta m^2 \cos 2\theta_v - \xi)^2 + (\Delta m^2)^2 \sin^2 2\theta_v}$$

and the mixing angle in matter, θ_m, is given by the equation

$$\tan 2\theta_m = \frac{\Delta m^2 \sin 2\theta_v}{\Delta m^2 \cos 2\theta_v - \xi} \tag{14}$$

where

$$\xi = 2V_W E \approx 1.526 \times 10^{-7}(Z/A)\rho E \ eV^2$$

(in this equation ρ is in g/cm^3 and the neutrino energy E is in MeV).

The behaviour of the two mass eigenvalues as a function of ξ is illustrated in Fig. 3.

Eq. (14) shows that, even if θ_v is very small, for $\xi = \Delta m^2 \cos 2\theta_v$ the denominator vanishes and the mixing angle in matter, θ_m, is equal to 45°, which corresponds to maximal mixing. This resonant behaviour was first noticed by Mikheyev and Smirnov [15] some years after Wolfenstein's original formulation of the theory of neutrino oscillations in matter. At the resonant value of ξ the difference between the two eigenstates is minimal and is equal to $\Delta m^2 \sin 2\theta_v$.

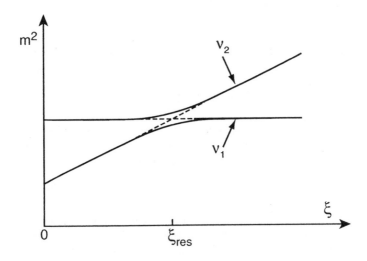

Figure 3: Neutrino mass eigenvalues in matter as a function of ξ for the case of small mixing angle in vacuum. On the right-hand side of the resonant value, ξ_{res}, ν_2 is mostly ν_e while on the left-hand side ν_2 is mostly ν_μ.

The oscillation length in matter, λ_m, is longer than in vacuum and is given by

$$\lambda_m = \lambda_v \frac{\Delta m^2}{\sqrt{(\Delta m^2 \cos 2\theta_v - \xi)^2 + (\Delta m^2)^2 \sin^2 2\theta_v}}$$

where λ_v is the oscillation length in vacuum given by Eq. (11). The maximum value of λ_m is reached at resonance, where $\lambda_m = \lambda_v / \sin 2\theta_v$.

The potential energy term V_W changes sign for $\bar{\nu}_e$ (see the Feynman graph of Fig. 2c). As a consequence, the difference between the two mass eigenvalues increases monotonically with density. There is no resonance, therefore, in the case of antineutrinos.

5 Solar neutrinos

5.1 Standard Solar Model

As all visible stars, the Sun was formed from the gravitational collapse of a cloud of gas consisting mostly of hydrogen and helium. This collapse produced an increase of the core density and temperature resulting in the ignition of nuclear fusion reactions. A state of hydrostatic equilibrium was reached when the kinetic and radiation pressure balanced the gravitational forces preventing any further collapse.

There are several nuclear fusion reactions occurring in the Sun core, all having the effect of transforming four protons into a He^4 nucleus:

$$4p \rightarrow He^4 + 2e^+ + 2\nu_e .$$

This reaction is followed by the annihilation of the two positrons with two electrons, so the average energy produced by this reaction and emitted by the Sun in the form of electromagnetic radiation is

$$Q = (4m_p - M_{He} + 2m_e)c^2 - < E(2\nu_e) > \approx 26.1 \text{ MeV}$$

where m_p, M_{He}, m_e are the proton, He^4 nucleus and electron mass, respectively, and $< E(2\nu_e) > \approx 0.59$ MeV is the average energy carried by the two neutrinos. The Sun luminosity is measured to be [1]

$$L_0 = 3.846 \times 10^{26} W = 2.400 \times 10^{39} \text{ MeV/s} .$$

From the values of Q and L_o it is possible to calculate the rate of ν_e emission from the Sun:

$$dN(\nu_e)/dt = 2L_o/Q \approx 1.8 \times 10^{38} \text{ s}^{-1}$$

from which one can calculate the solar neutrino flux on Earth using the average distance between the Sun and the Earth (1.496×10^{11}m):

$$\Phi_\nu \approx 6.4 \times 10^{10} \text{ cm}^{-2} \text{ s}^{-1} .$$

The Standard Solar Model (SSM), which has been developed and continuously updated by J.N. Bahcall during the past 20 years [16, 17], predicts the energy spectrum of the solar neutrinos. The main assumptions of the SSM are:

(i) hydrostatic equilibrium;

(ii) energy production by fusion;

(iii) thermal equilibrium (i.e., the thermal energy production rate is equal to the luminosity);

(iv) the energy transport inside the Sun is dominated by radiation.

Table 2 shows a list of Sun parameters.

Table 2: Sun parameters

Luminosity	3.846×10^{26} W
Radius	6.96×10^5 Km
Mass	1.989×10^{30} Kg
Core temperature T_c	15.6×10^6 K
Surface temperature T_s	5773 K
Hydrogen content in the core (by mass)	34.1%
Helium content in the core (by mass)	63.9%

The age of the Sun (4.6×10^9 years) is also known. The SSM calculations are performed by adjusting the initial parameters, by evolving them to the present day and by comparing the predicted and measured properties of the Sun. The initial composition of the Sun is taken to be equal to the present day measurement of the surface abundances. If the predicted properties disagree with the measured ones, the calculations are repeated with different initial parameters until agreement is found. These calculations require the knowledge of the absolute cross-sections for nuclear reactions in a very low energy region where little information is directly available from laboratory experiments. Another important ingredient in these calculations is the knowledge of the opacity versus radius which controls the energy transport inside the Sun and the internal temperature distribution.

There are two main nuclear reaction cycles in the Sun core:

(i) The pp cycle, responsible for 98.5% of the Sun luminosity. This cycle involves the following reactions:

$$p + p \rightarrow e^+ + \nu_e + d \tag{15}$$

$$p + d \rightarrow \gamma + \text{He}^3 \tag{16}$$

$$\text{He}^3 + \text{He}^3 \rightarrow \text{He}^4 + p + p \tag{17}$$

where the second He^3 nucleus in the initial state of reaction (17) is produced by another sequence of reactions (15) and (16).

Reactions (15) through (17) represent 85% of the pp cycle. In the remaining 15% reaction (17) is replaced by the following sequence of reactions:

$$\text{He}^3 + \text{He}^4 \rightarrow \gamma + \text{Be}^7 \tag{18}$$

$$e^- + \text{Be}^7 \rightarrow \nu_e + Li^7 \tag{19}$$

$$p + Li^7 \rightarrow \text{He}^4 + \text{He}^4 \,. \tag{20}$$

In approximately 1.9×10^{-3} of the cases reactions (19) and (20) are replaced by

$$p + Be^7 \to \gamma + B^8 \tag{21}$$

$$B^8 \to Be^8 + e^+ + \nu_e \tag{22}$$

$$Be^8 \to He^4 + He^4 . \tag{23}$$

Reaction (15) is replaced in 0.4% of the cases by the three-body fusion reaction

$$p + e^- + p \to d + \nu_e . \tag{24}$$

Finally, in an even smaller fraction of the cases ($\sim 2.4 \times 10^{-5}$), reaction (17) is replaced by

$$He^3 + p \to He^4 + e^+ + \nu_e . \tag{25}$$

It can be seen that in the pp cycle ν_e's are produced by the five reactions (15), (19), (22), (24) and (25). These neutrinos will be denoted as ν_{pp}, ν_{Be}, ν_B, ν_{pep} and ν_{hep}, respectively. While ν_{pp}, ν_B and ν_{hep} have a continuous energy spectrum, ν_{Be} and ν_{pep} are mono-energetic because they are produced in two-body final states.

(ii) The CNO cycle, which involves heavier elements. This cycle consists of the following chains of reactions:

$$p + N^{15} \to C^{12} + He^4$$

$$p + C^{12} \to \gamma + N^{13}$$

$$N^{13} \to e^+ + \nu_e + C^{13} \tag{26}$$

$$p + C^{13} \to \gamma + N^{14}$$

$$p + N^{14} \to \gamma + O^{15} \tag{27}$$

$$O^{15} \to e^+ + \nu_e + N^{15} \tag{28}$$

and

$$p + N^{15} \to \gamma + O^{16}$$

$$p + O^{16} \to \gamma + F^{17}$$

$$F^{17} \to e^+ + \nu_e + O^{17} \tag{29}$$

$$p + O^{17} \to N^{14} + He^4$$

followed by reactions (27) and (28). As for the pp cycle, the two chains of reactions in the CNO cycle have the overall effect of transforming four protons into a He^4 nucleus. Production of ν_e occurs in reactions (26), (28) and (29). These neutrinos will be denoted as ν_N, ν_O and ν_F, respectively.

Fig. 4 shows the ν_e flux as a function of energy, as predicted by the SSM for the different reactions. The ν_{pp} flux is the dominant component. However, neutrino cross-sections increase rapidly with energy (typically as E_ν^2 for energies well above threshold), so these neutrinos are not among the easiest ones to detect because of their low-energy. Fig. 4 displays also the energy threshold for the capture reaction

$$\nu_e + (A, Z) \rightarrow e^- + (A, Z + 1)$$

for a variety of nuclear isotopes. The SSM also makes predictions on neutrino production as a function of radius.

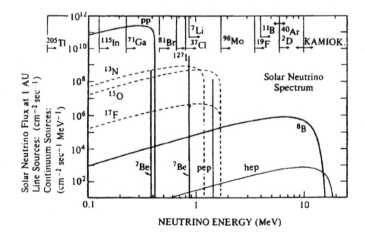

Figure 4: Solar neutrino energy spectrum as calculated from the SSM [16]. Energy thresholds for various neutrino detection processes are shown on top.

It must be finally pointed out that, while solar neutrinos arrive on Earth approximately 500 s after being produced, it takes of the order of 10^6 years for the energy produced in the same reactions to be transported from the Sun core to its surface. Thus the Sun luminosity which is measured at present is associated with neutrinos which reached the Earth $\sim 10^6$ years ago. This is not considered to be a problem for the SSM because the Sun is a star on the main sequence, with no appreciable change of properties over $\sim 10^8$ years.

5.2   Solar neutrino experiments

5.2.1 The Homestake experiment

Solar neutrinos were successfully detected for the first time in an experiment performed by Davis and collaborators [18] in the Homestake gold mine (South Dakota, U.S.A.). The method consists in measuring the production rate of ^{37}A from the capture reaction

$$\nu_e + {}^{37}\mathrm{Cl} \rightarrow e^- + {}^{37}\mathrm{A} \tag{30}$$

which occurs in a 390 m^2 tank filled with 615 tonnes of perchloroethylene (C_2Cl_4, a commonly used cleaning fluid). The isotope ^{37}Cl represents 24% of all natural chlorine, so there are approximately 125 tonnes of ^{37}Cl in the tank. The neutrino energy threshold for this reaction is 0.814 MeV, so this reaction is not sensitive to the ν_{pp} component (see Fig. 4).

The tank is installed deep underground in order to reduce ^{37}A production by cosmic rays. Every few months, Argon is extracted from the tank by N_2 flow. It is then separated, purified, mixed with natural Argon and used to fill a proportional counter. The presence of ^{37}A in the counter is then detected by observing its decay which occurs by electron capture with a half-life time $\tau_{1/2} = 34$d:

$$e^- + {}^{37}\mathrm{A} \rightarrow \nu_e + {}^{37}\mathrm{Cl} \ .$$

In this reaction an X-ray or an Auger electron emitted from the atomic transition to the orbital state left empty after electron capture is detected in the proportional counter. The extraction efficiency is measured by injecting a known amount of ^{37}A in the tank. On average, the ^{37}A production rate is of the order of 0.5 atoms/day.

It has become customary to express the solar neutrino capture rate in Solar Neutrino Units or SNU (1 SNU corresponds to 1 capture/s from 10^{36} nuclei). The result of the Homestake experiment averaged over more than 20 years of data taking, is [19]

$$R_{\mathrm{exp}}(^{37}\mathrm{C}) = 2.56 \pm 0.16 \pm 0.16 \tag{31}$$

where the first error is statistical and the second one represents the systematic uncertainties.

Table 3 shows the SSM predictions, as calculated by Bahcall et al. [17].

The total rate is predicted to be $R_{th}(^{37}\text{Cl}) = 7.7^{+1.2}_{-1.0}$ SNU, which disagrees with the measured value. An independent SSM calculation by Turck-Chièze et al. [20] predicts $R_{th}(^{37}\text{Cl}) = 6.4 \pm 1.4$ SNU which is again larger than the measured value.

Table 3: Solar neutrino contributions to reaction (30) as predicted by the SSM [17].

Solar neutrino component	^{37}A production rate (SNU)
ν_B	5.9
ν_{Be}	1.1
ν_O	0.4
ν_{pep}	0.2
ν_N	0.1
Total	$7.7^{+1.2}_{-1.0} \pm 3.0$

5.2.2 Gallium experiments

Two experiments, GALLEX and SAGE, have measured the rate of the reaction

$$\nu_e + {}^{71}\text{Ga} \to {}^{71}\text{Ge} + e^- \tag{32}$$

which has a neutrino energy threshold of 0.233 MeV and is sensitive, therefore, to the ν_{pp} contribution.

GALLEX, installed in the Gran Sasso underground laboratory (Italy), uses 30.3 tons of Gallium dissolved in HCl. SAGE, installed in the Baksan underground laboratory (Russia) uses 57 tons of metallic Gallium which is liquid at 40°C. The fraction of ^{71}Ga isotope in natural Gallium is 39.7%.

In both experiments ^{71}Ge is extracted every 3–4 weeks by means of physical and chemical methods and converted to GeH$_4$. This is a gaseous substance which is used to fill a proportional counter built from special, low radioactivity materials. This counter is used to detect the atomic X-rays emitted from ^{71}Ge decay which occurs by the electron capture reaction

$$e^- + {}^{71}\text{Ge} \to {}^{71}\text{Ga} + \nu_e \tag{33}$$

with a half-life of 11.43 d. The X-ray time and energy distributions provide evidence for ^{71}Ge production by solar neutrinos.

Both experiments have performed convincing checks of the ^{71}Ge extraction efficiency, which include the use of a very intense ^{51}Cr source producing mono-energetic neutrinos from the electron capture reaction

$$e^- + {}^{51}\text{Cr} \to \nu_e + {}^{51}\text{V}$$

with a half-life of 27.7 d. The neutrino energy from this source is 0.75 MeV and the initial flux at the detector corresponds to several times the solar neutrino flux.

In the GALLEX experiment the ^{71}Ge extraction efficiency has been measured directly by injecting known quantities of ^{71}As in the tank. This isotope decays to ^{71}Ge with a half-life of 2.72 d, either by β^+ decay or by electron capture.

The ^{71}Ge production rate from solar neutrinos is measured to be

$$R(^{71}\text{Ga}) = 77.5 \pm 6.2^{+4.3}_{-4.7} \text{ SNU}$$

by the GALLEX experiment [21]; and

$$R(^{71}\text{Ga}) = 66.6^{+6.8+3.8}_{-7.1-4.0} \text{ SNU}$$

by SAGE [22]. In both experiments the first error is statistical and the second one represents the systematic uncertainties.

After adding in quadrature the statistical and systematic errors, the weighted average of the two results is

$$R(^{71}\text{Ga}) = 72.7 \pm 5.7 \text{ SNU} . \tag{34}$$

The SSM predictions are shown in Table 4. Again, the measured ^{71}Ge production rate is much lower than the SSM predictions.

Table 4: Solar neutrino contributions to reaction (32), as predicted by the SSM

Solar neutrino component	^{71}Ge production rate (SNU)	
	Ref. [17]	Ref. [20]
ν_{pp}	69.6	70.6
ν_{pep}	2.8	2.8
ν_{Be}	34.4	30.6
ν_{B}	12.4	9.3
ν_{N}	3.7	3.9
ν_{O}	6.0	6.5
Total	129^{+8}_{-6}	124 ± 5

5.2.3 Super-Kamiokande

Super-Kamiokande is a real-time experiment which uses an underground detector installed in the Kamioka mine 350 km west of Tokyo.

The inner detector consists of a cylindrical tank filled with 32,000 tons of water. Approximately 40% of the tank surface are covered by 11,146 photo-multipliers with a diameter of 50 cm and pointing towards the liquid.

The inner detector is surrounded by an additional layer of water, with a thickness of 2 m and seen by 1,881 photomultipliers with a diameter of 20 cm. This outer detector is used to identify charged particles entering the detector from outside.

The inner detector is used as an imaging Čerenkov counter. Charged particles with $v/c \approx 1$ produce Čerenkov light at an angle of $\sim 41°$ to their direction of flight and the pattern of hit photomultipliers and their relative timing provide information on the particle direction and origin in the detector volume.

Solar neutrinos are detected by the scattering reaction

$$\nu + e^- \rightarrow \nu + e^- \tag{35}$$

which is suppressed by approximately 1/6 for ν_μ and ν_τ with respect to ν_e. The electron energy scale is calibrated by sending an electron beam of 5–16 MeV energy from a LINAC into the inner detector at various depths. The threshold for solar neutrino studies is as low as 5.5 MeV. Only events produced at a distance of at least 2 m from the inner detector walls are considered. This fiducial volume contains 22,500 tons of water.

The detected electron from reaction (35) has a very strong directional correlation with the incident neutrino. This property is used to demonstrate the solar origin of the events, as shown in Fig. 5 which displays the distribution of the angle between the electron direction and the Sun-to-Earth direction at the time of the event. The peak at $\cos \Theta_{sun} = 1$ is due to solar neutrinos.

The Super-Kamiokande experiment began data taking in May 1996 and has recently reported results from a run of 708 days [23]. The solar neutrino flux with an electron energy threshold of 6.5 MeV is measured to be

$$\Phi_\nu = (2.44 \pm 0.04 \pm 0.07) \times 10^6 \text{ cm}^{-2} \text{ s}^{-1} \tag{36}$$

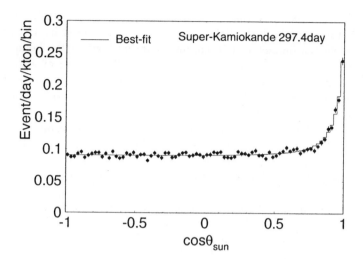

Figure 5: Distribution of the cosine of the angle between the electron direction and the Sun-to-Earth direction, as measured by Super-Kamiokande. The curve is a best fit to the data.

where the first error is statistical and the second one represents systematic uncertainties. This value is 47% of the SSM prediction [17], $\Phi_{SSM} = (5.15^{+0.98}_{-0.72}) \times 10^6$ cm^{-2} s^{-1}. Obviously with a detection threshold of 6.5 MeV the experiment is only sensitive to ν_B (see Fig. 4).

5.3 Interpretation of the solar neutrino data

The measured solar neutrino event rates are dominated by three components of the p–p cycle: ν_{pp}, ν_{Be} and ν_B.

We parametrize the deviations of the measured fluxes from the SSM predictions as follows:

$$
\begin{aligned}
x_{pp} &= \phi_m(\nu_{pp})/\phi_{SSM}(\nu_{pp}); \\
x_{Be} &= \phi_m(\nu_{Be})/\phi_{SSM}(\nu_{Be}); \\
x_B &= \phi_m(\nu_B)/\phi_{SSM}(\nu_B),
\end{aligned}
$$

where by ϕ_m we denote measured fluxes and by ϕ_{SSM} the predictions of Ref. [17]. With these parameters the result of the Homestake experiment (see Eq. 31) can be written as

$$1.2\, x_{Be} + 5.9 x_B + 0.6 = 2.56 \pm 0.23 \text{ SNU} \tag{37}$$

where the left term contains the SSM predictions for variable ν_{Be} and ν_B fluxes and the additional contribution (0.6 SNU) from all other solar neutrino components (see Table 3), while the right term is the experimental result with statistical and systematic uncertainties added in quadrature.

Similarly, with the help of Table 4 the combined result of the two Gallium experiments can be written as

$$69.6\, x_{pp} + 34.4\, x_{Be} + 12.4\, x_B + 12.6 = 72.7 \pm 5.7\ \text{SNU}\ . \tag{38}$$

Finally, Super-Kamiokande measures directly x_B:

$$x_B = 0.47 \pm 0.02\ . \tag{39}$$

The solution of the three equations is

$$x_{pp} = 1.03 \pm 0.14\ ;\ x_{Be} = -0.68 \pm 0.22\ .$$

If the contributions from all other neutrino components are set equal to zero in Eqs. (37, 38), the solution becomes

$$x_{pp} = 1.04 \pm 0.14\ ;\ x_{Be} = -0.18 \pm 0.22\ .$$

In both cases x_{Be} is consistent with zero.

The apparent absence of ν_{Be} in the solar neutrinos is a real puzzle, because neutrinos from $B^8(\nu_B)$ are observed by Super-Kamiokande (albeit at a rate which is 47% of the predicted one) and, as explained in Section 5.1, B^8 is formed from the fusion reaction

$$p + Be^7 \rightarrow \gamma + B^8$$

which implies that Be^7 must exist in the Sun. This in turn implies the occurence of the reaction

$$e^- + Be^7 \rightarrow Li^7 + \nu_e\ ,$$

which is responsible for ν_{Be} production, at a rate three orders of magnitude higher than the reaction responsible for B^8 formation which is strongly suppressed by Coulomb repulsion.

There are three possible explanations to this puzzle:

(i) At least two of the three measurements of the solar neutrino flux are wrong;

(ii) There is a basic flaw in the SSM, resulting in unreliable predictions of the solar neutrino flux (however, the SSM correctly predicts the results of helioseismological observations [24] which depend on the temperature profile of the Sun);

(iii) The ν_{Be}'s are produced as ν_e in the core of the Sun but are no longer ν_e when they reach the Earth.

This last explanation, which we assume to be the correct one, implies the occurrence of neutrino oscillations.

5.4 Vacuum oscillation solutions

The experimental results on solar neutrinos can be explained by oscillation parameters which strongly suppress the ν_{Be} component ($E_\nu = 0.861$ MeV) when these neutrinos reach the Earth at a distance $L \approx 1.5 \times 10^{11}$ m from the Sun. These are the so-called vacuum oscillation solutions, also nick-named 'just so' solutions because they require a precise mathematical relation between three unrelated physical quantities (Δm^2, L and the ν_{Be} energy).

These solutions, as recently derived by the Super-Kamiokande collaboration [23], are shown in Fig. 6a. The Δm^2 values are in the range $4 \times 10^{-11} - 5 \times 10^{-10}$ eV^2 and the mixing angle is large ($\sin^2 2\theta > 0.6$).

Two measurable effects are expected from these solutions:

(i) A seasonal variation of the measured solar neutrino flux exceeding the 6.7% solid angle variation associated with the excentricity of the Earth orbit around the Sun. The statistical precision of the present Super-Kamiokande results is insufficient to reach a definitive conclusion on this point (see Fig. 7a);

(ii) An energy dependent suppression of the solar neutrino flux. A significant energy dependence is indeed observed in the Super-Kamiokande experiment for electron energies above 13 MeV, in agreement with a vacuum oscillation solution (see Fig. 7b). However, it has been recently pointed out that the SSM prediction of the ν_{hep} component of the solar neutrino spectrum (see Fig. 4) is affected by very large theoretical uncertainties resulting from uncertainties on the cross-section for reaction (25). Hence one cannot exclude that the excess of events observed at high energies in the distribution of Fig. 7b could be due to a ν_{hep} flux exceeding the SSM prediction by an order of magnitude or more.

99% Confidence Level

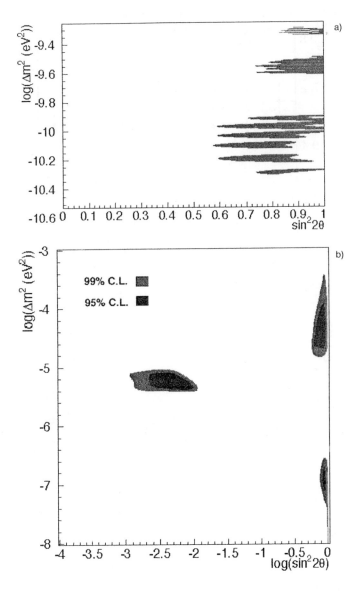

Figure 6: Vacuum oscillation (a) and MSW (b) solutions to the solar neutrino results under the assumption of two-neutrino mixing [23].

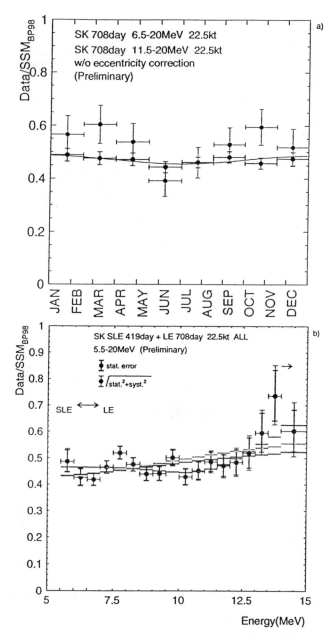

Figure 7: a) Yearly variation of the measured solar neutrino signal normalized to the average predicted signal. The curve represents the variation expected from the excentricity of the Earth orbit around the Sun; b) Ratio between measured and predicted solar spectrum. The curves show the expectations for the vacuum and MSW (small mixing angle) solution.

5.5 Matter enhanced solutions

Another class of oscillation solutions to the solar neutrino problem can be found in the framework of the theory of neutrino oscillations in matter described in Sec. 4.3. These are the so-called Mikheyev-Smirnov-Wolfenstein (MSW) solutions.

For neutrinos propagating through the Sun, the density ρ varies along the trajectory from a value higher than 100 g/cm^3 in the core to much less than 1 g/cm^3 in the outermost layers. The ratio Z/A also varies across the Sun because of the varying hydrogen abundance. Hence, in Eq. (12) the Hamiltonian depends on time. For a given set of mixing parameters m_1, m_2 and θ_v, Eq. (12) can be solved numerically with the condition that the initial neutrino state is a pure ν_e state, using the SSM prediction for the solar density, Z/A ratio and distribution of neutrino origins inside the Sun core.

The ideal case of constant density discussed in Sec. 4.3 represents a good approximation to a class of solutions of relatively short oscillation length for which the variation of the solar density over an oscillation length is negligible (the so-called adiabatic solutions):

$$\frac{1}{\rho}\frac{d\rho}{dr}\lambda_\mathrm{m} \ll 1$$

where r is the distance from the Sun centre. For such solutions the neutrino can be described as a superposition of mass eigenstates with slowly varying eigenvalues and mixing angle. In this case, if for a ν_e at production the condition $\xi > \Delta m^2 \cos 2\theta_v$ is satisfied, then θ_m is larger than 45° (see Eq. 14) and the dominant mass eigenstate is ν_2. If, furthermore, the adiabaticity condition is satisfied also at resonance, where λ_m is maximal, then the $\nu_2 \to \nu_1$ transition probability is negligible and the dominant mass eigenstate is still ν_2 when the neutrino emerges from the Sun. However, the ν_2 eigenstate in vacuum is mostly ν_μ because $\theta_\mathrm{v} < 45°$. Thus the Mikheyev-Smirnov resonance offers an elegant way to explain the solar neutrino problem even if the mixing angle in vacuum is small.

It must be pointed out that, in the case of small mixing angle, only the ν_e's produced with $\xi > \Delta m^2 \cos 2\theta_v$ may emerge from the Sun as ν_μ's as a result of the MSW effect. As ξ depends linearly on the neutrino energy E, this condition is satisfied only by neutrinos produced above a critical energy which depends on the mixing parameters.

The results from the latest analysis of the solar neutrino event rates in terms of matter enhanced oscillations [23] are shown in Fig. 6b. For each experiment the measured event rate corresponds to a region of allowed parameters in the $\sin^2(2\theta_v)$, Δm^2 plane. This region consists of a vertical band at large mixing angles, of a horizontal band at constant Δm^2 corresponding to adiabatic solutions and extending to small mixing angles, and of another band merging into the two previous ones for which the allowed values of $\sin 2\theta_v$ decrease with increasing Δm^2 (see Ref. [25]). Since the processes used to detect solar neutrinos have different energy thresholds, these regions do not coincide and the oscillation parameters which describe all available data are defined by their overlap. As shown in Fig. 6b there are three possible solutions, two with large mixing angles and $\Delta m^2 \approx 10^{-7}$ or 4×10^{-5} eV^2, and one with small mixing ($\sin^2 2\theta \approx 0.005$) and $\Delta m^2 \approx 6 \times 10^{-6}$ eV^2.

Two effects are expected from MSW solutions:

(i) a possible day-night variation of the solar neutrino event rate, with an increased rate at nights from matter enhanced oscillations for neutrinos crossing the Earth, resulting in an increase of the ν_e flux. The most recent measurement of the day-night asymmetry in the Super-Kamiokande experiment gives

$$\frac{R_d - R_n}{R_d + R_n} = -0.026 \pm 0.016 \text{ (stat.)} \pm 0.013 \text{ (syst.)} \,,$$

where R_d (R_n) is the day (night) event rate.

(ii) an energy dependent suppression of the electron energy spectrum (see Fig. 7b).

5.6 Future Solar Neutrino Experiments

5.6.1 SNO

The Sudbury Neutrino Observatory (SNO) is a solar neutrino detector installed in the Creighton mine near Sudbury, Ontario, at a depth of 2070 m (5900 m water equivalent) [26]. The detector consists of a spherical acrylic vessel with a 6 m radius containing $\sim$ 1000 tons of high purity heavy water $(D_2 0)$ surrounded by 7800 tons of ultra-pure water for shielding purposes. Čerenkov light produced in the heavy water is collected by 9456 photomultipliers with a diameter of 20 cm located on a concentric spherical surface at a radius of 9.5 m.

As for the SuperKamiokande experiment, solar neutrinos are detected by observing the elastic scattering reaction $\nu e^- \rightarrow \nu e^-$, which is dominated by the ν_e component of the solar flux and provides precise information on the incident neutrino direction. With an electron energy threshold set at 5 MeV the expected event rate is 1.4 d^{-1} assuming a 50% reduction of the ν_e flux.

However, in heavy water the charged current reaction

$$\nu_e + d \rightarrow p + p + e^-$$

also occurs, at a rate of 12.7 events/d for a 5 MeV threshold and for the same reduction of the ν_e flux. A measurable asymmetry with respect to the Sun position in the sky is also present in this reaction.

The main feature of SNO is its anticipated capability to detect the reaction

$$\nu + d \rightarrow p + n + \nu \tag{40}$$

which has the same cross-section for all three neutrino flavours and measures the total solar neutrino flux. Any significant difference between the neutrino flux measured from reaction (40) and that measured from charged-current reactions would provide, therefore, unambiguous proof of neutrino oscillations.

Two methods are used to detect reaction (40):

(i) $MgCl_2$ is added to the heavy water. In this case neutrons from reaction (40) undergo the capture process n + ^{35}Cl $\rightarrow$ ^{36}Cl + γ and the 8.5 MeV γ-ray converts in the heavy water.

(ii) He^3 proportional counters are inserted in the heavy water volume and detect the mono-energetic signal from the capture process n + He^3 $\rightarrow$ $H^3 + p$.

The event rate from reaction (40) is expected to be 5.5/d for a neutron detection efficiency of 40%.

The SNO detector is presently almost completely filled with heavy water and data taking is expected to start in the middle of 1999. After one year, it is planned to add $MgCl_2$ and to take data for another year. Then, after removal of $MgCl_2$ data taking will continue with an array of He^3 proportional counters in the heavy water volume.

5.6.2 Borexino

Borexino is an experiment presently under construction at the Gran Sasso National Laboratory [27]. The detector consists of a spherical acrylic vessel of 8.5 m diameter filled with very high purity, low activity liquid scintillator and viewed by an array of 1650 photomultipliers located on its surface. The relative timings of the photomultiplier signals provides information on the event position within the detector volume: only events occurring within a central fiducial volume corresponding to 100 tons of scintillator are considered. The entire detector is immersed in a cylindrical tank 16.5 m high with a 16.5 m diameter filled with high purity water and acting as a shield.

The aim of the experiment is to detect $\nu - e^-$ elastic scattering with an energy threshold as low as 0.25 MeV. If this is achieved, the experiment is sensitive to the ν_{Be} component of the solar flux (E = 0.861 MeV) which is expected to be strongly suppressed if neutrino oscillations are indeed the solution of the solar neutrino problem (see Sec. 5.3).

Fig. 8 shows the expected electron energy distribution. The recoil electrons from ν_{Be} elastic scattering on e^- at rest have a flat kinetic energy spectrum with the end-point at 0.664 MeV. In the absence of oscillation, and with a threshold as low as 0.25 MeV, one expects a contribution of 50 events/d from these neutrinos. Because of the scintillation light isotropy, the identification of the solar origin of the signal relies on the observation of its seasonal variation associated with the excentricity of the Earth orbit around the Sun.

Data taking is expected to start in the year 2000.

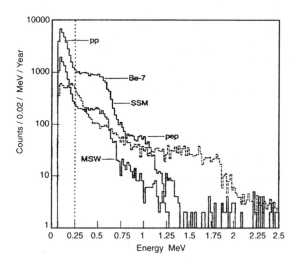

Figure 8: Simulated spectra of the signal expected in Borexino from the Standard Solar Model (upper line) and from the standard flux modulated by the neutrino oscillation expected from the MSW (small mixing angle) solution. The dotted line shows the expected background.

5.6.3 KAMLAND

The Kamioka Liquid Scintillator Anti Neutrino Detector (KAMLAND) is not a solar neutrino experiment. It is discussed here because it is sensitive to oscillation parameters which could explain the solar neutrino problem [28].

The detector is a transparent sphere with a diameter of 13 m filled with scintillating isoparaffin oil. This sphere is itself contained in a larger, concentric sphere (18 m diameter) filled with pure isoparaffin oil. Scintillation light from the inner sphere is collected by 1300 photomultipliers located on the surface of the outer sphere. The entire system is immersed in high purity water and installed in the Kamioka mine at a depth of 2700 m of water equivalent.

KAMLAND aims at detecting the $\bar{\nu}_e$ produced by five nuclear reactors located at distances between 150 and 210 km from the detector and producing a total thermal power of 127 GW. The $\bar{\nu}_e$, with an average energy of 3 MeV, are detected by measuring the e^+ signal from the reaction

$$\bar{\nu}_e + p \rightarrow e^+ + n \tag{41}$$

followed by the late γ signal from the neutron capture reaction $np \rightarrow d\gamma$ ($E_\gamma = 2.2$ MeV) which occurs after neutron thermalization.

When all reactors run at full power, the event rate is expected to be 3 d^{-1} with a signal-to-noise ratio of 10. The background is measured by observing the variation of the event rate with the reactor power.

Because of its large distance from the reactors and of the low $\bar{\nu}_e$ energy, KAMLAND is sensitive to $\Delta m^2 > 7 \times 10^{-6}$ eV^2 and $\sin^2 2\theta > 0.1$, a region which includes the large mixing angle, large Δm^2 MSW solution (see Fig. 6b).

KAMLAND will begin data taking in the year 2000.

6 Atmospheric neutrinos

6.1 Origin of atmospheric neutrinos

Since the total thickness of the atmosphere is $\sim 10^3$ g/cm^2, which is equivalent to ~ 10 interaction lengths, the interaction of a primary cosmic ray in the upper layers of the atmosphere results in the development of a hadronic shower leading to a flux of neutrinos from charged pion and muon decay. These neutrinos have energies ranging from ~ 0.1 GeV to several GeV. Their interaction rate is of the order of 100/y for a target mass of 1000 tons.

Since a ν_μ is produced from both $\pi^\pm$ and $\mu^\pm$ decay, and a ν_e from $\mu^\pm$ decay only, one expects the ratio between the ν_μ and ν_e fluxes on Earth to be of the order of 2 if both $\pi^\pm$ and $\mu^\pm$ decay in the atmosphere. This is indeed a very good approximation for neutrinos with energies lower than 3 GeV. At higher energies, this ratio increases because the $\mu^\pm$ decay path increases with energy and the fraction of $\mu^\pm$ decaying in the atmosphere and producing ν_e or $\bar{\nu}_e$ decreases.

Calculations of atmospheric neutrino fluxes [29] are affected by sizeable uncertainties which result from uncertainties on the composition and energy spectrum of the primary cosmic rays and on secondary particle distribution. In addition, these calculations ignore for simplicity the lateral shower development and treat the problem in one dimension only. The final uncertainty affecting the ν_μ and ν_e fluxes on Earth is estimated to be of the order of $\pm$ 30%. However, because of partial cancellations, the uncertainty on the predicted ν_μ/ν_e ratio is believed to be of the order of $\pm$ 5%.

6.2 ν_μ/ν_e flux ratio measurements

Six underground experiments have measured the atmospheric neutrino fluxes by detecting quasi-elastic interactions:

$$\nu_\mu\ (\nu_e) + n \to \mu^-(e^-) + p,$$
$$\bar{\nu}_\mu(\bar{\nu}_e) + p \to \mu^+(e^+) + n\ .$$

Three experiments (Kamiokande [30], IMB-3 [31] and Super-Kamiokande [32]) detect the Čerenkov light ring produced by relativistic particles in water. The other three experiments, FREJUS [33], NUSEX [34] and Soudan-2 [35] use calorimeters with high longitudinal and transverse segmentation. FREJUS and NUSEX took data in the 80's, while Soudan-2 is presently running.

Muons from quasi-elastic reactions appear in all these detectors as single, penetrating tracks. If the muon stops in the detector and decays, the decay electron can also be observed. The Super-Kamiokande experiment accepts also muon events from deep-inelastic neutrino interactions in the water if the muon exits the inner detector. These 'multi-ring' events are produced by ν_μ or $\bar{\nu}_\mu$ with an average energy of 15 GeV.

Electrons from quasi-elastic reactions produce single electromagnetic showers consisting of many short tracks which are easily identified in the calorimeters and result in diffuse Čerenkov light rings in the water detectors. For most experiments the electron-muon identification capabilities have been measured using test beams from accelerators. For the Kamiokande and Super-Kamiokande experiments the mis-identification probability is measured to be less than 2% [36].

The comparison between the measured and predicted ν_μ/ν_e ratio for the six experiments is shown in Table 5 which lists the values of the double ratio R defined as

$$R = \frac{(\nu_\mu/\nu_e)\ \text{measured}}{(\nu_\mu/\nu_e)\ \text{predicted}}\ .$$

With the exception of the values measured by FREJUS and NUSEX, all values of R are significantly lower than the expectation ($R = 1$). The NUSEX result is affected by a large statistical error, while the FREJUS result is not confirmed by Soudan-2 which uses the same detection technique. It can be concluded, therefore, that a small R value (of the order of 0.6) is now firmly established.

Table 5: Results on R. The momentum range of observed charged leptons (KamioKande, IMB, SuperKamiokande) or the visible energy range (Frejus, NUSEX, Soudan 2) is denoted by p_l.

Experiments	p_l(MeV/c)	Exposure (kt-yr)	R	Ref.
Kamiokande	e:100–1330 μ:200-1400	7.7	$0.60\pm^{0.06}_{0.05} \pm 0.05$	[30]
	e:1330– μ:1400–	6.0~8.2	$0.57\pm^{0.08}_{0.07} \pm 0.07$	[30]
IMB	e:100–1500 μ:300–1500	7.7	$0.54\pm0.05\pm0.12$	[31]
Frejus	e:200– μ:200–	2.0	$1.00\pm0.15\pm0.08$	[33]
NUSEX	e:200– μ:200–	0.74	$0.99\pm^{0.35}_{0.25}$	[34]
Soudan 2	e:150– μ:100–	2.83	$0.58\pm0.11\pm0.05$	[35]
Super-Kamiokande	e:100–1330 μ:200-1400	45.4	$0.668\pm^{0.024}_{0.023}\pm0.052$	[32]
	e:1330– μ:1400–	45.4	$0.663\pm^{0.044}_{0.041} \pm 0.079$	[32]

6.3 Zenith Angle Distribution

The flight path of atmospheric neutrinos from the production point to the detector, L, varies enormously with the zenith angle θ_Z. For example, neutrinos impinging on the detector from above ($\cos\theta_Z = 1$) are produced ~ 10 km above the detector, while upward going neutrinos ($\cos\theta_Z = -1$) have traversed the Earth and so have travelled for ~ 13000 km before reaching the detector. Also, the higher the neutrino energy, the better the outgoing charged lepton follows the incident neutrino direction. Hence the charged lepton zenith angle is a direct measurement of L. Because of the directionality of Čerenkov light the water detectors can be seen, therefore, as disappearance experiments with variable neutrino energies and path lengths. Measurements of the zenith angle distributions are a sensitive way to search for neutrino oscillations.

The Kamiokande experiment [37] published a dependence of R on θ_Z which disagreed from the expected shape at the level of ~ 2 standard deviations. The Super-Kamiokande θ_Z distributions [31], with a much larger event sample, are shown in Fig. 9. It is clear that there are less muon events in the upward direction (negative $\cos \theta_Z$) than expected, while the number of downard going muons is consistent with the expectation. For electrons, however, the distributions agree with expectations. The distributions $dR/d\cos\theta_Z$, also shown in Fig. 9, reflect the distortion seen in the muon distributions.

Fig. 10 shows the up-down asymmetry defined as $(U-D)/(U+D)$, where U(D) is the total number of events with $\cos \theta_Z < -0.2$ ($\cos\theta_Z > 0.2$), as a function of the charged lepton momentum. While for electron events the asymmetry is consistent with zero, for muon events its absolute value increases with momentum and reaches a value around –0.4 above 1 GeV.

6.4 Interpretation of the phenomenon

Figs. 9 and 10 clearly demonstrate the existence of a new phenomenon. Its most plausible interpretation is the occurrence of ν_μ oscillations. Since the ν_μ/ν_e ratio at production is equal to 2, or larger than 2, $\nu_\mu - \nu_e$ oscillations would induce a large up-down asymmetry for electrons as well, with more up-going than down-going electrons, in disagreement with the data. Hence the ν_μ predominantly oscillates to ν_τ or to a new type of 'sterile' neutrino, which we denote by ν_s.

Fig. 11 shows the region of $\nu_\mu - \nu_\tau$ oscillation parameters required to describe the Super-Kamiokande results under the assumption of two-neutrino mixing. The best fit values are $\Delta m^2 = 3.5 \times 10^{-3}$ eV^2, $\sin^2 2\theta = 1$, with $\chi^2 = 62.1$ for 68 degrees of freedom. The results from the fit are displayed in Figs. 9 and 10.

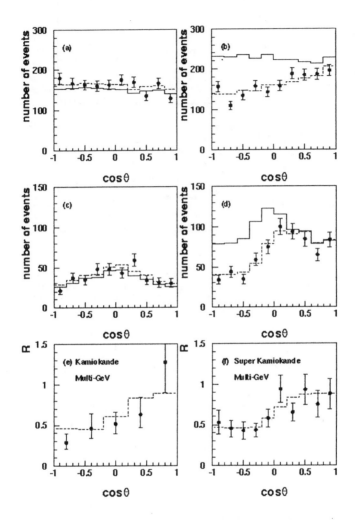

Figure 9: Zenith angle distributions, as measured in Super-Kamiokande: a) electron-like, sub-GeV events; b) muon-like, sub-GeV; c) electron-like, multi-GeV; d) muon-like, multi-GeV and partially contained events; e) ratio R for multi-GeV events as measured in Kamiokande [36]; f) ratio R for multi-GeV events, as measured in Super-Kamiokande [31]. The full line represents the expectation in the absence of neutrino oscillation. The dashed line is the expectation for $\nu_\mu - \nu_\tau$ oscillation with $\sin^2(2\theta) = 1$, $\Delta m^2 = 3.5 \times 10^{-3}$ eV^2.

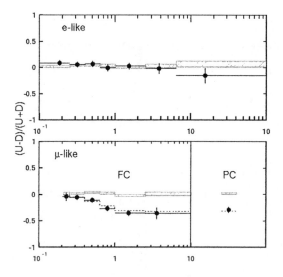

Figure 10: Up-down asymmetry vs energy, as measured in Super-Kamiokande [32]. The hatched bands represent the expectation in the absence of neutrino oscillation. The dashed line is the expectation for $\nu_\mu - \nu_\tau$ oscillation with $\sin^2(2\theta) = 1$, $\Delta m^2 = 3.5 \times 10^{-3}$ eV^2.

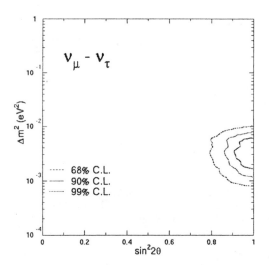

Figure 11: Region of $\nu_\mu - \nu_\tau$ oscillation parameters required to describe the Super-Kamiokande atmospheric neutrino results.

A $\nu_\mu - \nu_s$ oscillation also gives an acceptable fit with parameters similar to those required by the $\nu_\mu - \nu_\tau$ hypothesis. On the other hand, no oscillation gives $\chi^2 = 175$ for 69 degrees of freedom, corresponding to a probability of 5×10^{-9}. The hypothesis of $\nu_\mu - \nu_e$ oscillation is also in disagreement with the data: the best fit has $\chi^2 = 110$ for 67 degrees of freedom corresponding to a probability of 8×10^{-4}. This hypothesis is also rejected by the results of the Chooz experiment, as discussed in Sec. 7.

A $\nu_\mu - \nu_s$ oscillation can be distinguished from the oscillation to an 'active' neutrino by studying neutral-current interactions. Oscillations to an active neutrino do not change the rate of these events, whereas ν_s has neither charged- nor neutral-current interactions with matter.

In the Super-Kamiokande experiment events consisting of two electron-like rings are considered to be candidates for the reaction

$$\nu + N \rightarrow \nu + \pi^0 + N \tag{42}$$

and the two rings are identified as the photon showers from $\pi \rightarrow \gamma\gamma$ decay.

Fig. 12 shows the two-photon invariant mass distribution [38]. A peak containing ~ 270 events at the nominal π^0 mass is clearly visible. The rate of these events is compared with that of single-ring electron-like events, which is not affected by neutrino oscillations. This comparison is made by using the double ratio between the measured and predicted rates in the absence of oscillation. The result is

$$\frac{(\pi^0/e) \text{ measured}}{(\pi^0/e) \text{ predicted}} = 1.11 \pm 0.06 \pm 0.02 \pm 0.26$$

where the first error is statistical, the second one represents systematic uncertainties and the third, dominant error reflects the uncertainty on the cross-section for neutral-current events resulting in only one detectable π^0. A reduction of the order of 30% with respect to unity is expected from $\nu_\mu - \nu_s$ oscillations.

120

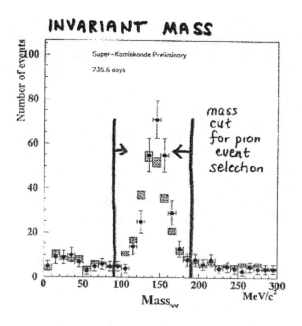

Figure 12: Two-photon invariant mass distribution as measured in SuperKamiokande [38].

Obviously, because of the size of the uncertainty on the π^0 cross-section no conclusion can be reached at present (as discussed in Sec. 9, the K2K experiment will greatly reduce this uncertainty in the near future). Similarly, the number of π^0 events is too small and their correlation to the incident neutrino direction too weak for a significant analysis of their zenith angle distribution.

A definitive proof that the observed anomalies (the low R value and the distorted muon zenith angle distribution) are the result of neutrino oscillations can only be obtained, in my opinion, by observing the expected oscillatory behaviour of the ν_μ signal as a function of L/E over at least a full oscillation cycle (see Eq. 10). For $\Delta m^2 = 3.5 \times 10^{-3}$ eV^2 signal minima are expected at $L/E = 708$ (n + 1/2) km/GeV, where $n = 0, 1, 2 ...$, thus L/E must be measured with a resolution significantly smaller than 708 km/GeV. We note that the distance between two minima (or two maxima) varies as $1/\Delta m^2$ and the required L/E resolution also varies in the same way.

In the Superkamiokande experiment the ν_μ energy in single-ring events can be measured accurately from the muon range (these events are dominated by quasi-elastic scattering for which the undetected recoil nucleon has a kinetic energy of typically less than 500 MeV). However, the detector measures accurately the muon zenith angle, and not that of the incident neutrino. Thus the neutrino direction is affected by an uncertainty which is as large as 30° at 1 GeV.

Fig. 13 shows the behaviour of the neutrino path length L as a function of its zenith angle θ_ν, as given by the equation

$$L = \sqrt{R_\oplus^2 \, \cos^2 \theta_\nu + 2hR_\oplus + h^2} - R_\oplus \cos\theta_\nu$$

where $R_\oplus$ is the Earth radius (6378 km) and h is the altitude at which the neutrino was produced in the atmosphere (typically of the order of 15 km with a large uncertainty).

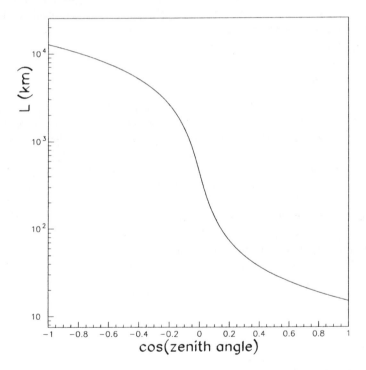

Figure 13: Neutrino path length L vs zenith angle, under the assumption that the neutrino was produced at an altitude of 15 km.

It is clear from Fig. 13 that downward-going neutrinos do not cover a full oscillation cycle in L/E. Furthermore, for these neutrinos L is poorly determined because of the large uncertainty on h.

122

In the θ_ν region near the horizontal direction ($\theta_\nu \approx 90°$) L varies rapidly with θ_ν and it requires a θ_ν resolution of $1° - 2°$ to determine L with the necessary precision. This is possible with neutrinos of very high energies which, however, are very few and produce muons which are generally not contained in the detector.

Finally, for upward-going neutrinos L can be determined with the required precision using multi-GeV muons but a neutrino energy resolution of the order of 10%, or less, is required in order to reach the resolution in L/E necessary to observe a full oscillation cycle for $\Delta m^2 \geq 10^{-3}$ eV^2.

A new atmospheric neutrino detector capable of measuring precisely the incident neutrino direction and energy is needed in order to observe the oscillatory L/E behaviour for $\Delta m^2 \geq 10^{-3}$ eV^2. The design of such a detector is not easy. As we shall see in Section 9, the problem is less difficult for long baseline experiments at accelerators, for which the neutrino path is fixed and very precisely known.

6.5 Upward-going muons

Charged-current interactions of upward-going ν_μ's in the rock produce upward-going muons which can either stop or traverse the underground detectors. The study of the zenith angle distribution of these muons in the interval $-1 < \cos\theta_z < 0$ has provided additional evidence for $\nu_\mu - \nu_x$ oscillations.

In the Super-Kamiokande detector, upward-going muons are identified by detecting the entrance and exit point and by requiring that their track length corresponds to an energy of at least 1.7 GeV [39]. Muons stopping in the detector are produced by ν_μ with $< E_\nu > \approx 10$ GeV, while muons traversing the detector have $< E_\nu > \approx 100$ GeV.

The measured zenith angle distribution of these muons is shown in Figs. 14 a,b. The data disagree with predictions in absence of oscillations, even if the overall normalization is adjusted to provide the best fit. However, good agreement is obtained under the assumption of $\nu_\mu - \nu_\tau$ oscillations with full mixing and Δm^2 values very similar to those required by the study of muons produced in the detector.

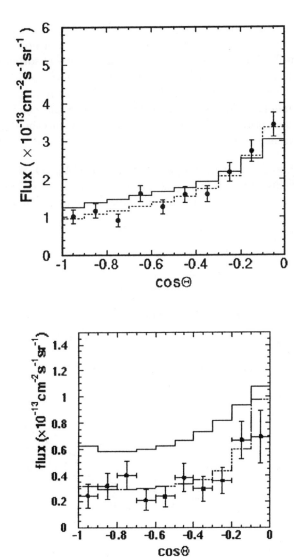

Figure 14: Zenith angle distribution of upward through-going muons (a) and upward-going stopping muons (b) as measured in Super-Kamiokande [39], compared with expectations from no neutrino oscillation (solid line) and $\nu_\mu - \nu_\tau$ oscillation (dashed line).

Upward-going muons have also been studied in the MACRO experiment at Gran Sasso [40]. The MACRO detector is a system of horizontal streamer tubes and scintillation counter planes interleaved with absorber plates (slabs of rock), 9.5 m high and with overall lateral dimensions 77 × 12 m. The muon direction is determined by time-of-flight. MACRO has a very good detection efficiency for muons near the vertical direction, but the efficiency decreases rapidly as $\cos\theta_Z$ approaches zero. The muon energy threshold is 1 GeV and the average parent ν_μ energy is ~ 50 GeV.

The zenith angle distribution, as measured in MACRO, is compared with expectations in Fig. 15. Again, an acceptable fit is obtained only if $\nu_\mu - \nu_\tau$ oscillations are assumed, with full mixing and Δm^2 close to the SuperKamiokande value.

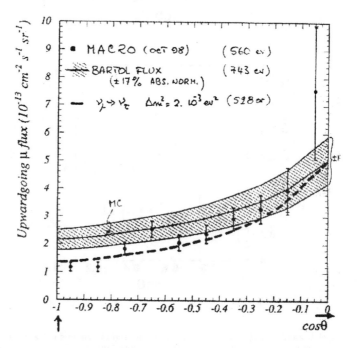

Figure 15: Zenith angle distribution of upward through-going muons, as measured in MACRO [40]. The hatched band represents the expectation from no neutrino oscillation with its uncertainty. The dashed line is the prediction for $\nu_\mu - \nu_\tau$ oscillation with full mixing and $\Delta m^2 = 2 \times 10^{-3}$ eV^2.

7 Long baseline experiments at nuclear reactors

The Chooz experiment [41] is a $\bar{\nu}_e$ disappearance experiment sensitive to oscillations with $\Delta m^2 > 9 \times 10^{-4}$ eV^2 and $\sin^2 2\theta > 0.1$. The detector is installed in an underground site under 115 m of rock at distances of 1114 and 998 m from the two 4.25 GW reactors of the Chooz nuclear power plant in France.

The detector consists of three concentric vessels. The innermost one contains 5 tons of Gadolinium-doped liquid scintillator (CH$_2$) which acts as the $\bar{\nu}_e$ target. The intermediate vessel, with a mass of 17 tons and also filled with CH$_2$ liquid scintillator, is used for containment. The outermost vessel, optically isolated from the two inner ones, contains 90 tons of liquid scintillator and acts as a veto counter.

The $\bar{\nu}_e$ are detected by measuring the prompt e^+ signal from reaction (41) followed by the delayed signal from neutron capture. Most neutrons are captured by Gadolinium, resulting in the emission of γ-rays with a total energy of ~ 8 MeV.

The event rate is 1.1 ± 0.3 d^{-1} with both reactors off and 25.5 ± 1.0 d^{-1} with both reactors at full power. The ratio between the measured and expected event rate is 0.98 ± 0.04 (stat.) ± 0.04 (syst.) in the absence of neutrino oscillation. Fig. 16 shows the region of $\bar{\nu}_e - \bar{\nu}_x$ oscillation parameters excluded by the Chooz experiment at the 90% confidence level. This region contains the parameter values required to describe the Super-Kamiokande results in terms of $\nu_\mu - \nu_e$ oscillations.

An experiment conceptually similar to the Chooz one has recently started data taking at the Palo Verde nuclear power plant in Arizona (three reactors with a total power of 10.9 GW). The detector is located at distances of 750 m from one reactor and of 890 m from the other two in an underground site under 16 m of earth. It consists of 12 tons of Gadolinium-loaded scintillator surrounded by 1 m thick water shield and by a liquid scintillator layer to reject cosmic rays. The central detector is segmented into 66 independent cells.

Because of the shallow site the background is much higher in the Palo Verde experiment than in the Chooz one. With all reactors off the event rate is 32.62 ± 1.02 d^{-1}, while it is 39.06 ± 1.00 d^{-1} when all reactors run at full power [42]. After ~ 40 days of data taking the experiment has not yet reached the sensitivity to oscillations of the Chooz experiment.

126

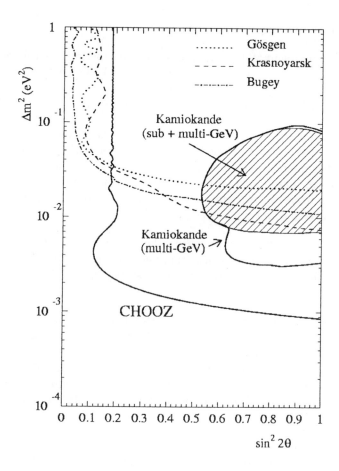

Figure 16: Boundary of the $\bar{\nu}_e - \bar{\nu}_x$ oscillation parameter region excluded by the Chooz experiment at the 90% confidence level. Also shown are the boundaries of the regions excluded by previous reactor experiments and the region of $\nu_\mu - \nu_e$ oscillation parameters allowed by the results of Ref. [37].

8   Neutrino Oscillation Searches at Accelerators

8.1 $\nu_\mu - \nu_e$, $\bar{\nu}_\mu - \bar{\nu}_e$ searches : LSND and KARMEN experiments

The Liquid Scintillator Neutrino Detector (LSND) [43] and the KArlsruhe-Rutherford Medium Energy Neutrino (KARMEN) experiment [44] use neutrinos produced in the beam stop of a proton accelerator. LSND has finished data taking at the Los Alamos Neutron Science Center (LANSCE) at the end of 1998, while KARMEN is still running at the ISIS neutron spallation facility of the Rutherford-Appleton Laboratory.

In these experiments neutrinos are produced by the following decay processes :

(i) $\pi^+ \to \mu^+ \nu_\mu$ (in flight or at rest);

(ii) $\mu^+ \to \bar{\nu}_\mu e^+ \nu_e$ (at rest);

(iii) $\pi^- \to \mu^- \bar{\nu}_\mu$ (in flight);

(iv) $\mu^- \to \nu_\mu e^- \bar{\nu}_e$ (at rest).

The relative $\bar{\nu}_e$ yield is very small (of the order of 4×10^{-4} with respect to $\bar{\nu}_\mu$) because π^- decaying in flight are a few % of all produced π^- and only a small fraction of μ^- stopping in heavy materials decays to $\nu_\mu e^- \bar{\nu}_e$ (π^- at rest are immediately captures by nuclei; most μ^- stopping in high-Z materials undergo the capture process $\mu^- p \to \nu_\mu n$). The neutrino energy distributions are shown in Fig. 17.

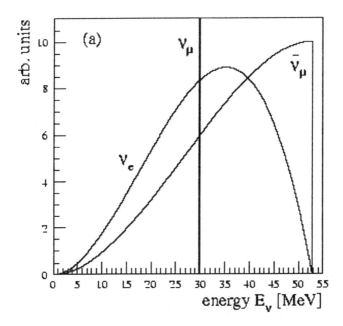

Figure 17: Neutrino spectra from the $\pi^+ \to \mu^+ \to e^+$ decay chain at rest.

Table 6 lists the main parameters of the two experiments. For the $\bar{\nu}_\mu - \bar{\nu}_e$ oscillation search, the $\bar{\nu}_e$ is detected by reaction (41) which gives a prompt e^+ signal followed by a delayed γ signal from neutron capture (the 2.2 MeV γ-ray from the reaction $np \rightarrow d\gamma$ and also, for KARMEN, the 8 MeV line from γ-rays emitted by neutron capture in Gadolinium, which is contained in thin layers of Gd_2O_3 placed between adjacent cells).

While the LANSCE beam is ejected in $\sim 500\mu s$ long spills 8.3 ms apart, the ISIS beam is pulsed with a time structure consisting of two 100 ns long pulses separated by 320 ns (this sequence has a repetition rate of 50 Hz). Thus it is possible to separate neutrinos from muon and pion decay from their different time distributions with respect to the beam pulse.

Table 7 lists preliminary results from LSND [45] and KARMEN [46], obtained after requiring space and time correlation between the prompt and delayed signal, as expected from $\bar{\nu}_e p \rightarrow e^+ n$ (and for KARMEN requiring also the time correlation between the e^+ signal and the beam pulse). The LSND result gives evidence for an excess of $\bar{\nu}_e$ events with a statistical significance of ~ 4.5 standard deviations.

Table 6: Parameters of the LSND and KARMEN experiments

	LSND	KARMEN
Proton beam kinetic energy	800 MeV	800 MeV
Proton beam current	1000 μA	200 μA
Detector	Single cylindrical tank; 1220 PMT's; collection of both scintillation and Čerenkov light.	512 cells filled with liquid scintillator; cell dim. $18 \times 18 \times 350$ cm
Detector mass	167 tons (mineral oil)	56 tons (mineral oil)
Event localization	timing	cell size
Distance from ν source	29 m	17 m
Angle between ν direction and proton beam	17°	90°
Data taking period	1993–98	Feb. 97–Feb. 99
Protons on target	1.8×10^{23}	2.9×10^{22}

Table 7: Preliminary results from LSND and KARMEN

	LSND	KARMEN
e^+ energy interval	20–60 MeV	16–50 MeV
N of observed events	70	8
Cosmic ray background	17.7 ± 1.0	1.9 ± 0.1
Total background	30.5 ± 2.7	7.82 ± 0.74
$\bar{\nu}_e$ signal (events)	39.5 ± 8.8	< 6.2 (90% C.L)
$\bar{\nu}_\mu - \bar{\nu}_e$ oscillation probability	$(3.3 \pm 0.9 \pm 0.5) \times 10^{-3}$	$< 4.2 \times 10^{-3}$ (90% C.L.)

Fig. 18 shows the preliminary e^+ energy distribution of the 70 events observed by LSND, together with the distributions expected from backgrounds and from $\bar{\nu}_\mu - \bar{\nu}_e$ oscillations for two different Δm^2 values. The region of oscillation parameters describing the LSND result is shown in Fig. 19, together with the region excluded by previous experiments and by the present KARMEN result. This figure shows that, if the LSND result is correct, the only allowed region of $\bar{\nu}_\mu - \bar{\nu}_e$ oscillation parameters is a narrow strip with Δm^2 between 0.2 and 2 eV^2 and $\sin^2 2\theta$ between 0.002 and 0.04.

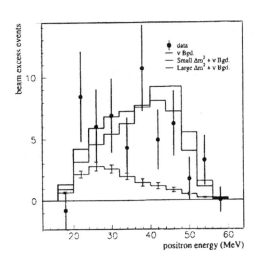

Figure 18: Preliminary e^+ energy distribution of the 70 events observed by LSND. Also shown are the distributions expected from backgrounds (histogram with error bars) and the expectations from $\bar{\nu}_\mu - \bar{\nu}_e$ oscillations for two different Δm^2 values.

130

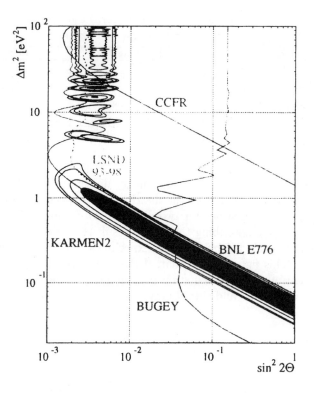

Figure 19: Region $\bar{\nu}_\mu - \bar{\nu}_e$ oscillation parameters allowed by the preliminary LSND results [45]. Also shown are the boundaries of the regions excluded by previous experiments and by the recent KARMEN results [46].

During the first three years of LSND data taking the target area of the LANSCE accelerator consisted of a 30 cm long water target located ~ 1 m upstream of the beam stop. This configuration enhanced the probability of pion decay in flight, allowing LSND to search for $\nu_\mu - \nu_e$ oscillations using ν_μ with energy above 60 MeV. In this case one expects to observe an excess of events from the reaction

$$\nu_e + C^{12} \rightarrow e^- + X$$

above the expected backgrounds. This reaction has only one signature (a prompt signal) but the higher energy, the longer track and the directionality of Čerenkov light help improving electron identification and measuring its direction.

In this search [47] LSND has observed 40 events to compared with 12.3 ± 0.9 events from cosmic ray background and 9.6 ± 1.9 events from machine-related (neutrino-induced) processes. The excess of events (18.1 ± 6.6 events) corresponds to a $\nu_\mu - \nu_e$ oscillation probability of $(2.6 \pm 1.0) \times 10^{-3}$, consistent with the value found from the study of the $\bar{\nu}_e p \rightarrow e^+ n$ reaction below 60 MeV.

8.2 Future searches for $\nu_\mu - \nu_e$ oscillations

The KARMEN experiment will finish data taking in the year 2001. By then, its sensitivity to $\bar{\nu}_\mu - \bar{\nu}_e$ oscillations will have improved by a factor of 1.7 with respect to the present value. However, in case of a negative result the exclusion region will not fully contain the region allowed by LSND. A new search for $\nu_\mu - \nu_e$ (or $\bar{\nu}_\mu - \bar{\nu}_e$) oscillations is needed, therefore, to unambiguously confirm or refute the LSND signal.

The Mini-BOONE experiment [48] is a first phase of a new, high sensitivity search for $\nu_\mu - \nu_e$ oscillations (BOONE is the acronym for Booster Neutrino Experiment). Neutrinos are produced using an 8 GeV, high intensity proton beam from the Fermilab Booster Synchrotron. The beam consists mainly of ν_μ from π^+ decay with a small contamination ($\sim 0.3\%$) of ν_e, with a broad energy distribution from 0.3 to 2 GeV.

The Mini-BOONE detector will be installed at a distance of 500 m from the neutrino source. It consists of a 6 m radius spherical tank filled with mineral oil. Čerenkov light produced in the oil is collected by ~ 1500 photomultiplier tubes located on the surface of the sphere. The detector is surrounded by anticoincidence counters and will use the different pattern of Čerenkov light expected for muons, electrons and π^0 to identify these particles.

For a fiducial mass of 445 tons, Mini-BOONE expects to detect $\sim 5 \times 10^5$ $\nu_\mu \, C^{12} \rightarrow \mu^- X$ events and ~ 1700 $\nu_e C^{12} \rightarrow e^- X$ events in one year. A $\nu_\mu - \nu_e$ oscillation probability of 0.003 will result in an excess of ~ 1500 $\nu_e C^{12} \rightarrow e^- X$ events.

If no oscillation signal is observed, Mini-BOONE will exclude a region of oscillation parameters which extends to $\sin^2 2\theta \approx 4 \times 10^{-4}$ at large Δm^2 and to $\Delta m^2 \approx 0.02$ eV^2 at full mixing, thus completely excluding the region presently allowed by LSND. If, however, a signal is observed, it should be possible to measure precisely the oscillation parameters, possibly by using a second detector at a different distance.

Mini-BOONE will begin taking data in the year 2002.

8.3 $\nu_\mu - \nu_\tau$ searches at CERN

Two experiments searching for $\nu_\mu - \nu_\tau$ oscillations have recently completed data taking at CERN. They both used the wide-band neutrino beam from the CERN

450 GeV proton synchrotron (SPS). The method adopted by both experiments consists in detecting τ^- production with a sensitivity corresponding to a ν_τ/ν_μ ratio much larger than the value expected from conventional ν_τ sources in the beam (the main ν_τ production process is D_s production by the primary protons, followed by the decay $D_s \to \tau\nu_\tau$). The observation of τ^- could only result, therefore, from $\nu_\mu - \nu_\tau$ oscillations.

These experiments are sensitive to Δm^2 values above a few eV^2. According to the so-called 'see-saw' model [49], the neutrino masses obey one of the two relations

$$m_1 : m_2 : m_3 = m_e^2 : m_\mu^2 : m_\tau^2;$$
$$m_1 : m_2 : m_3 = m_u^2 : m_c^2 : m_t^2 ,$$

involving either the charged lepton masses $(m_e, \; m_\mu, \; m_\tau)$ or the $Q = 2/3$ quark masses $(m_u, \; m_c, \; m_t)$. In both cases one has $m_1 \ll m_2 \ll m_3$.

Assuming that the solar neutrino problem is the result of $\nu_e - \nu_\mu$ oscillations with small mixing, Δm^2 is equal to $[m(\nu_\mu)]^2$ to a very good approximation, giving $m(\nu_\mu) \approx 3 \times 10^{-3}$ eV for $\Delta m^2 = 10^{-5}$ eV^2. Then, from the see-saw relations given above the range of values for the ν_τ mass (between ~ 1 and ~ 30 eV) is such that the ν_τ could be, at least partially, an important component of dark matter and $\nu_\mu - \nu_\tau$ oscillations can be observed using ν_μ beams from high-energy proton accelerators and baselines of the order of 1 km if the mixing angle is not too small.

The two experiments, CHORUS and NOMAD, are installed one behind the other at a distance of ~ 820 m from the proton target. A pair of pulsed magnetic lenses located after the target produces an almost parallel wide-band beam of positive hadrons. Neutrinos from π or K decay reach the detectors, while iron and earth shielding absorb surviving hadrons and range out decay muons. The distance between the proton target and the end of the decay tunnel is 414 m. Fig. 20 shows the expected neutrino energy spectrum and Table 8 lists its mean energies and relative abundances [50]. The ν_τ 'natural' abundance is estimated to be $\sim 5 \times 10^{-6}$ [51].

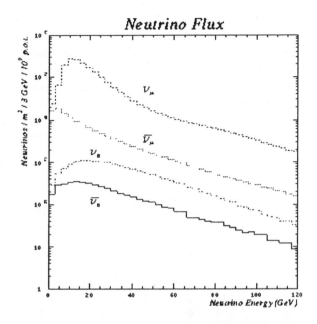

Figure 20: Neutrino fluxes from the CERN wide-band beam.

Table 8: Mean energies and relative abundances for ν fluxes and CC interactions at the NOMAD detector [50]. The integrated ν_μ flux is 1.11×10^{-2} ν_μ per proton on target

ν type	Flux		Charged-current interactions	
	$<E_\nu>$ GeV	Relative abundance	$<E_\nu>$ GeV	Relative abundance
ν_μ	23.5	1.00	42.6	1.00
$\bar{\nu}_\mu$	19.2	0.061	41.0	0.0249
ν_e	37.1	0.0094	56.7	0.0148
$\bar{\nu}_e$	31.3	0.0024	53.6	0.0016
ν_τ	~ 35	$\sim 5 \times 10^{-6}$		

8.3.1 CHORUS

CHORUS (Cern Hybrid Oscillation Research apparatUS) aims at detecting the decay of the short-lived τ lepton in nuclear emulsion. This technique provides a spacial resolution of ~ 1 μm, well matched to the average τ^- decay length of 1 mm.

134

The apparatus [52] is shown in Fig. 21. It consists of an emulsion target with a total mass of $\sim$ 770 kg, followed by an electronic tracking detector made of scintillating fibres, an aircore hexagonal magnet, electromagnetic and hadronic calorimeters and a muon spectrometer consisting of magnetized iron toroids interleaved with drift chambers. The hexagonal magnet provides a field of 0.1 T, over a length of 0.75 m, oriented along the sides of an hexagon with no radial dependence. It is used to determine the charge and momentum of low-energy particles with a resolution $\sigma(p)/p \sim 20\%$ for momenta between 2 and 10 GeV.

CHORUS

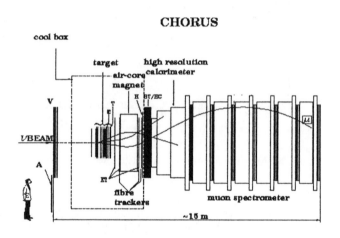

Figure 21: Layout of the CHORUS detector.

Neutrino events with a μ^- or a negatively charged hadron are selected and the tracks are followed back to the exit point from the emulsion target. The method is illustrated in Fig. 22. It relies on special emulsion sheets mounted between the target and the fibre tracker (these sheets are replaced every few weeks during the run). With the reconstruction accuracy of the fibre tracker, the track position on the special sheet is predicted within an area of 360 μm $\times$ 360 μm. In this area, because of the short exposure time, one finds, an average, 5 muon tracks which are rejected by angular measurement. The search is then continued in an area of 20 μm $\times$ 20 μm into the emulsion target, with negligible background despite the long exposure time of the target (2 years).

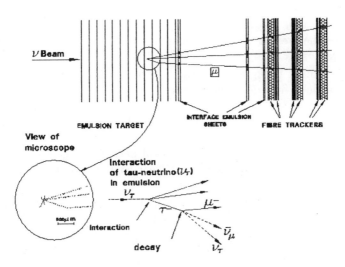

Figure 22: Expected configuration of a typical ν_τ charged-current interaction in the CHO-RUS emulsion. In this example the τ^- decays to $\mu^- \nu_\tau \bar{\nu}_\mu$.

Events in which the neutrino interaction point is found in the emulsion target are analysed to search for τ^- decay, which is identified by the presence of a change in direction (kink) of a negatively charged track, as expected from one-prong decays. No other charged leptons must be observed at the primary vertex and the transverse momentum (p_T) of the selected particle with respect to the τ candidate direction is required to be larger than 0.25 GeV (to eliminate strange particle decays). Negative tracks with impact parameters larger than ~ 8 μm are also considered as possible secondary particles from τ^- decay.

Negative muons with momentum $p < 30$ GeV or negative hadrons with $1 < p < 20$ GeV are considered as possible τ^- decay products. If observed, the kink is required to be within 3.95 mm from the interaction point for muon tracks, and 2.37 mm for hadron tracks. The emulsion scanning procedure to locate the interaction point and to reject events incompatible with τ^- decays is fully automatic. The remaining events undergo a computer assisted eye scan to confirm the presence of a τ decay. This step is necessary because the development process deforms the emulsion and the tracks become distorted. It must be verified, therefore, that the decay kink or the large impact parameter are genuine and are not the result of local distortions. This is achieved by scanning and measuring background tracks close to the event under consideration (in the CHORUS experiment these tracks are mostly straight-through muons from SPS beams).

CHORUS took data between May 1994 and the end of 1997. Table 9 summarizes the present status of the analysis [53]. No τ^- candidate has been observed yet.

The dominant background in the one-muon channel is the production of charmed particles from $\bar{\nu}_\mu$ interactions

$$\bar{\nu}_\mu \ N \to \mu^+ D^- X$$

followed by the decay $D^- \to \mu^- +$ neutral particles. These events are dangerous only if the μ^+ is not identified. Their contribution to the data sample analysed so far is 0.197 ± 0.040 events.

In the muonless channel the background from charm production amounts to 0.075 ± 0.015 events. A more serious background is the interaction of negative hadrons with nuclei producing only one outgoing negatively charged particle with no evidence for nuclear break-up (these interactions are called 'white kinks'). The rate of such events is affected by a large uncertainty and is estimated to be 0.656 ± 0.656 events.

Table 9: Status of the CHORUS analysis [53].

	One-muon events	Muonless events
Expected number of events	458.6×10^3	116×10^3
Fraction scanned so far	54%	47%
Events with identified ν interaction	102.8×10^3	16.7×10^3
N_τ for $< P_{\mu\tau} > = 1$	4023	1137

For a particular τ decay channel, i, the expected number of events is given by

$$N_\tau^i = BR^i \int \Phi_{\nu_\mu}(E) P_{\mu\tau} \sigma_\tau \ A_\tau^i \ \epsilon_\tau^i \ dE$$

where: BR^i is the decay branching ratio; $\phi_{\nu_\mu}(E)$ is the ν_μ energy spectrum; $P_{\mu\tau}$ is the oscillation probability; σ_τ is the cross-section for τ^- production; A_τ^i is the acceptance and reconstruction efficiency (including the vertex-finding efficiency); and ϵ_τ^i is the efficiency of the decay search.

Similarly, the expected number of one-muon events from ν_μ charged-current interactions is given by

$$N_\mu = \int \phi_{\nu_\mu}(E) \sigma_\mu \ A_\mu \ dE \ .$$

At large Δm^2 values $P_{\mu\tau}$ is constant and one can use average values:

$$N_\tau^i = N_\mu P_{\mu\tau} \, BR^i \frac{<\sigma_\tau> <A_\tau^i>}{<\sigma_\mu> <A_\mu>} <\epsilon_\tau^i> \, . \qquad (43)$$

The ratio $<\sigma_\tau> / <\sigma_\mu>$ has the value 0.53. For the $\tau \to \mu$ decay channel $<A_\tau^\mu> / <A_\mu>$ is 1.075 because of the requirement $P_\mu < 30$ GeV which suppresses ν_μ charged-current interactions more than $\tau \to \mu$ decays. The efficiency of the decay search is estimated by simulations and its value, $\epsilon_\tau^\mu \approx 0.37$, is verified experimentally using $\mu^-\mu^+$ events from $\nu_\mu N \to \mu^- D^+ X$ followed by $D^+ \to \mu^+$ decay.

Muonless τ^- decays include the $\tau^- \to h^-$ and $\tau^- \to e^-$ channels, where h^- is a charged hadron, and also $\tau^- \to \mu^-$ events in which the μ^- was not identified. To allow an easy combination of the results from the one-muon and muonless channels, an 'equivalent number of muonic events' is defined for the muonless sample using the relation

$$N_\mu^{eq} = N_\mu \frac{\sum_i <A_\tau^i> <\epsilon_\tau^i> BR^i}{<A_\tau^\mu> <\epsilon_\tau^\mu> BR^\mu} \, ,$$

where the sum extends to all muonless τ decay channels and N_μ is the corresponding number of one-muon events from ν_μ charged-current interactions. With this trick the oscillation probability $P_{\mu\tau}$ can be determined for muonless events using Eq. (43) with $i = \mu$ and replacing N_μ by N_μ^{eq}.

The expected values of N_τ, as obtained by setting $P_{\mu\tau} = 1$ in Eq. (43) are given in the last row of Table 9. The 90% confidence upper limit on $P_{\mu\tau}$ is

$$P_{\mu\tau} < 2.38/N_\tau = 0.46 \times 10^{-3}$$

where 2.38 is the Poisson upper limit of a null result at the 90% confidence level taking into account a systematic uncertainty of 17% on N_τ [54]. The region of oscillation parameters excluded by this result under the assumption of two-neutrino mixing is shown in Fig. 23 together with the results from previous experiments [55].

138

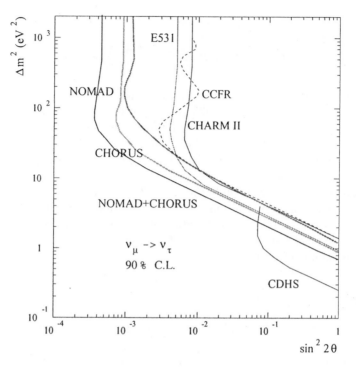

Figure 23: The $\Delta m^2 - \sin^2(2\theta)$ plane for $\nu_\mu - \nu_\tau$ oscillation. The regions excluded by CHORUS and NOMAD, and by their combined results [56], are shown together with the results of previous experiments [55].

A more stringent limit, $P_{\mu\tau} < 0.35 \times 10^{-3}$, would be obtained [56] using the method recently proposed by Feldman and Cousins [57].

CHORUS is expected to reach the upper limit $P_{\mu\tau} < 10^{-4}$ if no event is found after the completion of the analysis.

8.3.2 NOMAD

NOMAD (Neutrino Oscillation MAgnetic Detector) is designed to search for $\nu_\mu - \nu_\tau$ oscillations by observing τ^- production using kinematical criteria [58], which require a precise measurement of secondary particle momenta. The main detector components are [59] (see Fig. 24):

- drift chambers (DC) used to reconstruct charged particle tracks and also acting as the neutrino target (fiducial mass ~ 2.7 tons, average density 0.1 g/cm^3, radiation length ~ 5 m);

- nine independent transition radiation detectors (TRD) for electron identification, interleaved with additional drift chambers;

- an electromagnetic calorimeter (ECAL) located behind a 'preshower' detector (PRS);

- a hadronic calorimeter (HCAL);

- large-area muon chambers.

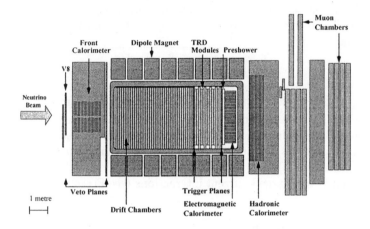

Figure 24: Side view of the NOMAD detector.

DC, TRD and ECAL are located inside a uniform magnetic field of 0.4 T perpendicular to the beam direction.

The NOMAD experiment aims at detecting τ^- production by observing both leptonic and hadronic decay modes of the τ^-. The decay $\tau^- \to \nu_\tau e^- \bar{\nu}_e$ is particularly attractive because the main background results from ν_e charged-current interactions which are only $\sim 1.5\%$ of the total number of neutrino interactions in the target fiducial volume and have an energy spectrum quite different from that expected from $\nu_\mu - \nu_\tau$ oscillations (see Fig. 20 and Table 8). The selection of this decay relies on the presence of an isolated electron in the final state and on the correlation among the lepton transverse momentum ($\vec{p}_T^{\,e}$), the total transverse momentum of the hadronic system ($\vec{p}_T^{\,H}$) and the missing transverse momentum ($\vec{p}_T^{\,m}$) (only the momentum components perpendicular to the beam direction can be used because the incident neutrino energy is unknown). In the case of ν_e charged-current events $\vec{p}_T^{\,e}$ is generally opposite to $\vec{p}_T^{\,H}$ and $|\vec{p}_T^{\,m}|$ is small (it should be exactly zero if the momenta of all secondary

140

particles were measured precisely and the target nucleon were at rest). On the contrary, in $\tau^- \to \nu_\tau e^- \bar{\nu}_e$ decays there is a sizeable $|\vec{p}_T^m|$ associated with the two outgoing neutrinos. Furthermore, in a large fraction of events $\vec{p}_T^m$ is at opposite azimuthal angles to $\vec{p}_T^H$, in contrast with ν_e charged-current interactions for which large values of $|\vec{p}_T^m|$ result mostly from hadrons escaping detection (in these cases the azimuthal separation between $\vec{p}_T^H$ and $\vec{p}_T^m$ is small).

The selection of hadronic τ^- decays relies on the observation of a hadron, or of a collimated system of hadrons, consistent with τ^- decay and well isolated from the other hadrons in events containing no primary, isolated lepton. A powerful variable to reject neutral-current events is the transverse momentum of the candidate hadron(s) from τ^- decay with respect to the total visible momentum:

$$Q_T = \sqrt{(p_h^\tau)^2 - \frac{(\vec{p}_h^\tau \cdot \vec{P})^2}{P^2}} \,,$$

where $\vec{p}_h^\tau$ is the momentum of the candidate hadron(s) from τ^- decay and $\vec{P}$ is the total visible momentum. For neutral-current events $\vec{P}$ is the total momentum of the hadronic jet and Q_T is the transverse momentum of a particle (or system of particles) in the jet. This is generally much smaller than the value expected for hadron(s) from τ^- decay which are well isolated from the hadronic jet produced in ν_τ charged-current interactions (see Fig. 25).

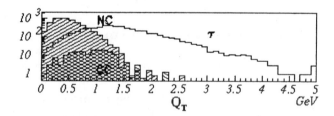

Figure 25: Expected NOMAD distribution of the variable Q_T (see text) for simulated ν_μ neutral-current (NC), charged-current (CC) interactions and $\tau^- \to \nu_\tau \pi^-$ decays.

A particularly dangerous background in the search for hadronic τ^- decays results from ν_μ or ν_e charged-current interactions in which the primary, high-p_T lepton was not identified. This background is reduced to a tolerable level by rejecting all the events containing a high-p_T negative particle which could not reach the subdetectors providing lepton identification (TRD, PRS, ECAL, muon chambers).

In practice, for most τ^- decay channels the method to separate the signal from backgrounds uses ratios of likelihood functions. These functions are approximated by products of probability density functions of kinematic variables. These are obtained from large samples of simulated events, after corrections to take into account differences between simulated and real data, as described in Ref. [60].

For the $\tau^- \to e^-$ decay channel the analysis uses two likelihood ratios, λ_{e1} and λ_{e2}, which are used to separate the signal from neutral-current and ν_e charged-current background, respectively. Fig. 26 shows the event distribution in the $\ln\lambda_{e1}$, $\ln\lambda_{e2}$ plane for backgrounds and signal. The framed region in the upper-right corner (the 'signal box') contains a sizeable fraction of $\tau^- \to e^-$ events and little background. For the data sample collected between 1995 and 1997 the predicted background amounts to $6.3^{+1.6}_{-1.0}$ events [61].

The analysis is not allowed to look at the data in the signal box until a robust background prediction has been provided. When the signal box is opened, 5 events are found (see Fig. 26d), in agreement with the predicted background.

Table 10 shows a summary of backgrounds and efficiencies for all analyses of deep-inelastic scattering events (DIS), as reported in Ref. [61], and for the analysis of low-multiplicity (LM) events reported in Ref. [60] (DIS events are defined by the requirement $p^H > 1.5$ GeV). For all channels there is good agreement between the observed number of events and the background prediction. The resulting 90% confidence level upper limit using the method of Ref. [57] is [61]

$$P_{\mu\tau} < 0.6 \times 10^{-3} ,$$

which corresponds to the exclusion region shown in Fig. 23.

The CHORUS and NOMAD limits can be combined using the method of Ref. [57]. The combined limit is [56]

$$P_{\mu\tau} < 0.23 \times 10^{-3} .$$

The corresponding exclusion region is outlined in Fig. 23.

142

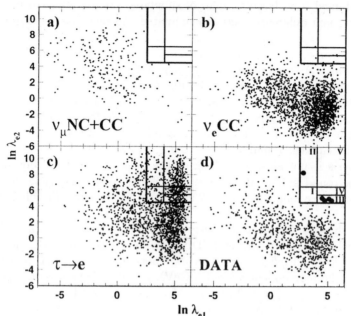

Figure 26: Scatter plot of ln λ_{e2} vs ln λ_{e1} for (a) simulated ν_μ neutral-current interactions; (b) simulated ν_e charged-current interactions; (c) simulated $\tau^- \to e^- \nu_\tau \bar{\nu}_e$ events; (d) NOMAD data.

Table 10: Summary of NOMAD results [61]. N_τ is the number of expected τ^- events for $P_{\mu\tau} = 1$

Analysis	$\epsilon(\%)$	$Br(\%)$	N_τ	Number of events Est. Bkgnd	Obs.
$\tau \to e$ DIS	3.5	17.8	2818	$6.3^{+1.6}_{-1.0}$	5
$\tau \to h(n\pi^0)$ DIS	0.78	49.8	1727	5.0 ± 1.2	5
$\tau \to \rho$ DIS	1.6	25.3	1891	$5.0^{+1.7}_{-0.9}$	5
$\tau \to 3\pi(\pi^0)$ DIS	2.9	15.2	1180	6.5 ± 1.1	5
$\tau \to e$ LM	3.4	17.8	218	$0.5^{+0.6}_{-0.2}$	0
$\tau \to \pi(\pi^0)$ LM	1.5	37.3	198	$0.1^{+0.3}_{-0.1}$	1
$\tau \to 3\pi(\pi^0)$LM	2.0	15.2	108	$0.4^{+0.6}_{-0.4}$	0

9   Long baseline experiments at accelerators

Long baseline experiments at accelerators extend the sensitivity of searches for $\nu_\mu - \nu_x$ oscillations to Δm^2 as low as 10^{-3} eV^2 using ν_μ beams of well known properties which can be monitored and varied if needed. The main goal of these experiments is to verify that the atmospheric neutrino results discussed in Sec. 6 are indeed associated with oscillations, to establish the nature of the oscillation and to measure its parameters.

Table 11 shows a list of parameters for the three existing projects.

Table 11: Long baseline projects

Project	Accelerator	Location of far detector	Distance (Km)	$< E_\nu >$ GeV	Status
K2K	KEK 12 GeV proton synchrotron	Kamioka mine	250	1.4	Start April 1999
NuMI	Fermilab 120 GeV Main Injector (MI)	Soudan mine	730	16 or lower	Start 2002
NGS	CERN 450 GeV SPS	Gran Sasso Lab	732	30 or lower	not yet approved

9.1   K2K

The K2K project [62] uses neutrinos from the decay of π and K mesons produced by the KEK 12 GeV proton synchrotron and aimed at the SuperKamiokande detector at a distance of 250 km. The beam consists mainly of ν_μ, with $\bar{\nu}_\mu$ and ν_e contaminations of 4% and 1%, respectively. The ν_μ energy spectrum is shown in Fig. 27.

For a run of 10^{20} protons on target (3 years, corresponding to an effective data taking time of 12 months) the total number of events in the SuperKamiokande fiducial volume (22.5 Ktons of water) is expected to be 345 ν_μ charged-current, 120 neutral current and 4 ν_e charged-current interactions in the absence of neutrino oscillations. The beam energy is below threshold for τ^- production ($E_\nu \approx 3.5$ GeV), so no search for τ^- appearance is possible.

144

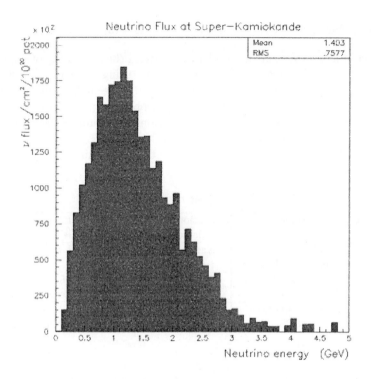

Figure 27: Expected ν_μ energy spectrum at the SuperKamiokande detector in the absence of neutrino oscillation.

Fig. 28a shows the expected distortion of the ν_μ flux at 250 km for a ν_μ oscillation with $\Delta m^2 = 3.5 \times 10^{-3}$ eV^2 and full mixing. Such a distortion can be detected by comparing the energy distribution of beam-associated muon-like events in SuperKamiokande with the distribution measured in a similar, 1 Kton water detector located at 300 m from the neutrino source on the KEK site (see Fig. 29). No oscillation effects are expected at this distance and the number of events is much larger ($\sim 4 \times 10^5 \nu_\mu$ charged-current interactions in a fiducial volume of 21 tons). Fig. 30 shows the region of oscillation parameters which is excluded at the 90% confidence level if no significant difference is observed between the near and far detector.

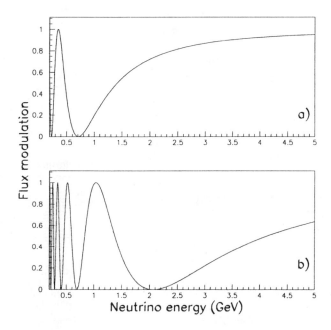

Figure 28: Expected flux modulation from ν_μ disappearance in the K2K (a) and NUMI project (b), for a two-neutrino oscillation with $\Delta m^2 = 3.5 \times 10^{-3}$ eV^2 and full mixing. The energy values where minima and maxima occur are proportional to Δm^2.

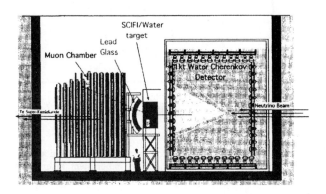

Figure 29: Layout of the near detector of the K2K project.

146

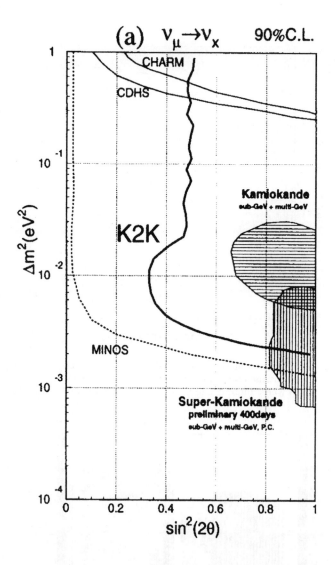

Figure 30: Region of $\nu_\mu - \nu_\tau$ or $\nu_\mu - \nu_s$ oscillation paramaters excluded at the 90% confidence level if no oscillation signal is detected by the K2K experiment after three years of data taking.

The near detector includes a system of scintillating fibres in water, a lead glass calorimeter and a muon range telescope to monitor and measure precisely the ν_μ, $\bar{\nu}_\mu$ and ν_e energy spectra and space distributions. The near detector will also measure precisely the cross-section for π^0 production in neutral-current interactions. As discussed in Sec. 6.4 a precise knowledge of this quantity is crucial to understand if the apparent disappearance of atmospheric ν_μ's is the result of oscillations to an active or to a sterile neutrino.

In the K2K project the expected number of single-π^0 events in Super-Kamiokande is only 25 for a three-year run. This number is too small to provide any significant information on the issue of active versus sterile neutrinos from the K2K experiment itself.

K2K has started data taking in April 1999.

9.2 NuMI and the MINOS experiment

The NuMI project uses neutrinos from the decay of π and K mesons produced by the new Fermilab Main Injector (MI), a 120 GeV proton synchrotron capable of accelerating 5×10^{13} protons with a cycle time of 1.9 s. The expected number of protons on target is 3.6×10^{20}/y. The decay pipe is 675 m long.

The neutrino beam will be aimed at the Soudan mine in Minnesota (an inactive iron mine) at a distance of 730 km from the proton target. The beam will consist primarly of ν_μ, with 0.6% ν_e. Fig. 31 shows the expected energy distributions of ν_μ charged-current events for three different neutrino beams which correspond to different tunes and locations of the focusing elements. For the high energy beam the total number of events is ~ 3000/y for a detector mass of 1000 tons.

The expected distortion of the ν_μ flux for an oscillation with $\Delta m^2 = 3.5 \times 10^{-3}$ eV^2 and full mixing is shown in Fig. 28b. It is clear that such a distortion can be best detected using the lowest energy beam of Fig. 31.

The MINOS experiment [63] will use two detector, one (the 'near detector') located at Fermilab, the other (the 'far detector') located in a new underground hall to be built at the Soudan site at a depth of 713 m (2090 m of water equivalent). Both detectors are iron-scintillator sandwich calorimeters with a toroidal magnetic field in the iron plates.

The far detector (Fig. 32) has a total mass of 5400 tons and a fiducial mass of 3300 tons. It consists of magnetized octagonal iron plates, 2.54 cm thick, interleaved with active planes of 4 cm wide, 8 m long scintillator strips providing both calorimetric and tracking information.

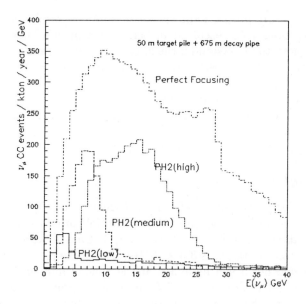

Figure 31: Neutrino interaction energy spectra predicted for different focusing conditions at the Soudan location of the NuMI project. 'Perfect focusing' is the ideal case of all secondary π^+ and K^+ being focused into a pencil beam.

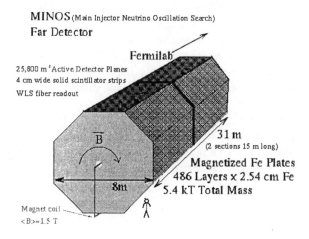

Figure 32: Sketch of the MINOS detector.

The near detector has a total mass of 920 tons and a fiducial mass of 100 tons. It will be installed 250 m dowstream from the end of the decay pipe.

The comparison of ν_μ charged-current event rate and energy distribution in the two detectors will be sensitive to oscillations which can be detected with a statistical significance of at least four standard deviations over the full parameter space currently suggested by the atmospheric neutrino results (see Fig. 33). The oscillation parameters will be measured precisely over most of this region.

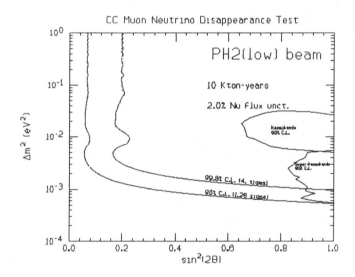

Figure 33: Excluded region (90% confidence) and 4σ discovery region for a 10 kton × y exposure of the MINOS experiment from a comparison of the ν_μ CC event spectra in the far and near detector.

The measurement of the ratio between neutral- and charged-current event rate (NC/CC) is very important because it will be used to discriminate between $\nu_\mu - \nu_\tau$ and $\nu_\mu - \nu_s$ oscillations. In the former case NC/CC is larger in the far detector, while for $\nu_\mu - \nu_s$ oscillations it has the same value in the near and far detector.

The MINOS experiment should begin data taking at the end of the year 2002.

9.3 The NGS project

The NGS project (Neutrinos to Gran Sasso) has not yet been approved. It consists in aiming a neutrino beam from the CERN 450 GeV SPS to the Gran Sasso National Laboratory in Italy at a distance of 732 km. The three existing underground halls at Gran Sasso, under ~ 4000 m of water equivalent, are already oriented towards CERN and ICARUS [64], a 600 ton detector suitable for oscillation searches, will start operation in the year 2000 to search for proton decay and to study atmospheric and solar neutrinos.

If approved before the end of 1999, the NGS beam will be operational in the year 2005. It will be used for ν_τ appearance experiments, for which a detector in a 'near' location should not be necessary. The rate of ν_τ charged-current interactions from $\nu_\mu - \nu_\tau$ oscillations is given by

$$N_\tau = A \int \phi_{\nu_\mu}(E) \, P_{\mu\tau}(E) \, \sigma_\tau(E) \, dE \tag{44}$$

where A is a normalization constant proportional to the detector mass, $\phi_{\nu_\mu}(E)$ is the ν_μ energy spectrum at the detector, $P_{\mu\tau}(E)$ is the $\nu_\mu - \nu_\tau$ oscillation probability and $\sigma_\tau (E)$ is the cross-section for ν_τ charged-current interactions. The integration lower limit is set by the energy threshold for τ^- production, 3.5 GeV.

Under the assumption of two-neutrino mixing, $P_{\mu\tau}(E)$ is given by Eq. (10). For a large fraction of the Δm^2 interval suggested by the atmospheric neutrino results the condition $1.27 \, \Delta m^2 \, L/E < 1$ holds for $L = 732$ km and $E > 3.5$ GeV. In this case Eq. (44) can be approximated as

$$N_\tau = 1.61 \, A \, \sin^2(2\theta) \, (\Delta m^2)^2 \, L^2 \int \phi_{\nu_\mu}(E)\sigma_\tau(E)\frac{dE}{E^2} \; . \tag{45}$$

In this approximation N_τ varies as $(\Delta m^2)^2$. In addition, because at large distances $\phi_{\nu_\mu}(E)$ decreases as L^{-2}, N_τ does not depend on L. However, the background from conventional neutrino events is proportional to L^{-2}, so the ratio between the τ^- signal and the background varies as L^2.

In the approximation of Eq. (45), the neutrino beam must be designed with the goal of maximizing the integral of Eq. (45) which does not depend on Δm^2. A preliminary beam design, based on a 1000 m long decay tunnel, is described in Ref. [65]. The design has been improved [66] to optimize the τ^- production rate. The ν_μ mean energy is 17 GeV and the rate of ν_μ charged-current interactions is ~ 2448/y for a detector with a mass of 1000 tons (this value corresponds to 4.5×10^{19} protons on target, which is a realistic figure for a one-year run after the shut-down of LEP).

The rates of other neutrino events relative to ν_μ events are 0.007, 0.02 and 0.0007 for ν_e, $\bar{\nu}_\mu$ and $\bar{\nu}_e$, respectively. Table 12 lists the expected yearly rates of τ^- events [66] for three different Δm^2 values.

Table 12: Yearly rate of τ^- events vs Δm^2 for a 1000 ton detector [66] (1 year $= 4.5 \times 10^{19}$ protons on target).

Δm^2 (eV^2)	N_τ
10^{-3}	2.48
3×10^{-3}	21.7
5×10^{-3}	58.5

ICARUS [64] is a new detector concept based on a liquid Argon Time Projection Chamber (TPC) which allows three-dimensional reconstruction of events with spatial resolution of the order of 1 mm.

Primary ionisation electrons drift in very pure liquid Argon over distances of the order of 1.5 m and are collected by wire planes which provide two of the three coordinates and measure the ionisation, while the third coordinate along the drift direction is determined by measuring the drift time.

Liquid Argon has a density of 1.4 g/cm^3, a radiation length of 14 cm and an interaction length of 84 cm. ICARUS is also an excellent calorimeter, with an expected resolution $\sigma_E/E = 3\%/\sqrt{E}$ and $13\%/\sqrt{E}$ for electromagnetic and hadronic showers, respectively (E in GeV).

A 600 ton ICARUS module is presently being constructed. The cryostat cold volume (534 m^3) is 19.6 m long and 4.2 m high. Three additional modules will be built if the operation of the first module is successful.

ICARUS will search for τ appearance using kinematical criteria similar to those used in the NOMAD experiment (see Sec. 8.3.2). However, for the $\tau^- \to e^-$ decay channel a background rejection power in excess of 10^4 was needed in NOMAD, while in ICARUS a rejection power of $\sim 10^2$ is sufficient because of the much smaller number of events. This value requires looser selection criteria and the detection efficiency for $\tau^- \to e^-$ events becomes $\sim 50\%$. With four modules one expects ~ 10 τ^- events/y for a $\nu_\mu - \nu_\tau$ oscillation with $\Delta m^2 = 5 \times 10^{-3}$ eV^2 and full mixing, to be compared with a background of 0.25 events.

The τ identification from other τ decay channels is presently under study. Since the τ^- production rate at low Δm^2 is proportional to $(\Delta m^2)^2$, a detector consisting of 4 ICARUS modules should be sensitive to oscillations with $\Delta m^2 \geq 2 \times 10^{-3}$ eV^2 after a running time of four years.

Another interesting detector concept for a ν_τ appearance search is OPERA [67]. The detection of one-prong τ decays is performed by measuring the τ^- decay kink in space, as determined by two track segments measured with very high precision in nuclear emulsion.

Fig. 34 illustrates the OPERA concept. The main components of the target are 1 mm thick Pb plates where most neutrinos interact. One-prong τ decays occurring in the 3 mm gap between the emulsion detectors ES1, ES2 are expected to result in observable kinks. ES1 and ES2 consist of two 50 μm thick emulsion layers glued to a 100 μm plastic foil. The 3 mm gap between ES1 and ES2 is filled with a very low density spacer to which ES1 and ES2 are glued to ensure that their relative positions are stable and precisely known.

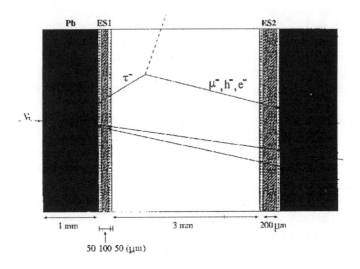

Figure 34: Schematic structure of the OPERA target.

The OPERA detector is arranged in 'bricks', each consisting of $\sim$ 30 Pb-ES1-ES2 layers with transverse dimensions 15 $\times$ 15cm^2, for a total mass of $\sim$ 800 tons. Magnetized iron toroids and conventional trackers will be used to identify μ^- and to minimize the background from the decay of charm particles produced by ν_μ charged-current interactions. Conventional trackers are located after each plane of bricks to identify the brick where the neutrino interaction took place. This brick is immediately removed from the detector for scanning and measurement of the emulsion layers.

The global τ^- detection efficiency is estimated to vary between 0.29 at low Δm^2 and 0.33 at $\Delta m^2 \approx 10^{-2}$eV^2. For a run of one year (4.5 $\times 10^{19}$ protons on target) the background is expected to be 0.4 events, mostly from charm production and decay in events in which the primary μ^- was not identified. The number of expected τ^- events for full mixing is listed in Table 13 for different Δm^2 values.

Table 13: Number of detected τ^- in OPERA per 4.5×10^{19} protons on target (1 year) (from Ref. [66])

Δm^2 (eV^2)	Detected N^-/y
1.0×10^{-3}	0.5
3×10^{-3}	4.5
5.0×10^{-3}	12.0

The OPERA concept is being also considered at Fermilab in the frame of the NuMI project. For the highest energy version of the NuMI beam (see Fig. 31) the yearly τ^- production rate in a 1000 ton detector is a factor of ~ 2 higher than in the NGS project, because of the much higher number of protons on target (3.6×10^{20} for NuMI, 4.5×10^{19} for NGS). However, the high energy neutrino beam is not the best choice for MINOS if Δm^2 is below $\sim 5 \times 10^{-3}$ eV^2. It is not clear, therefore, if the two experiment will be compatible.

10 Summary and Conclusions

As discussed in the previous sections, studies of solar and atmospheric neutrinos and oscillation searches at accelerators have provided evidence or hints for neutrino oscillations. If these results are interpreted in terms of two-neutrino mixing, the following regions of oscillation parameters in the ($\sin^2 2\theta$, Δm^2) plane are suggested:

- several 'islands' with Δm^2 in the range $4 \times 10^{11} - 5 \times 10^{-10}$ eV^2 and large mixing angles (the 'vacuum oscillation' solutions to the solar neutrino problem);

- two islands with $\Delta m^2 \approx 10^{-7}$ eV^2 or 4×10^{-5} eV^2 and large mixing angles, or one with $\Delta m^2 \approx 6 \times 10^{-6}$ eV^2 and small mixing angle (the MSW solutions to the solar neutrino problem);

- an island with Δm^2 in the range $10^{-3} - 10^{-2}$ eV^2 and large mixing angle, dominated by $\nu_\mu - \nu_\tau$ or $\nu_\mu - \nu_s$ oscillation (the solution to the atmospheric neutrino problem);

- a narrow strip with Δm^2 between 0.2 and 2 eV^2 and $\sin^2 2\theta$ between 0.002 and 0.04, required to describe the $\bar{\nu}_\mu - \bar{\nu}_e$ oscillation signal claimed by LSND.

From this information one can draw the following conclusions:

- four neutrino states are needed to describe simultaneously the solar and atmospheric neutrino measurements and the LSND result, because with three neutrinos there are only two independent Δm^2 values. If the need for a fourth neutrino is confirmed, then this neutrino must be sterile (no coupling to W and Z bosons);

- unless neutrinos are degenerate in mass, Δm^2 is equal to the square of the mass of the heavier neutrinos to a very good approximation. In this case the heaviest neutrino has a mass of at most 1.4 eV. Hence neutrinos are not the main component of dark matter in the Universe.

By the middle of the next decade, more data from experiments presently running and results from experiments just beginning or in preparation will provide answers to several crucial questions. In particular, we expect to learn if the solar neutrino problem is due to oscillations and to know the oscillation parameters with much less uncertainties than the present ones.

We expect also to know rather precisely the oscillation parameters responsible for the atmospheric neutrino problem, either from further data on atmospheric neutrinos, or from long baseline experiments at accelerators (or from both). We should also know if the dominant oscillation is $\nu_\mu - \nu_\tau$ or $\nu_\mu - \nu_s$ from measurements of neutral-current interactions or from the observation of ν_τ appearance.

Finally, we shall definitely know if the $\bar{\nu}_\mu - \bar{\nu}_e$ oscillation signal observed by LSND is real.

In the longer term, if neutrino oscillations are confirmed, the elements of the mixing matrix need to be measured. This task requires neutrino beams at least two orders of magnitude more intense than present ones. The most promising idea to achieve this goal is based on high-energy muon storage rings with long straight sections pointing to a neutrino detector. The advantage of such 'neutrino factories' is that they provide beams of precisely known composition (50% ν_μ, 50% $\bar{\nu}_e$ from μ^- decay, or 50% $\bar{\nu}_\mu$, 50% ν_e from μ^+ decay), and also precisely calculable fluxes and energy spectra [68]. With such beams, backgrounds are very low for some appearance searches, such as $\nu_\mu - \nu_\tau$ oscillations detected by observing the $\tau^- \to e^-$ decay channel. In addition it will be possible to measure CP violation in the neutrino sector.

The construction of neutrino factories represents a challenge for accelerator technology.

Acknowledgements

I wish to express my sincere thanks to Faqir Khanna, Bruce Campbell and Manuella Vincter, organizers of the Lake Louise Winter Institute, for inviting me to give these lectures and for organizing a very successful and stimulating meeting. I am grateful to Audrey Schaapman and Lee Grimard, LLWI secretaries, for their help. Finally, last but not least, I thank Geneviève Prost for typing and editing these lectures.

References

1. Particle Data Group, Eur. Phys. J. **C3**, 122, (1998).
2. For a review see D.O Caldwell, Neutrino Dark Matter, hep-ph/9902219.
3. Particle Data Group, Eur. Phys. J. **C3**, 313, (1998).
4. V.M. Lobashev, Direct Search for the Neutrino Mass, in Proc. of the XVIII Physics in Collision Conference, Frascati (Italy), June 17-19, 1998.
5. J.F. Wilkerson, Nucl. Phys. (Proc. Suppl.) **B31**, 32, (1993).
6. K. Assamagan *et al.*, Phys. Rev. **D53**, 6065, (1996).
7. B. Jeckelmann, P.F.A. Goudsmit and H.J. Leisi, Phys. Lett. **B335**, 326, (1994).
8. For a review see F.J.M. Farley and E. Picasso, Ann. Rev. Nucl. Particle Science **29**, 243, (1979).
9. R. Barate *et al.*, Eur. Phys. J. **C2**, 395, (1998).
10. L. Baudis *et al.*, Limits on the Majorana neutrino mass in the 0.1 eV range, hep-ex/9902014.
11. For a review see M.K. Moe, Nucl. Phys. (Proc. Suppl.) **B38**, 36, (1995).
12. B. Pontecorvo, Sov. Phys. JETP **26**, 984, (1968).
13. Z. Maki, M. Nakagawa and S. Sakata, Prog. Theor. Phys. **28**, 870, (1962).
14. L. Wolfenstein, Phys. Rev. **D17**, 2369, (1978).
15. S.P. Mikheyev and A.Yu. Smirnov, Nuovo Cim. **9C**, 17, (1986).
16. J.N. Bahcall, Rev. Mod. Phys. **50**, 881, (1978); Rev. Mod. Phys. **59**, 505, (1987);
J.N. Bahcall and R.K. Ulrich, Rev. Mod. Phys. **60**, 297, (1988);
J.N. Bahcall and M. Pinsonneault, Rev. Mod. Phys. **67**, 781, (1995).
17. J.N. Bahcall, S. Basu and M. Pinsonneault, Phys. Lett. **B433**, 1 (1998);
J.N. Bahcall, P.I. Krastev and A.Yu. Smirnov, Phys. Rev. **D58**, 6016, (1998).
18. R. Davis, Prog. Part. Nucl. Phys. **32**, 13, (1994).
19. B.T. Cleveland *et al.*, Astrophys. J. **496**, 505, (1998).
20. S. Turck-Chièze and I. Lopez, Astrophys. J. **408**, 347, (1993).

21. W. Hampel *et al.*, Phys. Lett. **B447**, 127, (1999).
22. V.N. Gavrin *et al.*, Results from SAGE, Proc. XVIII Int. Conf. on Neutrino Physics and Astrophysics (Neutrino 98), Takayama, Japan, 4–9 June 1998 (to be published in Nucl. Phys. **B.** (Proc. Suppl.)).
23. M.B. Smy, Solar Neutrino Energy Spectrum with SuperKamiokande, presented at DPF99, Los Angeles, January 5-9, 1999.
24. J.N. Bahcall, M. Pinsonneault, S. Basu and J. Christensen-Dalsgaard, Phys. Rev. Lett. **78**, 171, (1997).
25. N. Hata and P. Langacker, Phys. Rev. **D49**, 3622, (1994).
26. A. Mac Donald, Status of the SNO Project, Proc. XVIII Int. Conf. on Neutrino Physics and Astrophysics (Neutrino 98), Takayama, Japan, 4-9 June 1998 (to be published in Nucl. Phys. **B.** (Proc. Suppl.)).
27. L. Oberauer, Status of Borexino, Proc. XVIII Int. Conf. on Neutrino Physics and Astrophysics (Neutrino 98), Takayama, Japan, 4-9 June 1998 (to be published in Nucl. Phys. **B.** (Proc. Suppl.)).
28. A. Suzuki, KAMLAND Project, Proc. XVIII Int. Conf. on Neutrino Physics and Astrophysics (Neutrino 98), Takayama, Japan, 4-9 June 1998 (to be published in Nucl. Phys. **B.** (Proc. Suppl.)).
29. G. Barr, T.K. Gaisser and T. Stanev, Phys. Rev. **D39**, 3532, (1989); V. Agrawal *et al.*, Phys. Rev. **D53**, 1314, (1996); M. Honda *et al.*, Phys. Lett. **B248**, 193, (1990); Phys. Rev. **D52**, 4985, (1995).
30. K.S. Hirata *et al.*, Phys. Lett. **B205**, 416, (1988); Phys. Lett. **B280**, 146, (1992).
31. D. Casper *et al.*, Phys. Rev. Lett. **66**, 2561, (1991); R. Becker-Szendy *et al.*, Phys. Rev. **D46**, 3720, (1992).
32. Y. Fukuda *et al.*, Phys. Lett. **B433**, 9, (1998); Phys. Lett. **B436**, 33, (1998); M. Messier, Atmospheric Neutrinos at SuperKamiokande, presented at DPF99, Los Angeles, January 5-9, 1999.
33. Ch. Berger *et al.*, Phys. Lett. **B227**, 489, (1989); Phys. Lett. **B245**, 305, (1990); K. Daum *et al.*, Z. Phys. **C66**, 417, (1995).
34. M. Aglietta *et al.*, Europhys. Lett. **8**, 611, (1989).
35. W.W.M. Allison *et al.*, Phys. Lett. **B391**, 491, (1997); E.A. Peterson, Contained Events in Soudan-2, Proc. XVIII Int. Conf. on Neutrino Physics and Astrophysics (Neutrino 98), Takayama, Japan, 4-9 June 1998 (to be published in Nucl. Phys. **B.** (Proc. Suppl.)).
36. S. Kasuga *et al.*, Phys. Lett. **B374**, 238, (1996).
37. Y. Fukuda *et al.*, Phys. Lett. **B335**, 237, (1994).

38. K. Scholberg, Atmospheric neutrinos in Super-Kamiokande, presented at the VIII Int. Workshop on Neutrino Telescopes, Venice, Italy, February 23-26, 1999.
39. Y. Fukuda *et al.*, Phys. Rev. Lett. **82**, 2644, (1999); Y Totsuka, Evidence for Neutrino Oscillation Observed in Super-Kamiokande, presented at the Nobel Symposium, Stockholm, 20-25 August 1998.
40. M. Spinetti, Atmospheric Neutrino Results from MACRO, presented at the VIII Int. Workshop on Neutrino Telescopes, Venice, Italy, February 23-26, 1999.
41. M. Apollonio *et al.*, Phys. Lett. **B240**, 397, (1998).
42. Y. Wang, Recent Results from Palo Verde and Future Prospects, presented at the XXXIV Rencontres de Moriond on Electroweak Interactions and Unified Theories, Les Arcs, France, 13-20 March 1999.
43. C. Athanassopoulos *et al.*, Nucl. Intrum. Methods Phys. Res. **A388** (1997).
44. B. Zeitnitz, Prog. Part. Nucl. Phys. **32**, 351, (1994).
45. G. Mills, Results from LSND, presented at the XXXIV Rencontres de Moriond on Electroweak Interactions and Unified Theories, Les Arcs, France, 13-20 March 1999.
46. T. Jannakos, Latest Results from the Search for $\bar{\nu}_\mu \to \bar{\nu}_e$ Oscillations with KARMEN2, presented at the XXXIV Rencontres de Moriond on Electroweak Interactions and Unified Theories, Les Arcs, France, 13-20 March 1999.
47. C. Athanassopoulos *et al.*, Phys. Rev. **C58**, 2489, (1998).
48. E. Church *et al.*, A proposal for an experiment to measure $\nu_\mu - \nu_e$ oscillations and ν_μ disappearance at the Fermilab Booster: BOONE, 7 December 1997 (unpublished).
49. T. Yanagida, Prog. Theor. Phys. **B135**, 66, (1978); M. Gell-Mann, P. Ramond and R. Slansky, in Supergravity, eds. P. van Nieuwenhuizen and D. Freedman (North Holland, Amsterdam, 1979) 315.
50. G. Collazuol *et al.*, Neutrino Beams: Production Models and Experimental Data, CERN-OPEN-98-032.
51. M.C. Gonzales-Garcia and J.J. Gomez-Cadenas, Phys. Rev. **D55**, 1297, (1997); B. Van de Vijver and P. Zucchelli, Nucl. Instr. and Methods **A385**, 91, (1997).
52. E. Eskut *et al.*, Nucl. Instr. Methods Phys. Res. **A401**, 7, (1997).
53. M. Messina, Latest results from the CHORUS Neutrino Oscillation Experiment, presented at the XXXIV Rencontres de Moriond on Elec-

troweak Interactions and Unified Theories, Les Arcs, France, 13-20 March 1999.

54. R.D. Cousins and V.L. Highland, Nucl. Instr. Methods **A320**, 331, (1992).

55. E531: N. Ushida *et al.*, Phys. Rev. Lett. **57**, 2897, (1986);
    CHARM-II: M. Gruwé *et al.*, Phys. Lett. **B309**, 463, (1993);
    CCFR: K.S McFarland *et al.*, Phys. Rev. Lett. **75**, 3993, (1995);
    CDHS: F. Dydak *et al.*, Phys. Lett. **B134**, 281, (1984).

56. R. Petti, private communication.

57. G.J. Feldman and R.D. Cousins, Phys. Rev. **D57**, 3873, (1998).

58. C. Albright and R. Shrock, Phys. Lett. **B84**, 123, (1979).

59. J. Altegoer *et al.*, Nucl. Instr. Methods **A404**, 96, (1998).

60. J. Altegoer*et al.*, Phys. Lett. **B431**, 219, (1998).

61. P. Astier *et al.*, Phys. Lett. **B453**, 169, (1999).

62. K. Nishikawa, Status of the K2K experiment, Proc. XVIII Int. Conf. on Neutrino Physics and Astrophysics (Neutrino 98), Takayama, Japan, 4-9 June 1998 (to be published in Nucl. Phys. B. (Proc. Suppl.)).

63. S. Wojcicki, Long Baseline Neutrino Oscillation Program in the U.S.A., Proc. XVIII Int. Conf. on Neutrino Physics and Astrophysics (Neutrino 98), Takayama, Japan, 4–9 June 1998 (to be published in Nucl. Phys. B. (Proc. Suppl.)).

64. A. Bettini *et al.*, Nucl. Instr. Methods **A332**, 395, (1993);
    J.P. Revol *et al.*, A search program for explicit neutrino oscillations at long and medium baselines with ICARUS detectors, ICARUS-TM-97/01 (5 March 1997).

65. G. Acquistapace *et al.*, The CERN neutrino Beam to Gran Sasso, CERN 98–02 (19 May 1998).

66. A. Ereditato *et al.*, Towards the optimization of the NGS neutrino beam for $\nu_\mu - \nu_\tau$ appearance experiments, ICARUS-TM/98-13, OPERA 980722-01 (August 10, 1998);
    J.L. Baldy *et al.*, The CERN Neutrino beam to Gran Sasso (NGS), CERN-SL/99-034 (May 1999).

67. K. Kodama *et al.*, The OPERA ν_τ appearance experiment in the CERN-Gran Sasso neutrino beam, CERN SPSC 98-25 (October 9, 1999).

68. S. Geer, Phys. Rev. **D57**, 6989, (1998).

PHYSICS BEYOND THE STANDARD MODEL

R.D. PECCEI

Department of Physics and Astronomy, UCLA, Los Angeles, CA 90095-1547

Abstract

These lectures describe why one believes there is physics beyond the Standard Model and review the expectations of three alternative explanations for the Fermi scale. After examining constraints and hints for beyond the Standard Model physics coming from experiment, I discuss, in turn, dynamical symmetry breaking, supersymmetry and extra compact dimension scenarios associated with the electroweak breakdown.

1 Introduction

The Standard Model (SM) [1], gives an excellent theoretical description of the strong and electroweak interactions. This theory, which is based on an $SU(3) \times SU(2) \times U(1)$ gauge group, has proven extraordinarily robust. As shown in Fig. 1, all data in the electroweak sector to date appears to be in perfect agreement with the SM predictions, and there are just a few (quite indirect) hints for physics beyond the SM. Nevertheless, there are theoretical aspects of the SM which suggest the need for new physics. In addition, there are certain open questions within the SM whose answers can only be found by invoking physics beyond the SM.

In the last year, the observation by the SuperKamiokande [3] collaboration of neutrino oscillations provided the first experimental indication that some new physics exists which causes a large splitting among the leptonic doublets. While in the quark sector $m_u/m_d \sim O(1)$, it appears that in the leptonic sector $m_{\nu_\ell}/m_\ell \lesssim 10^{-8}$. As we shall see, the most natural explanation for this phenomena is the existence of a new scale far above the electroweak scale.

In these Lectures I will try to explore some of the new physics scenarios which are motivated by theoretical considerations and try to confront and constrain them with what we know experimentally, both from the indirect hints coming from the electroweak sector as well as from the more direct hints coming from neutrino oscillations.

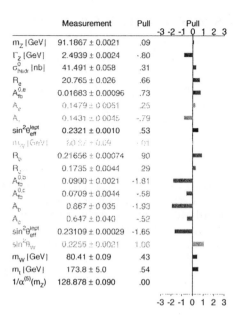

Figure 1: Standard Model fit, from Ref.2.

2 Theoretical Issues in the Standard Model

The Standard Model Lagrangian can be written as the sum of four pieces. Schematically, one has

$$\mathcal{L}_{\text{SM}} = -\sum_f \bar{\psi}_f \gamma^\mu \frac{1}{i} D_\mu \psi_f - \frac{1}{4} \sum_i F_i^{\mu\nu} F_{i\mu\nu} + \mathcal{L}_{\text{SB}} + \mathcal{L}_{\text{Yukawa}} . \tag{1}$$

The first two terms in the above Lagrangian contain the interactions of the fermions in the theory with the gauge fields and the self-interactions of the gauge fields. The precision electroweak measurements, as well as most QCD tests, essentially have checked these pieces of the SM. In fact, for the electroweak tests, all that the symmetry-breaking piece, $\mathcal{L}_{\text{SB}}$ and the Yukawa piece, $\mathcal{L}_{\text{Yukawa}}$, provide are a **renormalizable cut-off** M_H and a **large** fermion mass m_t, respectively. Of course, $\mathcal{L}_{\text{SB}}$ also allows for the spontaneous generation of mass for the W and Z bosons. The masses of these excitations are given by the formulas

$$M_W^2 = \frac{1}{4} g^2 v_F^2 \; ; \quad M_Z^2 = \frac{1}{4}(g^2 + g'^2) v_F^2 \tag{2}$$

which involve the $SU(2)[g \equiv g_2]$, and $U(1)[g']$, coupling constants as well as a mass scale, v_F, arising from $\mathcal{L}_{SB}$. This scale—the Fermi scale—is related to the Fermi constant G_F and sets the scale of the electroweak interactions:

$$v_F = (\sqrt{2}G_F)^{-1/2} \simeq 250 \text{ GeV} . \tag{3}$$

It is important to note that neither the pure gauge field piece of $\mathcal{L}_{SM}$ nor the fermion-gauge piece of this Lagrangian contain explicit mass terms. A mass term for the gauge fields

$$\mathcal{L}_{\text{mass}}^{\text{Gauge}} = -\frac{1}{2}\sum_i m_i^2 A_i^\mu A_{\mu i} \tag{4}$$

is forbidden explicitly by the local $SU(3) \times SU(2) \times U(1)$ symmetry. However, masses for the gauge fields [cf. Eq. 2] can arise after the symmetry breakdown $SU(2) \times U(1) \to U(1)_{\text{em}}$. Similarly, fermion mass terms of the form

$$\mathcal{L}_{\text{mass}}^{\text{Fermions}} = -\sum_f m_f(\bar{\psi}_{fL}\psi_{fR} + \bar{\psi}_{fR}\psi_{fL}) \tag{5}$$

are forbidden by the $SU(2) \times U(1)$ assignments of the fermions. This follows since all left-handed fermions are part of $SU(2)$ doublets, while all right-handed fermions are $SU(2)$ singlets.

Because of these circumstances, mass generation in the Standard Model is intimately connected to the spontaneous breakdown of the electroweak $SU(2) \times U(1)$ symmetry. As a result, not only are the gauge boson masses M_W and M_Z proportional to the Fermi scale v_F, but so are the masses for all the fermions as well as the mass of the Higgs boson H

$$m_f \sim v_F ; \quad M_H \sim v_F . \tag{6}$$

The difference between Eq. 2 and Eq. 6 is that, in the latter case, the proportionality constants are **not** known. Or, better said, they are related to phenomena we have not yet seen.

There are, however, two important exceptions to the pattern given by Eq. 6. First of all, the masses of hadrons are not simply related to the masses of quarks. Thus they depend on another scale besides v_F. This scale, Λ_{QCD}, is a **dynamical scale** whose magnitude can be inferred from the running of the $SU(3)$ coupling constant. A convenient definition is to take Λ_{QCD} to be the scale where $\alpha_3(q^2) = g_3^2(q^2)/4\pi$ becomes of $O(1)^a$. Then Λ_{QCD} serves to set

---

[a] There is no equivalent dynamical scale for the weak $SU(2)$ group since its coupling becomes strong at scales much below the scale v_F, where the $SU(2)$ group breaks down.

the mass scale of the light hadrons which receive the bulk of their mass from QCD dynamical effects[b]. Hadrons containing heavy quarks, on the other hand, get most of their mass from the mass of the heavy quark. Thus, less of their mass depends on QCD dynamics and Λ_{QCD}.

The second exception to Eq. 6 is provided by neutrinos. It is clear that for any particle carrying electromagnetic charge the only allowed mass term must involve particles and antiparticles, as detailed in Eq. 5. Lorentz invariance, however, allows one to write down mass terms involving two particle fields, or two antiparticle fields. Such mass terms, called Majorana mass terms, are allowed for neutrinos. In particular, since the right-handed neutrinos have no $SU(2) \times U(1)$ quantum numbers, one can write down an $SU(2) \times U(1)$ invariant mass term for these states of the form

$$\mathcal{L}_{\text{mass}}^{\text{Majorana}} = -\frac{1}{2}(\nu_R^T M_R \tilde{C} \nu_R + \bar{\nu}_R \tilde{c} M_R^\dagger \bar{\nu}_R^T) \ . \tag{7}$$

Here $\tilde{C}$ is a charge conjugation matrix needed for Lorentz invariance [5]. The right-handed neutrino mass matrix M_R contains mass scales which are totally **independent** from v_F. We will return to this point later on in these lectures.

Ignoring these more detailed questions, one of the principal issues which remains open in the Standard Model is the nature of the Fermi scale v_F. The role of symmetry breakdown as a generator of mass scales is familiar in superconductivity. In that case, the formation of an electron number violating Cooper pair $\langle e^\uparrow e^\downarrow \rangle$ [6] sets up a mass gap between the normal and the superconducting ground states. The Fermi scale v_F plays an analogous role in the electroweak theory. It is the scale of the order parameter which is responsible for the breakdown of $SU(2) \times U(1)$ down to $U(1)_{\text{em}}$.

Although the size of v_F ($v_F \sim 250$ GeV) is known, its precise origin is yet unclear. Two possibilities have been suggested for the origin of v_F:

i) The Fermi scale is associated with the vacuum expectation value (VEV) of some **elementary scalar** field, or fields $\langle \Phi_i \rangle$.

ii) The Fermi scale is connected with the formation of some **dynamical condensates** of fermions of some underlying deeper theory, $\langle \bar{F}F \rangle$.

Roughly speaking, the above two alternatives correspond to having $\mathcal{L}_{SB}$ being described either by a weakly coupled theory or by a strongly coupled theory.

The nature and origin of the Fermi scale, of course, is not the only unanswered theoretical question in the SM. Equally mysterious is the physics which

---

[b]The mass squared of the pseudoscalar octet is an interesting exception. Since these states are quasi-Nambu Goldstone bosons their mass squared is proportional to the light quark masses. In fact, one has $m_\pi^2 \sim v_F \Lambda_{QCD}^4$.

gives rise to $\mathcal{L}_{\text{Yukawa}}$—the piece of the SM Lagrangian which is responsible for the masses of, and mixing among, the elementary fermions in the theory. In contrast to v_F, however, here one does not have directly a scale to associate with this Lagrangian. It could well be that the **flavor problem**—the origin of the fermion masses and of fermion mixing—is the result of physics operating at scales which are much larger than v_F. Indeed, as we will see, trying to generate $\mathcal{L}_{\text{Yukawa}}$ itself from physics at a scale of order v_F is fraught with difficulties.

In view of the above, I will concentrate for now on only on the symmetry breaking piece of the SM Lagrangian. In this context, it proves useful to begin by examining the simplest example of $\mathcal{L}_{\text{SB}}$ in which this Lagrangian involves just one complex doublet Higgs field: $\Phi = \begin{pmatrix} \phi^o \\ \phi^- \end{pmatrix}$:[c]

$$\mathcal{L}_{\text{SB}} = -(D_\mu \Phi)^\dagger (D^\mu \Phi) - \lambda \left[\Phi^\dagger \Phi - \frac{1}{2} v_F^2 \right]^2 . \tag{8}$$

In the above λ is an, arbitrary, coupling constant which, however, must be positive to guarantee a positive definite Hamiltonian.

The Fermi scale v_F enters directly as a scale parameter in the Higgs potential

$$V = \lambda \left[\Phi^\dagger \Phi - \frac{1}{2} v_F^2 \right]^2 . \tag{9}$$

The sign in front of the v_F^2 term is chosen appropriately to guarantee that V will be asymmetric, with a minimum at a non-zero value for $\Phi^\dagger \Phi$. This fact is what triggers the breakdown of $SU(2) \times U(1)$ to $U(1)_{\text{em}}$, since it forces the field Φ to develop a non-zero VEV[d].

$$\langle \Phi \rangle = \frac{1}{\sqrt{2}} \begin{pmatrix} v_F \\ 0 \end{pmatrix} . \tag{10}$$

Because v_F is an internal scale in the potential V, in isolation, it clearly makes no sense to ask what physics fixes the scale of v_F to be any given particular number. This question, however, can be asked if one considers the SM in a larger context. For instance, one can imagine that the SM is an effective theory valid up to some very high cut-off scale Λ, where new physics comes in. An obvious candidate for Λ is the Planck scale $M_P \sim 10^{19}$ GeV, the scale associated with gravity, embodied in Newton's constant $G_N = \frac{1}{M_P^2}$. In

---

[c]Often, one associates the nomenclature Standard Model to the electroweak theory in which $\mathcal{L}_{\text{SB}}$ is precisely given by this simplest option.
[d]With only one Higgs doublet one can always choose $U(1)_{\text{em}}$ as the surviving $U(1)$ in the breakdown. So the choice $\langle \phi^o \rangle \neq 0$; $\langle \phi^- \rangle = 0$ is automatic.

this broader context then it makes sense to ask what is the relation of v_F to the cut-off Λ. In fact, because the $\lambda\Phi^4$ theory is trivial[7], with the only consistent theory being one where $\lambda_{\text{ren}} \to 0$, considering the scalar interactions in $\mathcal{L}_{\text{SB}}$ without some high energy cut-off is not sensible. Let me explain.

One can readily compute the evolution of the coupling constant λ as a function of q^2. One finds that $\lambda(q^2)$ evolves in an **opposite** way to the way in which the QCD coupling constant $\alpha_3(q^2)$ evolves, growing as q^2 gets larger. This can be seen immediately from the Renormalization Group equation (RGE) for λ

$$\frac{d\lambda}{d\ln q^2} = +\frac{3}{4\pi^2}\lambda^2 + \ldots \tag{11}$$

This equation, in contrast to the QCD case, has a positive rather than a negative sign in front of its first term. As a result, if one solves the above RGE, including only this first term, one finds a singularity at large q^2 which is a reflection of this growth

$$\lambda(q^2) = \frac{\lambda(\Lambda_o^2)}{1 - \frac{3\lambda(\Lambda_o^2)}{4\pi}\ln\frac{q^2}{\Lambda_o^2}} \, . \tag{12}$$

This singularity is known as the Landau pole, since Landau was the first to notice this anomalous kind of behaviour[8].

One cannot really trust the location of the Landau pole derived from Eq. 12

$$\Lambda_c^2 = \Lambda_o^2 \exp\left[\frac{4\pi^2}{3\lambda(\Lambda_o^2)}\right] \, , \tag{13}$$

since Eq. 12 stops being valid when λ gets too large. When this happens, of course, one should not have neglected the higher order terms in Eq. 11. Nevertheless, once the cut-off Λ_c is fixed, one can predict $\lambda(q^2)$ for scales q^2 sufficiently below the cut-off. Indeed, the $\lambda\Phi^4$ theory is perfectly sensible as long as one restricts oneself to $q^2 \ll \Lambda_c^2$. If one wants to push the cut-off to infinity, however, one sees from (13) that $\lambda(\Lambda_o^2) \to 0$. This is the statement of triviality[7], within this simplified context.

In the case of the SM, one can "measure" where the cut-off Λ_c is in $\mathcal{L}_{\text{SB}}$ from the value of the Higgs mass. Using the potential (9) one finds that

$$M_H^2 = 2\lambda(M_H^2)v_F^2 \, . \tag{14}$$

Obviously, as long as the Higgs mass is light with respect to v_F the coupling λ is small and the cut-off is far away. Indeed, using Eqs. 13 and 14 one finds that even if $M_H \sim 200$ GeV, then the cut-off is very large still, of order of the

Planck mass $\Lambda_c \sim M_P$! So, as long as M_H is that light, or lighter, the effective theory described by $\mathcal{L}_{\text{SB}}$ is very reliable, and **weakly coupled**, with $\lambda \leq 0.3$. In these circumstances it is meaningful to ask the question whether the large hierarchy

$$v_F \ll \Lambda_c \tag{15}$$

is a stable condition. This question, following 't Hooft [9], is often called the problem of **naturalness**.

If, on the other hand, the Higgs mass is heavy, of order of the cut-off ($M_H \sim \Lambda_c$), then it is pretty clear that $\mathcal{L}_{\text{SB}}$ as an effective theory stops making sense. The coupling λ is so strong that one cannot separate the particle-like excitations from the cut-off itself. Numerical investigations on the lattice [10] have indicated that this occurs when

$$M_H \sim \Lambda_c \sim 700 \text{ GeV} . \tag{16}$$

In this case, it is clear that $\langle \Phi \rangle$, as the order parameter of the symmetry breakdown, must be replaced by something else.

Before discussing this latter point, let me first return to the light Higgs case. Here one must worry about the naturalness of having the Fermi scale v_F be so much smaller that the Planck mass M_P, which is clearly a physical cut-off. It turns out, in general, that the hierarchy $v_F \ll M_P$ is **not** stable. This is easy to see since radiative effects in a theory with a cut-off destabilize any pre-existing hierarchy. Indeed, this was 't Hooft's original argument [9]. Quantities that are **not protected** by symmetries suffer quadratic mass shifts. This is the case for the Higgs mass. This mass, schematically, shifts from the value given in Eq. 14 to

$$M_H^2 = 2\lambda v_F^2 + \alpha \Lambda_c^2 . \tag{17}$$

It follows from Eq. 17 that if $\Lambda_c \sim M_P \gg v_F$, the Higgs bosons cannot remain light. Or, saying it another way, if one wants the Higgs to remain light, one needs an enormous amount of fine tuning of parameters to guarantee that, in the end, it remains a light excitation. This kind of fine-tuning is really unacceptable, so one is invited to look for some protective symmetry to guarantee that the hierarchy $v_F \ll M_P$ is stable.[e]

Such a protective symmetry exists—it is supersymmetry (SUSY)[11]. SUSY is a boson-fermion symmetry in which bosonic degrees of freedom are paired with fermionic degrees of freedom. If supersymmetry is exact then the masses of the fermions and of their bosonic partners are the same. In a supersymmetric version of the Standard Model all quadratic divergences cancel. Thus

---

[e]Note that a stable hierarchy $v_F \ll M_P$ does not explain why one has such a hierarchy to begin with. This is a much harder question to answer.

parameters like the Higgs boson mass will not be sensitive to a high energy cut-off. Roughly speaking, via supersymmetry, the Higgs boson mass is kept light naturally since its fermionic partner has a mass which is protected by a chiral symmetry and is of $O(v_F)$.

Because one has not seen any of the SUSY partners of the states in the SM yet, it is clear that if a supersymmetric extension of the SM exists then the associated supersymmetry must be broken. Remarkably, even if SUSY is broken the naturalness problem in the SM is resolved, provided that the splitting between the fermion-boson SUSY partners is itself of $O(v_F)$. For example, the quadratic divergence of the Higgs mass due to a W-loop is moderated into only a logarithmic divergence by the presence of a loop of Winos, the spin-1/2 SUSY partners of the W bosons. Schematically, in the SUSY case, Eq. 17 gets replaced by

$$M_H^2 = 2\lambda v_F^2 + \alpha(\tilde{M}_W^2 - M_W^2) \ln \Lambda_c/v_F \ . \tag{18}$$

So, as long as the masses of the SUSY partners (denoted by a tilde) are themselves not split away by much more than v_F, radiative corrections will not destabilize the hierarchy $v_F \ll \Lambda_c$.

Let me recapitulate. Theoretical considerations regarding the nature of the Fermi scale have brought us to consider two alternatives for new physics associated with the $SU(2) \times U(1) \to U(1)_{\rm em}$ breakdown and $\mathcal{L}_{\rm SB}$:

i) $\mathcal{L}_{\rm SB}$ is the Lagrangian of some elementary scalar fields interacting together via an asymmetric potential, whose minimum is set by the Fermi scale v_F. The presence of non-vanishing VEVs triggers the electroweak breakdown. However, to guarantee the naturalness of the hierarchy $v_F \ll M_P$, both $\mathcal{L}_{\rm SB}$ and the whole Standard Model Lagrangian must be augmented by other fields and interactions so as to enable the theory (at least approximately) to be supersymmetric. Obviously, if this alternative is true, there is plenty of new physics to be discovered, since all particles have superpartners of mass $\tilde{m} \simeq m + O(v_F)$.

ii) The symmetry breaking sector of the SM has itself a dynamical cut-off of $O(v_F)$. In this case, it makes no sense to describe $\mathcal{L}_{\rm SB}$ in terms of strongly coupled scalar fields. Rather, $\mathcal{L}_{\rm SB}$ describes a dynamical theory of some new strongly interacting fermions F, whose condensates cause the $SU(2) \times U(1) \to U(1)_{\rm em}$ breakdown. The strong interactions which form the condensates $\langle \bar{F}F \rangle \sim v_F^3$ also identify the Fermi scale as the dynamical scale of the underlying theory, very much analogous to $\Lambda_{\rm QCD}$. If this alternative turns out to be true, then one expects in the future to see lots of new physics, connected with these new strong interactions, when one probes them at energies of $O(v_F)$.

In the past year, a third very speculative alternative has been suggested besides the two possibilities above[12]. This alternative is based on the idea that, perhaps, in nature there could exist some extra "largish" compact dimensions of size R. [13] In such theories, the fundamental scale of gravity in $(d + 4)$-dimensions could well be quite different than the Planck scale. In particular, it may well be that

$$[M_P]_{d+4} \sim v_F \ . \tag{19}$$

That is, at short distances $(r < R)$ the scale of gravity could well be different than the Planck scale, the usual scale of 4-dimensional gravity valid at large distances $(r > R)$. Indeed, this short-distance scale could be identical to the Fermi scale. The relationship between these two gravity scales depends both on R and on the number of extra dimensions d: [12]

$$M_P = ([M_P]_{d+4})^{\frac{d+2}{2}} R^{\frac{d}{2}} \sim v_F [v_F R]^{\frac{d}{2}} \tag{20}$$

where the second (approximate) equality holds if Eq. 19 holds.

Obviously, if Eq. 19 were to be true, then there is no naturalness issue—the Fermi scale is the scale of gravity in the "true" extra-dimensional theory! From Eq. 20 it follows that, if $d = 2$, then the scale of the compact dimensions needed is quite large $R \sim 10^{-1}$ cm! On the other hand, if $d = 6$, as string theory suggests, then $R \sim 10^{-13}$ cm. These distances are small enough that perhaps one would not have noticed the modifications implied for the gravitational potential at distances $r < R$.

Although these theories do not suffer from any naturalness problem, and thus are perfectly consistent with a single weakly-coupled Higgs field, they do predict the existence of other phenomena beyond the SM. In particular, if this alternative is correct, one would expect copious production of gravitons at energies of order $\sqrt{s} \sim v_F$, as one begins to excite the compact dimensions. Thus, also here there is spectacular new physics to find!

In these Lectures, I will try to illustrate some of the consequences of all the three alternatives for v_F alluded to above. In addition, to try to divine which of the above ideas is most likely to be correct, I want to explore in some depth some of the points which come from experiment suggesting possible traces of physics beyond the Standard Model. In the next section, I will try to describe in more detail what these hints of physics beyond the Standard Model are and what is their likely origin.

3 Constraints and Hints for Beyond the Standard Model Physics

There are four different experimental inputs which help shed some light on possible physics beyond the Standard Model. I will discuss them in turn.

3.1 Implications of Standard Model Fits

One of the strongest constraints on physics beyond the SM is that the SM gives an excellent fit to the data, as we already illustrated in Fig. 1. In practice, since all fermions but the top are quite light compared to the scale of the W and Z-bosons, all quantities in the SM are specified as functions of 5 parameters: g; g'; v_F; M_H; and m_t. It proves convenient to trade the first three of these for another triplet of quantities in the SM which are better measured: α; M_Z; and G_F. This trade-off has become the common practice in the field. Once one has adopted a set of **standard parameters** then all physical measurable quantities can be expressed as a function of this "standard set". For example, the W-mass in the SM is given as a function of these parameters as:

$$M_W|_{\text{SM}} = M_W(\alpha; M_Z; G_F; m_t; M_H) \ . \tag{21}$$

Because α, M_Z, and G_F, as well as $m_t{}^f$ are rather accurately known, all SM fits essentially serve to constrain only **one** unknown–the Higgs mass M_H. This constraint, however, is not particularly strong because all radiative effects depends on M_H only logarithmically. That is, radiative corrections give contributions of $O\left(\frac{\alpha}{\pi} \ln M_H/M_Z\right)$.

The result of the SM fit of all precision data gives for the Higgs mass the result:[15]

$$M_H = \left(76^{+85}_{-47}\right) \text{ GeV} \tag{22}$$

or

$$M_H < 262 \text{ GeV} \quad (95\% \text{ C.L.}) \ . \tag{23}$$

These results for the Higgs mass are compatible with limits on M_H coming from direct searches for the Higgs boson in the process $e^+e^- \to ZH$ at LEP 200. The limit presented at the 1998 ICHEP in Vancouver was[16]

$$M_H > 89.8 \text{ GeV} \quad (95\% \text{ C.L.}) \ . \tag{24}$$

However, preliminary results presented at the 1999 Winter Conferences have raised this bound to the neighborhood of 95 GeV.

It is particularly gratifying that the SM fits indicate the need for a light Higgs boson, since this "solution" is what is internally consistent. Let me illustrate how this emerges, for example, from studies of the Z-leptonic vertex. The axial coupling of the Z

$$\mathcal{L}_{\text{eff}} = \frac{eZ^\mu}{2\cos\theta_W \sin\theta_W} \bar{e}[\gamma_\mu g_V - g_A \gamma_\mu \gamma_5]e \tag{25}$$

---

fThe top mass is quite accurately determined now. The combined value obtained by the CDF and DO collaborations fixes m_t to better than 3%: $m_t = (174.8 \pm 5.0)$ GeV.[14]

gets modified by radiative corrections to

$$g_A^2 = \frac{1}{4}[1 + \Delta\rho] . \tag{26}$$

The shift in the ρ-parameter, $\Delta\rho$, gets its principal contribution from m_t. However, it has also a (weak) dependence on M_H.[17]

$$\Delta\rho|_{\text{Higgs}} = -\frac{3 G_F M_W^2}{4\pi^2 \sqrt{2}} \tan^2 \theta_W \ln \frac{M_H}{M_Z} + \ldots \simeq -10^{-3} \ln \frac{M_H}{M_Z} . \tag{27}$$

The SM fit gives [15]

$$\Delta\rho = (3.7 \pm 1.1) \times 10^{-3} , \tag{28}$$

with the Higgs contribution giving, for $M_H = 300$ GeV, $\Delta\rho = -1 \times 10^{-3}$. Obviously, if M_H were to be very large, the Higgs contribution could have even changed the sign of $\Delta\rho$. The value emerging from the SM fit instead is perfectly compatible with having a rather light Higgs mass. In fact, one nice way to summarize the result of the SM fit is that, approximately, this fit constrains

$$\ln \frac{M_H}{M_Z} \lesssim 1 . \tag{29}$$

That is, there are **no** large logarithms associated with the symmetry breaking sector.

I should remark that a good SM fit does not necessarily exclude possible extensions of the SM involving either new particles or new interactions, provided that these new particles and/or interactions give only small effects. Typically, the effects of new physics are small if the excitations associated with this new physics have mass scales several times the W-mass.

One can quantify the above discussion in a more precise way by introducing a general parametrization for the vacuum polarization tensors of the gauge bosons and the $Z b \bar{b}$ vertex. These are the places where the dominant electroweak radiative corrections occur and therefore are the quantities which are probably the most sensitive new physics [18]. I do not want to enter into a full discussion of this procedure here since it is already well explained in the literature [18]. However, I want to talk about one example, connected to modifications of the gauge fields vacuum polarization tensors, because precision electroweak data serves to provide a strong constraint on dynamical symmetry breaking theories-excluding theories which are QCD-like.

There are four distinct vacuum polarization contributions $\Sigma_{AB}(q^2)$, where the pairs $AB = \{ZZ, WW, \gamma\gamma, \gamma Z\}$. For sufficiently low values of the momentum transfer q^2 ($q^2 \simeq M_W^2$) it obviously suffices to expand $\Sigma_{AB}(q^2)$ only up to $O(q^2)$. Thus, approximately, one needs to consider 8 different parameters associated with these contributions:

$$\Sigma_{AB}(q^2) = \Sigma_{AB}(0) + q^2 \Sigma'_{AB}(0) + \dots , \tag{30}$$

with the remaining corrections being terms of $O\left(\frac{q^4}{\Lambda^2}\right)$ with Λ being the scale of the new physics. In fact, there are not really 8 independent parameters since electromagnetic gauge invariance requires that

$$\Sigma_{\gamma\gamma}(0) = \Sigma_{\gamma Z}(0) = 0 . \tag{31}$$

Of the 6 remaining parameters one can fix 3 combinations of coefficients in terms of G_F, α and M_Z. Hence, in a most general analysis, the gauge field vacuum polarization tensors (for $q^2 \lesssim M_W^2$) only involve 3 arbitrary parameters. The usual choice [18], is to have one of these contain the main quadratic m_t-dependence, leaving the other two essentially independent of m_t.

I will proceed with my discussion in terms of the parametrization of Altarelli and Barbieri [18], where these three parameters are chosen to be

$$\epsilon_1 = \Delta\rho = \left[\frac{\Sigma_{ZZ}(0)}{M_Z^2} - \frac{\Sigma_{WW}(0)}{M_W^2}\right] \overset{\text{SM}}{=} \frac{3G_F m_t^2}{8\pi^2\sqrt{2}} - \frac{3G_F M_W^2}{4\pi^2\sqrt{2}} \tan^2\theta_W \ln\frac{M_H}{M_Z} \tag{32}$$

$$\epsilon_2 = \left[\Sigma'_{11}(0) - \Sigma'_{33}(0)\right] \overset{\text{SM}}{=} -\frac{G_F M_W^2}{2\pi^2\sqrt{2}} \ln\frac{m_t}{M_Z} + \dots \tag{33}$$

$$\epsilon_3 = \left[\Sigma'_{3\gamma}(0) - \Sigma'_{3Z}(0)\right] \overset{\text{SM}}{=} \frac{G_F M_W^2}{12\pi^2\sqrt{2}} \ln\frac{M_H}{M_Z} - \frac{G_F M_W^2}{6\pi^2\sqrt{2}} \ln\frac{m_t}{M_Z} + \dots \tag{34}$$

Here, as usual, $Z = 3 - \sin^2\theta_W\gamma$ and $W^\pm = \frac{1}{\sqrt{2}}[1 \mp i2]$. In the above, I have displayed also the leading dependence on m_t and M_H of the ϵ_i in the SM.

The interesting parameter here is ϵ_3, whose experimental value turns out to be [15]

$$\epsilon_3 = (3.9 \pm 1.1) \times 10^{-3} . \tag{35}$$

One can estimate ϵ_3 in a dynamical symmetry breaking theory, if one assumes that the spectrum of such a theory, and its dynamics, is QCD-like [19]. From its definition, one sees that ϵ_3 involves the difference between the spectral functions of vector and axial vector currents

$$\epsilon_3 = \frac{1}{2} \left[\Sigma'_{VV}(0) - \Sigma'_{AA}(0)\right] . \tag{36}$$

This difference has two components in a dynamical symmetry breaking theory. There is a contribution from a heavy Higgs boson ($M_H \sim$ TeV) characteristic of such theories, plus a term detailing the differences between the vector and axial vector spectral functions. This second component reflects the resonances with these quantum numbers in the spectrum of the underlying theory which gives rise to the symmetry breakdown. The first piece is readily estimated from the SM expression, using $M_H \sim$ TeV. The second piece, in a QCD-like theory, can be deduced by analogy to QCD, modulo some counting factors associated with the type of underlying theory one is considering. One finds [19]

$$
\begin{aligned}
\epsilon_3 &= \epsilon_3|_{M_H \simeq 1 \text{ TeV}} + \frac{\alpha}{12\pi \sin^2 \theta_W} \int_0^\infty \frac{ds}{s} [R_V(s) - R_A(s)] \\
&= 6.65 \times 10^{-3} + N_D \left(\frac{N_{TC}}{4}\right) \left[\frac{2\alpha\pi}{\sin^2 \theta_W} \frac{f_\pi^2}{m_\rho^2}\right] \\
&= \left[6.65 \pm 3.4 N_D \left(\frac{N_{TC}}{4}\right)\right] \times 10^{-3} .
\end{aligned}
\tag{37}
$$

The second line above follows if the underlying theory is QCD-like, so that the resonance spectrum is saturated by ρ-like and A_1-like, resonances. Here N_D is the number of doublets entering in the underlying theory and N_{TC} is the number of "Technicolors" in this theory[g]. Using Eq. 34 and using $N_{TC} = 4$, as is usually assumed, one sees that

$$
\epsilon_3 = \begin{cases} 10.05 \times 10^{-3} & N_D = 1 \\ 20.25 \times 10^{-3} & N_D = 4 \end{cases}
\tag{38}
$$

These values for ϵ_3 are, respectively, 5.5σ and 15σ away from the best fit value of ϵ_3, obtained from fitting all the high precision electroweak data. Obviously, one cannot countenance anymore a dynamical symmetry breaking theory which is QCD-like!

Provided the superpartners are not too light, nothing as disastrous occurs instead if one considers a supersymmetric extension of the SM. Fig. 2, taken from a recent analysis of Altarelli, Barbieri, and Caravaglios,[20] shows a typical fit, scanning over a range of parameters in the MSSM—the minimal supersymmetric extension of the SM. Although the MSSM can improve the χ^2 of the fit over that for the SM (which is already very good!), these improvements are small. In effect, the MSSM radiative corrections fits are all slightly better than the SM fits. This is not surprising, since these latter fits contain more parameters. Interestingly, however, these fits do not provide better bounds on sparticles than the bounds obtained by direct searches. Of course, for certain cases there are constraints. For instance, there cannot be too large a stop-

---

[g] For QCD, of course, $N_D = 1$ and $N_{TC} = 3$.

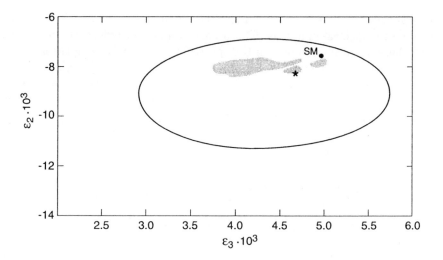

Figure 2: Comparison of SM and MSSM fits in the $\epsilon_2 - \epsilon_3$ plane, from Ref. 20. The ellipse is the 1-σ range determined by the data. The shaded region is the result of a scan over a range of SUSY parameters, with the star marking the lowest χ^2 point.

sbottom splitting because such a splitting would give too large a value for $\epsilon_1 = \Delta\rho$.

3.2 Hints of Unification

Although the SM coupling constants are very different at energies of order of the Fermi scale,[h] these couplings can become comparable at very high energies because they evolve differently with q^2. [21] Indeed, it is quite possible that the SM couplings unify into a single coupling at high energy, reflecting an underlying Grand Unified Theory (GUT) which breaks down to the SM at a high scale. If G denotes the GUT group, then one can imagine the sequence of spontaneous breakings

$$G \xrightarrow{M_X} SU(3) \times SU(2) \times U(1) \xrightarrow{v_F} SU(3) \times U(1)_{\text{em}} , \qquad (39)$$

with $M_X \gg v_F$.

To test this assumption one can compute the evolution of the SM coupling constants using the Renormalization Group Equations (RGE) and see if, indeed, these coupling constants unify. To leading order, the evolution of

---

[h] One has, for instance, $\alpha_3(M_Z^2) \simeq 0.12$, while $\alpha_2(M_Z^2) = \alpha(M_Z^2)/\sin^2\theta_W \simeq 0.034$.

each coupling constant can be evaluated separately from the others, since they decouple from each other:

$$\frac{d\alpha_i(\mu^2)}{d\ln\mu^2} = -\frac{b_i}{4\pi}\alpha_i^2(\mu^2) . \tag{40}$$

These equations imply a logarithmic change for the inverse couplings

$$\alpha_i^{-1}(q^2) = \alpha_i^{-1}(M^2) + \frac{b_i}{4\pi}\ln\frac{q^2}{M^2} . \tag{41}$$

The rate of change of the coupling constants with energy is governed by the coefficients b_i which enter in the RGE. In turn, these coefficients depend on the **matter content** of the theory—which matter states are "active" at the scale one is probing. In general, one has [22]

$$b_i = \sum_{\text{states}} \left\{ \frac{11}{6}\ell_i^{\text{vector}} - \frac{1}{3}\ell_i^{\text{chiral fermion}} - \frac{1}{12}\ell_i^{\text{real scalar}} \right\} \tag{42}$$

with the ℓ_i being group theoretic factors. For $SU(N)$ groups [22] $[\ell_i]_{\text{Adjoint}} = 2N$, while for fields transforming according to the fundamental representation, $[\ell_i]_{\text{Fundamental}} = 1$. For $U(1)$ groups, $\ell_i = 2Q^2$, where Q is a "property normalized" charge. That is, a charge which allows the possibility of unifying the $U(1)$ groups with the other non-Abelian groups. Let me explain this last point further.

For non-Abelian groups, the generators in the fundamental representation are conventionally normalized [22] so that

$$\text{Tr } t_a t_b = \frac{1}{2}\delta_{ab} . \tag{43}$$

For Abelian groups one can always rescale the charge. Thus no similar convention exists for this case. For example, for the electroweak group, instead of the usual hypercharge Y, one can define a new charge Q related to Y by a constant:

$$Y = \xi Q . \tag{44}$$

Obviously, the conventional hypercharge coupling can be turned into a Q-coupling, by rescaling the $U(1)$ coupling constant:

$$g'Y = g'\xi Q = g_1 Q . \tag{45}$$

For unification, one wants a $U(1)$ charge that is normalized in the same way as the non-Abelian generators, when one sweeps over all quarks and leptons

$$\sum_{q+\ell} \text{Tr } t_a^2 = \sum_{q+\ell} \text{Tr } Q^2 \equiv \frac{1}{\xi^2}\sum_{q+\ell} \text{Tr } Y^2 . \tag{46}$$

Table 1: Coefficients b_i in the SM and in the SUSY SM

Coefficient	SM	SUSY SM
b_1	$-41/10$	$-33/5$
b_2	$+19/6$	-1
b_3	$+7$	$+3$

Because $\sum\limits_{q+\ell} \mathrm{Tr}\ t_a^2 = 2$, while $\sum\limits_{q+\ell} \mathrm{Tr}\ Y^2 = 10/3$ it follows that $\xi = \sqrt{5/3}$. So,

$$Q = \sqrt{\frac{3}{5}} Y \ ; \quad g_1 = \sqrt{\frac{5}{3}} g' \ . \tag{47}$$

Using Eq. 42, it is straightforward to compute the coefficients b_i in the SM. For example, for the QCD coupling, one finds

$$b_3 = \left\{ \frac{11}{6} \cdot 6 - \frac{1}{3} \cdot 12 \right\} = 7 \ , \tag{48}$$

where the first factor above is the contribution of the gluonic degree of freedom and the second factor above comes from the 6 species of left-handed quarks plus the 6 species of right-handed quarks. Obviously, if there were to be supersymmetric matter, all the coefficients b_i would be modified at scales at or above where this matter starts to be produced. For example, in the supersymmetric QCD case, the gluons are now accompanied by spin 1/2 gluinos (which are chiral fermions) and each of the quarks of a given helicity has two real spin zero squark partners. For SUSY QCD then, the coefficient b_3 becomes

$$b_3 = \left\{ \frac{11}{6} \cdot 6 - \frac{1}{3} \cdot 12 - \frac{1}{3} \cdot 6 - \frac{1}{12} \cdot 24 \right\} = 3 \ . \tag{49}$$

Table 1 compares the predictions for the coefficients b_i of the SM and the SUSY extension.[i]

Using the result of this table, along with input data for $\alpha_i(M_Z^2)$ [31], one can compute the evolution of the coupling constants in both models [23]. As is shown in Fig. 3, for the SM there is a **near unification** of the couplings around $M_X \simeq 10^{15}$ GeV. However, rather remarkably, in the supersymmetric extension of the SM, the presence of the SUSY matter, by altering the evolution, appears to give a **true unification** of the coupling constants at $M_X \simeq 10^{16}$ GeV.

---

[i]Note that the SUSY SM has, by necessity, 2 Higgs doublets, while the SM is assumed to have only 1 Higgs doublet.

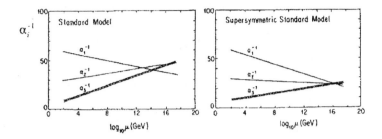

Figure 3: Evolution of couplings without and with SUSY matter.

The unification of the couplings in the SUSY SM case is quite spectacular. However, *per se*, this is only suggestive. It is not either a "proof" that a low energy supersymmetry exists, nor does it mean that there exists some high energy GUT! The proof of the former really requires the discovery of the predicted SUSY partners, while for GUTs one must find typical phenomena which are associated with these theories–like proton decay. This said, however, one can gather additional ammunition in favor of this picture from some of the properties of the top quark. I turn to this point next.

3.3   Implications of a Large Top Mass

Of all the quarks and leptons, only top has a mass which is of order of the Fermi scale, $v_F \simeq 250$ GeV. In this sense, top is unique among all the fundamental particles, since it has a mass whose value is basically set by the value of the order parameter responsible for the electroweak breakdown: $m_t \sim v_F$. All other elementary excitations are related to v_F by constants which are much less than unity.

If one is permitted a perturbative analysis, having a large top mass, in turn, gives further constraints. This is particularly true for the case of the SM, where a large top mass influences what Higgs masses are allowed. However, interesting consequences also arise in the (theoretically more pristine) SUSY SM. In both cases, rather than dealing with the "physical" top quark mass determined experimentally by CDF and DO:[14]

$$m_t = (173.8 \pm 5.0) \text{ GeV} , \tag{50}$$

it is more convenient to consider instead the running mass

$$m_t(m_t) \simeq \frac{m_t}{1 + \frac{4}{3\pi}\alpha_3(m_t^2)} = (165 \pm 5) \text{ GeV} . \tag{51}$$

176

Table 2: Coefficients entering the RGE for λ_t for the SM and the SUSY SM

Coefficient	SM	SUSY SM
a_t	9/2	6
a_b	3/2	1
a_τ	1	0
c_1	17/20	17/5
c_2	9/4	3
c_3	8	16/3

This is because $m_t(m_t)$ is directly related to the diagonal couplings of the top to the VEV of the Higgs boson, H_u:[j]

$$m_t(m_t) = \lambda_t(m_t)\langle H_u \rangle . \tag{52}$$

The Yukawa coupling λ_t also obeys a RGE. Keeping only the dominant 3rd generation couplings, this equation reads:[24]

$$\frac{d\lambda_t(\mu)}{d\ln\mu^2} = \frac{1}{32\pi^2}\left[a_t\lambda_t^2(\mu) + a_b\lambda_b^2(\mu) + a_\tau\lambda_\tau^2(\mu) - 4\pi c_i\alpha_i(\mu^2)\right]\lambda_t(\mu) . \tag{53}$$

Here λ_b and λ_τ are, respectively, the Yukawa couplings of the b-quark and the τ-lepton to the (corresponding) Higgs field. The coefficients a_t, a_b, a_τ and c_i in Eq. 53 again depend on the matter content of the theory. Table 2 details them both for the SM and its SUSY extension.

Because the coefficient $a_t > 0$, it follows that also the top coupling $\lambda_t(\mu)$ will have a Landau pole at large values of the scale μ. Of course, just as for the case of the Higgs coupling λ discussed earlier, the location of this singularity is not to be trusted exactly since Eq. 53 breaks down in its vicinity. Nevertheless, there are significant differences between where the Landau pole for λ_t is in the SM and where it is in the SUSY SM.

SM Case

Because one assumes that there is only one Higgs boson in the Standard Model, it follows that $\langle H_u \rangle = \frac{1}{\sqrt{2}}v_F \simeq 174$ GeV. This implies, in turn, a precise value for $\lambda_t(m_t)$ from Eq. 52:

$$\lambda_t(m_t) = 0.95 \pm 0.03 . \tag{54}$$

[j] In the SM there is only one Higgs boson, so the subscript u is unnecessary. For the SUSY SM, H_u is the Higgs field which couples to the right-handed up-quarks (while H_d couples to the right-handed down-quarks).

Further, since $m_t \gg m_b, m_\tau$ and $c_3 \alpha_3 \gg c_2 \alpha_2, c_1 \alpha_1$, to a good approximation the RGE (53) reduces to

$$\frac{d\lambda_t(\mu)}{d \ln \mu^2} = \frac{1}{32\pi^2} \left[\frac{9}{2} \lambda_t^2(\mu) - 32\pi\alpha_3(\mu^2) \right] \lambda_t(\mu) . \tag{55}$$

Using $\alpha_3(m_t^2) \simeq 0.118$, one sees that the above square bracket is **negative** at $\mu = m_t$. Thus in the SM, $\lambda_t(\mu)$ **decreases** as μ increases above m_t, at least temporarily. However, for very large μ, eventually $\lambda_t(\mu)$ will begin growing and eventually it will diverge at some scale–the Landau pole.

Eq. 55 and its companion for $\alpha_3(\mu^2)$, Eq. 40, can be solved in closed form [24]. One finds for $\lambda_t(\mu^2)$ the expression

$$\lambda_t^2(\mu) = \frac{\eta(\mu)\lambda_t^2(m_t)}{\left[1 - \frac{9}{16\pi^2}\lambda_t^2(m_t)I(\mu) \right]} , \tag{56}$$

where the functions $\eta(\mu)$ and $I(\mu)$ contain information on the running of the strong coupling constant:

$$\eta(\mu) = \left[\frac{\alpha_3(m_t^2)}{\alpha_3(\mu^2)} \right]^{-c_3/b_3} = \left[\frac{\alpha_3(\mu^2)}{\alpha_3(m_t^2)} \right]^{8/7}$$

$$I(\mu) = \int_{\ln m_t}^{\ln \mu} d\ln \mu' \eta(\mu') \tag{57}$$

Using Eq. 56 it is easy to check that the Yukawa coupling $\lambda_t(\mu)$ decreases well beyond the Planck scale, with $\lambda_t(M_P) \simeq 0.65 < \lambda_t(m_t)$, so that the top sector is perturbative throughout the region of interest. In fact, $\lambda_t(\mu)$ does not begin to get large until $\mu \sim 10^{30}$ GeV, with the Landau pole occuring around $\mu = 10^{32}$ GeV—scales well beyond the Planck scale.

Even though the top Yukawa coupling is below unity for $\mu < M_P$, this coupling is large enough to affect the Higgs sector of the Standard Model. Our discussion of the Higgs self-coupling λ in Section II was based on the RGE in which **only** terms involving λ were retained. In fact, at higher order, the RGE equation for λ is influenced both by the top Yukawa coupling and the electroweak gauge couplings. The full RGE for λ, rather than Eq. 11, reads [25]

$$\frac{d\lambda(\mu)}{d \ln \mu^2} = \frac{3}{4\pi^2} \left[\lambda^2(\mu) - \frac{1}{4}\lambda_t^4(\mu) + \frac{\pi}{128}[3 + 2\sin^2\theta_W + \sin^4\theta_W]\alpha_2(\mu^2) \right] . \tag{58}$$

The important point to notice in this equation is the **negative** contribution coming from the top coupling. This contribution, just like the α_3 contribution in Eq. 55, can cause $\lambda(\mu)$ to **decrease** at first. Indeed, if the Higgs coupling $\lambda(M_H)$ is not large enough, because the Higgs boson is light, the relatively large contribution coming from the λ_t^4 term can drive $\lambda(\mu)$ **negative** at some scale μ. This cannot really happen physically, because for $\lambda < 0$ the Higgs potential is unbounded!

To avoid this vacuum instability below some cut-off Λ_c–typically $\Lambda_c \sim M_P$– one needs to have $\lambda(M_H)$, and therefore the Higgs mass, sufficiently large. Hence, these considerations give a lower bound for the Higgs mass. Taking $\Lambda_c = M_P$, this lower bound is [26]

$$M_H \geq 134 \text{ GeV} . \tag{59}$$

Lowering the cut-off Λ_c, weakens the bound on M_H. Interestingly, to have a SM Higgs as light as 100 GeV—which is the region accessible to LEP 200— requires a very low cut-off, of order $\Lambda_c \sim 100$ TeV. [27] So, finding such a Higgs may, very indirectly, point to a scenario with extra compact dimensions where such low-cut-offs are allowed. Parenthetically, I should note that these kinds of vacuum stability bounds cease to be valid in models with more than one Higgs doublet, like in the SUSY SM case.

SUSY SM

The situation is quite different for λ_t if there is supersymmetric matter. Because supersymmetry necessitates two Higgs doublets, $\lambda_t(m_t)$ is no longer fixed solely by the value of the top mass. The vacuum expectation values of the two Higgs bosons involve a further parameter, $\tan\beta$, besides v_F:

$$\langle H_u \rangle = \frac{1}{\sqrt{2}} v_F \sin\beta ; \quad \langle H_d \rangle = \frac{1}{\sqrt{2}} v_F \cos\beta . \tag{60}$$

Thus, one has, instead of Eq. 54,

$$\lambda_t(m_t) = \frac{0.95 \pm 0.03}{\sin\beta} . \tag{61}$$

Keeping again only the leading terms, the RGE for $\lambda_t(\mu)$ in the presence of SUSY matter reads now:

$$\frac{d\lambda_t(\mu)}{d\ln\mu^2} = \frac{1}{32\pi^2} \left[6\lambda_t^2(\mu) + \lambda_b^2(\mu) - \frac{64\pi}{3}\alpha_3(\mu^2) \right] \lambda_t(\mu) . \tag{62}$$

Because of the different coefficients that SUSY matter implies, it is no longer necessarily true that the square bracket above is negative at $\mu = m_t$, as was the case in the SM. Instead, since $64\pi\alpha_3(m_t^2)/3 \simeq 7.9$, the square bracket above can actually vanish in two regions of parameter space. The first is the region of large $\tan\beta$ where, for large scales μ, $\lambda_b(\mu) \simeq \lambda_t(\mu) \simeq 1$ [Yukawa unification region]. The second is a region where $\tan\beta \sim O(1)$, so that $\lambda_b(m_t) \ll \lambda_t(m_t)$, with the top contribution cancelling directly that coming from the $SU(3)$ corrections [in detail, this requires $\sin\beta = 0.83 \pm 0.03$].

In either of the two regions above, the presence of SUSY matter forces λ_t to an infrared fixed point [28] as μ becomes of order m_t

$$\frac{d\lambda_t(\mu)}{d\ln\mu^2}\bigg|_{\mu\sim m_t} \simeq 0 \,. \tag{63}$$

This is a very interesting possibility, since such a condition essentially serves to drive quite different values of $\lambda_t(\mu)$ at high scales μ down to the same fixed point value λ^*, at scales $\mu \simeq m_t$. This behavior is illustrated in Fig. 4. Asking that Eq. 63 holds, one sees that the fixed point λ^* is given by

$$\lambda^* = \frac{32\pi}{9}\alpha_3(m_t^2)\left[\frac{1}{1+\frac{m_b^2(m_t)\tan^2\beta}{6m_t^2(m_t)}}\right] \simeq 1.3\left[\frac{1}{1+\frac{m_b^2(m_t)\tan^2\beta}{6m_t^2(m_t)}}\right]\,. \tag{64}$$

Two remarks are in order

i) The fixed point behavior for $\lambda_t(m_t)$ does, indeed, need supersymmetry. In the SM with only ordinary matter, one would get a fixed point behavior from the RGE for λ_t only if m_t would have been approximately 250 GeV!

ii) If one were to assume that near the Planck mass $\lambda_t(\mu)$ were to be large then, if there is SUSY matter, the fixed point behavior for λ_t essentially serves to predict the correct value for the top mass $m_t(m_t) \simeq m_t^* \simeq 170$ GeV seen by experiment.

3.4 Neutrino Oscillations

Although hints of neutrino oscillations have been around for some time, notably connected with the solar neutrino puzzle [29], real evidence for oscillatioins only emerged last year from data on atmospheric neutrinos studied by the large underground water Cerenkov detector, SuperKamiokande. In June 1998, the SuperKamiokande Collaboration [3] reported a pronounced zenith angle dependence for the flux of multi-GeV atmospheric ν_μ events, but no such dependence for the atmospheric ν_e flux. The collaboration interpreted the large up-down asymmetry seen [139 up-going ν_μ's versus 256 down-going ν_μ's] as evidence

180

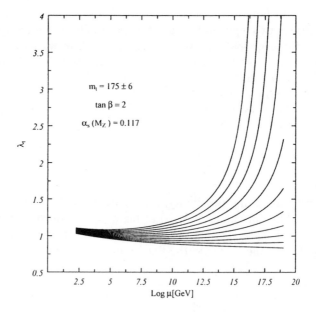

Figure 4: Focusing of the Yukawa couplings as $\lambda_t \to \lambda^*$.

for $\nu_\mu \to \nu_X$ oscillations, with ν_X being either a ν_τ or, possibly, a new sterile neutrino ν_s.[k]

In the usual 2-neutrino mixing formalism [5], the weak interaction eigenstates $|\nu_\mu\rangle$ and $|\nu_X\rangle$ are linear combinations of two mass eigenstates $|\nu_1\rangle$ and $|\nu_2\rangle$:

$$\begin{pmatrix} |\nu_\mu\rangle \\ |\nu_X\rangle \end{pmatrix} = \begin{pmatrix} \cos\theta & \sin\theta \\ -\sin\theta & \cos\theta \end{pmatrix} \begin{pmatrix} |\nu_1\rangle \\ |\nu_2\rangle \end{pmatrix} . \tag{65}$$

The probability that after traversing a distance L a ν_μ neutrino of energy E emerges as a ν_X neutrino is then given by the well known formula

$$P(\nu_\mu \to \nu_X; L) = \sin^2 2\theta \, \sin^2 \frac{\Delta m^2 L}{4E} , \tag{66}$$

---

[k] A sterile neutrino is one that has no $SU(2) \times U(1)$ interactions. Because ν_s does not have any couplings to the Z-bosons it does not contribute to the Z width, so a light ν_s is not excluded by the precise neutrino counting result from LEP: $N_\nu = 2.991 \pm 0.011$.[31]

with $\Delta m^2 = m_2^2 - m_1^2$ being the difference in mass squared between the two neutrino eigenstates. From an analysis of their results, the SuperKamiokande collaboration [3] deduced that the observed up-down asymmetry could be explained by neutrino oscillations if the mixing was nearly maximal, $\sin^2 2\theta \simeq 1$, and if $\Delta m^2 \simeq 2.5 \times 10^{-3}$ eV^2.

The SuperKamiokande results provide a **lower bound** on neutrino masses. Since

$$\Delta m^2 = m_2^2 - m_1^2 \simeq 2.5 \times 10^{-3} \text{ eV}^2 \ , \tag{67}$$

it follows that at least one neutrino has a mass larger than

$$m_2 \geq 5 \times 10^{-2} \text{ eV} \ , \tag{68}$$

with the bound being satisfied if $m_2 \gg m_1$. Such small masses, compared to the masses of quarks and leptons, suggests that new physics is at work. Indeed, the simplest way to understand why neutrinos have tiny masses is through the see-saw mechanism [30], which involves new physics at a scale much larger than the Fermi scale. Let me briefly discuss this reasoning.

Because neutrinos have no charge, as we remarked earlier, the most general mass term for these states can contain also particle-particle terms, besides the usual particle-antiparticle contribution [5]. One has, considering one species of neutrinos for simplicity,

$$\begin{aligned}
\mathcal{L}^\nu_{\text{mass}} &= -m_D[\bar{\nu}_L \nu_R + \bar{\nu}_R \nu_L] - \frac{m_R}{2}[\bar{\nu}_R \tilde{C} \bar{\nu}_R^T + \nu_R^T \tilde{C} \nu_R] \\
&\quad - \frac{m_L}{2}[\nu_L^T \tilde{C} \nu_L + \bar{\nu}_L \tilde{C} \bar{\nu}_L^T] \ .
\end{aligned} \tag{69}$$

In the above the different mass terms conserve/violate different symmetries. To wit:

The Dirac mass m_D: conserves lepton number L, but violates $SU(2) \times U(1)$.

The Majorana mass m_R: violates lepton number L, but conserves $SU(2) \times U(1)$.

The Majorana mass m_L: violates both lepton number L and $SU(2) \times U(1)$.

Because the Dirac mass term has the same form as the usual quark and lepton masses, it is sensible to imagine that m_D should be of the same order of magnitude as these masses. Hence, schematically, one expects

$$m_D \sim m_\ell \sim v_F \tag{70}$$

182

where the proportionality constant to the Fermi scale may, indeed, be quite small. Thus, if one wants the physical neutrinos to have very small masses, this must be the result of the presence of the Majorana mass terms in Eq. 69.

There are two simple ways of achieving this goal, depending on whether one assumes that a right-handed neutrino ν_R exists or not. If one does not involve a ν_R, the simplest effective interaction one can write using only ν_L which preserves $SU(2) \times U(1)$ is [32]

$$\mathcal{L}_{\text{eff}} = \frac{1}{M}(\nu_\ell \; \ell)_L^T \tilde{C} \vec{\tau} \begin{pmatrix} \nu_\ell \\ \ell \end{pmatrix}_L \cdot \Phi^T C \vec{\tau} \Phi + \text{h.c.} \tag{71}$$

Here M is some, presumably large, scale which is associated with these lepton number violating processes. This term, when $SU(2) \times U(1)$ breaks down to $U(1)_{\text{em}}$, generates a mass for the neutrino

$$m_\nu \equiv m_L = \frac{v_F^2}{M} . \tag{72}$$

One sees that to get neutrino masses of the order of those inferred from SuperKamiokande one requires $M \sim 10^{15}$ GeV—a scale of the order of the GUT scale!

One can get a similar result if one includes right-handed neutrinos in the theory. In this case, it is convenient to rewrite the general neutrino mass terms of Eq. 69 in terms of both neutrino fields, ν, and their charged conjugate, ν^c. Since [5]

$$\nu^c = \tilde{C}\bar{\nu}^T \; ; \quad \overline{\nu^c} = \nu^T \tilde{C} , \tag{73}$$

Eq. 69 takes the form

$$\mathcal{L}^\nu_{\text{mass}} = -\frac{1}{2}\left[\left(\overline{(\nu_L)^c} \; \overline{\nu_R}\right)\right]\begin{pmatrix} m_L & m_D \\ m_D & m_R \end{pmatrix}\begin{pmatrix} \nu_L \\ (\nu_R)^c \end{pmatrix} + \text{h.c.} . \tag{74}$$

If one neglects m_L altogether ($m_L = 0$) and $m_R \gg m_D$, then it is easy to see that the eigenvalues of the neutrino mass matrix

$$\mathcal{M} = \begin{pmatrix} 0 & m_D \\ m_D & m_R \end{pmatrix} \tag{75}$$

have a large splitting, producing a heavy neutrino and one ultralight neutrino:

$$m_{\text{heavy}} \sim m_R \; ; \quad m_{\text{light}} \sim \frac{m_D^2}{m_R} . \tag{76}$$

This is the famous see-saw mechanism [30].

In the see-saw mechanism the light neutrino state, n, is mostly ν_L, while the heavy neutrino state, N, is mostly ν_R:

$$\begin{aligned}
\nu_L &\simeq n_L + \frac{m_D}{m_R} N_L \\
\nu_R &\simeq N_R - \frac{m_D}{m_R} n_R \ .
\end{aligned} \tag{77}$$

Assuming that SuperKamiokande has observed $\nu_\mu \to \nu_\tau$ oscillations, and that the light neutrino has a mass given by Eq. 76, then

$$m_\nu = \frac{m_D^2}{m_R} \sim 5 \times 10^{-2} \text{ eV} \ . \tag{78}$$

If $m_D \simeq m_\tau$, one requires $m_R \sim 10^{11}$ GeV. If $m_D \simeq m_t$, on the other hand, one requires $m_R \sim 10^{15}$ GeV. Irrespective of the choice, again one sees that to obtain neutrino masses in the sub-eV range via the see-saw mechanism one needs to involve new scales, connected to m_R, which are much above v_F.

These two examples make it clear that the neutrino oscillations detected by SuperKamiokande are definitely signs of new physics. However, it is quite likely that this new physics is disconnected from the precise mechanism which causes the $SU(2) \times U(1)_{em}$ breakdown. This is certainly the case if the light neutrinos involved in the oscillations are produced by the see-saw mechanism, since the parameter m_R is an $SU(2) \times U(1)$ singlet. Thus the scale m_R has nothing at all to do with v_F. This is likely to be true also if the light neutrinos are generated by effective interactions of the type shown in Eq. 71. Although light neutrino masses in this case arise only after $SU(2) \times U(1)$ breaking, the physics that gives origin to these masses is the physics associated with the scale M (likely some GUT physics) characterizing the effective interactions.

Because of the above, neutrino oscillations are unlikely to give much information on the nature of the physics which gives rise to the Fermi scale v_F. For this reason, in what follows, I shall not pursue this interesting topic further, preferring to concentrate instead on the dynamics of electroweak symmetry breaking.

4 Promises and Challenges of Dynamical Symmetry Breaking

The idea behind a dynamical origin for the Fermi scale v_F is rather simple [33]. One imagines that there exists an underlying strong interaction theory which confines and that the fundamental fermions F of this theory carry also $SU(2) \times U(1)$ quantum numbers. If the confining forces acting on F allow the formation of $\langle \bar{F} F \rangle$ condensates then, in general, these condensates will cause the breakdown of $SU(2) \times U(1)$, since $\bar{F}F$ also carries non-trivial $SU(2) \times U(1)$ quantum numbers. The dynamical scale Λ_F associated with the underlying strongly interacting theory is then, *de facto*, the Fermi scale:

$$\langle \bar{F}F \rangle \sim \Lambda_F^3 \sim v_F^3 \ . \tag{79}$$

There are two generic predictions of such theories:

i) Because of the strongly coupled nature of the underlying theory, there should be no light Higgs boson in the spectrum.

ii) Just like in QCD, this underlying theory should have a rich spectrum of bound states which are singlets under the symmetry group of the underlying theory. These states, typically, should have masses

$$M \sim \Lambda_F \sim v_F \ . \tag{80}$$

Among these states there should be a heavy Higgs boson.

It has become conventional to denote the underlying strong interaction theory responsible for the breakdown of $SU(2) \times U(1) \rightarrow U(1)_{\text{em}}$ as Technicolor, which was the name originally used by Susskind [33]. Just as there are two generic predictions for Technicolor theories, there are also two necessary requirements for these theories coming from experiment. These are

iii) The underlying theory must lead naturally to the connection between W and Z masses embodied in the statement that $\rho = 1$, up to radiative corrections. That is $M_W^2 = M_Z^2 \cos^2 \theta_W$. As we shall see, this obtains if the underlying theory has some, protective, $SU(2)$ global symmetry.

iv) The Technicolor spectrum must be such that the parameter ϵ_3 defined in Sec. 3 is small, as seen experimentally. For this to be so, one needs that

$$\int_0^\infty \frac{ds}{s} [R_V(s) - R_A(s)] < 0 \ . \tag{81}$$

The first requirement above is easy to achieve in most Technicolor theories (but, eventually, quite constraining). The second requirement is much harder to implement, since it requires understanding some unknown strong dynamics! Let me begin by discussing how one can guarantee that the underlying theory give $\rho = 1$. For this purpose, it proves useful to examine how this happens in the SM. There $\rho = 1$ emerges as a result of an **accidental symmetry** in the Higgs potential. If one writes out the complex Higgs doublet Φ in terms of real fields

$$\Phi = \frac{1}{\sqrt{2}} \left(\begin{array}{c} \phi_1 + i\phi_2 \\ \phi_3 + i\phi_4 \end{array} \right) , \tag{82}$$

it is immediately clear that the potential

$$V = \lambda \left(\Phi^\dagger \Phi - \frac{v_F^2}{2} \right)^2 \tag{83}$$

has a bigger symmetry than $SU(2) \times U(1)$, namely $O(4)$. The VEV of Φ is, in the notation of Eq. 82, the result of ϕ_1 getting a VEV: $\langle \phi_1 \rangle = v_F$. Obviously, this VEV causes the breakdown of $O(4) \to O(3)$. It is the remaining $O(3)$ symmetry, after the spontaneous breakdown, which forces $\rho = 1$. Indeed, this symmetry requires the 11, 22 and 33 matrix elements of the weak-boson mass matrix to have all the same value:

$$M^2 = \frac{1}{4}v_F^2 \left[\begin{array}{cccc} g_2^2 & 0 & 0 & 0 \\ 0 & g_2^2 & 0 & 0 \\ 0 & 0 & g_2^2 & g_2 g' \\ 0 & 0 & g_2 g' & g'^2 \end{array} \right] , \tag{84}$$

giving $\rho = 1$.

To guarantee $\rho = 1$ in Technicolor models one must build in the same custodial $O(3) \sim SU(2)$ symmetry present in the Higgs potential. Such a custodial $SU(2)$ symmetry in fact exists in QCD with just 2 flavors. Neglecting the u- and d-quark masses the QCD Lagrangian has an $SU(2)_\mathrm{L} \times SU(2)_\mathrm{R} \times U(1)_\mathrm{L+R}$ global symmetry[l]. However, only the vectorial piece of the $SU(2)_\mathrm{L} \times SU(2)_\mathrm{R} \sim SU(2)_V \times SU(2)_A$ symmetry survives as a good symmetry (Isospin) of QCD, since the formation of $\langle \bar{u}u \rangle = \langle \bar{d}d \rangle \neq 0$ condensates breaks the $SU(2)_A$ symmetry spontaneously.

Given the circumstances described above, the simplest way to guarantee that $\rho = 1$ in Technicolor models is to make these models look very much like QCD. Indeed, this was the strategy adopted originally by Susskind and Weinberg[33]. One organizes the underlying Technifermions F in doublets

$$\left(\begin{array}{c} U_i \\ D_i \end{array} \right) \quad i = 1, \ldots, N_D \tag{85}$$

---

[l]The $U(1)_\mathrm{L-R}$ symmetry present at the Lagrangian level is not preserved at the quantum level, because of the nature of the QCD vacuum[34].

but assumes, just like in QCD, that the left- and right-handed components of these states transform differently under $SU(2) \times U(1)$:

$$\begin{pmatrix} U_i \\ D_i \end{pmatrix}_{\mathrm{L}} \sim 2 \; ; \quad U_{i\mathrm{R}} \sim 1; \; D_{i\mathrm{R}} \sim 1 \; . \tag{86}$$

Neglecting the electroweak interactions, if the Technifermions are massless, then the Technicolor theory has a large global chiral symmetry $SU(2N_D)_{\mathrm{L}} \times SU(2N_D)_{\mathrm{R}} \supset SU(2)_{\mathrm{L}} \times SU(2)_{\mathrm{R}}$. The condensates which one assumes form due to the strong Technicolor forces, and which break $SU(2) \times U(1) \to U(1)_{\mathrm{em}}$,

$$\langle \bar{U}_{i\mathrm{L}} U_{i\mathrm{R}} \rangle = \langle \bar{D}_{i\mathrm{L}} D_{i\mathrm{R}} \rangle \neq 0 \; , \tag{87}$$

also break this global symmetry down. In particular, $SU(2)_{\mathrm{L}} \times SU(2)_{\mathrm{R}} \to SU(2)_{\mathrm{L+R}}$ and this custodial $SU(2)$ symmetry serves to guarantee that in the gauge boson spectrum $M_W^2 = M_Z^2 \cos^2 \theta_W$.

If one pushes the QCD-Technicolor analogy a bit more, one can infer something about the scale of the Technicolor mass spectrum. Both QCD and Technicolor have an approximate $SU(2)_{\mathrm{L}} \times SU(2)_{\mathrm{R}}$ global symmetry broken down to $SU(2)_{\mathrm{L+R}}$. For QCD such a breakdown gives rise to the pions, $\vec{\pi}$, as Nambu-Goldstone bosons. Technicolor, analogously will have also three Technipions $\vec{\pi}_T$. However, in contrast to the pions which are real states, the Technipions, when $SU(2) \times U(1)$ is gauged, become the longitudinal components of the $W^{\pm}$ and Z gauge bosons. Nevertheless, from the $\vec{\pi} - \vec{\pi}_T$ analogy one can estimate the importance of Technicolor interactions and their associated spectra. More specifically $\pi - \pi$ scattering in QCD should tell us something about $\pi_T - \pi_T$ scattering in Technicolor.

For $\pi - \pi$ scattering one can write a partial wave expansion for the scattering amplitude $\mathcal{A}$ of the form

$$\mathcal{A} = 32\pi \sum_J (2\pi + 1) P_J(\cos\theta) a_J(s) \; . \tag{88}$$

The $SU(2) \times SU(2)$ chiral symmetry of QCD allows one to calculate the s-wave scattering amplitudes a_o at threshold, and one finds [35]

$$a_o^{I=0} = \frac{s}{16\pi f_\pi^2} \; ; \quad a_o^{I=2} = -\frac{s}{32\pi f_\pi^2} \tag{89}$$

where f_π is the pion decay constant. Unitarity requires the partial wave amplitudes to be bounded $|a_J| \leq 1$, with strong interactions signalled by these amplitudes saturating the bound. Using Eq. 89, naively one sees that this occurs when the energy squared $s \sim 16\pi f_\pi^2$. Indeed, at these energies $\pi - \pi$

scattering already is dominated by resonance formation. For the Technicolor theory, by analogy, one should have similar formulas with f_π replaced by v_F. Hence, if the analogy holds, one should expect Technicolor resonances to appear at an energy scale of order $4\sqrt{\pi}v_F \sim 1.7$ TeV. If one trusts this estimate, the physics of the underlying Technicolor theory will be hard to see, even at the LHC!

The analogy between a possible Technicolor theory and QCD, however, cannot be pushed too far. Indeed, we know from our discussion of the precision eletroweak tests that the spectrum of vector and axial resonances in the Technicolor theory must be quite **different** than QCD, since what is required is that Eq. 81 hold—which has the **opposite** sign of what obtains in QCD! Thus Technicolor, in some fundamental aspects, must be quite different than QCD. This also emerges from a different set of considerations, connected with the mass spectrum of quarks and leptons. I turn to this issue now.

Although strange, and difficult to implement in practice, it is possible to imagine that a Technicolor theory exists in which, as far as the electroweak radiative corrections go, the presence of a heavy Higgs state around a TeV and a Technicolor spectrum in the few TeV range combine to mimic the effects of a single light ($M_H \sim 100$ GeV) Higgs, giving a tiny ϵ_3. A much harder task is to ask that this theory also generate a **realistic** mass spectrum for the quarks and leptons. In my view, this latter problem is the principal difficulty of Technicolor theories.

In weakly coupled theories, where the Fermi scale is a parameter put in by hand, one can easily generate quark and lepton masses through Yukawa couplings. In these theories there is no reason that the physics which is associated with v_F be connected to the physics which served to produce the Yukawa couplings. Indeed, it is likely that this latter physics is one associated with scales much larger than v_F. This freedom of decoupling the origin of the quark and lepton mass spectrum from the Fermi scale does not exist for theories where v_F is generated through a strongly coupled theory. In these theories one is forced to try to understand fermion mass generation at scales of order of the Fermi scale v_F, or just one or two orders of magnitude higher. This complicates life immensely.

To generate quark and lepton masses in Technicolor theories at all, one must introduce some communication between these states—which I shall denote collectively as f—and the Technifermions F, whose condensates cause the electroweak breakdown. This necessitates, in general, introducing yet **another** strongly coupled underlying theory, which has been dubbed extended Technicolor (ETC) [36]. Spontaneous ETC breakdown, in conjunction with Technicolor-induced electroweak breakdown, is at the root of the quark and

188

lepton mass spectrum. However, since at least one state in this spectrum, top, has a mass of $O(v_F)$, the scale associated with the ETC breakdown cannot be very much larger than v_F. Let me discuss this in a bit more detail.

The ETC interactions couple the ordinary fermions f to the Technifermions F. As a result, the exchange of an ETC gauge boson between pairs of fF states, when ETC spontaneously breaks down, generates an effective interaction which, schematically, reads

$$\mathcal{L}_{\text{eff}}^{\text{ETC}} = \frac{1}{\Lambda_{\text{ETC}}^2} \bar{F}_{\text{L}} F_{\text{R}} \bar{f}_{\text{R}} f_{\text{L}} + \text{h.c.} \tag{90}$$

Such an interaction generates a mass term for the ordinary fermions, as the result of the formation of the $SU(2) \times U(1)$ breaking Technifermion condensate $\langle \bar{F}_{\text{L}} F_{\text{R}} \rangle \sim v_F^3$. Thus one finds

$$m_f = \frac{\langle \bar{F}_{\text{L}} F_{\text{R}} \rangle}{\Lambda_{\text{ETC}}^2} \sim \frac{v_F^3}{\Lambda_{\text{ETC}}^2} \ . \tag{91}$$

If one tries to use this formula to produce a top mass, since m_t itself is of $O(v_F)$ one sees, as alluded to above, that the ETC scale Λ_{ETC} cannot be large—typically, $\Lambda_{\text{ETC}} \sim (1-10)$ TeV. Such a low ETC scale, however, is troublesome because it generally leads to too large flavor changing neutral currents (FCNC). For instance, as discussed long ago by Dimopoulos and Ellis [37], the box graph containing both ETC and Technifermion exchanges gives a contribution to the $\bar{K}^o - K^o$ mass difference which is far above that coming from the weak interactions, unless $\Lambda_{\text{ETC}} \gtrsim 100$ TeV.

Although the FCNC $\leftrightarrow m_t$ conundrum is not the only problem of Technicolor/ETC models,[m] its theoretical amelioration has been a principal target for partisans of dynamical symmetry breaking [39]. Fortunately, as I will discuss below, one of the more interesting "solutions"—dubbed Walking Technicolor (WTC) [40]—involves theories that are rather different from QCD dynamically. It is perhaps not unreasonable to hope that for such theories the constraint (81) actually might hold. Some arguments in favor of this contention actually exist [41].

The essence of how WTC models ameliorate the $m_t \leftrightarrow$ FCNC conundrum can be readily appreciated by noting that the Fermi scale v_F and the Technifermion condensate $\langle \bar{F} F \rangle$, in general, are sensitive to rather **different** parts

---

[m] For instance, because of the large global symmetries present in the Technicolor sector, condensate formation leads to the appearance of many more (pseudo) Nambu-Goldstone bosons than just the Technipions, $\tilde{\pi}_T$. Unless other interactions can generate sufficiently large masses for these extra states, their presence in the Technicolor spectrum invalidates these theories, since no trace of these states has yet been seen experimentally [38].

of the self-energy of Technifermions. Since v_F measures the strength of the matrix element of the spontaneously broken currents between a Technipion state and the vacuum, v_F^2 is proportional to an integral over the square of the Technifermion self-energy. Schematically, this result, gives for v_F^2 the formula[42]

$$v_F^2 \sim \int \frac{d^4p}{(p^2)^2} \Sigma^2(p) \sim \Sigma^2(0) , \qquad (92)$$

where the second approximation is valid up to logarithmic terms. The condensate $\langle \bar{F}F \rangle$, on the other hand, just involves an integral over the Technifermion self-energy. Again, schematically,

$$\langle \bar{F}F \rangle \sim \int \frac{d^4p}{p^2} \Sigma(p) \sim \Lambda_{\rm ETC}^2 \Sigma(\Lambda_{\rm ETC}) . \qquad (93)$$

The second term above, again to logarithmic accuracy, recognizes that this integral is dominated by the largest scales in the theory. In our case, since one integrates up to the ETC scale, this is $\Lambda_{\rm ETC}$.

One can understand how WTC theories work from Eqs. 92 and 93, augmented by a result of Lane and Politzer[43] detailing the asymptotic behavior of fermionic self-energies in theories with broken global symmetries. Lane and Politzer showed that

$$\Sigma(q^2) \stackrel{q^2 \to \infty}{\sim} \frac{\Sigma^3(0)}{q^2} \sim \frac{\Lambda^3}{q^2}, \qquad (94)$$

where the last line recognizes that the self-energy $\Sigma(0)$ scales according to the dynamical scale Λ of the theory in question. For ordinary Technicolor theories, by the time one has reached the ETC scale, the Technifermion self-energy should have already reached the asymptotic form (94). Thus

$$\langle \bar{F}F \rangle \sim \Lambda_{\rm ETC}^2 \Sigma(\Lambda_{\rm ETC}) \sim \Sigma^3(0) \sim v_F^3 \qquad (95)$$

and the condensate $\langle \bar{F}F \rangle$ indeed scales as v_F^3, as we have assumed. This, however, is not the case for WTC theories. In these theories, the evolution with q^2 of the WTC coupling constant, as well as of the Technifermion self-energy, is very slow—hence the name, Walking Technicolor. In particular, the Technifermion self-energy at the ETC scale is assumed to be nowhere near the asymptotic form (94) so that,

$$\Sigma(\Lambda_{\rm ETC}) \gg \frac{\Lambda_{\rm TC}^3}{\Lambda_{\rm ETC}^2} . \qquad (96)$$

In this case, there is a large disparity between the condensate $\langle \bar{F}F \rangle$ and v_F^3:

$$\langle \bar{F}F \rangle \sim \Lambda_{\rm ETC}^2 \Sigma(\Lambda_{\rm ETC}) \gg \Lambda_{\rm TC}^3 \sim v_F^3 . \qquad (97)$$

Although Walking Technicolor theories are quite interesting dynamically, and some WTC theories have been constructed which are semirealistic[44], there are still many difficulties in practice, notably with top itself. For instance, to really get the top mass large enough, one has to really have very slow "walking". Since we want $m_t \sim v_F$ and m_t is given by the formula

$$m_t \sim \frac{\langle \bar{F}F \rangle}{\Lambda_{\mathrm{ETC}}^2} \sim \Sigma(\Lambda_{\mathrm{ETC}}) \tag{98}$$

one needs

$$\Sigma(\Lambda_{\mathrm{ETC}}) \sim \Sigma(0) . \tag{99}$$

Such "slow walking" again is only realistic if Λ_{ETC} is relatively near to v_F. However, then FCNC problems re-emerge, even in the WTC context! Furthermore, the large ETC effects that are used to boost the top mass up cause other problems. In particular, as Chivukula, Selipsky and Simmons pointed out[45], the same graphs that give rise to the top mass produce a rather large anomalous $Zb\bar{b}$ vertex. In the simplest WTC model, this gives an unacceptably large shift for the ratio R_b—the ratio of the rate of $Z \to b\bar{b}$ to that of hadrons—$(\delta R_b)_{\mathrm{ETC}} \simeq -0.01$.

To avoid the problems alluded to above, as well as other problems[39], lately some hybrid models have been developed. These, so called, topcolor-Technicolor models[46] get the masses for all first and second generation quarks and leptons from a WTC/ETC theory. However, the top (and bottom) masses come from yet a **third** underlying strong interaction theory–topcolor[47]–which produces 4-Fermi effective interactions involving these states. So top, effectively, gets its mass from the presence of a top-quark condensate $\langle \bar{t}t \rangle$, formed as a result of the topcolor theory. Because $\langle \bar{t}t \rangle$ also breaks $SU(2) \times U(1)$, in these topcolor-Technicolor models, the Fermi scale arises both as the result of these condensates and of the usual Technifermion condensates $\langle \bar{F}F \rangle$.

Although these theories have some attractive features, and some interesting predictions of anomalies in top interactions,[48] one has moved a long way away from the simple idea that the electroweak breakdown is much like the BCS theory of superconductivity! Of course, ultimately, only experiment will tell if these rather complicated ideas have merit or not. From the point of view of simplicity, however, the weak coupling SM alternative of invoking the existence of low energy supersymmetry seems a more desirable route to follow. I turn to this topic next.

5 The SUSY Alternative

The procedure for constructing a supersymmetric extension of the SM is straight-forward, and rather well known by now [49]. For completeness, let me outline the principal steps here. They are:

i) One associates scalar partners to the quarks and leptons (squarks and sleptons) and fermion partners to the Higgs scalar(s) (Higgsinos), building chiral supermultiplets

$$\{\Psi\} = \{\tilde\psi, \psi\} \,, \qquad (100)$$

composed of complex scalars $\tilde\psi$ and Weyl fermions ψ. For instance the left-handed electron chiral supermultiplet

$$\{\Psi_e\} = \{\tilde e_L, e_L\} \qquad (101)$$

contains a (left-handed) selectron $\tilde e_L$ and a left-handed electron.

ii) One associates spin-1/2 partners to the gauge fields (gauginos), building vector supermultiplets

$$\{V\} = \{V^\mu, \lambda\} \,, \qquad (102)$$

composed of a gauge field V^μ and a Weyl fermion gaugino λ.

iii) One supersymmetrizes all interactions. For example, the ordinary Yukawa coupling of the Higgs H_d to the quark doublet Q_L and the quark singlet d_R is now accompanied by two other vertices involving $\tilde H_d\, Q_L\, \tilde d_R$ and $\tilde H_d\, \tilde Q_L\, d_R$.[n]

In addition to the above three points, supersymmetry imposes some constraints on which kind of interactions are allowed [49]. For our purposes here, two of these constraints are most significant:

iv) Interactions among chiral superfields are derivable from a **superpotential** $W(\{\Phi_i\})$ which involves only $\{\Phi_i\}$ and not $\{\Phi_i^*\}$.

v) The scalar potential $V(\phi_i)$ follows directly from the superpotential, plus terms—the, so called, D-terms—arising from gauge interactions [49]

$$V(\phi_i) = \sum_i \left|\frac{\partial W}{\partial \phi_i}\right|^2 + \frac{g_a}{2}\,|\phi_i^* t_a \phi_i|^2 \,. \qquad (103)$$

Here t_a is the appropriate generator matrix for the scalar field ϕ_i for the symmetry group whose coupling is g_a.

---

[n] Here $\tilde H_d$ is the spin-1/2 SUSY partner of H_d and $\tilde Q_L, \tilde d_R$ are the spin-0 partners of Q_L, d_R, respectively.

I remark that point (iv) above is the reason one needs two different Higgs fields in supersymmetry. Even though the scalar field H_d^* has the same quantum numbers as H_u, a superpotential term W_3 ($\mathbf{H_d^*}$, $\mathbf{Q_L}$, $\mathbf{u_R}$) is not allowed. There is another way to understand why a second Higgs supermultiplet is needed in a supersymmetric extension of the SM, involving anomalies. The Higgsino $\tilde{H}_d$ has opposite charges to the Higgsino $\tilde{H}_u$. Omitting one of these two fields in the theory would engender a chiral anomaly in the $SU(2) \times U(1)$ theory, since for $SU(2) \times U(1)$ to be anomaly free it is necessary that [50]

$$\text{Tr} \, [Q]_{\text{fermions}} = 0 \, . \tag{104}$$

Although Eq. 104 holds for the quarks and leptons, it would fail if only $\tilde{H}_d$ was included and not $\tilde{H}_u$. So anomaly consistency requires two Higgs bosons in a SUSY extension of the SM.

With this brief precis of the SUSY SM in hand, let me examine what are the implications of supersymmetry for the issue of $SU(2) \times U(1)$ breaking. If one ignores for the moment any Yukawa couplings, the only superpotential term one is allowed to write down for the SUSY SM is

$$W(\mathbf{H_u}, \mathbf{H_d}) = \mu \mathbf{H_u H_d} \, . \tag{105}$$

This superpotential gives the following scalar potential for the SUSY SM [cf. Eq. 103]

$$\begin{aligned} V(H_u, H_d) &= \mu^2 [H_u^\dagger H_u + H_d^\dagger H_d] + \frac{1}{8} g'^2 [H_u^\dagger H_u - H_d^\dagger H_d]^2 \\ &\quad + \frac{1}{8} g_2^2 [H_u^\dagger \vec{\tau} H_u + H_d^\dagger \vec{\tau} H_d] \cdot [H_u^\dagger \vec{\tau} H_u + H_d^\dagger \vec{\tau} H_d] \, . \end{aligned} \tag{106}$$

However, because all the coefficients in the above are positive, it is clear that the potential V **cannot** break $SU(2) \times U(1)$!

This is really not a disaster, since a SUSY SM is not a realistic theory without including some breaking of the supersymmetry. Once one adds some SUSY breaking terms to Eq. 106, then it is quite possible that these terms can cause the $SU(2) \times U(1) \to U(1)_{\text{em}}$ breakdown. However, not any type of supersymmetry breaking terms are allowed. To preserve the solution to the hierarchy problem which supersymmetry provided (namely, logarithmic sensitivity to the cutoff—$\ln \Lambda_c$—and not quadratic sensitivity—Λ_c^2) one must ask that the SUSY-breaking Lagrangian has **only** terms of dimensionally $d < 4$. Thus, schematically, $\mathcal{L}_{\text{SUSY breaking}}$ has the form [51]

$$\mathcal{L}_{\text{SUSY breaking}} = -\sum_{ij} \mu_{oij}^2 \phi_i^\dagger \phi_j - \sum_i m_i \lambda_i \lambda_i + \sum_j A_j W_{3j}(\phi_i) + B W_2(\phi_i) \, . \tag{107}$$

In the above the ϕ_i fields are scalars and the λ_i fields are gauginos. The coefficients μ_{oij}^2 are scalar mass terms for the ϕ_i, ϕ_j scalars; m_i are possible gaugino masses; and A_i and B are scalar coefficients which multiply the trilinear and bilinear couplings of scalar fields which follow from the form of the superpotential.

Including SUSY breaking terms, the Higgs potential is modified to

$$V(H_u, H_d) = (H_u^\dagger \; H_d^\dagger)\mathcal{M}^2 \left(\begin{array}{c} H_u \\ H_d \end{array} \right) + \frac{1}{8}g'^2[H_u^\dagger H_u - H_d^\dagger H_d]^2$$

$$+ \frac{1}{8}g_2^2[H_u^\dagger \vec{\tau} H_u + H_d^\dagger \vec{\tau} H_d] \cdot [H_u^\dagger \vec{\tau} H_u + H_d^\dagger \vec{\tau} H_d] \, , \quad (108)$$

where the mass squared $\mathcal{M}^2$ is given by

$$\mathcal{M}^2 = \left(\begin{array}{cc} \mu^2 + \mu_{11}^2 & -B\mu + \mu_{12}^2 \\ -B\mu + \mu_{12}^2 & \mu^2 + \mu_{22}^2 \end{array} \right) \equiv \left(\begin{array}{cc} m_1^2 & m_3^2 \\ m_3^2 & m_2^2 \end{array} \right) . \quad (109)$$

Obviously, a breakdown of $SU(2) \times U(1) \to U(1)_{em}$ is now possible **provided**

$$\det \mathcal{M}^2 < 0 \, . \quad (110)$$

Note that the Higgs potential in the SUSY SM, even though it involves 2 Higgs fields, is considerably more restricted than that of the SM. In particular, the quartic terms of the potential are **not** arbitrary. Because they arise from the D-terms, these couplings are fixed by the strength of the gauge interactions themselves.

If $SU(2) \times U(1) \to U(1)_{em}$, the spectrum of the potential V of Eq. 108 contains 5 physical scalars: $h; H; A;$ and $H^\pm$. The first three of these are neutrals, with two scalars h and H and one pseudoscalar A.[o] The masses of all 5 of these states are functions of the 5 independent parameters entering in V, namely $(m_1^2, m_2^2, m_3^2, g'^2, g_2^2)$ or, more physically, the set of three masses (M_W^2, M_Z^2, M_A^2) and two mixing angles $(\tan\beta, \cos\theta_W)$. However, since $M_W^2 = M_Z^2 \cos^2\theta_W$, because of doublet Higgs breaking, and because M_W and M_Z are well measured experimentally, in effect the Higgs spectrum in the SUSY SM has only two unknowns: $\tan\beta$ and M_A^2.

A straightforward calculation, using the potential of Eq. 108 yields the following results[52]

$$M_{H^\pm}^2 = M_A^2 + M_W^2$$

$$M_{H,h}^2 = \frac{1}{2}(M_A^2 + M_Z^2) \pm \frac{1}{2}\left[(M_A^2 + M_Z^2)^2 - 4M_Z^2 M_A^2 \cos^2 2\beta\right]^{1/2} \quad (111)$$

---

[o]By convention $M_h < M_H$.

It is easy to see from Eq. 111 that there is always one **light Higgs** in the spectrum:

$$M_h \leq M_Z |\cos 2\beta| \leq M_Z \ . \tag{112}$$

However, the bound of Eq. 112 is not trustworthy, as it is quite sensitive to radiative effects which are enhanced by the large top mass [53]. Fortunately, the magnitude of the radiative shifts for M_h^2 can be well estimated, by either direct conputation [54] or via the renormalization group [55].

It is useful to illustrate the nature of these radiative shifts for M_h^2 using the RGE in the special limit in which $M_h \to (M_H)_{SM}$, but where $(M_H)_{SM}$ is fixed to be at the Z-mass. This limit [55] obtains as M_A gets very large ($M_A \to \infty$) and $|\cos 2\beta| \to 1$. In the above limit, the only remaining light field in the theory is M_h and, since $M_h = M_Z$ at tree level, the theory is just like the SM except that the quartic Higgs coupling λ, in the SM, is fixed to

$$\lambda_{SM} = \frac{1}{8}(g_2^2 + g'^2) \ . \tag{113}$$

Recalling the RGE [Eq. 58] for the evolution of λ, with its large **negative** contribution due to λ_t, one sees that, approximately

$$\lambda(\mu) = \lambda(M_h) - \frac{3}{8\pi^2}\lambda_t^4 \ln \frac{\mu}{M_h} \ . \tag{114}$$

Using Eq. 114 one can readily estimate the radiative shift from the tree level equation $M_h = M_Z$. One has

$$M_h^2 = 2\lambda(M_h)v_F^2 = 2\lambda(\mu)v_F^2 + \frac{3}{8\pi^2}\lambda_t^4 \ln \frac{\mu}{M_h} \ . \tag{115}$$

Using Eq. 113 and taking μ to be the characteristic scale of the SUSY partners- $\tilde{m}$- the, relatively, slow running of the gauge couplings allows one to write, approximately

$$2\lambda(\tilde{m})v_F^2 = \frac{1}{4}(g_2^2 + g'^2)\big|_{\tilde{m}} v_F^2 \simeq \frac{1}{4}(g_2^2 + g'^2)\big|_{M_Z} v_F^2 = M_Z^2 \ . \tag{116}$$

Hence Eq. 115 yields the formula [55]

$$M_h^2 = M_Z^2 + \frac{3\alpha m_t^4}{2\pi \sin^2\theta_W M_W^2} \ln \frac{\tilde{m}^2}{M_Z^2} \ . \tag{117}$$

This is quite a large shift since, for the scale of the SUSY partners $\tilde{m} \simeq 1$ TeV, numerically one finds $\Delta M_h \simeq 20$ GeV.

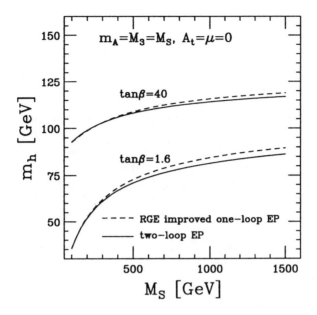

Figure 5: Plots of M_h as a function of $\tilde{m} \equiv M_S$ for two values of $\tan\beta$, from Ref 56.

Eq. 117 was obtained in a particular limit ($|\cos 2\beta| \to 1$), but an analogous result can be obtained for all $\tan\beta$. It turns out that for small $\tan\beta$ the shifts are even larger than those indicated in Eq. 117. However, for these values of $\tan\beta$ the tree order contribution is also smaller, since $M_h|_{\text{tree}} < M_Z \cos 2\beta$. To illustrate this point, the expectations for M_h, plotted as a function of $\tilde{m}$ is shown in Fig. 5 for two values of $\tan\beta$.

The results shown in this figure [56] neglect any details in the SUSY spectrum, since they have all been subsumed in the average parameter $\tilde{m}$. It turns out that the most important effect of the SUSY spectrum for ΔM_h arises if there is an incomplete cancellation between the top and the stop contributions, due to large $\tilde{t}_L - \tilde{t}_R$ mixing [55]. At their maximum these effects can cause a further shift of order $(\Delta M_h)_{\text{mixing}} \simeq 10$ GeV.

One can contrast these predictions of the SUSY SM with experiment. At LEP 200, the four LEP collaborations have looked both for the process $e^+e^- \to hZ$ and $e^+e^- \to hA$. The first process is analogous to that used for searching for the SM Higgs, while hA production is peculiar to models with two (or more) Higgs doublets. One can show that these two processes are complementary, with one dominating in a region of parameter space where the other is small, and *vice versa*[52]. LEP 200 has established already rather strong bounds for M_h and M_A setting the 95% C.L. bounds (for $\tan\beta > 0.8$)[57]

$$M_h > 77 \text{ GeV} ; \quad M_A > 78 \text{ GeV} \tag{118}$$

As Fig. 6 shows, if there is not much $\tilde{t}_{\rm L} - \tilde{t}_{\rm R}$ mixing the low $\tan\beta$ region $[0.8 < \tan\beta < 2.1]$ is also already excluded.

Although the SUSY SM is rather predictive when it comes to the Higgs sector, beyond this sector the spectrum of SUSY partners and possible allowed interactions is quite model dependent. Most supersymmetric extensions of the SM considered are assumed to contain a discrete symmetry, R-parity, which is conserved. This assumption simplifies considerably the form of the possible interactions one has to consider. In fact, R-parity conservation provides an essentially unique way to generalize the SM since R, defined by

$$R = (-1)^{Q+L+2J} , \tag{119}$$

with Q being the quark number, L the lepton number and J the spin, turns out simply to be $+1$ for all particles and -1 for all sparticles.

Obviously, R parity conservation implies that SUSY particles enter in vertices always in pairs, and hence sparticles are always pair produced. This last fact implies, in turn, the **stability** of the lightest supersymmetric particle (LSP), even in the presence of supersymmetry breaking interactions. Although supersymmetry must be broken, since we do not observe multiplets of particles and sparticles of the same mass, SUSY breaking interactions are quite restrictive and do not end up by violating the stability of the LSP. Let me discuss the issue of SUSY breaking in a little more detail, since the manner in which one breaks supersymmetry is the principal source of model-dependence for the SUSY SM.

In general[58], one assumes that SUSY is spontaneously broken at some scale Λ in some **hidden sector** of the theory. This sector is coupled to ordinary matter by some **messenger** states of mass M, with $M \gg \Lambda$, and all that obtains in the visible sector is a set of soft SUSY breaking terms—terms of dimension $d < 4$ in the Lagrangian of the theory.[p] Ordinary matter contains supersymmetric states with masses $\tilde{m} \sim$ TeV, with $\tilde{m}$ given generically by

$$\tilde{m} \sim \frac{\Lambda^2}{M} . \tag{120}$$

Within this general framework, two distinct scenarios have been suggested which differ in what one assumes are the messengers that connect the hidden SUSY breaking sector with the visible sector. In supergravity models[59]

---

[p]Terms of $d = 4$ would re-introduce the hierarchy problem.

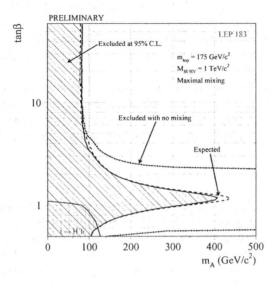

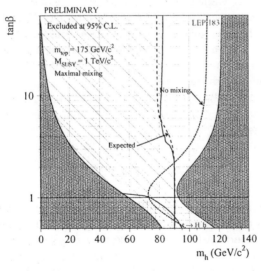

Figure 6: LEP200 limits for M_h and M_A as a function of $\tan\beta$, from Ref. 57.

(SUGRA), the messengers are gravitational interactions, so that $M \sim M_{\text{Planck}}$. Then, because of Eq. 120 and the demand that $\tilde{m} \sim$ TeV, the scale of SUSY breaking in the hidden sector is of order $\Lambda \sim 10^{11}$ GeV. In contrast, in models where the messengers are gauge interactions [60] (Gauge Mediated Models) with $M \sim 10^6$ GeV, then the scale of spontaneous breaking of supersymmetry is around $\Lambda \sim 10^3$ TeV.

In both cases one assumes that the supersymmetry is a local symmetry, gauged by gravity [61]. Then the massless fermion which originates from spontaneous SUSY breaking, the goldstino, is absorbed and serves to give mass to the spin-3/2 gravitino—the SUSY partner of the graviton. This mass is of order

$$m_{3/2} \sim \frac{\Lambda^2}{M_{\text{P}}} . \tag{121}$$

Obviously, in SUGRA models the gravitino has a mass of the same order as all the other SUSY partners ($\tilde{m} \sim$ TeV). However, in SUGRA models, in general, one does not assume that the gravitino is the LSP. However, in Gauge Mediated Models, since $\Lambda \ll 10^{11}$ GeV, the gravitino is definitely the LSP.

Besides the above difference, the other principal difference between SUGRA and Gauge Mediated Models of supersymmetry breaking is the assumed form of the soft breaking terms. In SUGRA models, to avoid FCNC problems, one needs to assume that the soft breaking terms are **universal**. This assumption is unnecessary in Gauge Mediated Models, where in fact one can explicitly compute the form of the soft breaking terms and show that they do not lead to FCNC. Let me discuss this a little further.

The basic point is the following. As I alluded to earlier [cf. Eq. 107], as a result of supersymmetry breaking one ends up, in general, with nondiagonal mass terms for the scalar fields ϕ_i:

$$\mathcal{L}_{\substack{\text{soft} \\ \text{mass}}} = -\sum_{ij} \mu_{oij}^2 \phi_i^\dagger \phi_j . \tag{122}$$

These terms, which connect states of the same charge, can lead to large FCNC through effective mass diagonal couplings of gluinos to squarks and quarks. The scalar mass insertion (122) gives rise to a $\tilde{g} d_L \tilde{s}_L$ vertex. Such a vertex can generate a very large $K^o - \bar{K}^o$ mixing term, through a gluino-squark box graph. With SUSY masses in the TeV range, this would be a total disaster. Hence, effectively, only diagonal soft mass terms can be countenanced. In the SUGRA models, this circumstance forces one to consider only **universal** soft breaking terms, with

$$\mu_{oij}^2 = \mu_o^2 \delta_{ij} . \tag{123}$$

In the case of Gauge Mediated Models, this flavor blindness arises more naturally since SUSY breaking is the result of gauge interactions which are flavor diagonal. As a result, the soft mass terms can only couple diagonally and the soft masses will be proportional to the gauge couplings squared. One can show that the soft mass for the ith scalar field is given by [60]

$$\mu_i^2 = \sum_{a=1}^{3} \frac{\alpha_a^2}{(4\pi)^2} C_i \frac{\Lambda^4}{M^2} , \qquad (124)$$

where C_i is the appropriate Casimir factor for the scalar field in question. It follows from this equation that in Gauge Mediated Models, in general, squarks are heavier than sleptons since the latter do not have strong interactions. In SUGRA models both squarks and sleptons are assumed to have the same mass at large scales ($\mu \gtrsim \Lambda$). However, as a result of the evolution of couplings these universal masses can become quite different at scales of order 100 GeV. Thus SUGRA models at low scales turn out to be not so dissimilar in their spectrum to Gauge Mediated Models.

This point is particularly germane for gauginos, where the differences between SUGRA and Gauge Mediated Models are quite small. For Gauge Mediated Models, the analogous equation to Eq. 124 for gaugino masses is only quadratically dependent on the gauge couplings [60]

$$m_a = \frac{\alpha_a}{4\pi} \frac{\Lambda^2}{M} . \qquad (125)$$

Hence the ratio of the $SU(3)$, $SU(2)$ and $U(1)$ gaugino masses scale with the α_i

$$m_1 : m_2 : m_3 = \alpha_1 : \alpha_2 : \alpha_3 . \qquad (126)$$

In SUGRA models, although at high scales one assumes a universal gaugino mass $m_{1/2}$, the masses of the individual gauginos evolve in the same way as the squared gauge coupling constants

$$\frac{dm_i}{d\ln\mu^2} = -\frac{1}{16\pi^2} b_i m_i . \qquad (127)$$

Whence, at low scales, Eq. 126 holds again. As a result, the most important difference between these two SUSY breaking scenarios is that in Gauge Mediated Models the gravitino is the LSP, while in SUGRA models, the LSP is, in general, thought to be a neutralino—a spin-1/2 partner to the neutral gauge and Higgs bosons.

In Gauge Mediated Models, because the gravitino is the LSP, an important role for phenomenology is played by the "next lightest" SUSY state—the NLSP.

In general, the NLSP in these models is either a slepton or a neutralino[60] and, because it is not the lightest, this NLSP is unstable. The decay

$$\text{NLSP} \to \text{ordinary state} + \text{gravitino} \tag{128}$$

has a lifetime which scales as

$$\tau_{\text{NLSP}} \to \frac{\Lambda^4}{M_{\text{NLSP}}^5} . \tag{129}$$

Depending on whether the NLSP decays occur within, or outside, the detector, the phenomenology of the Gauge Mediated Models will be similar, or rather different, to that of SUGRA models. This is because the gravitino acts essentially as missing energy, which is the signal associated with the SUGRA LSP.

I will not explicitly discuss here what are the expected signals for either the SUGRA or the Gauge Mediated scenarios, since the expected phenomenology is both involved and quite dependent on the actual spectrum of supersymmetric states assumed [62]. I note here only that the present Tevatron and LEP 200 lower bounds on SUSY-states, although somewhat model dependent, are in the neighborhood of 100 GeV. More precisely, the weakest bounds are for the neutralinos ($M_{\chi^\circ} \gtrsim 30$ GeV), followed by charginos ($M_{\chi^\pm} \gtrsim 70$ GeV), sleptons ($M_{\tilde{l}} \gtrsim 90$ GeV) and squarks and gluinos ($m_{\tilde{q}}, m_{\tilde{g}} \gtrsim 200$ GeV). [62]

Before closing this Section, I would like to make two final points concerning the SUSY alternative. These are

i) Not only does supersymmetry allow for a stable hierarchy between the Planck mass and the Fermi scale, but supersymmetry breaking itself can help trigger the $SU(1) \times U(1) \to U(1)_{\text{em}}$ breaking. The Higgs mass matrix $\mathcal{M}^2$ can start with det $\mathcal{M}^2 > 0$ at high scales but can evolve to det $\mathcal{M}^2 < 0$ at the Fermi scale. Indeed m_2^2, the mass term associated with the H_u Higgs, is strongly affected by the large top Yukawa coupling and can change sign as one evolves to low q^2. This radiative breaking of $SU(2) \times U(1)$, triggered by SUSY breaking[63], is a very attractive feature of the SUSY SM. For it to produce the Fermi scale, however, one must inject already at very high scales a SUSY breaking mass parameter also of order v_F. Why this should be so remains a mystery.

ii) In SUGRA models the neutralino LSP, which is a linear combination of all the neutral weak gaugino and Higgsino fields

$$\tilde{\chi} = \alpha_\gamma \tilde{\gamma} + \alpha_Z \tilde{Z} + \alpha_u \tilde{H}_u + \alpha_d \tilde{H}_d \tag{130}$$

is an excellent candidate for cold dark matter [64]. In the Gauge Mediated case, the gravitino could provide a warm dark matter candidate [65], provided that

$$m_{3/2} \sim \frac{\Lambda^2}{M_{\mathrm{P}}} \sim \mathrm{KeV} \; , \tag{131}$$

which requires $\Lambda \sim 10^3$ TeV. The presence of possible candidates for the dark matter in the Universe, is another quite remarkable feature of low energy SUSY models and speaks in favor of the SUSY alternative.

6 Are there Extra Dimensions of Size Greater than a $(\mathrm{TeV})^{-1}$?

To conclude these lectures, I want to make some remarks on the phenomenological consequences of imagining that the Fermi scale is associated with the Planck scale of $(d+4)$-dimensional gravity. That is

$$v_F \sim (M_{\mathrm{P}})_{d+4} \equiv M \; . \tag{132}$$

The extra d-dimensions in this theory are assumed to be compact and of size R. By comparing the Newtonian forces in $(d+4)$-dimensions with that in four dimensions, one can interrelate M and the usual Planck mass $M_{\mathrm{P}} = G_N^{-1/2} \simeq 10^{19}$ GeV. [12] One has

$$F_4 = \frac{1}{M_{\mathrm{P}}^2} \frac{m_1 m_2}{r^2} \; ; \quad F_{4+d} = \frac{1}{(M)^{2+d}} \frac{m_1 m_2}{r^{d+2}} \; . \tag{133}$$

Then, using Gauss' law, it follows that [12]

$$M_{\mathrm{P}} = M(MR)^{d/2} \; . \tag{134}$$

Obviously, demanding that M be of the order of the Fermi scale v_F, fixes the size R of the compact dimensions as a function of the number of these extra dimensions d. One finds that for $d = 2$, $R \sim 10^{-1}$ cm, while for $d = 6$, $R \simeq 10^{-13}$ cm. For distances r below the size R of the compact dimensions one will begin to register the fact that gravitational forces do not follow the familiar r^{-2} law. Thus the first test that this idea is not nonsensical is to look for possible modifications of the usual Newtonian potential at short distances. For $r \sim R$, the potential will feel Yukawa-like correctione of strength $2d$: [66]

$$V(r) = -\frac{G_N m_1 m_2}{r} \left\{ 1 + 2de^{-r/R} + \dots \right\} \; . \tag{135}$$

There have been a variety of tests of the Newtonian potential at (relatively) short distances. For $r > 100\mu$, the most stringent limits for deviation from the

Newtonian expectations [67] allow $R \sim 10^{-1}$ cm for strengths $\alpha \equiv 2d$ of $O(1)$. Below 100μ, the best bounds on non-Newtonian forces come from Casimir experiments but, as shown in Fig. 7, these experiments only put some constraints on modifications which have a strength α much below unity. This figure also shows the limits which might be achievable in a proposed experiment employing 1 KHz mechanical oscillators as test masses. This experiment [67] should be able to push α to $\alpha \sim 10^{-1}$ for compact dimensions $R \sim 10^{-1}$ cm. Thus it could **directly test** for deviations from Newtonian gravity to a level where an effect might be seen.

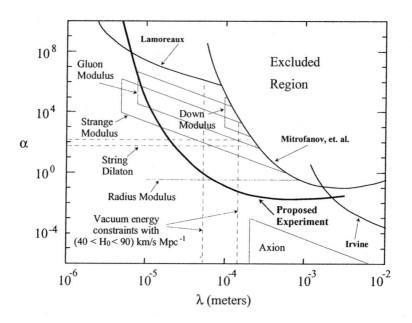

Figure 7: Experimental status of gravitational strength forces at short distances, from Ref. 67.

At high energy, the presence of extra compact dimensions of finite size R can be detected because these theories allow abundant graviton production to occur. Obviously, to see any effects one needs experiments at energies $(\sqrt{s})_{\text{parton}} \sim M \sim v_{F^-}$ energies which will be attainable at the LHC. The presence of compact dimensions of size R allows gravitons to get produced, at energies of order $(\sqrt{s})_{\text{parton}} \sim M$, with a probability proportional to M^{-2} **not**

M_P^{-2}. However, the graviton production is rather soft, with the softness being greatest the larger the number of compact dimensions d is.

This can be understood qualitatively as follows. In the theories under discussion, the SM fields exist essentially on a 4-dimensional hypersurface, with gravity acting both on this hypersurface, as well as on a set of compact dimensions of size R. From the point of view of the 4-dimensional theory, a graviton with its momentum p^α acting in the compact dimensions is equivalent to having a particle of mass

$$m_{\text{eff}} \sim \frac{i}{R} \, , \tag{136}$$

where i is an integer associated with the level of excitation. It follows that, for given total energy E, one can access many states in the compact dimensions. Using (136), the spacing among these states is of order $\Delta E \sim R^{-1}$. Thus, for processes of total energy, E, the total number of states probed in the compact dimensions is

$$N \sim \left(\frac{E}{\Delta E} \right)^d \sim (RE)^d \, . \tag{137}$$

The probability of gravitational production at $(\sqrt{s})_{\text{parton}} = E$ is inversely proportional to the Planck mass M_P, but directly proportional to the number of states N excited in the compact dimensions. If N is large, which will occur for $E \gg R^{-1}$, then, effectively, the probability of producing gravitational radiation is much greater than the classical expectations. Indeed, as alluded to above, this probability scales like M^{-2} not M_P^{-2}. One has

$$\text{Probability} \sim \frac{1}{M_P^2} N \sim \frac{(ER)^d}{M_P^2} = \frac{(ER)^d}{M^{2+d} R^d} = \frac{1}{M^2} \left(\frac{s}{M^2} \right)^{d/2} \, . \tag{138}$$

Note, however, that even though the probability indeed scales like M^{-2} there is an additional soft infrared multiplying factor of $(s/M^2)^{d/2}$. Therefore, one learns that strong graviton production comes on slowly for $s < M^2$. Thus the importance of these effects for the LHC, even if these theories were to be true, is crucially dependent on the value of M.

Mirabelli, Perelstein and Peskin [66], as well as others [68], have studied graviton production at LEP and Tevatron energies to try to obtain bounds on M from present day data. These authors have looked for the processes $e^+e^- \to \gamma G$ and $p\bar{p} \to$ jet G, with the graviton experimentally being manifested as missing energy. Analyses of the LEP data for the process $e^+e^- \to \gamma\nu\bar{\nu}$ and of the CDF/DO bounds for monojet production yield the bounds:

$$M \gtrsim 600\text{GeV} \ (d = 6); \quad M \gtrsim 750\text{GeV} \ (d = 2). \tag{139}$$

Obviously the LHC will be sensitive to much greater values of M. However, at least for $d = 2$, astrophysics already puts a strong constraint on M. Basically, if M is too low graviton emission will cool off supernovas too quickly. The fact that SN 1987a does not show any anomalous cooling then allows one to set a bound of $M \gtrsim 30$ TeV, for $d = 2$.[69]

7 Concluding Remarks

In these lectures I have discussed the theoretical arguments for believing that there is physics beyond the electroweak theory, related to the origin and magnitude of the Fermi scale v_F. Each of the suggestions for new physics which I examined: dynamical symmetry breaking, supersymmetry and extra compact dimensions, had its own characteristic signals and its own set of theoretical puzzles to resolve. A priori, it is clearly not possible to select among these alternatives purely theoretically. However, from the few meager experimental hints for physics beyond the SM that we have today, supersymmetry does appear to be favored. What is exciting is that we should know quite soon which of these alternatives, if any, are correct. If one is lucky, the answer may come perhaps already from experiments at LEP200 or the Tevatron. However, the origin of the Fermi scale should certainly become clear once the LHC starts running.

Acknowledgements

I am grateful to Bruce Campbell and Fakir Khanna for the wonderful hospitality at Chateau Lake Louise. This work was supported in part by the Department of Energy under contract No. DE-FG03-91ER40662, Task C.

References

1. For an introduction to the Standard Model see, for example, my lectures, and those of others, at TASI 90, **Testing the Standard Model**, Proceedings of the 1990 Theoretical Advanced Study Institute in Elementary Particle Physics, eds. M. Cvetic and P. Langacker (World Scientific, Singapore 1991). A subset of the original references includes: S.L. Glashow, Nucl. Phys. **B22**, 579, (1961); S. Weinberg, Phys. Rev. Lett. **19**, 1264, (1967); A. Salam, in **Elementary Particle Theory**, ed. N. Svartholm (Amquist and Wiksels, Stockholm 1969); H. Fritzsch, M. Gell-Mann and H. Leutwyler, Phys. Lett. **47**, 365, (1973); H.D. Politzer, Phys. Rev. Lett. **30**, 1346, (1973); D. Gross and F. Wilczek, Phys. Rev. Lett. **30**, 1343, (1973); S. Weinberg, Phys. Rev. Lett. **31**, 494, (1973).

2. M.W. Grünewald, in the Proceedings of the 29th Conference on High Energy Physics, ICHEP98, Vancouver, Canada, July 1998, eds. A. Astbury, D. Axen and J. Robinson (World Scientific, Singapore 1999) p569.

3. SuperKamiokande Collaboration, Y. Fukuda *et al.*, Phys. Rev. Lett. **81**, 1562, (1998).

4. For a discussion see, for example, R.D. Peccei in **Concepts and Trends in Particle Physics**, eds. H. Latal and H. Mutter, in Proceedings of the XXV Int. Universität Wochen für Kernphysik, Schladming, Austria, 1986 (Springer-Verlag, Berlin 1987).

5. For a discussion see, for example, R.D. Peccei, hep-ph/9906509, to be published in the Proceedings of the VIII Escuela Mexicana de Particulas y Campos, Oaxaca, Mexico, 1998.

6. For a discussion see, for example, A.L. Fetter and J.D. Walecka, **Quantum Theory of Many Particle Systems** (McGraw Hill Book Company, New York 1971).

7. M. Aizenman, Comm. Math. Phys. **86**, 1, (1982); J. Frölich, Nucl. Phys. **B200**, [FS4] 281, (1982).

8. L. Landau in **Niels Bohr and the Development of Physics**, ed. W. Pauli (McGraw Hill, New York 1955).

9. G. 't Hooft, in **Recent Developments in Gauge Theories**, Proceedings of the Nato Advanced Study Institute, Cargese, France, 1979, eds G. 't Hooft *et al.* (Plenum Press, New York 1980)

10. M. Luescher and P. Weisz, Nucl. Phys. **B290**, 25, (1987); **B295**, 65, (1988); **B318**, 705, (1989); J. Kuti, L. Lin and Y. Shen, Phys. Rev. Lett. **61**, 678, (1988); A. Hasenfratz *et al.*, Phys. Lett. **199B**, 531, (1987); Nucl. Phys. **B317**, 81, (1989); G. Bhanot *et al.*, Nucl. Phys. **B343**, 467, (1990); **B353**, 551, (1991).

11. For an introduction see, for example, J. Wess and J. Bagger, **Supersymmetry and Supergravity** (Princeton University Press, Princeton 1992) 2nd Edition.

12. N. Arkani-Hamed, S. Dimopoulos and G. Dvali, Phys. Lett. **B429**, 263, (1998); Phys. Rev. **D59**, 086004, (1999); I. Antoniadis, S. Dimopoulos and G. Dvali, Nucl. Phys. **B516**, 70, (1998); I. Antoniadis, N. Arkani-Hamed, S. Dimopoulos and G. Dvali, Phys. Lett. **B436**, 257, (1998).

13. I. Antoniadis, Phys. Lett. **B246**, 377, (1990); Proceedings of the PASCOS-91 Symposium, Boston, Mass. (World Scientific, Singapore 1991); see also, J. Lykken, Phys. Rev. **D54**, 3693, (1996).

14. The data from the Tevatron is summarized by E. Barberis and W.-M. Yao, in the Proceedings of the 29th Conference on High Energy Physics, ICHEP98, Vancouver, Canada, July 1998, eds. A. Astbury, D. Axen and

J. Robinson (World Scientific, Singapore 1999) p1098 and p1093.

15. The fits to the SM of precision electroweak data are summarized by D. Karlen, in the Proceedings of the 29th Conference on High Energy Physics, ICHEP98, Vancouver, Canada, July 1998, eds. A. Astbury, D. Axen and J. Robinson (World Scientific, Singapore 1999) p47; see, also, Ref. 31.

16. The data from LEP200 is summarized by P.A. McNamara, in the Proceedings of the 29th Conference on High Energy Physics, ICHEP98, Vancouver, Canada, July 1998, eds. A. Astbury, D. Axen and J. Robinson (World Scientific, Singapore 1999) p1303.

17. M. Veltman, Acta Phys. Pol. **B8**, 475, (1977). The $O(\alpha^2)$ corrections to ρ have been calculated by J. Van der Bij and M. Veltman, Nucl. Phys. **B231**, 205, (1984) and are proportional to M_H^2.

18. M.E. Peskin and T. Takeuchi, Phys. Rev. Lett. **65**, 964, (1990); Phys. Rev. **D46**, 381, (1992); B. Holdom and J. Terning, Phys. Lett. **B247**, 88, (1990); D.C. Kennedy and P. Langacker, Phys. Rev. Lett. **65**, 2967, (1990); G. Altarelli and R. Barbieri, Phys. Lett. **B253**, 161, (1990); G. Altarelli, R. Barbieri and S. Jadach, Nucl. Phys. **B369**, 3, (1992); G. Altarelli, R. Barbieri and F. Caravaglios, Nucl. Phys. **B405**, 3, (1993); Phys. Lett. **B349**, 145, (1995).

19. M.E. Peskin and T. Takeuchi, Phys. Rev. **D46**, 381, (1992).

20. G. Altarelli, R. Barbieri and F. Caravaglios, Int. J. Mod. Phys. **A13**, 1031, (1998).

21. H. Georgi, H.R. Quinn and S. Weinberg, Phys. Rev. Lett. **33**, 451, (1974).

22. R. Slansky, Phys. Reports **79C**, 1, (1981).

23. See, for example, U. Amaldi, W. de Boer and H. Fürstenau, Phys. Lett. **B260**, 447, (1991).

24. See, for example, M. Olechowski and S. Pokorski, Phys. Lett. **B257**, 388, (1991).

25. See, for example, B. Schrempp and M. Wimmer, Progr. in Part. and Nucl. Phys. **37**, 1, (1996).

26. G. Altarelli and G. Isidori, Phys. Lett. **B337**, 141, (1994).

27. M. Lindner, Z. Phys. **C31**, 295, (1986).

28. B. Pendelton and G.G. Ross, Phys. Lett. **B98**, 291, (1981); C.T. Hill, Phys. Rev. **D24**, 691, (1981).

29. For a recent discussion see, for example, J.N. Bahcall, P.I. Krastev and A. Yu. Smirnov, Phys. Rev. **D58**, 096016, (1998).

30. T. Yanagida, in Proceedings of the Workshop on the Unified Theories and Baryon Number in the Universe, Tsukuba, Japan 1979, eds. O.

Sawada and A. Sugamoto, KEK Report No. 79-18; M. Gell-Mann, P. Ramond and R. Slansky, in **Supergravity**, Proceedings of the Workshop at Stony Brook, NY, 1979, eds. P. van Nieuwenhuizen and D. Freedman (North-Holland, Amsterdam 1979).

31. See, for example, the compilation of the latest electroweak data from LEP and the SLC in the report of the LEP Electroweak Working Group, CERN-EP/99-15, February 1999.

32. See, for example, S. Weinberg, in Proceedings of the XXIII International Conference on High Energy Physics, Berkeley, California 1986, ed. S.C. Loken (World Scientific, Singapore 1987).

33. L. Susskind, Phys. Rev. **D20**, 2619, (1979); S. Weinberg, Phys. Rev. **D13**, 974, (1976).

34. See, for example, R.D. Peccei, in **Broken Symmetries**, eds. L. Mathelitsch and W. Plessas, Lecture Notes in Physics 521 (Springer Verlag, Berlin 1999).

35. See, for example, J.F. Donoghue, E. Golowich and B.R. Holstein, **Dynamics of the Standard Model** (Cambridge University Press, Cambridge, UK 1992).

36. S. Dimopoulos and L. Susskind, Nucl. Phys. **B155**, 237, (1979); E. Eichten and K. Lane, Phys. Lett. **B90**, 125, (1980).

37. S. Dimopoulos and J. Ellis, Nucl. Phys. **B182**, 505, (1981).

38. See, for example, E. Fahri and L. Susskind, Phys. Rept. **74C**, 277, (1981); see, also, Ref. 38.

39. For an extensive review, see R.S. Chivukula, hep-ph/9803219, lectures presented at 1997 Les Houches Summer School, Les Houches, France.

40. B. Holdom, Phys. Rev. **D24**, 1441, (1981); K. Yamawaki, M. Bando, and K. Matsumoto, Phys. Rev. Lett. **56**, 1335, (1986); T. Akiba and T. Yanagida, Phys. Lett. **169B**, 432, (1986); T. Appelquist, D. Karabali and L. Wijewardhana, Phys. Rev. Lett. **57**, 982, (1986).

41. R. Sundrum and S.D.H. Hsu, Nucl. Phys. **B391**, 127, (1993).

42. H. Pagels, Phys. Rept. **16C**, 219, (1975).

43. K. Lane, Phys. Rev. **D10**, 2605, (1974); H.D. Politzer, Nucl. Phys. **B117**, 397, (1976).

44. See, for example, B. Holdom, Phys. Lett. **B246**, 169, (1990); see, also, B. Holdom in **Dynamical Symmetry Breaking**, ed. K. Yamawaki (World Scientific, Singapore 1992); R. Sundrum, Nucl. Phys. **B395**, 60, (1993).

45. R.S. Chivukula, S.B. Selipsky and E.H. Simmons, Phys. Rev. Lett. **69**, 575, (1992).

46. C.T. Hill, Phys. Lett. **B345**, 483, (1995); E. Eichten and K. Lane, Phys.

Lett. **B352**, 382, (1995); K. Lane, Phys. Rev. **D54**, 2204, (1996).

208

Lett. **B352**, 382, (1995); K. Lane, Phys. Rev. **D54**, 2204, (1996).

47. Y. Nambu, in Proceedings of the 1988 International Workshop on New Trends in Strong Coupling Gauge Theories, Nagoya, Japan, eds. M. Bando, T. Muta and K. Yamawaki; V.A. Miransky, M. Tanabashi and K. Yamawaki, Mod. Phys. Lett. **A4**, 1043, (1989); W.A. Marciano, Phys. Rev. Lett. **62**, 2793, (1989); W.A. Bardeen, C.T. Hill and M. Lindner, Phys. Rev. **D41**, 1647, (1990).

48. E. Eichten, K. Lane and J. Womersley, Phys. Lett. **B405**, 305, (1997).

49. For reviews see, for example, H.P. Nilles, Phys. Rept. **110C**, 1, (1984); H.E. Haber and G.L. Kane, Phys. Rept. **117C**, 75, (1985).

50. C. Bouchiat, J. Iliopoulos and Ph. Meyer, Phys. Lett. **B38**, 519, (1972); D.J. Gross and R. Jackiw, Phys. Rev. **D6**, 477, (1972).

51. E. Cremmer, S. Ferrara, L. Girardello and A. Van Proyen, Nucl. Phys. **B212**, 506, (1983); L. Girardello and M. Grisaru, Nucl. Phys. **B194**, 65, (1984).

52. J.F Gunion, H.E. Haber, G. Kane and S. Dawson, **The Higgs Hunter's Guide** (Addison-Wesley Publishing Company, Reading, Mass 1990).

53. H.E. Haber and R. Hempfling, Phys. Rev. Lett. **66**, 1815, (1991); Y. Okada, M. Yamaguchi and T. Yanagida, Prog. Theo. Phys **85**, 1, (1991); J. Ellis, G. Ridolfi and F. Zwirner, Phys. Lett. **B257**, 83, (1991); **B262**, 477, (1991).

54. S. Heinemeyer, W. Hollik and G. Weiglein, Phys. Rev. **D48**, 091701, (1998); Phys. Lett. **B440**, 296, (1998); **B455**, 179, (1999); see, also, R. Hempfling and A.H. Hoang, Phys. Lett. **331**, 99, (1994); R.-J. Zhang, Phys. Lett. **B447**, 89, (1999).

55. For a review see, for example, H.E. Haber, hep-ph/9901365, to appear in the Proceedings of the 4th International Symposium on Radiative Corrections (RADCOR 98), Barcelona, Spain, 1998.

56. R.-J. Zhang, Phys. Lett. **B447**, 89, (1999).

57. The data from LEP is summarized by K. Desch, in the Proceedings of the 29th Conference on High Energy Physics, ICHEP98, Vancouver, Canada, July 1998, eds. A. Astbury, D. Axen and J. Robinson (World Scientific, Singapore 1999) p1309.

58. For a discussion see, for example, M.E. Peskin, hep-ph/9705479, lectures presented at the 1996 European School of High Energy Physics, Carry-le Rouet, France.

59. For reviews see, for example, P. Nath, R. Arnowitt and A.H. Chamseddine, **Aplied N=1 Supergravity** (World Scientific, Singapore 1984); R. Arnowitt and P. Nath, Proceedings of the VII J.A. Swieca Summer School, ed. E. Eboli (World Scientific, Singapore 1994).

209

60. M. Dine and A.E. Nelson, Phys. Rev. **D48**, 1277, (1993); M. Dine, A.E. Nelson and Y. Shirman, Phys. Rev. **D51**, 1362, (1995); M. Dine, A.E. Nelson, Y. Nir and Y. Shirman, Phys. Rev. **D53**, 2658, (1996); for a review, see G.F. Giudice and R. Rattazzi, hep-ph/9801271.
61. S. Ferrara, D.Z. Freedman and P. van Nieuwenhuizen, Phys. Rev. **D13**, 3214, (1976).
62. For a review of SUSY searches, see D. Treille, in the Proceedings of the 29th Conference on High Energy Physics, ICHEP98, Vancouver, Canada, July 1998, eds. A. Astbury, D. Axen and J. Robinson (World Scientific, Singapore 1999) p87.
63. L. Ibañez and G.G. Ross, Phys. Lett. **B110**, 215, (1982); L. Alvarez-Gaume, M. Clauson and M.B. Wise, Nucl. Phys. **B207**, 96, (1982).
64. See, for example, A. Bottino, F. Donato, G. Mignola, S. Scopel, P. Belli and A. Incicchiti, Phys. Lett. **B402**, 103, (1997); see also A. Bottino, astro-ph/9611137, invited talk at the International Workshop on the Identification of Dark Matter, Sheffield, UK, 1996.
65. See, for example, S. Borgani, A. Masiero, and M. Yamaguchi, Phys. Lett. **356B**, 189, (1996).
66. E.A. Mirabelli, M. Perelstein and M.E. Peskin, Phys. Rev. Lett. **82**, 2236, (1999).
67. For a recent review, see J.C. Long, H.W. Chan and J.C. Price Nucl. Phys. **B539**, 23, (1999).
68. T. Han, J.D. Lykken and R.-J. Zhang, Phys. Rev. **D59**, 105006, (1999); G.F. Giudice, R. Rattazzi and J.D. Wells, Nucl. Phys. **B544**, 3, (1999).
69. N. Arkani-Hamed, S. Dimopoulos and G. Dvali, Phys. Rev. **D59**, 086004, (1999).

ELECTROWEAK RESULTS FROM e^+e^- COLLIDERS

JAMES L. PINFOLD

Centre for Subatomic Research,
Physics Department,
University of Alberta,
Edmonton,
Alberta T6G 2N5,
E-mail: pinfold@phys.ualberta.ca

The measurements performed at LEP and SLC over the past nine years have substantially improved the precision of the test of the Standard Model of electroweak interactions. The precision of these tests is such that they are now sensitive to purely weak radiative corrections. This allows us to indirectly determine the W-boson mass ($M_W = 80.367 \pm 0.029$GeV), the top-quark mass ($M_t = 161^{+8.2}_{-7.1}$ GeV) and to set an upper limit on the Higgs boson mass of 262 GeV at the 95% confidence level. Direct measurement of M_W at LEP-II, SDC and D0 as well at M_t at SDC and D0 are in good agreement with the indirect determinations.

1 Introduction

The LEP/SLC program provides a window on the vast panorama of electroweak physics. A vista that spans over 100 GeV in energy with cross-sections ranging from approximately 40 nb at the hadronic peak of the Z^0 to around a picobarn for Z^0Z^0 production and slightly less than 20 pb for W^+W^- production.

Precision measurements of the mass, width and partial widths of the Z^0 boson and of the asymmetries in its production and decay, allow us to see past the tree level predictions for the first time. These tests of the electroweak theory have been advanced by the direct measurements of the properties of the W boson at LEP-2, and Fermi National Accelerator Laboratory (FNAL), as well as the direct measurement of the top quark mass at FNAL.

The top quark mass has been constrained to $161^{+8.2}_{-7.1}$ GeV, using LEP-1 data, W mass measurements[1] and νN scattering results[2]. The direct measurement of the top quark mass made at by the CDF and D0 Collaborations[3], $M_t = 173.8 \pm 5.0$ GeV, is in good agreement with the indirect determination at LEP and SLC. The Higgs mass, the last unknown quantity in the Standard Model, can also be constrained by measurements made at LEP, SLC, and FNAL.

The results presented here are those available for the 1998 summer conferences, unless otherwise specified. Electroweak results from around the Z^0 pole were obtained from the LEP and SLD Electroweak Working Group.

2 LEP and its Detectors

The Large Electron Positron collider (LEP) is a storage ring with circumference of 27.4 km made up of eight 2.8 km long arcs linked by eight straight sections. It has four interaction regions housing the LEP experiments ALEPH, DELPHI, L3, and OPAL. The beam is maintained in its orbit by 3400 dipole magnets and is focussed by 800 quadrapoles and 500 sextapoles.

In the first phase of LEP (LEP-1), that began in 1989, electrons and positrons were brought into collision at a centre of mass energy at, or around, the mass of the Z^0 . The luminous volume produced at the collision points is approximately $0.30 \times .06 \times 2.00$ mm^3, where the three dimensions are measured along the bending radius, the perpendicular to the bending plane, and along the beam. The luminosity of an e^+e^- collider is given by the formula:

$$L = \frac{k_b f N^+ N^-}{A} \qquad (1)$$

where k_b is the number of bunches, f is the bunch rotation frequency (11 kHz), $N^{+(-)}$ the number of particles per $e^+(e^-)$ bunch $(\mathcal{O}(10^{11}))$ and A is the cross-sectional area of the beams assuming they overlap completely.

The beam energy was maintained by copper radio-frequency (RF) cavities. Starting in 1995 the RF was upgraded for the second phase of LEP (LEP-2) with the installation of superconducting RF (SCRF) cavities. This upgrade was finished in 1998. The peak luminosity achieved in this first phase was $2.4 \times 10^{31} cm^2 s^{-1}$. By the end of LEP-I, in 1995, the four LEP experiments had recorded around 15 million hadronic Z^0 decays and approximately 1.6 million leptonic Z^0 decays. About 80% of integrated luminosity of ~ 200 pb^{-1} per experiment, was delivered at a centre-of-mass energy within 100 MeV of the Z^0 mass. The remaining 20% was used to scan the resonance.

The second phase of LEP began in 1995 when the centre of mass energy reached 130/140 GeV, during this phase (LEP-1.5) LEP provided a luminosity of $\sim 5 pb^{-1}$ per experiment. This and subsequent beam energy increases were allowed by the the process of replacing conventional with SCRF cavities. In 1996 the W^+W^- production threshold was reached and LEP-2 began. During this phase each experiment obtained a luminosity of $\sim 20 pb^{-1}$ whilst running at two centre-of-mass energies: 161 and 172 GeV.

In 1997 LEP-2 achieved the centre-of-mass energy of 183 GeV and provided a luminosity of approximately $55 pb^{-1}$ per experiment. In 1998 LEP-2, operating at a centre-of-mass energy of 189 GeV, obtained an instantaneous luminosity of $\sim 10^{32} cm^2 s^{-1}$ whilst delivering $\sim 190 pb^{-1}$ integrated luminosity to each experiment. A summary of the performance of the LEP accelerator from 1993 until 1998 is given in Fig. 1

212

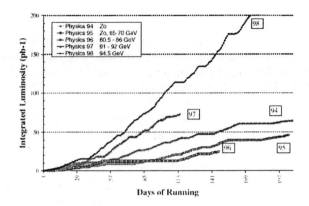

Figure 1: A summary of LEP performance in terms of integrated luminosity versus time is shown for 1993 until the end of data taking in 1998.

The current goals of the LEP program are to provide to each experiment $500pb^{-1}$ of data above the $e^+e^- \to W^+W^-$ threshold and to achieve a centre of mass energy of ~ 200 GeV, or more.

2.1 The LEP Detectors

Each of the LEP detectors has a cylindrical symmetry, a solenoidal magnetic field and a design that follows a basic pattern. We can illustrate that pattern by considering one of the detectors, OPAL[4]. A schematic cross-sectional view of OPAL is given in Fig. 2.

A particle exiting the beampipe first encounters the silicon microstrip vertex detector a cylindrical arrangement of ladders of silicon detectors two layers deep. Each layer provides an accurate measurement of the $r - \phi$, utilizing readout strips every $50\mu m$ and $r - z$ position, from readout strips every $100\mu m$, of the traversing track. The next detector encountered is the vertex chamber, a cylindrical drift chamber of radius ~ 25 cm, that makes 12 $r - \phi$ measurements and 6 $r - z$ " stereo" measurements. The impact resolution obtained with this detector is $\sim 50\mu m$ for high momentum tracks.

The central tracking detector of OPAL is a cylindrical "jet" chamber of radius approximately 2m, allowing 159 $r - \phi$ measurements of a track that fully traverses the detector. The outer radius of the jet chamber is panelled with planar drift chambers that make 6 $r - \phi$ position measurements. A 0.435 Tesla solenoidal field is provided by a conventional water cooled warm coil that surrounds the tracking system. The overall momentum resolution obtained by

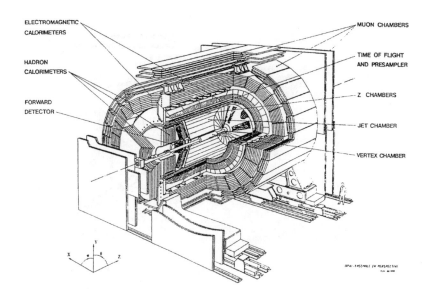

ELECTROMAGNETIC
CALORIMETERS

HADRON
CALORIMETERS

FORWARD
DETECTOR

MUON CHAMBERS

TIME OF FLIGHT
AND PRESAMPLER

Z CHAMBERS

JET CHAMBER

VERTEX CHAMBER

Figure 2: A cut-away view of the OPAL detector.

OPAL is $\sigma(p_t)/p_t = [0.02^2 + (0.0015p_t)^2]^{1/2}$ (p_t in GeV).

The barrel region OPAL electromagnetic calorimeter lies outside of the field region and consists of 9440 lead-glass blocks with depth $\sim 22X^0$ arranged in a pointing geometry. Hermiticity is maintained by endcap leadglass EM calorimeters each consisting of 1132 lead-glass blocks. The energy resolution of the barrel EM calorimetry is $\sim 20\%/\sqrt{E} + 2\%$. A magnet return yoke instrumented with limited streamer tubes forms the hadronic calorimetry in the barrel region.

The layout of the other LEP detectors follows the same basic design as the OPAL detector. ALEPH, like OPAL, is a general purpose detector[5]. Tracking is performed by a 1.8 m Time Projection Chamber (TPC). A fine grain EM calorimeter surrounds the TPC. Both the TPC and the EM calorimeter are immersed in a 1.5 Tesla superconducting solenoidal field. Muons are identified by external muon chambers and by the instrumented hadron calorimeter.

A TPC also forms the main tracking detector of the DELPHI detector[6]. A Ring Imaging Cerenkov Counter (RICH) surrounds the TPC. The High-density Projection Chamber (HPC) provides fine grain EM calorimetery. All of these detector elements are bathed in a 1.2 Tesla solenoidal field provided by a superconducting solenoid housed in a cryostat of radius 2.6 m.

214

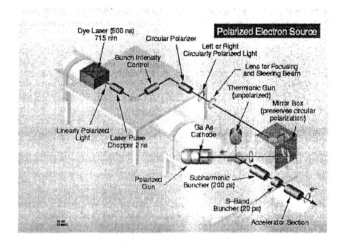

Figure 3: A schematic view of the SLC polarized electron source.

The central tracker of the L3 detector [7] is a 1 m radius Time Expansion Chamber (TEC). The surrounding precision electromagnetic calorimetry is comprised of ~ 8000 Bismuth Germanium Oxide (BGO) crystals. Jet energy measurement is performed with a hadron calorimeter with Uranium absorber plate structure. Outer muon chambers perform precise measurements of muon momentum since the muons are tracked over 4 m inside a 0.5 Tesla magnetic field provided by a magnetic coil that encloses the whole detector

2.2 The Stanford Linear Collider

The Stanford Linear Collider (SLC) is a 3.2 km long linear e^+e^- collider that started operation in 1989. The SLD [8] detector was installed at the single interaction point of SLC in 1990.

The SLC was designed to operate at the Z^0 and has a maximum beam energy of ~ 50 GeV. The luminous volume, with transverse dimension of only a few microns, of SLC is much smaller than that of LEP. Despite the small beam area the luminosity of SLC, which has reached 3×10^{30} cm^{-2} s^{-1} (300 Z's per hour), is somewhat smaller than that of LEP. This is due to the comparatively small beam currents and low collision frequency.

Although the SLC operates at a smaller luminosity than LEP it has the great advantage of a highly polarized electron beam. In 1993 a new source of polarized electrons was commissioned [9]. It was based on an electron gun

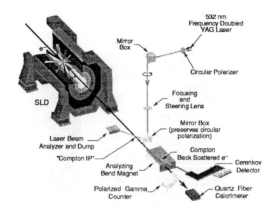

Figure 4: A schematic view of the SLC polarization measurement system based on a Compton Polarimeter.

and a GaAs strained cathode providing electrons with around 90% polarization at a rate of 120 Hz. The polarized electron source of the SLC is shown in Fig. 3. The electron polarization is measured downstream of the Interaction Point (IP) with a Compton polarimeter - a Cerenkov detector array at several angles measures the asymmetry of recoil electrons that scatter of the circularly polarized laser light provided by a YAG laser. The average polarization measured is approximately 80%. Fig. 4 shows the main features of the Compton polarimeter system.

2.3 The SLD Detector

The SLD detector[8] is a general purpose solenoidal detector with design, functionality and coverage similar to a LEP detector. It is designed to study physics at the Z^0 scale. A quadrant view of the SLD detector is shown in Fig. 5. Charged-particle tracking is performed using a CCD-based vertex chamber and a central drift chamber together with a set of endcap drift chambers. The overall momentum resolution is: $\sigma(p_t)/p_t = [0.0095 + 0.0026p_t]$ (P_t is GeV). The $300M$ channel CCD system is arranged in three overlapping layers to allow precision measurement of heavy flavour decays vertices. The solenoidal magnet employs a conventional warm aluminium coil of diameter 5.9 m, to produce a 0.6T field.

Particle energy is measured by three calorimeters. First, a Liquid Argon Calorimeter (LAC) with a lead absorber structure to measure EM and hadronic energy. The energy resolution for the LAC has been measured to be

The SLD Detector

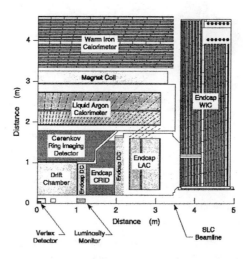

Figure 5: A cross-sectional view of one quadrant of the SLD detector.

$15\%\sqrt{E}$ for the EM and $60\%\sqrt{E}$ for the hadronic section. Third, a warm iron hadronic calorimeter instrumented with Iarocci tubes. Lastly, a luminosity monitor formed from silicon tungsten sampling calorimeters. Charged particle identification is provided by a RICH Detector.

3 The LEP Energy Calibration

A precise knowledge of the centre-of-mass energy at the LEP collision points is required for an accurate measurement of the mass and width of the Z^0. Energy scans were performed in 1993 and 1995 and in both years luminosity was collected at three scan points: at the peak, as well as ~ 1.79 GeV above and below the peak.

Using these scan points the errors on M_Z and Γ_Z depend approximately on the sum and difference of the centre-of-mass energies (E_{CM}'s) at the two off peak points:

$$\Delta M_Z \simeq 0.5\Delta(E_{+2} + E_{-2}); \quad \Delta\Gamma_Z \simeq \frac{\Gamma_Z\Delta(E_{+2} - E_{-2})}{(E_{+2} - E_{-2})} = 0.71\Delta(E_{+2} - E_{-2}).$$
(2)

where $E_{\pm 2}$ are the luminosity weighted E_{CM}'s at the two off-peak points. In order to match the statistical precision of the measurement of M_Z and Γ_Z, $E_{\pm 2}$ have to be known with an error of $\sim$ 1MeV.

3.1 The Beam Energy of LEP

The energy loss of about 125 MeV per turn (at $\sqrt{s} \approx M_Z$), due to synchrotron radiation in the LEP arcs, is compensated by acceleration in the Radio Frequency (RF) cavities placed symmetrically either side of the OPAL and L3 intersection regions. There is a Gaussian energy distribution of beam particles in each bunch with spread (rms) of approximately 40 MeV at LEP-1. The mean bunch energy at the interactions points is equal to the beam energy. Deviations due to misalignment errors of the RF stations are well understood and give rise to corrections which are known to a precision of a fraction of an MeV. Small differences in the orbits followed by the electron and positron bunches due to synchrotron energy loss and lattice imperfections can give energy differences $\mathcal{O}(0.1 \text{ MeV})$.

To first approximation $< E_{CM} >$ is equal to the sum of the average beam energies at the interaction point. A shift in E_{CM} will be induced if there is a correlation between the energy of the particles in the bunch and their transverse position. At the collision point this correction is proportional to the difference of the dispersions of the two bunches and the offset of their centres.

3.2 Measuring Beam Energy Using Resonant Depolarization

The LEP beams are vertically polarized by the emission of synchrotron radiation in the vertical bending field. A polarization greater than $\sim 10\%$ can be obtained when the LEP beams are not in collision. A Compton polarimeter is used to measure the degree of vertical polarization of the beam with an accuracy of $\sim 2\%$. The precession frequency of the polarization vector can be precisely measured by inducing a resonant depolarization of the beam with a radial depolarizing field provided by a coil. If the spin precession is in phase with the perturbation caused by the radial field then the spin rotations add coherently with each turn and the vertical polarization is eventually destroyed. It takes around a second (or 10^4 turns) to turn the polarization so that it is all in the radial plane. The average value of the beam energy can be measured to about 1 MeV using the method of resonant depolarization [10].

3.3 Variation with Time of the Collision Energy

The causes of time variation of the LEP beam energy can be divided into three main categories, those effects changing: the dipole field; the vertical quadrapole

218

Table 1: The size and error of the the main causes of time variation in the LEP E_{CM} .

Effect	Size (MeV)	Error (MeV)
Temperature Variation	~3	0.3
Rise per fill	~3	1.0
Horizontal corrector settings	~1	0.4
Tidal effects (daily)	~10	0.1
Geological shifts (weekly)	~10	0.3
RF corrections	~ 10	0.5
Vertical collision effects	< 1	0.3

field experienced in an orbit; and, the energy at the collision point. The causes are summarized in Table 1.

Unfortunately, the beam energy cannot be monitored using resonant depolarization whilst beams are in collision. A model based on a set of monitored quantities (magnet currents, field measurements, RF status, etc.) can be used to follow the evolution of the energy as a function of time [12]. Magnetic field measurements are made using two main methods. The first relies on Nuclear magnetic Resonance (NMR) probes, at LEP there are 16 probes for 3200 dipoles. The probes can be readout continuously. But, they are prone to radiation damage. The precision of the measurements effected using NMR probes is of order 10^{-6}. The second method uses flux loops which are cycled every few weeks. The precision of this method is a few time 10^{-4}.

Using NMR probes measurements it was established that there is a rise of the dipole magnetic field during a fill due to stray currents flowing along the beampipe. These current spikes disturb the fields of the dipole magnets, causing a unidirectional walk along their hysteresis curves resulting in an increase of the field that saturates within several hours. As can be seen from Fig. 6 these effects are correlated to return currents of electrical trains, such as the TGV, that do not return to the power supply along the tracks.

The LEP orbit length is fixed by the RF frequency which is stable to one part in 10^{10}. Strain in the Earth's crust arising from tidal forces [13] causes the radius of the LEP ring to change by $\sim 100\mu$m. The orbit is will thus be off centre in the quadrupoles. The extra deflection of the beams in the quadrupoles will change the beam energy by approximately 10 MeV. This whole effect takes place over the timescale of a day. Fig. 7 shows a comparison of a detailed tidal model [11] with the change in the beam energy over a period a day as measured using the technique of resonant depolarization.

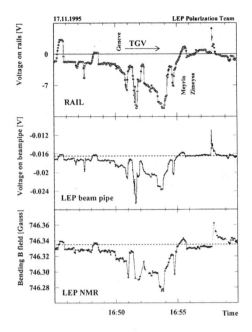

Figure 6: The railroad track potential (top), the LEP beam pipe potential (middle) and the dipole magnetic field measured by NMR (bottom).

4 Parameters of the Electroweak Standard Model

The present theory of the electroweak (EW) interaction is a gauge invariant field theory with the symmetry group $SU(2)_L \times U(1)_Y$ spontaneously broken by the Higgs mechanism. A common parameterization of the EW Standard Model that can be used to predict the observables measured at LEP and SLC is based on three precisely known input quantities. The first is the fine structure constant, α. It is precisely measured as $\alpha^{-1} = 137.0359395(61)$ at $q^2 \sim 0$. It determines the strength of QED part of the neutral current. Second, the Fermi constant, $G_F = 1.16639(2) \times 10^{-5} GeV^{-2}$, obtained from muon decay measurements. G_F determines the strength of the charged current. Third, the mass of the Z^0 vector boson, $M_Z = 91.1867(21)$ GeV, as measured at LEP.

Any process can be predicted at tree level using these three quantities with the following caveats. First, the Cabibbo-Kobayashi-Maskawa (CKM) mass-mixing matrix is required for tree level predictions of charge current processes involving flavour tagged hadronic final states. Second, that the phase space effects caused by the finite mass of the final state fermions can be neglected.

220

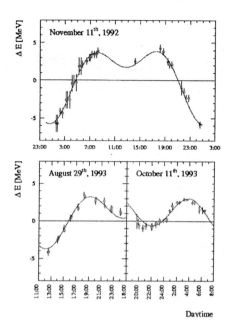

Figure 7: The change in beam energy measured over the period of a day using the technique of resonant depolarization. The Cartwright-Taylor-Edder tidal model fits the data well.

The Higgs mechanism generates a nondiagonal mass matrix for the gauge bosons. Diagonalizing this matrix gives:

$$Z_\mu = cos\theta_W W_\mu^3 + sin\theta_W B_\mu; \quad A_\mu = -sin\theta_W W_\mu^3 + cos\theta_W B_\mu. \quad (3)$$

where $W_\mu^{(1,2,3)}$ are the gauge fields associated with the $SU_L(2)$ group and the gauge field B_μ is associated with the $U(1)$ group. The Weinberg mixing angle (θ_W) is related to the $U(1)$ and $SU_L(2)$ couplings g_1 and g_2, via the relations:

$$g_1 cos\theta_W = g_2 sin\theta_W = e = \sqrt{4\pi\alpha} \quad (4)$$

where e is the positron charge and α is the fine structure constant.

The masses of the fundamental vector bosons are related at the Born level by:

$$sin^2\theta_W = 1 - \frac{M_W^2}{M_Z^2} \quad (5)$$

The Standard Model specifies the coupling constants between the gauge bosons and the fermions(f). These couplings are summarized in Fig. 8. In

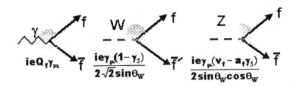

Figure 8: The coupling for the EM, charged and neutral current weak interactions is specified, in all cases the coupling is proportional to the positron charge.

particular, the strength of the weak neutral current and M_W are given by:

$$\sqrt{2}G_F M_Z^2 = \frac{\pi\alpha}{s_W^2 c_W^2}; \quad M_W^2 = \frac{\pi\alpha}{\sqrt{2}G_F s_W^2} \quad (6)$$

where $s^2\theta_W$ is used to represent $sin^2\theta_W$ The neutral current vector (v) and axial-vector (a) couplings are also specified by the theory:

$$v_f = I_3^f - 2Q_f sin^2\theta_W, \quad a_f = I_3^f, \quad (7)$$

where Q_f is the charge of the fermion. The Standard Model predicts a universal coupling for the three generations , where the left handed (L) and right handed (R) couplings are not equal:

$$L_f = a_f + v_f \neq R_f = v_f - a_f \quad (8)$$

In order to predict accurately the observables measured at LEP and the SLC one needs to perform calculations to higher orders than tree level. These calculations require three additional input parameters. First, M_f, the fermion masses, are required. The fermion masses appear in weak boson self energy corrections of the form $[(m_i^2 - m_j^2)/M_Z^2]$ (i and j are two fermions of the same isodoublet) and in vertex corrections of the form $[M_f/M_Z^2]^2$. Such corrections are in general small and well known except for the case of the top quark. The fermion masses also contribute to the photon self energy. This correction can be taken into account by replacing the fine structure as $q^2 = 0$ with that at $q^2 = M_Z^2$.

Second, α_s, the strong coupling constant (at $q^2 = M_Z^2$) appears in corrections, due to gluon radiation, of any weak process involving quarks. It is known to $\sim 3\%$ and only has a minor effect in processes where quarks contribute only through radiative corrections. Third, M_H, the Higgs boson mass which appears in higher order terms as a logarithmic correction of the form $log_e(M_H/M_Z)$ and thus has only a small effect on Standard Model predictions.

Any electroweak observable can be calculated within the framework of the Standard Model – with sufficient accuracy to match experimental precision – as a function of the parameters: $\alpha(M_Z^2), G_f, M_Z, M_f$ and α_s. Indeed, the accuracy of the predictions of the electroweak Standard Model are limited mainly by the precision with which the parameters M_t, M_H and α_s are known.

4.1 Cross-section for the Process $e^+e^- \rightarrow f\bar{f}$

The basic physics process at LEP-1 is $e^+e^- \rightarrow f\bar{f}$. This reaction has contributions from: Z^0 exchange; γ exchange; and, the interference between the weak and EM contributions. Considering only non-electron final states (the inclusion of Bhabha scattering, $e^+e^- \rightarrow e^+e^-$, requires s and t channel diagrams to be added which are a complicating factor) the differential cross-section is given by:

$$d\sigma_{e^+e^- \rightarrow f\bar{f}} = d\sigma_{Z-exch.} + d\sigma_{\gamma-exch.} + d\sigma_{Z\gamma-int.} \qquad (9)$$

The terms on the right hand side of this equation are,

$$d\sigma_{\gamma-exch.} = \frac{\pi\alpha^2}{2s} Q_f^2 D(1 + cos^2\theta), \qquad (10)$$

$$d\sigma_{Z\gamma-int.} = \frac{\alpha Q_f D G_f M_Z^2}{2\sqrt{2}} \frac{(s - M_Z^2)}{(s - M_Z^2)^2 + M_Z^2\Gamma_Z^2}[v_e v_f(1 + cos^2\theta) \qquad (11)$$
$$+ 2a_e a_f cos\theta],$$

$$d\sigma_{Z-exch.} = \frac{D G_f^2 M_Z^4}{16\pi} \frac{s}{(s - M_Z^2)^2 + M_Z^2\Gamma_Z^2}[(v_e^2 + a_e^2)(v_f^2 + a_f^2)(1 + cos^2\theta)$$
$$+ 8v_e a_e v_f a_f cos\theta]. \qquad (12)$$

where Q_f is the fermion charge (D=1 for leptons and 3 for quarks), and θ is the scattering angle.

5 Radiative Corrections

Radiative corrections at LEP/SLC can be divided into four categories as summarized in Fig. 9. The treatment of radiative corrections used here follows that of Ref. [15]. QED corrections (Fig. 9(a)) involving the radiation of a real photon or the exchange of a virtual photon are the largest. For example, they give rise to a 30% reduction in the Z^0 pole cross-section.

Vacuum polarization corrections (Fig. 9(b)), or oblique corrections, are important for three reasons. First, they give rise to corrections that are second only in numerical importance to the QED corrections. Second, they are universal in that they do not depend on the final state. Third, they are sensitive to new physics. The $\mathcal{O}(10\%)$ corrections due to the $\gamma\gamma$ propagator can be renormalized by using the a running QED coupling constant $(\alpha^2_{M_z})$. On the other hand the ZZ and γZ propagators can have order $\sim 1\%$ effects on observables at the Z^0 pole.

Vertex corrections (Fig. 9(c)) depend on the final state. They are large only for the $Z^0 \rightarrow b\bar{b}$ vertex. These corrections are also sensitive to new physics. Box diagrams (Fig. 9(d)) are almost negligible at LEP1/SLC energies and will not be considered further.

The oblique corrections have a *non-decoupling* nature. Essentially, this property means that the effect of particles with masses much larger than the electroweak scale does not disappear as M_Z/M_X to some power, where M_X is the large particle mass associated with a large coupling constant. Thus, the heavy degrees of freedom (DOF) do not *decouple* allowing their contributions to the vacuum polarization corrections to grow with M_X. An example of such a heavy particle is the top quark. Its contribution to the Z^0 self energy is proportional to $(M_t/M_Z)^2$. Non-decoupling is also seen at the $Z \rightarrow b\bar{b}$ vertex. The contribution of the vertex correction is shown in Fig. 9. It too is proportional to $(M_t/M_Z)^2$

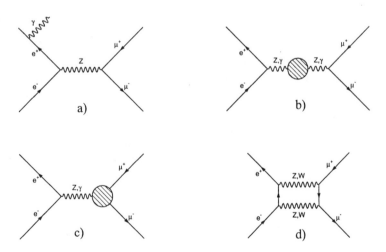

Figure 9: The Feynman diagrams showing the one loop level corrections to the process $e^+e^- \rightarrow \mu^+\mu^-$. Diagram (a) shows the QED corrections, (b) the vacuum polarization corrections, (c) vertex corrections, and (d) box corrections.

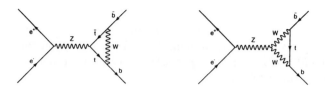

Figure 10: Vertex corrections to the $Z \to b\bar{b}$ vertex involving the top quark.

5.1 Radiative Corrections to the $e^+e^- \to f\bar{f}$ Cross-section

The vacuum polarization and vertex corrections to the process $e^+e^- \to f\bar{f}$ can be implemented using an amplitude with the same structure as the tree level amplitude, where these radiative corrections are absorbed into complex energy dependent couplings [16], [17], [18]. This approach is called the Improved Born Approximation (IBA). It is useful procedure since the vacuum polarization and vertex corrections at the one-loop level are naturally separated from the corrections due to QED, they in fact form a gauge invariant subset. Box diagrams involving ZZ and WW can also be separated in a similar way. A detailed discussion can be found in Ref.[17].

In the IBA the amplitude for the process $e^+e^- \to f\bar{f}$ contains the running complex couplings, $\overline{\alpha}(s)$, $\overline{s}_W^2(s)$ and $\overline{\rho}(s)$.

Vacuum Polarization Corrections

These corrections can be divided into three categories. The first is the photon self energy diagram correction, that is one involving photon lines to both vertices. The correction due to this source is absorbed into the effective coupling constant:

$$\overline{\alpha}(s) = \frac{\alpha(0)}{1 + \Pi^\gamma(s)} = \frac{\alpha(0)}{1 - \Delta\alpha(s)}, \tag{13}$$

where $\Delta\alpha(s)$ is a complex quantity and $\Pi^\gamma(s)$ is defined as, $\Pi^\gamma(s) = \Sigma_\gamma(s)/s$, where $\Sigma_\gamma(s)$ stands for the photon self energy function [17].

The second correction is due to vacuum polarization diagrams involving γ and Z lines to the vertices. According to Ref.[17] corrections arising from this source can be absorbed into an effective mixing angle $\overline{s}_W^2(s)$

$$\overline{s}_W^2(s) \equiv s_W^2(1 + \Delta\kappa(s)), \tag{14}$$

where the complex quantity $\Delta\kappa(s)$ is:

$$\Delta\kappa(s) = -\frac{c_W}{s_W} \frac{\Pi^{\gamma Z}(s)}{1 + \Pi^\gamma(s)}. \tag{15}$$

The third type of correction is that due to Z self energy diagram, involving Z lines to the vertices. Following again the procedure described in Ref.[17] the real part of Z self-energy is absorbed into the effective parameter $\bar{\rho}(s)$ defined by:

$$\sqrt{2}G_F M_Z^2 \bar{\rho}(s) = \frac{e^2}{4s_W^2 c_W^2} \frac{1}{1 + \Pi^Z(s)}, \tag{16}$$

where

$$\Pi^Z(s) = \frac{\Re[(\Sigma_Z(s)]}{s - M_Z^2} \tag{17}$$

Using the Optical Theorem the imaginary part of the Z self energy can be expressed [17] as,

$$\Gamma_Z(s) = \frac{\Im[\Sigma_Z(s)]}{1 + \Pi^Z(s)} \cong \Gamma_Z(M_Z^2) \frac{s}{M_Z^2}, \tag{18}$$

where $\Gamma_Z(s)$ is the Born total decay width. At tree level $\bar{\rho}(s)$ is unity. In order to manifest higher order corrections to the tree level approach it is useful to define the quantity $\Delta\rho(s)$:

$$\bar{\rho}(s) \equiv [1 + \Delta\rho(s)]. \tag{19}$$

The quantities $\bar{\rho}(s)$ and $\Delta\rho(s)$ include only the real part of the Z vacuum polarization.

Vertex Corrections

The flavour dependent vertex corrections can be absorbed into effective parameters giving rise to a set of effective parameters for each $f\bar{f}$ final state. Following again Ref.[17] the one loop vertex corrections to the Z amplitude can be introduced via the complex form factors F_{VZf} and F_{AZf} which modify the Born vector and axial couplings to the Z current:

$$C_\mu^Z = [\gamma_\mu(Q^f + F_{VZf}(s) - F_{VZf}(s)\gamma_5)]. \tag{20}$$

The couplings and vacuum polarization corrections can be redefined to incorporate vertex corrections by defining flavour dependent complex effective weak mixing angles in the following way:

$$sin^2\theta_{eff}^f(s) \equiv s_W^2[1 + \Delta\kappa(s) + \Delta\kappa_f(s)]. \tag{21}$$

The flavour dependent compled effective ρ parameters can be defined as:

$$\rho^f(s) \equiv [1 + \Delta\rho(s) + \Delta\rho^f(s)]. \tag{22}$$

The quantities $\Delta\kappa_f(s)$ and $\Delta\rho^f(s)$ are given in terms of the vertex functions ($F_{VZf}(s)$ and $F_{AZf}(s)$) mentioned previously:

$$\Delta\kappa_f(s) = -\frac{1}{Q_f}\frac{c_W}{s_W}(F_{VZf}(s) - \frac{v_f}{a_f}F_{AZf}(s)), \tag{23}$$

$$\Delta\rho_f(s) = (1 + \frac{F_{AZf}(s)}{a_f})^2 - 1, \tag{24}$$

where

$$a_f = \frac{I_3^f}{2c_W s_W}; \quad v_f = \frac{I_3^f - 2Q_f s_W^2}{2c_W s_W}. \tag{25}$$

The Z-exchange amplitude can now be rewritten in terms of the complex effective vector and axial couplings defined as:

$$g_{Vf} = \sqrt{\rho^f(s)}[I_3^f - 2Q_f sin^2\theta_{eff}^f(s)]; \quad g_{Af} = \sqrt{\rho^f(s)}I_3^f. \tag{26}$$

The Z decay width can also be rewritten in terms of these couplings:

$$\Gamma_Z = \sum_f N_c^f \frac{G_F M_Z^3}{6\sqrt{2}\pi}(g_{Af}^2 M_Z^2 + g_{Vf}^2 M_Z^2). \tag{27}$$

where, $N_c^f = 1$ for leptons and $3(1 + \alpha_s/\pi)$ for quarks. All small mass terms have been neglected.

Although the effective parameters described above are complex functions of s, their imaginary parts are small. So to is their energy dependence, which can be ignored near to the Z^0 pole. At the present level of precision of the determination of the electroweak variables the only imaginary part that cannot be neglected is that of $\overline{\alpha}(s)$.

The Photon Self Energy

The term $\Delta\alpha$ is sensitive to fermion mass through $log_e(s/m_f^2)$ arising from the photon self energy, $\Pi^\gamma(s)$. Consequently, $\Delta\alpha$ is sensitive to light fermion masses.

The contributions of the leptons and the top quark to the photon self energy can be calculated using an analytical expression given in Ref.[19]. The contribution of the light quarks can be computed by relating, through unitarity, the cross-section $e^+e^- \to q\bar{q}$ via one photon exchange to the imaginary part of the Π^γ function. Using relatively recent experimental data [20] one obtains:

$$\Delta\alpha_h(M_Z^2) \equiv -\Re[\Pi_h^\gamma(s)] = 0.0280 \pm 0.0007. \tag{28}$$

the real part of the photon self-energy is obtained by adding to $\Delta\alpha_h$ the real parts of the top and leptonic contributions giving, $\Delta\alpha = 0.0632 \pm 0.0007$ [15].

Radiative Corrections to M_W

Radiative correction modify the standard expression,

$$G_F = \frac{\pi \alpha}{\sqrt{2} s_W^2} \frac{1}{M_W^2} \ [q^2 \approx 0], \tag{29}$$

which becomes:

$$M_W^2 = \frac{\pi \alpha}{\sqrt{2} s_W^2 G_f}(1 + \Delta r), \tag{30}$$

where,

$$\Delta r = \Delta \alpha - \frac{c_W^2}{s_W^2}\Delta \rho + \Delta r_{rem}, \tag{31}$$

$\Delta \alpha$ is the photon vacuum polarization and $\Delta \rho$ is the radiative correction to the ρ parameter. The third term incorporates remaining corrections. At tree level, $\rho = M_W^2/(c_W^2 M_Z^2) = 1$. That is, the ρ parameter is the ratio of the neutral current amplitude to the charged current amplitude at $q^2 \approx 0$. It is equivalent to the parameter $\overline{\rho}$ introduced earlier. If one loop corrections are included, $\rho = 1 + \Delta \rho$. The leading term of $\Delta \rho$ is:

$$\Delta \rho = \frac{G_F \sqrt{2}}{16 \pi^2} \sum_f N_c^f \Delta m_f^2 + \ldots, \tag{32}$$

where Δm_f^2 is the isodoublet mass splitting and the sum extends over all fermion isodoublets (f). Obviously, the largest contribution to $\Delta \rho$ is provided by the $top - bottom$ iso-doublet:

$$\Delta \rho \sim \Delta \rho_{t-b} \sim 3\sqrt{(2)}\frac{G_F M_t^2}{16 \pi^2} = 0.0099 \left(\frac{M_t}{175 GeV}\right)^2. \tag{33}$$

In the Standard Model there is an accidental $SU(2)_R$ symmetry in the Higgs sector that at the one loop level kills any quadratic Higgs boson mass (M_H) dependence of $\Delta \rho$. However, there is still a leading logarithmic dependence on M_H:

$$\Delta r^{Higgs} \approx \sqrt{2}\frac{G_F M_W^2}{16 \pi^2}\left[\frac{11}{3}(log_e \frac{M_H^2}{M_W^2} - \frac{5}{6})\right] = 0.0025 \left(log_e \frac{M_H^2}{M_W^2} - \frac{5}{6}\right). \tag{34}$$

At $M_H = 300$ GeV this is $\Delta r^{Higgs} = 0.0045$.

The prediction for Δr can be written as [15]:

$$\Delta r \approx \Delta \alpha - \frac{c_W^2}{s_W^2}\Delta \rho_{t-b} + [\Delta_{rem}^{quarks} + \Delta_{rem}^{top} + \Delta_{rem}^{leptons} + .. + \Delta r^{Higgs}]. \tag{35}$$

228

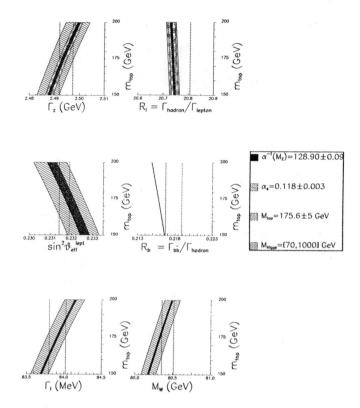

Figure 11: The top mass dependence for six electroweak variables. The vertical dotted lines indicate $\pm 1\sigma$ of experimental accuracy available in the winter of 1998. The bands show the error on the prediction due to uncertainties in the input parameters. The hatching code is shown to the right.

Putting in numerical values [15]:

$$\Delta r \approx 0.0632 - 0.0349 + 0.0124 + .. + 0.0025(log_e \frac{M_H^2}{M_W^2} - \frac{5}{6}). \qquad (36)$$

The above expression for Δr^{Higgs} assumes that $M_H^2 \gg M_W^2$. However, current data indicates that the Higgs boson is light and that this is not a good approximation. An exact for for Δr^{Higgs} can be found in Ref.[21].

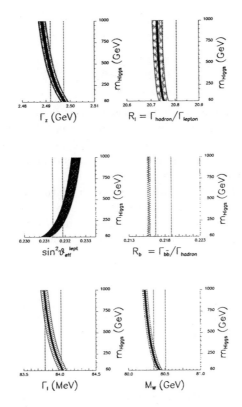

Figure 12: The Higgs mass dependence for six electroweak variables. The vertical dotted lines indicate $\pm 1\sigma$ of experimental accuracy available in the winter of 1998. The bands show the error on the prediction due to uncertainties in the input parameters. The hatching code is shown in the previous figure.

6 Input Parameter Dependence of Standard Model Predictions

The input parameter dependence of the electroweak Standard Model can be illustrated using six important observables. Fig. 11 and Fig. 12 [15] show the predictions for these variables, as a function of the the input value of M_H and M_t, respectively. The value of M_t is arbitrarily allowed to range around the known value: from 150 to 200 GeV. The value of M_H, which is only now starting to be constrained by the data, is allowed to range from 60 to 1000 GeV.

There are four numerically important loop corrections: the three vacuum polarization corrections, Δr, $\Delta \kappa$, $\Delta \rho$; and, the $Z \to b\bar{b}$ vertex correction. The leading terms in the vacuum polarization corrections all depend on $\Delta\rho$ and consequently all have similar leading M_t dependence:

$$\Pi^Z \to 1 + \Delta\rho + ...; \quad \Pi^W \to 1 + \Delta\alpha - \frac{c_W^2}{s_W^2}\Delta\rho + ...; \quad \Pi^{\gamma Z} \to 1 - \frac{c_W^2}{s_W^2}\Delta\rho + ... \quad (37)$$

The $\Delta\alpha$ correction can be ignored for now because it does not depend on M_t or M_H. The M_t dependence is only important for the $Z^0 \to b\bar{b}$ vertex because of the large CKM coupling (V_{tb}). The M_H dependence of the vertex corrections can be ignored because of the small mass dependent coupling of the Higgs boson to light fermions.

Thus, the predictions for the total width of the Z^0 (Γ_Z), the leptonic width of the Z^0 (Γ_l) and the mass of the W boson (M_W) share a similar dependence on M_t and M_H. The prediction for the $sin^2\theta^l_{eff}$ is opposite in sign but otherwise has a similar dependence. The prediction for $R_b (= \Gamma_{b\bar{b}}/\Gamma_{had})$ depends on the M_t because of the vertex correction but at the one loop level it is independent of M_H. It does not depend appreciably on vacuum polarization corrections since they mostly cancel in the ratio. It is independent of M_H. The prediction for R_l ($= \Gamma_{had}/\Gamma_{lept}$) has only a small dependence of M_t and M_H but it is affected by α_s via Γ_{had}. At present precision $sin^2\theta^l_{eff}$ shows the most sensitivity to M_t and M_H.

7 The Lineshape of the Z^0

The Z^0 lineshape is described by four almost uncorrelated variables: M_Z, Γ_Z, R_l and the total cross-section for the process, $e^+e^- \to Z^0 \to q\bar{q}$ (σ_H). The Z^0 lineshape is fitted using a formula that depends on the lineshape parameters. The absolute Z^0 mass scale is determined by a precise calibration of the centre-of-mass energy at the collision points. As QED radiative corrections can be as large as 30% near the Z^0 peak, as shown in Fig. 13, the effects of Initial State Radiation (ISR) must be included in any lineshape parameterization.

The cross-section for the process $e^+e^- \to f\bar{f}$ ($f \neq e$), corrected for 1st order ISR is according to Ref.[22]:

$$\sigma(s) = \sigma_0(s)(1 + \delta_1 + \beta log_e x_0) + \int_{x_0}^1 \beta(\frac{1}{x} - 1 + \frac{x}{2})\sigma_0(s')dx, \quad (38)$$

where x is the ISR photon energy in units of beam energy, and $s'(= s(1 - x))$ is the square of E_{cm} after ISR. The term δ_1 ($\sim$ 9%) accounts for the portions of the soft or virtual corrections that are independent of the infrared singularity. The symbol x_0 represents a cut-off parameter that is used to divide the corrections between those due to hard photon emission and those due to soft and virtual photons. The definition of β is:

$$\beta \equiv \frac{2\alpha}{\pi}[log_e(\frac{s}{m_e^2}) - 1],\tag{39}$$

where β is approximately 0.11 at LEP energies. The integral term in this expressions is the convolution of the cross-section at the centre of mass energy remaining after ISR radiation, with the photon spectrum.

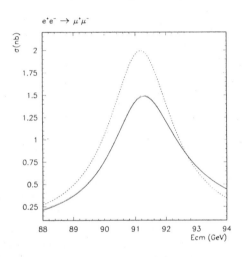

Figure 13: Effect of QED ISR corrections on muon-pair production at and around the Z pole. The dashed line represents the cross-section without ISR. The dotted line shows the cross-section curve with $\mathcal{O}(\alpha)$ exponentiated ISR. The solid line depicts the cross-section curve for $\mathcal{O}(\alpha^{\epsilon})$ exponentiated ISR.

As $\sigma_0(s)$ falls away rapidly from the Z pole, σ may be written as:

$$\sigma(s) \sim \sigma_0(s)[1 + \delta_1 + \beta log_e(\frac{\Gamma_Z}{M_Z})],\tag{40}$$

for centre of mass energies around M_Z^2. The overall correction described by the above relation is $\sim -30\%$. Well below the Z resonance the dominant correction is δ_1. The QED radiation corrections around the Z pole are shown in Fig. 13. Around the pole hard photons from ISR are strongly suppressed. One can see that well above the Z pole the ISR corrected cross-section is larger than the non-ISR corrected cross-section. This is due to events that can "radiate" down to the Z pole. One can also see, well below the Z pole, the domination of δ_1.

The measured cross-sections are fitted using a formula in which a reduced cross-section $\sigma^R_{f\bar{f}}$ is convoluted with a radiator function[23] $H(s, s')$ in which all initial state radiative corrections are included:

$$\sigma_{f\bar{f}}(s) = \int_{4m_f^2}^{s} H(s, s')\sigma^R_{f\bar{f}}(s')ds'. \tag{41}$$

The reduced cross-section is:

$$\sigma^R_{f\bar{f}}(s) = \sigma^{peak}_{f\bar{f}} \cdot \frac{s\Gamma^2_Z}{(s - M_Z^2)^2 + (s\Gamma_Z/M_Z)^2} + \text{``}(\gamma - Z)\text{''} + \text{``}|\gamma|^{2\text{''}}, \tag{42}$$

where

$$\sigma^{peak}_{f\bar{f}} = \frac{12\pi}{M_Z^2} \frac{\Gamma_e\Gamma_f}{\Gamma_Z^2} \cdot \frac{1}{1 + 3\alpha/4\pi}. \tag{43}$$

The formula for $\sigma^R_{f\bar{f}}(s)$ given above consists of three terms. The first term includes a relativistic Breit-Wigner distribution describing Z-exchange. The second term describes the "$(\gamma - Z)$" interference which is much smaller and of course zero when $s = M_Z^2$. In the fit this correction is usually fixed to the Standard Model prediction. If the interference term is allowed to be a free parameter in the fit it the error on M_Z is substantially increased. The photon exchange term "$|\gamma|^2$" is fixed in the fit to the QED prediction. It is only a few percent of the Z exchange term. In the case of Bhabha scattering the contributions from additional t-channel Z and γ exchange diagrams are set to the Standard Model prediction.

The determination the total leptonic width of the Z can easily be divided into a study of the three partial leptonic widths. The assumption of lepton universality will increase the sensitivity of the fit. In the case of the hadronic decays of the Z all flavours of final state are counted together. If lepton universality is not assumed six parameters are needed in the fit: M_Z, Γ_Z, σ_h^0, Γ_e, Γ_μ, and Γ_τ. If lepton univerality is assumed four parameters are needed: M_Z, Γ_Z, σ_h^0, and $R_l = \Gamma_h/\Gamma_1$.

The above described prescription for performing the lineshape fit is implemented in computer programs. At LEP there are two main programs. The first is ZFITTER[24] which is used by the LEP collaborations DELPHI, L3 and OPAL. The second is MIZA[25] which is used by ALEPH. The overall precision of the description of ISR in the calculation of the total cross-section is around 0.06% in both programs.

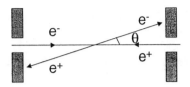

Figure 14: Schematic depiction of the low angle detectors used for the determination of luminosity at LEP.

7.1 Determining the Cross-section

The cross-section, σ_j for a given process j is given by the following formula:

$$\sigma_j = \frac{(N_j - N_j^{bk})}{\epsilon_j L}. \tag{44}$$

where N_j is the number of events of type j, N_j^{bk} is the contribution from background events, ϵ_j is the event selection efficiency, and L is the integrated luminosity.

Determining the Luminosity

At LEP Bhabha scattering is used to determine the luminosity because it is a well understood QED process, dominated by t-channel photon exchange, with a large cross-section. In this case the luminosity is given by the following expression:

$$L = \frac{N_{bh} - N_{bh}^{bk}}{\epsilon_{bh} \sigma_{bh}^{th.}}. \tag{45}$$

As can be seen from the above formula to obtain a precise measurement of the luminosity the detection efficiency, ϵ^{bh}, has to be well known and large, σ_{bh}^{th} has to be calculated precisely, and N_{bh}^{bk} has to be small and well determined. Prior to LEP it was the determination of the absolute efficiency that formed the main contribution to the overall uncertainty on the luminosity.

A typical LEP luminosity detector consists of two cylindrical calorimeters deployed at small angles on either side of the interaction point as indicated schematically in Fig. 14.

The Bhabha cross-section can be approximated as:

$$\sigma_{bh}^{th} \sim \frac{16\pi\alpha^2}{s} \left(\frac{1}{\theta_{min}^2} - \frac{1}{\theta_{max}^2} \right), \tag{46}$$

where θ_{min} and θ_{max} are polar angles defining the inner and outer acceptance of the luminosity detectors. To a first approximation:

$$\Delta\sigma/\sigma \approx 2\Delta\theta_{min}/\theta_{min} \approx 2\Delta R_{min}/R_{min}. \tag{47}$$

Thus, to compute the efficiency precisely one needs to know well the position of the inner edge of the detector, R_{min}, with respect to the beam. Therefore LEP luminometers have either tracking devices in conjunction with their calorimetry or else very fine grain calorimetry.

All LEP experiments upgraded their luminometers between 1992 and 1994. A precision of the order of 20 μm was achieved in the knowledge of the position of the inner edge of the detectors. Typically, $R_{min} \sim$ 60mm for these detectors. Thus, the uncertainty on the luminosity of these detectors is around 7×10^{-4}. This is a big improvement on the uncertainty on the luminosity at the start of LEP, which was around 2%.

Random coincidences of off-momentum electrons and positrons form the main background of the Bhabha process. This background, which is at the per-mil level, can be well determined from data samples obtained from prescaled low-energy triggers and by studying acoplanar coincidences. Consequently, the contribution of the backgrounds to the systematic error on the luminosity is very small. Usually, it is the knowledge of the positions of the mechanical structure of the luminometer that gives rise to the largest uncertainty. The overall experimental error lies in the range 0.07% → 0.1%, depending on the experiment. The combined LEP error is around 0.05%. The theoretical error on the calculation of the Bhabha cross-section is $\sim$ 0.11% [26] and is due to missing sub-leading $\mathcal{O}(\alpha^2, log_e[-t/m_e^2])$) corrections.

Hadronic Cross-Section

Hadronic events, which form approximately 70% of the decays of the Z, carry most of the weight in the lineshape fit. The high multiplicity, large acceptance (greater than 97%) and large visible energy of these events result in clear and efficient event selection. Typically, the LEP experiments select hadronic events requiring the number of calorimeter clusters to be greater than $\sim$ 10 with a minimum summed energy of about 10 → 20% of E_{cm} [27]. ALEPH and DELPHI have independent charged track based selections on multiplicity (around five) and energy (approximately 10% of E_{cm}). Backgrounds to the hadronic event selection are typically $\sim$ 1%, mostly from two-photon and tau-pair events.

Each experiment accumulated around four million hadronic Z decays resulting in statistical errors less than one per mil at the Z peak. The main systematic errors in the determination of the hadronic cross-section are due to the knowledge of the two-photon backgrounds. Overall systematic errors range from 0.5 → 1.6 per mil, depending on the experiment. The hadronic cross-section measured by the ALEPH collaboration is shown in Fig. 15 [28].

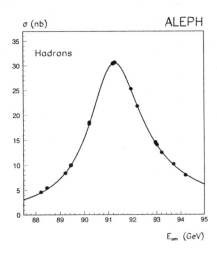

Figure 15: The cross-section for $e^+e^- \to hadrons$ as a function of E_{cm} as measured by the ALEPH Collaboration. The solid line depicts the Standard Model fit.

The Charged Lepton Cross-section

Although, charged lepton pairs are only $\sim$ 10% of the decay width of the Z they allow a precise measurement of the electroweak couplings of lepton species to the Z. Selection of leptonic events at LEP is typically based on criteria of large visible energy/momentum and low multiplicity. In the case of an electronic final state the EM calorimetry is also utilized. Muon pair events have additional selection criteria based on the penetration of the tracks through the hadronic calorimetry to the muon walls of each experiment. The tau pair event selection uses, among other criteria, the missing energy of the event.

The geometrical acceptance for leptonic final states of around 85% is mainly determined by a cut on the production angle. Selection efficiencies are high ranging from around 85% for tauonic final states to greater than 95% for electrons and muons. The main backgrounds are other leptonic final states of different species. Although, in the case of taus the two-photon background is relevant. Between 1989 and 1995 each LEP experiment accumulated be-

236

tween 300 and 400 thousand events arising from the decay of the Z to charged leptons.

In the case of electrons and muons the reliability of Monte carlo simulations of the detector response forms the greatest source of systematic error ($\sim 0.5\%$). However, for tauonic final states the knowledge of the background contamination is the main source of systematic error($\sim 0.5\%$). The cross-section measured by DELPHI for the three lepton species is given in Fig. 16 [29].

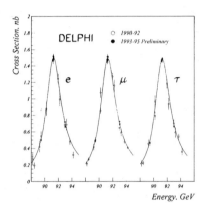

Figure 16: The cross-section for $e^+e^- \to l^+l^-$ ($l = e, \mu, \tau$) as a function of E_{cm} as measured by the DELPHI Collaboration. The solid lines depict the Standard Model fit.

8 The Z^0 Lineshape Results

The lineshape parameters, M_Z, Γ_Z, σ_h^0 and $R_l = \Gamma_h/\Gamma_1$ are determined by the fitting procedure described previously. The fit utilizes the precise calibration of the LEP centre-of-mass energy. The parameters obtained by each experiment are averaged taking into account sources of error that the experiments have in common [30]. The largest correlation between the four Lineshape Parameters is $\sim 15\%$. The data used in determining the lineshape parameter results given below is that presented at the summer conferences in 1998.

8.1 The Z^0 Mass and Width

The mass of the Z boson is the most precise single measurement made at LEP. The results obtained by the LEP experiments together with the LEP ̇average are shown in Fig. 17. The measurement is limited by the systematic error due mainly to a common uncertainty in the absolute energy scale of LEP (± 1.5 MeV).

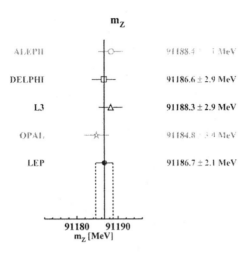

Figure 17: The Z mass as determined by the LEP experiments as well as the LEP average. The errors bars do not include the common errors which are included in the numerical values given to the right of the figure.

The dependence of the Z^0 width on the vacuum polarization corrections makes it an important observable. The experimental measurement of the Z^0 width and its LEP average are shown in Fig. 18. The main systematic error arises from the energy calibration as is the case for the measurement of the Z^0 mass. Another important source of systematic error is the knowledge of the background due to non-resonant processes, for example quasi-real photon-photon collisions. The error due to this source is ~ 1 MeV but uncorrelated between experiments. The theoretical error associated with the determination of the width is less than 1 MeV.

8.2 The Ratio R_l

The results from the LEP experiments on the measurement of R_l are given in Fig. 19. This variable is interesting because it depends on the strong coupling constant but does not depend strongly on M_t or M_H. Thus allowing $\alpha_s(M_Z^2)$ to be determined with the minimum of theoretical uncertainty. This can be done by assuming the Standard Model and using it to calculate the ratio of the couplings of the Z to quarks and leptons.

238

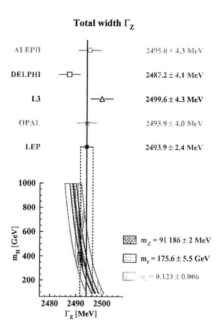

Figure 18: The Z width as determined by the four LEP experiments, as well as its LEP average. A comparison is made with the Standard Model value.

The experimental systematic error is dominated by the precision with which the efficiencies and backgrounds of the leptonic selections are known. The error due to this source is about 0.5% per experiment. The lack of a full $\mathcal{O}(\alpha^2)$ event generator for Bhabha scattering gives rise to a "theoretical" error in the t−channel correction in the electron channel. The common error from this source is approximately 0.1%.

8.3 The Hadronic Peak Cross-section

The results of the measurement of σ_h^0 by the LEP experiments are shown in Fig. 20. The Standard Model prediction of this cross-section is not very sensitive to M_H, M_t or to $\alpha_s(M_Z^2)$. This makes σ_h^0 a good variable to look for deviations from the Standard Model without the uncertainty related to the precision with which the variables M_H, M_t and $\alpha_s(M_Z^2)$ are known.

There are two main sources of experimental systematic error. First, the knowledge of the background and efficiency of the hadron selection. This contributes an uncorrelated error of around $0.10 \rightarrow 0.15\%$ per experiment. Second,

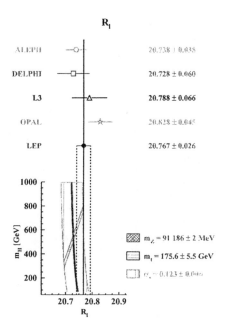

Figure 19: The ratio of the hadronic to the leptonic partial widths of the Z^0 measured by the LEP experiments, as well as its LEP average. A comparison is made with the Standard Model prediction.

the uncertainty on the absolute luminosity measurement which contributes between 0.07% and 0.15%. There is a common theoretical systematic error due to the knowledge of the of the small angle Bhabha scattering cross-section. This error is around 0.11%.

8.4 Quantities Derived from the Lineshape Fit

If lepton universality is not assumed there are six free parameters in the lineshape fit. As we have seen these parameters include the three leptonic widths: Γ_e, Γ_μ, and Γ_τ. The LEP values for these widths are given in Fig. 21. As can be seen the three results are consistent with being the same and are also in agreement with the fitted leptonic width assuming lepton universality. As the leptonic width of the Z does not depend on α_s it provides an important constraint on vacuum polarization corrections.

The width of the Z to neutrino pairs, called the invisible width, can be

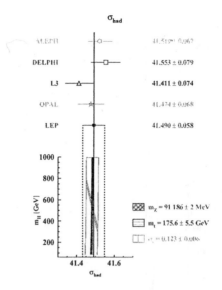

Figure 20: The hadronic peak cross-section as measured by the four LEP experiments along with the average LEP value. A comparison is made with the Standard Model expectation.

obtained from the sum:

$$\Gamma_Z = \Gamma_h + 3\Gamma_l + \Gamma_{inv} \qquad (48)$$

The ratio of the invisible width to the leptonic width is the number of light neutrino families, N_ν assuming that the invisible width arises only from neutrino-pair final states. Thus, the following relation can be written:

$$N_\nu \cdot \frac{\Gamma_\nu}{\Gamma_l} = \frac{\Gamma_{inv.}}{\Gamma_l} = \frac{\Gamma_Z}{\Gamma_l} - R_l - 3. \qquad (49)$$

The ratio, Γ_ν/Γ_l, is obtained from the Standard Model prediction. The value for this ratio has only a small error since the M_H and M_t dependence largely cancels out in the ratio. Fig. 22 shows that the LEP data is in very good agreement with the Standard Model stipulation that there are three families of light neutrinos.

9   The Decay Widths of Z^0 for Heavy Flavours

No picture of the Standard Model would be complete without precision tests of the quark sector since electroweak observables involving quarks are sensitive

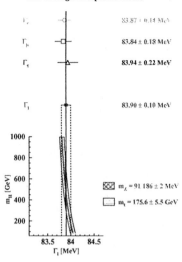

LEP averages of leptonic widths

Γ_e 83.87 ± 0.14 MeV

Γ_μ 83.84 ± 0.18 MeV

Γ_τ 83.94 ± 0.22 MeV

Γ_l 83.90 ± 0.10 MeV

m_Z = 91 186 ± 2 MeV

m_t = 175.6 ± 5.5 GeV

Figure 21: The three average charged leptonic partial widths obtained by the four LEP experiments. The comparison to the fit where lepton universality is assumed is shown. The Standard Model prediction is also indicated.

to Standard Model parameters. Heavy flavour tagging is an invaluable tool in this endeavour since it permits the precise determination of the partial widths $Z \rightarrow b\bar{b}$ and $Z \rightarrow c\bar{c}$. One can remove most of the dependence the vacuum polarization corrections and α_s by forming the ratios:

$$R_b \equiv \frac{\Gamma_{b\bar{b}}}{\Gamma_{had}}, \;\; R_c \equiv \frac{\Gamma_{c\bar{c}}}{\Gamma_{had}}. \tag{50}$$

As has been mentioned previously R_b receives large vertex corrections and is consequently dependent mostly on M_t. The ratio R_c is only slightly dependent on M_t, due to the contribution of $Z \rightarrow b\bar{b}$ decays to the denominator.

It seems to be a rule of nature that free quarks cannot be observed in the laboratory. Thus, electroweak measurements involving quark final states have to be made once hadronization has taken place. Also, it is not easy to tag the flavour of the initial quark giving rise to the hadronic final state. However, it is possible to obtain enriched samples of $Z \rightarrow b\bar{b}/c\bar{c}$ events at LEP and SLC due primarily to the large mass and finite lifetime of b and c flavoured hadrons. These properties allow a number of effective tagging techniques.

242

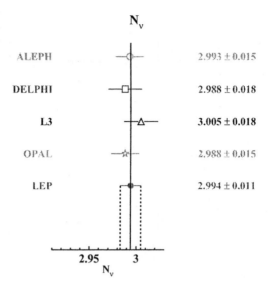

Figure 22: The number of light generations determined by the four LEP experiments along with the LEP average.

9.1 Lifetime Tagging of Heavy Flavour

Lifetime tagging relies on the finite lifetime of b-hadrons which is around 1.5ps. Also b hadrons carry away a substantial fraction of the total energy of the event, typically ~ 30 GeV at LEP-1. At LEP energies the resulting Lorentz boosts give b-hadrons decay lengths of typically a few millimetres.

The finite decay length of the b-hadron gives rise to tracks with a finite distance of closest approach, or impact parameter (IP). The IP for b-hadron tracks can be measured with an experimental resolution varying between 20μm and 70μm depending on the momentum of the particle. At LEP-1 the IP for the products of B-hadron decay is typically 300μm. The IP can be signed. It is taken to be positive when the track in question crosses the jet axis (assumed to be the b-quark direction) before the primary vertex. The definition of IP is summarized in Fig. 23. Using impact parameter as a method of tagging b-hadrons high purity samples ($\sim 90\%$) can be obtained with high relative efficiency ($\sim 25\%$).

At LEP the beamspot has transverse size of (xy) of $\sim 200 \times 10$ microns with a length in Z of around 1 cm. The primary vertex can be determined by fitting the charged particels comprising the event to a common vertex using

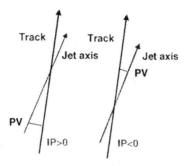

Figure 23: The definition and signing of the Impact Parameter (IP) of a track.

the beam-spot as a constraint. At LEP the beam spot is typically determined with a resolution of around 50μm, in x, 10μm in y and 100μm in z. At the SLC the beamspot constraint is more stringent with an xy size of $\sim 1\mu$m which can be estimated with a resolution of $\sim 10\mu$m.

The secondary vertex can be determined and the decay length estimated. One can use a "buildup" or a "teardown" method to fit tracks to a vertex. The decay length significance can be estimated by dividing the decay length by the error on the estimate of the decay length. An example of an OPAL measurement of decay length significance is shown in Fig. 24[32]. Similar purities to the IP case have been achieved using decay length significance as a method to tag heavy flavour. However, the efficiency ($\sim 20\%$) is usually less than in the IP case for the same purity.

9.2 Tagging of Heavy Flavour Using prompt Leptons

The three main sources of prompt leptons are the primary semi-leptonic decays, $b, c \to l + X$ and the chain decay $b \to c \to l$. Typical $p \times p_t$ distributions of prompt leptons for these decays are given in Fig. 25.

The branching ratios to prompt leptons are around 10% per species with lepton identification efficiencies ranging between 60% to 90% at LEP. The large mass for $b(c)$-hadrons combined with the fact that they tend to take a large share of the available energy – typically around 70% of the beam energy at LEP-1 – explains why prompt leptons from $b(c)$ decays tend to have large transverse momentum (p_t) with respect to the jet axis.

The most commonly used approach to jet finding at LEP is the JADE algorithm[33]. In this approach particles are combined until they form an invariant mass that is typically around the b-hadron mass. It becomes easier

244

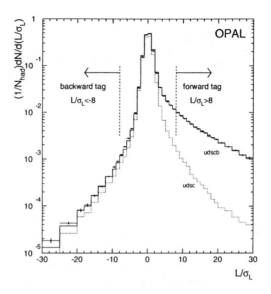

Figure 24: The decay length significance for tracks for *udsc* and *udscb* events as determined by the OPAL collaboration.

to distinguish $b \to l$ decays from the chain decay $b \to c \to l$ if the transverse momentum is measured with respect to the jet axis reconstructed without the momentum contribution of the prompt lepton. Purities of $\sim 80\%$, selecting $\sim 25\%$ of the primary semi-leptonic decays, have been obtained using high p_t lepton b-flavour tagging.

9.3 Using the Event Shape to Tag Heavy Flavour

The process $Z^0 \to b\bar{b}$ tend to have a characteristic shape. This is due to the "hard" fragmentation of the b-quark where there is a tendency for less energy to be carried away by gluon radiation and thus more energy available for the heavy flavour hadron decay products. The b-hadron decay products show high p_t with respect to the jet axis. Finally, single particles carrying a large fraction of the momentum are not as likely in events containing heavy flavour.

Tagging of heavy flavour using event shape has poor efficiency and purity. However, multivariate techniques such as Artificial Neural Networks have successfully utilized event shape variables, in combination with other variables based on lifetime and prompt lepton tagging, to provide the highest possible selection efficiency for heavy flavour events[34].

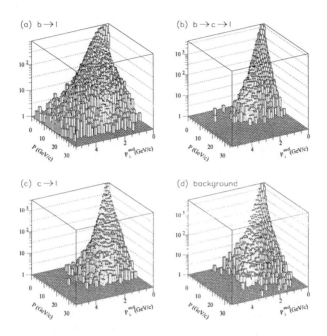

Figure 25: Momentum against transverse momentum distributions for leptons from: 1) primary b-hadrons decays; 2) the chain decay $b \to c \to l$; 3) primary c decays; 4) lepton fakes.

9.4 Charm Tagging Using D^*'s

The distinctive kinematics of $D^{*\pm}$ decays can be used to tag charm hadron decays[37]. The decay chain of interest is, $D^{*+} \to D^0\pi^+ \to K^-\pi^+\pi^+$ with branching ratio $\sim 2.7\%$ and Q value ~ 6 MeV. A plot of $\Delta M = M(K\pi\pi) - M(K\pi)$ yields, for the process in question, a narrow peak above a small background as shown in Fig. 26 [36].

Since primary D^*'s carry away a larger fraction of the original beam energy, a cut on $X = E_{D^*}/E_{beam}$ can be used to select $Z \to c\bar{c}$ samples [37]. By using X and opposite hemisphere tagging it is possible to improve the separation of the separate contributions of $Z \to c\bar{c}$ and $Z \to b\bar{b}$ events[38].

Another method of tagging $Z \to c\bar{c}$ events is based on the inclusive properties of the slow π^+ associated with the $D^{*+} \to D^0\pi^+$ decay. The longitudinal momentum of this pion is kinematically constrained to be less than 3 GeV and its P_T with respect to the original c quark direction is much lower than the typical 300 MeV of kinetic energy of fragmentation tracks at LEP-1.

246

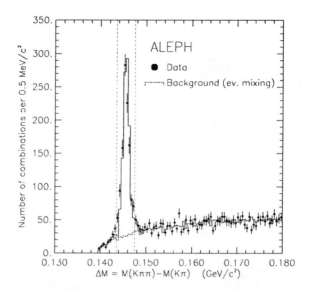

Figure 26: The mass difference $\Delta M = M(K\pi\pi) - M(K\pi)$ measured by the ALEPH Collaboration utilizing the channel, $D^{*+} \rightarrow D^0\pi^+, D^0 \rightarrow K^-\pi^+$ The data depicted by points with error bars are compared with data (solid line). The estimated background is depicted by the dashed line.

10 Measurement of R_b and R_c

At LEP-1 b-hadrons from the decay $Z^0 \rightarrow b\bar{b}$ are usually boosted into opposite hemispheres, where the axis of the hemisphere is defined by the thrust axis. The thrust of an event is defined by the following relation:

$$T = \frac{\sum_i |\overline{p_i}.\overline{n}|}{\sum_i |\overline{p_i}|}, \tag{51}$$

where $\overline{n}$ is the vector that maximizes the quantity T, p_i is the momentum of particle i and the sum runs over all reconstructed particles in the event.

One can write expressions for the fraction of hemispheres that receive a b−tag (F_1) and the number of events in which both hemispheres receive a b−tag (F_2):

$$F_1 = R_b\epsilon_b + R_c\epsilon_c + \epsilon_{uds}(1 - R_b - R_c), \tag{52}$$

$$F_2 = R_b.C_b.\epsilon_b^2 + R_c.\epsilon_c^2 + \epsilon_{uds}^2(1 - R_b - R_c), \tag{53}$$

where ϵ_b, ϵ_c, and ϵ_{uds} are the tag efficiencies on b, c, and uds events respectively and C_b is the correlation coefficient between b-tag efficiencies in each hemisphere. In the early experimental determinations of R_b it was assumed that the two hemispheres are uncorrelated, that is $C_b = 1$.

The double tagging method allows R_b and ϵ_b to be determined experimentally. For example, if $\epsilon_c = \epsilon_{uds} = 0$, then:

$$R_b \approx \frac{C_b F_1^2}{F_2}, \ \epsilon_b = \frac{F_2}{C_b . F_1} \tag{54}$$

More careful analyses showed that there are correlations between the hemispheres and that C_b is not exactly one. These correlations arise from three main sources. First, detector effects where, for example, one has correlated loss of acceptance due to the beam hole. Second, physics effects where, for example, gluon radiation pushes both $b-$jets into the same hemisphere. Third, algorithm effects, for instance a bias in the vertex position due to the inclusion of displaced $b-$hadron decay tracks.

The uncertainty on ϵ_{uds} and particularly ϵ_c is another source of systematic error. The uncertainty on ϵ_c is evaluated with the aid of a Monte Carlo program tuned to c hadron properties. A major uncertainty in the efficiency for light quarks is hard gluon splitting to $b\bar{b}$. Experimental measurements of $g \to c\bar{c}$ [39] and theoretical estimates [40] can be used to estimate this source of uncertainty. Another source of systematic error in the estimation of the charm background is the value of R_c. Usually, the Standard Model value for R_c is used in the analysis.

The results from the LEP and SLD experiments on the measurement of R_b are given in Fig. 27 along with the LEP/SLC average and Standard Model expectation. The latest analyses in this area use several b-tagging methods [41]. These multiple b-tag analyses allow a better determination of the efficiencies for charm and light quark tagging from data.

10.1 Measurement of R_c

As has been described previously D^*'s can be used to tag $Z \to c\bar{c}$ events. The resulting sample of tagged events can be used to measure R_c. Since all D^{*+} states have to eventually decay to D^0 or D^+ mesons the $D^0 + D^+$ yield can be used to improve the efficiency of the R_c measurement [43]. Initially, LEP analyses, using this method to tag charm, assumed that at LEP energies the hadronization of c quarks gives the same fraction of D^* as in lower energy experiments [42]. The full statistics of LEP-1, in combination with double tagging, allow us to drop this assumption [43,44]. Another method of tagging charm is the charm particle counting method. In this case one uses the fact that the yield of charmed mesons and baryons is directly proportional to R_c [45]. The LEP

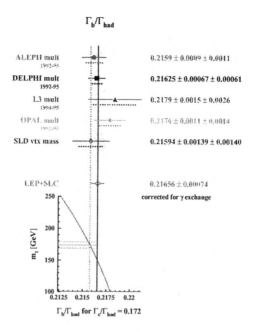

Figure 27: The results from the LEP and SLD experiments on the measurement of R_b. The Standard Model prediction is given as function of M_t.

and SLD measurements for R_c together with the Standard Model expectation is shown in Fig. 28.

10.2 R_b and R_c Versus the Standard Model

The improvement in the understanding and handling of systematic error, as well as increased statistics, has resulted in the average values for R_b and R_c moving into good agreement with the Standard Model expectation. This agreement is illustrated in Fig. 29[65] where the Standard Model expectation is shown for a value of the top mass of 175.6 ± 5.5 GeV.

11   Asymmetries at the Z^0 Pole

The different coupling of the Z to right-handed and left-handed fermions is the cause of parity violation in the weak neutral current. Assuming unpolarized beams, the Z boson in the process $e^+ e^- \to Z \to f\bar{f}$ is produced polarized along the direction of the beams. The degree of polarization (A_e) is dependent

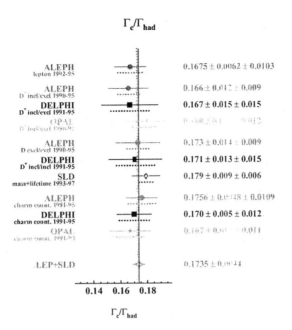

Γ_c/Γ_{had}

ALEPH lepton 1992-95		$0.1675 \pm 0.0062 \pm 0.0103$
ALEPH D incl/excl 1990-95		$0.166 \pm 0.012 \pm 0.009$
DELPHI D* incl/excl 1991-95		$0.167 \pm 0.015 \pm 0.015$
OPAL D incl/excl 1990-95		$0.168 \pm 0.011 \pm 0.012$
ALEPH D excl/excl 1990-95		$0.173 \pm 0.014 \pm 0.009$
DELPHI D* incl/incl 1991-95		$0.171 \pm 0.013 \pm 0.015$
SLD mass+lifetime 1993-97		$0.179 \pm 0.009 \pm 0.006$
ALEPH charm count. 1991-95		$0.1756 \pm 0.0048 \pm 0.0109$
DELPHI charm count. 1991-95		$0.170 \pm 0.005 \pm 0.012$
OPAL charm count. 1991-95		$0.167 \pm 0.005 \pm 0.011$
LEP+SLD		0.1735 ± 0.0044

0.14   0.16   0.18

Γ_c/Γ_{had}

Figure 28: The results from the LEP and SLD experiments, together with the LEP/SLD average, of the measurement of R_c.

on the ratio of the vector and axial coupling constants to the electron. As the decay $Z \to f\overline{f}$ violates parity, the fermions will be emitted in a preferential direction with respect to the Z spin. If the Z is fully polarized the asymmetry is $(3/4)A_f$. The conservation of angular momentum results in a correlation between the direction of spin of the Z and the helicity of the fermion.

The angular dependence of the process, $e^+e^- \to Z^0 \to f\overline{f}$ can be expressed as:

$$\frac{1}{\sigma C(cos\theta)} \frac{d\sigma}{dcos\theta} = \frac{3}{8}(1 + cos^2\theta) + A_{FB}cos\theta, \qquad (55)$$

where θ is the centre-of-mass scattering angle, $C(cos\theta)$ is the acceptance function and A_{FB} is the forward-backward asymmetry, which can be determined using the relation:

$$A_{FB} = \frac{\sigma_{forward} - \sigma_{backward}}{\sigma_{tot}}. \qquad (56)$$

This asymmetry at the Z peak can be expressed as:

$$A_{FB}^f = \frac{3}{4}A_e A_f, \qquad (57)$$

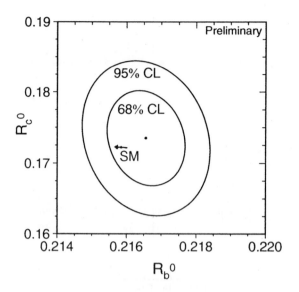

Figure 29: The contour formed from R_b plotted against R_c. The Standard Model expectation is shown for a top mass range of 175.6 ± 5.5 GeV.

where,

$$A_f = \frac{2g_{v_f}g_{a_f}}{g_{v_f}^2 + g_{a_f}^2}. \tag{58}$$

The asymmmetries provide a direct measurement of:

$$sin^2\theta_{eff}^{lept} = \frac{1}{4}(1 - \frac{g_{v_l}}{g_{a_l}}) \tag{59}$$

Assuming lepton universality the ratios of the coupling constants of the Z to the charged leptons are equal.

$Z^0 \to f\overline{f}$ events can be characterized by the helicity and direction if the fermion f. If we define the forward hemisphere to be the one into which the electron beam is pointing these events can be separated into the following categories: FR, forward right-handed; BR, backward right-handed; FL, forward left-handed; and, BL for backward left-handed. Using these definitions three types of asymmetry can be defined:

$$A_{pol} = \frac{\sigma_{FR} + \sigma_{BR} - \sigma_{FL} - \sigma_{BL}}{\sigma_{tot}} = \frac{\sigma_R - \sigma_L}{\sigma_{tot}}, \tag{60}$$

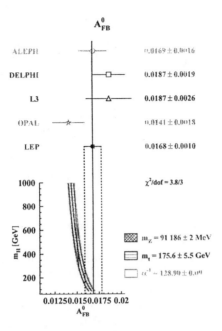

Figure 30: The LEP wide determinations of A_{FB}^l as well as the LEP average. The variation of the Standard Model prediction for this quantity with Higgs mass is also shown.

$$A_{pol}^{FB} = \frac{\sigma_{FR} + \sigma_{BL} - \sigma_{BR} - \sigma_{FL}}{\sigma_{tot}}, \qquad (61)$$

$$A_{FB} = \frac{\sigma_{FR} + \sigma_{FL} - \sigma_{BR} - \sigma_{BL}}{\sigma_{tot}} = \frac{\sigma_F - \sigma_B}{\sigma_{tot}}. \qquad (62)$$

The polarization asymmetry, A_{pol}, depends only on the helicity of the fermions emitted in the Z-decay. The Forward-Backward polarization asymmetry, A_{pol}^{FB} measures the polarization of the Z^0 and does not depend on f-flavour. Both of these asymmetries require the fermion helicity to be measured. Practically, this is only possible in a final state containing taus. The Forward-Backward asymmetry only needs the fermion charge and direction. Thus, it can be measured for all flavour final states that can be tagged. The FB asymmetry is proportional to the product of A_{pol} and A_{pol}^{FB} and it can be determined without measuring the polarization of the final state.

The asymmetries measured at LEP and SLC (corrected for ISR, FSR,γ-exchange and $Z - \gamma$-interference) can be expressed in terms of ratios of effective coupling of fermions to the neutral current:

$$A_{pol} = -A_f; \; A_{pol}^{FB} = -\frac{3}{4}A_e; \; A_{FB} = \frac{3}{4}A_eA_f; \; A_{LR} = A_e; \; A_{FB}^{pol} = \frac{3}{4}A_f. \quad (63)$$

11.1 Lepton Forward-Backward Asymmetry

The lepton Forward-Backward asymmetry can be determined from the angular distribution of the final state leptons. Information contained in the angular distribution is extracted by either, for example, counting:

$$A_{FB}^{\mu} = \frac{N_F^{\mu^-} - N_B^{\mu^-}}{N_F^{\mu^-} + N_B^{\mu^-}}. \quad (64)$$

Or, it can be measured by fitting the odd term of the angular distribution (θ) of (say) the negative lepton.

The corrected asymmetries, $A_{FB}^0(l = e, \mu, \tau)$ are found by fitting the measured A_{FB}'s to a model independent formula that incorporates radiative corrections and the energy dependence of the asymmetry. This fit is done simultaneously with the lineshape data to account for the effect of the energy uncertainty. Fig. 30 shows the LEP wide determinations of A_{FB}^{ll} as well as the LEP average. The variation of the Standard Model prediction for this quantity with Higgs mass is also shown in the figure.

11.2 The Tau Polarization Asymmetry

The helicities of the taus in the final state of the decay $Z^0 \to \tau^+\tau^-$ are nearly 100% anticorrelated. The average tau polarization is given by:

$$P_\tau = \frac{\sigma_R - \sigma_L}{\sigma_R + \sigma_L} \quad (65)$$

where, for example, σ_L is the cross-section to make a left-handed τ^-. Thus, A_{pol} is the tau-polarization measured over the entire $cos\theta$ range.

A_{pol}^{FB} can be found by comparing the measured polarization in the forward and backward direction. One can also fit the measured dependence of the polarization as a function of θ:

$$P_\tau(cos\theta) = \frac{A_{pol}(1 + cos^2\theta) + \frac{8}{3}A_{pol}^{FB}cos\theta}{(1 + cos^2\theta) + A_{FB}cos\theta}. \quad (66)$$

The measurement of tau polarization asymmetry is dominated by the channels: $\tau^\pm \to \pi^\pm\nu_\tau$ and $\tau^\pm \to 2\pi\nu_\tau$. Leptonic channels have smaller sensitivities

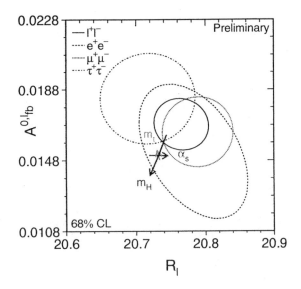

Figure 31: A plot of R_l against A^l_{FB} around the Z^0 pole. The solid contour depicts the 68% confidence level assuming lepton universality. The Standard Model expectation is shown for the parameter ranges: $M_t = 173.8 \pm 5.0$ GeV; $M_H = 300^{+700}_{-200}$ GeV; and, $\alpha_s = 0.119 \pm 0.002$.

because of the two neutrinos involved. Tau decay channels have selection criteria based on: multiplicity, particle identification, photon counting, and visible invariant mass.

The main source of background to the $\pi^\pm\nu$ channel arises from $2\pi\nu$ final states where the π^0 is not identified. The purity of the pionic channel ranges from 84% to 94%, depending on the LEP experiment. In the $2\pi\nu$ channels the main source of backgrounds is from tau decays with more than one π^0. The selection purity for this channel is similar to that of the single pion channel. The decay modes with more than three pions in the final state, with an integrated branching ratio of $\sim 10\%$, are usually not used because of the large model dependencies and low selection purity.

As we have seen the measurement of A_{pol} and A^{FB}_{pol} for taus can be interpreted as measurements of A_e and A_τ after a correction is applied to take into account the effects of photon exchange, $Z - \gamma$ interference and ISR/FSR. The LEP wide values of A_e and A_τ obtained from tau polarization measurements are in good agreement with the Standard Model. The LEP averages are[65]:

$$A_\tau = 0.1431 \pm 0.0045, \ A_e = 0.1479 \pm 0.0051. \tag{67}$$

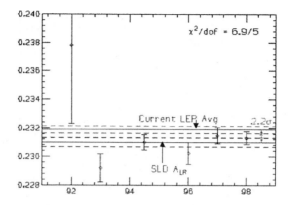

Figure 32: A plot of the variation of the value of $sin^2\theta_W^{eff}$ estimated from the SLD value of A_{LR} as a function of time in years. The result is compared with the LEP average.

11.3 Lepton Universality

Lepton universality requires that the vector g_V^l and axial g_A^l couplings of the neutral current to the various flavours of leptons are equal. The sum of the squares of the couplings is determined by a measurement of Γ_l. The coupling ratio is determined by A_{FB}^l. Thus, a good test of lepton universality is a plot of R_l versus A_{FB}^l around the Z^0 pole, as shown in Fig. 31. As can be seen the data is completely consistent with lepton universality.

11.4 Beam Polarization and Asymmetries

The beam polarization of SLC allows the cross-section for Z^0 production to be measured with left-handed (σ_l) and right-handed (σ_r) polarized beams. This information can be summarized in a variable called the left-right asymmetry, or A_{LR}, which is defined as:

$$A_{LR} = \frac{1}{P}\frac{\sigma_l - \sigma_r}{\sigma_l + \sigma_r}, \tag{68}$$

where P is the luminosity weighted average value of the beam polarization. One can also measure the Forward-Backward asymmetry when the polarization is switched:

$$A_{FB}^{pol} = \frac{1}{P}\frac{(\sigma_{F,r} - \sigma_{F,l}) - (\sigma_{B,r} - \sigma_{B,l})}{\sigma_l + \sigma_r}. \tag{69}$$

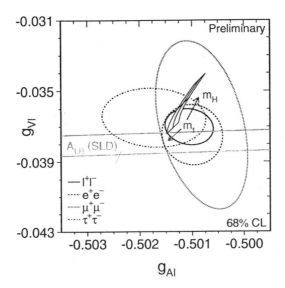

Figure 33: The 68% contours for g_{V_l} versus g_{A_l} ($l = e, \mu, \tau$). The solid contour shows the result assuming lepton universality. The Standard Model expectation is calculated for $M_t = 173.8 \pm 5.0\text{GeV}$ and $M_H = 300^{+700}_{-210}$.

Measurement of A_{LR} at SLD

The helicity of the SLC e^- beam is chosen randomly pulse-to-pulse. To measure A_{LR} one counts hadronic events that are produced with left- and right-handed beams. The beam polarization P_i associated with each Z is used to form the luminosity weighted average:

$$P = (1 + \epsilon)\frac{1}{N_Z}\sum_i^{N_Z} P_i \tag{70}$$

where ϵ is a small correction ($0.001 \to 0.002$) for chromatic and beam transport effects.

The systematic uncertainties on the polarization scale are running at around 1% with contributions from uncertainties in: laser polarization (0.10% in 1998); detector linearity (0.50%); AP calibration (0.30%); electronic noise (0.2%); interchannel consistency (0.8%).

The asymmetry is measured using hadronic Z decays that are selected with high purity using the following event level criteria: there must be at least 22 GeV of energy in the calorimeters; events must have at least four tracks;

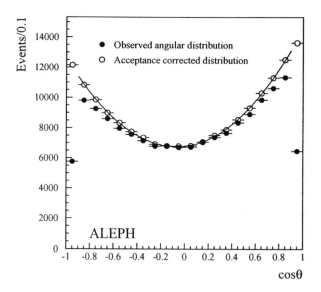

Figure 34: The Forward-Backward b asymmetry determined from inclusive leptons as observed by the ALEPH Collaboration. The curves with and without acceptance correction are shown.

and, the ratio of the "vector" to scalar energy sum in the calorimeter is less than 0.6.

The events produced with left-handed (N_L) and right-handed (N_R) polarization are counted and in conjunction with the luminosity weighted polarization the initial value of A_{LR} can be determined, using:

$$A_{LR} = \frac{1}{P}\frac{(N(L) - N(R))}{(N(L) + N(R))}. \tag{71}$$

A correction is then applied to account for: residual backgrounds to the Z sample; residual, left-right asymmetries in the luminosity, polarization, beam energy. etc. Finally, Electroweak interference is taken into account in order to extract A_{LR} at the Z^0 pole.

Fig. 32 shows the variation of the value of $sin^2\theta_W^{eff}$ estimated from the SLD value of A_{LR} as a function of time. The result is compared with the LEP average. Note the $\sim 2\sigma$ difference between the LEP and SLD average value for $sin^2\theta_W^{eff}$.

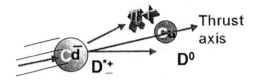

Figure 35: The charge correlation between a prompt D^* and the c quark charge in $Z \to c\bar{c}$ decays.

11.5 Universality of the Neutral Current Couplings

It is interesting to revisit the contour plots for g_{V_l} versus g_{A_l} for the various leptons flavours (l) with the additional information obtained from the SLD determination of A_{LR}. This plot is shown in Fig. 33. The measurements are in good agreement with lepton universality as expressed by the Standard Model. The allowed range of g_{V_l} versus g_{A_l} determined from A_{LR} is also consistent with lepton universality.

12 The Quark Forward Backward Asymmetry

The sensitivity of fermion forward-backward asymmetries to the mixing angle $sin^2\theta^l_{eff}$ is particularly enhanced for quarks. Another advantage is that the quark asymmetry dependence on the E_{cm} is weaker than in the case of lepton asymmetries because of the smaller quark electric charge. A_q must be known before A_e and $sin^2\theta^l_{eff}$ can be determined from quark Forward-Backward asymmetries. It can be obtained by evaluating:

$$\frac{g_{v_q}}{g_{a_q}} = 1 - \frac{2Q_q}{I^3_q}(sin^2\theta^l_{eff} + C_q),$$ (72)

where C_q is the residual vertex correction which can be calculated within the framework of the Standard Model. The correction depends on the top quark mass and for $Z \to b\bar{b}$ events it is 0.014, for $M_t = 175$ GeV [46]. In the case of u, d, s, c quarks C_q is small with little dependence on the Standard Model parameters.

In order to measure the quark asymmetry the quark charge must be determined. There are four basic methods with which this can be done. First the use of prompt leptons from semileptonic heavy flavour decays. Second, particularly in the case of $Z \to c\bar{c}$, exploiting D^* tagging. Third, using heavy flavour tagging plus jet charge determination. Fourth, utilizing jet charge from inclusive hadrons

258

12.1 Heavy Quark Asymmetries Using Leptons

In primary semilptonic decays of b-hadrons the sign of the lepton correlates with the quark-antiquark nature of the produced quark. A good definition of the quark direction is the thrust axis which can be signed by lepton charge (Q) in the hemisphere into which the thrust points. Thus, one obtains the following signed estimator of the b quark direction.

$$cos\theta_b = -Qcos\theta_{thrust} \qquad (73)$$

In the case of semileptonic b-decays A_{FB} can be determined from a fit to the angular distribution. An example of an acceptance corrected angular distribution measured by the ALEPH collaboration is shown in Fig. 34[47].

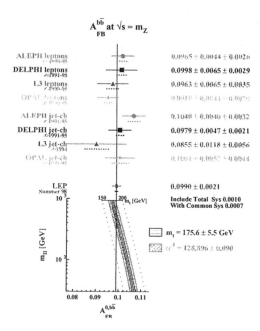

Figure 36: The measured asymmetry $A_{FB}^{b\bar{b}}$ for the four LEP experiments using two leptons and jet-charge to ascertain the quark charge. The LEP average is also given along with a comparison with the Standard Model expectation.

Even for perfectly pure samples of semileptonic b decays there are two principle factors that result in the dilution of the observed asymmetry. The first reduction factor is due to the presence of $B^0\overline{B}^0$ oscillations. The second dilution factor is due to cascade decays $b \to c \to l$. Such decays yield the wrong

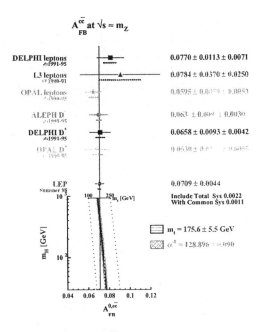

$A_{FB}^{c\bar{c}}$ at $\sqrt{s} \approx m_Z$

DELPHI leptons z:1991-95	$0.0770 \pm 0.0113 \pm 0.0071$
L3 leptons (?) 1990-91	$0.0784 \pm 0.0370 \pm 0.0250$
OPAL leptons z:1990-95	$0.0595 \pm 0.06^{..} \pm 0.0053$
ALEPH D* z:1991-95	$0.063 \pm 0.06^{..} \pm 0.0030$
DELPHI D* z:1991-95	$0.0658 \pm 0.0093 \pm 0.0042$
OPAL D* 1990-95	$0.0638 \pm 0.0^{..} \pm 0.0055$
LEP Summer 98	0.0709 ± 0.0044
	Include Total Sys 0.0022 With Common Sys 0.0011

$m_t = 175.6 \pm 5.5$ GeV
$\alpha^{-1} = 128.896 \pm 0.090$

Figure 37: The measured asymmetry $A_{FB}^{c\bar{c}}$ for the four LEP experimnents using two leptons and D^*'s to ascertain the quark charge. The LEP average is also given along with a comparison with the Standard Model expectation.

charge for the lepton. There is also a small correction from the background asymmetry. By calculating the expected dilution factors, using a Monte Carlo estimate for the background asymmetry and using an independent determination of the c asymmetry, it is possible to arrive at a value of the b-asymmetry. The use of an independent measurement of the c asymmetry can be avoided if a fit to the lepton distribution in the (p, p_t) plane is performed. This allows the c asymmetry to become a free parameter of the fit.

The precision of LEP measurements of the b asymmetry is limited primarily by statistics [48]. The main source of uncertainty arise from the lack of knowledge of the $B^0\overline{B}^0$ mixing parameter.

12.2   Using D^*'s to Determine the c Quark Asymmetry

The charge correlation between the prompt D^* and the c quark can be used to sign the thrust axis, as illustrated in Fig. 35. In this way the c quark asymmetry can be determined in way similar to the b asymmetry. In this case the main discrimination variable between the b and c quark is the variable, $X_E = E_{D^*}/E_{beam}$. For example, a cut of $X_E > 0.5$ gives a purity $\sim 80\%$ for the selection of the process $Z \to c\bar{c}$. The effect of the remaining b asymmetry, arising from the charge correlation in the decay $b \to D^*$, can be corrected for in the following ways. The effect of the b asymmetry can be subtracted using an independent measurement of A_{FB}^b [49]. Another method is to fit the shape of the distribution of X_E with a b and c component [50]. Whatever method is chosen the effect of mixing on the b asymmetry must be properly taken into account.

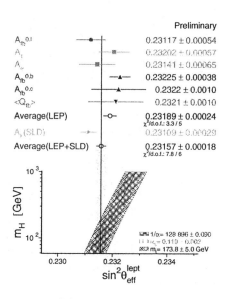

Figure 38: Determinations of $sin^2\theta^l_{eff}$ from asymmetry measurements at LEP and SLC compared with Standard Model predictions.

12.3 Asymmetries Utilizing Jet Charge

Lifetime tagging can be used to select $Z \to b\bar{b}$ decays. However, this tagging method does not give any information on the original quark charge. Despite the complexity of the hadronization process, the fragmentation and decay products retain some memory of the original quark charge. An estimator called the jet charge can be built up from the remnants of information on the original quark charge contained in the particles comprising the hadronic jet.

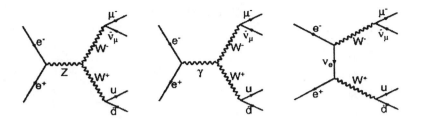

Figure 39: The three main Feynman diagrams describing W^+W^- production at LEP.

The jet charge is defined as:

$$Q_h = \frac{\Sigma_i |p_{L_i}|^\kappa q_i}{\Sigma_i |p_{L_i}|^\kappa}, \tag{74}$$

where q_i is the charge of particle i in one hemisphere defined by the thrust axis, p_{L_i} is the longitudinal component of particle i with respect to the thrust axis and the parameter κ is an adjustable weighting parameter.

The average of the difference between the jet charges estimated for the 'forward' and 'backward' hemispheres is related to the b asymmetry through the formula:

$$< Q^b_{FB} > \equiv < Q_F - Q_B > = \delta_b A^b_{FB} \tag{75}$$

where the quantity δ_b is called the charge separation. Of course, there is always some contamination from lighter quarks and a more general relation than the one above must be employed:

$$< Q_F - Q_B > = \sum_{f=u,d,..}^{b} P_f C_f \delta_f A^f_{FB}. \tag{76}$$

where C^f are the acceptance factors for each quark flavour and P_f are the purities for each flavour ($P_{u,d,s,c} \ll 1$). The determination of the charge separation is described in detail in Ref.[51].

The jet-charge technique sketched above can be used to evaluate $A^0_{FB}(b)$ and $A^0_{FB}(c)$. Just as in the case of lepton Forward-Backward asymmetries corrections have to be applied for QED ISR and FSR and for the effects of photon-exchange and $Z - \gamma$ interference. As we are dealing with heavy quarks there are additional corrections due to the strong interaction. At the present level of precision one loop QCD corrections are sufficient[52]. The LEP results for $A^{b\bar{b}}_{FB}(b)$ and $A^{c\bar{c}}_{FB}(c)$ are shown in Fig. 36 and Fig. 37. As can be seen

OPAL Preliminary

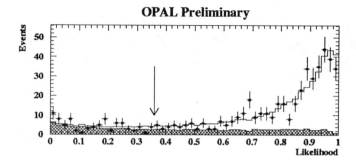

Figure 40: The distribution of the likelihood discriminant used to select events of the type $W^+W^- \rightarrow q\bar{q}q\bar{q}$ for the preselected events. The points indicate the data, and the histogram represents the Monte Carlo expectation, where the hatched area shows the estimated contribution of the total background. The selection cut is indicated by the arrow.

from the figures these results are in good agreement with the Standard Model expectation.

12.4 Asymmetry Measurements and The Effective Electroweak Mixing Angle

A compilation of the various measurements of $sin^2\theta^l_{eff}$ determined from the various asymmetry measurements described is shown in Fig. 38. As can be seen from the figure the value of effective mixing angle derived from the LEP b Forward-Backward asymmetry is high compared to the average while the measurement of A^0_{LR} from SLD, shows the opposite behaviour. In fact these two values are separated by $\sim 2\sigma$.

There are two types of measurements that contribute here. The first type is based on leptonic asymmetries and contains: $A^0_{FB}(l)$; the measurements of A_e and A_τ from tau polarization; and, A^l_{LR} from SLC. The second type contains the other three measurements that require a knowledge of A_q terms in order to derive the sine. These terms, containing A_q have only a weak dependence on $sin^2\theta^l_{eff}$ in the Standard Model.

13 W-Physics at LEP-2

As we have seen LEP-1 and SLC have a sensitivity to M_W, as well as M_t, M_Z and M_H, through electroweak radiative corrections. However, in the summer of 1996 LEP crossed the W^+W^- production threshold for the first time, allowing the LEP experiments to directly measure W boson properties.

The integrated luminosity above the W-pair production threshold, up to the end of summer 1998, is $\sim 250\ pb^{-1}$ per experiment. This period can be divided into four parts. First, a run a 161 GeV in 1996 in which each LEP experiment took around 10 pb^- of data. Second, a run at 172 GeV, again in 1996, in which each experiment took about $\sim 10\ pb^{-1}$ of data. Third, a run at 183 GeV in 1997, where $\sim 56\ pb^{-1}$ per experiment was delivered. Fourth, a long run in 1998 where approximately 180 pb^{-1} of data per experiment was delivered at a centre of mass energy of 189 GeV. It is envisaged that each LEP experiment will record approximately 8 thousand W^+W^- pairs before the end of the LEP program. This should allow a LEP-2 combined error on M_W to reach ~ 35 MeV.

W pairs at LEP are produced via three main (CC03) diagrams, as shown in Fig. 39. However, the Standard Model four fermion diagrams with the same final state, that can interfere with the three main channels, are much more numerous. In order to obtain CC03 cross-sections, the measurements have to be corrected for the effects of four fermion processes. The ALEPH and OPAL analyses treat four fermion effects as additional background to the CC03 processes that has to be subtracted (or even added). In the case of DELPHI and L3 multiplicative correction factors are applied in order to go from measured four fermion cross-sections to CC03 cross-sections. Both procedures obtain the same results to within $\sim 1\%$.

The final states for W-pair decays are as follows: BR$(W^+W^- \to qqqq) = 45.6\%$; BR$(W^+W^- \to qql\nu) = 14.6\%(l = e, \mu, \tau)$; and, BR$(W^+W^- \to ll\nu\nu) = 10.6\%$. The WW decays can be classified in 3 topologies: four jets; two jets a lepton and missing energy; and, two leptons plus missing energy.

13.1 W Cross-sections and Branching Ratios

$WW \to qqqq$ (BR $\sim 46\%$)

The signature for this channel is simply 4 jets with similar energy and very little missing energy. The main backgrounds are $e^+e^- \to q\bar{q}(\gamma) + n(gluons)$ and ZZ production.

In the $E_{cm} = 183$ GeV analysis, all LEP experiments begin with some preselection. ALEPH then uses an Artificial Neural Network (ANN) analysis on the remaining events. The variables that discriminate most are: the angles between jets; the Fox Wolfram moments; and the minimum and maximum invariant masses. The L3 collaboration also uses a neural network approach to differentiate between QCD background and the WW signal. Their ANN has 10 input variables. The most discriminating variables are: the Y_{34} variable from the Durham jet-finder[54]; minimum jet-jet angle; and, minimum and maximum energies. In both analyses the cross-section was obtained from a fit to the ANN output distribution.

W Leptonic Branching Ratios

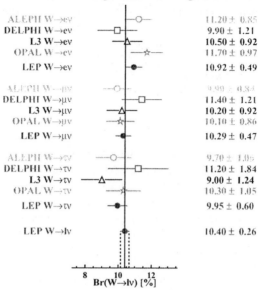

Figure 41: The leptonic branching ratios determined by each LEP experiment for each lepton flavour. The LEP average for each lepton flavours is also given. The Standard Model value for the leptonic branching ratio of the W is $\sim 10.8\%$.

The OPAL collaboration chose a maximum likelihood approach to the $WW \rightarrow qqqq$ analysis. Four input variables were used: $log(y_{45})$ from the Durham jet finder; the QCD matrix element; a modified Nachtmann-Reiter angle; and, sphericity. These variables are used to construct a likelihood discriminator which is transformed into event weights. The cross-section was obtained by cutting on the likelihood discriminator. The distribution of the likelihood parameter obtained by OPAL[55] is given in Fig. 40.

The DELPHI analysis exploited the difference in jet energies and jet-jet angles between the signal and QCD background. They summarized this in one variable:

$$D = \frac{E_{min}}{E_{max}} \cdot \frac{\Theta_{min}}{E_{max} - E_{min}}, \quad (77)$$

where Θ_{min} is the minimum angle between any jet-pair in the event.

All four experiments obtain selection efficiencies $\sim 85\%$ while reducing backgrounds to between $1.0 \rightarrow 1.8$ pb. At $E_{cm} = 183$ GeV the four LEP experiments selected between 391 → 475 events each.

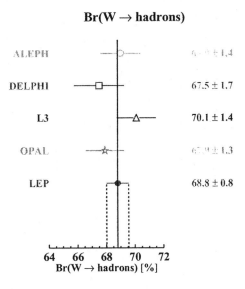

Figure 42: The hadronic branching ratios determined by each LEP experiment. The LEP average is also given. The Standard Model value for the hadronic branching ratio of the W is ∼ 67.5%.

$WW \to qql\nu$ (BR ∼ 44%)

The signature for this important channel is two separated jets with one isolated energetic lepton, or tau jet, plus missing momentum not pointing along the beam pipe. The neutrino is identified with the missing momentum vector. The main backgrounds arise from: $e^+e^- \to q\bar{q}(\gamma)$; and, the four fermion processes, $e^+e^- \to q\bar{q}l^+l^-$. The ALEPH[53] and OPAL analyses build a probability/likelihood that the event is of the type qql^+l^- from lepton energy, isolation and missing momentum.

DELPHI and L3 adopt a different approach, they select events by cutting on the above mentioned variables. At a centre-of-mass-energy of 183 GeV the LEP experiments each selected between 288 → 362 events in this channel. Selection efficiencies ranged between 75 → 90% for e, μ and 50 → 65% for final states containing a tau.

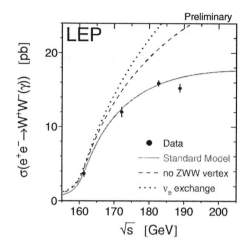

Figure 43: A plot of the W-pair cross-section against the centre-of-mass-energy. The dots show the data. The solid line shows the Standard Model fit to the data. The dashed line gives a fit to the data when the theory does not include a ZWW vertex. The dotted line gives a fit to the data where the neutrino exchange diagram is the only contributor to the W-pair cross-section.

$WW \rightarrow l\nu l\nu$ (**BR** $\sim 11\%$)

The main signatures for this process are two acoplanar energetic leptons (or tau jets) plus missing momentum that is not pointing along the beam pipe. The important backgrounds are: $e^+e^- \rightarrow Z + \gamma \rightarrow l^+l^-\gamma$; Bhabha events; and, two-photon processes. The typical analysis is an OR of a topological two lepton search and dedicated final state selections. At the centre-of-mass-energy of 183 GeV roughly $54 \rightarrow 78$ events per experiment were selected in this channel. Typical selection efficiencies were $\sim 50 \rightarrow 60\%$.

Combined Maximum Likelihood Fit

Measurements of the total W-pair cross-sections and branching ratios are obtained via a combined likelihood fit to all $WW \rightarrow ffff$ channels using:

$$L = \prod_i P(N_i, \mu_i), \qquad (78)$$

where P is the Poissonian probability distribution, N_i is the number of observed events in channel (i) and μ_i is the number of expected events in channel

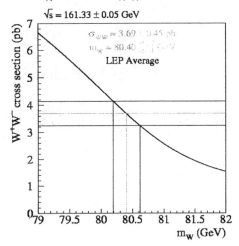

Figure 44: A plot of the WW cross-section as a function of W boson mass. The grey band gives the cross-section range allowed by W boson mass range, according to the Standard Model.

(i). The expected number of events takes into account the background (σ_i^{bg}) and efficiency matrix (ϵ_{ij}):

$$\mu_i = L.(\sum_j \epsilon_{ij}\sigma_j + \sigma_i^{bg}), \qquad (79)$$

where j is the process and L is the integrated luminosity. Using this method all known systematic effects are taken into account and overlaps in the selected events are avoided.

The resulting values of the leptonic branching ratio for the W boson determined by each LEP experiment for each lepton flavour together with the LEP average for each lepton flavour, are summarized in Fig. 41. The Standard Model value for the leptonic branching ratio of the W is $\sim 10.8\%$ The W boson hadronic branching ratio of the W boson determined by each LEP experiments is given in Fig. 42.

It is possible to determine the CKM matrix element $|V_{cs}|$ from this branching ratio, in the following way. To a very good degree of approximation:

$$\frac{Br(W \to q\bar{q})}{1 - Br(W \to q\bar{q})} = \left(1 + \frac{\alpha_s(M_W^2)}{\pi}\right) \sum_{i=u,c;j=d,s,b} |V_{ij}|^2. \qquad (80)$$

268

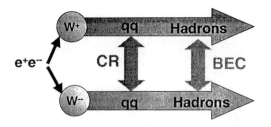

Figure 45: A schematic view of color reconnection and Bose Einstein correlation between the W's in the final state of the process: $e^+e^- \to W^+W^-$.

Utilizing the $BR(W \to q\bar{q})$ from LEP combined results and $\alpha_s = 0.118 \pm 0.003$ from the (then) current world average, one obtains: $\Sigma|V_{ij}|^2 = 2.131 \pm 0.076$ (unitarity $\equiv 2$). The value of $|V_{cs}|$ can be extracted by fixing the other CKM matrix elements to their measured values, to obtain: $|V_{cs}| = 1.04 \pm 0.04$, where the assumption of unitarity has not been used.

The LEP average WW cross-section plotted as a function of centre-of-mass-energy is shown in Fig. 43. The fact that the Standard Model provides the best fit and that the theory without a ZWW vertex does not, indicates that the data is consistent with the non-Abelian nature of the Standard model.

13.2 Estimate of M_W from Cross-section Measurements near the WW Threshold

Assuming the validity of the Standard Model the W boson mass can be extracted from the cross-section. The sensitivity of this method is maximal near to the threshold as shown in Fig. 44 [15]. For a mass of $M_W = 80.40^{+0.22}_{-0.21}$ GeV the W-pair cross-section is $\sigma_{WW} = 3.69 \pm 0.45$ pb.

13.3 The Direct Measurement of the W Boson Mass

At higher energies M_W is extracted fitting the invariant mass distribution of the W decay distribution. Two main methods are used: convolution; and Monte Carlo re-weighting.

The Monte-Carlo reweighting technique is the primary method used by all LEP collaborations except DELPHI. In this approach a large number of Monte Carlo events a correspondence between generated and constructed masses, to be established. These events are generated with a definite M_W value an analytical code is used to reweight them according to the mass that best fits the data.

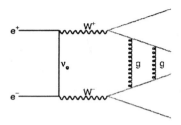

Figure 46: A diagram showing the QCD interaction between partons in the final state of the W-pair decay.

The convolution technique is the baseline method employed by the DEL-PHI analysis. The theoretical W line shape curve is convoluted with an analytical function describing detector effects. All data are then compared with the theory plus convolution curve to obtain a likelihood dependent on M_W. This method allows the use of different event weights depending on the resolution thus improving the precision. This method is described in detail at this meeting [56].

Using both reweighting and convolution methods it is possible to perform two parameter fits to extract M_W and Γ_W. The correlation between the two measurements is small.

13.4   Colour Reconnection and Bose Einstein Effects in W^+W^- Decay

W decay vertices in WW decays are separated by distances of the order of 0.1 fm. However, the hadronic scale is of the order of 1 fm. Obviously there is a large spacetime overlap in which interconnections between the colour singlets may be imagined. This scenario is illustrated in Fig. 45.

Colour Reconnection Effects in $e^+e^- \rightarrow W^+W^-$

QCD interactions between partons, shown in Fig. 46, can be studied by investigating event shapes and particle multiplicities since colour reconnection would influence the final event topology. Although perturbative effects are expected to be small, non-perturbative effects may be large. Hence one must rely on models to describe possible reconnection effects. One such is the Ellis-Geiger model [57] which predicts large multiplicity effects. Colour reconnection effects are not only restricted to WW production but appears for example also in top-quark production at hadron colliders. Additionally, reconfiguration of colour flow is also possible at LEP-1 for example in $q\bar{q}gg$ events.

270

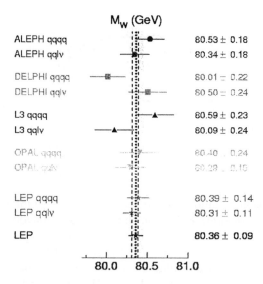

Figure 47: The value of M_W determined by the LEP experiments for $qqqq$ and $qql\nu$ final states. The LEP average of M_W for both types of final states and for all final states, is shown.

Various Monte Carlo models such as PYTHIA[58], HERWIG[59] and ARIADNE[60] incorporate various colour reconnection effects. For example, both PYTHIA and ARIADNE implement the Gustafson-Hakkinen model, but in different ways. The charged multiplicity for WW final states as measured by, for example, the OPAL collaboration[61] can be compared with the predictions for various Monte Carlo techniques. In this analysis the ARIADNE model has the worst agreement with the data.

Colour reconnection effects might affect some event topologies rather than others. For example, the Ellis-Geiger model predicts that such effects would increase as the angle between two outgoing jets in W-pair decays decreases. However, with current statistics the properties of the final state of $WW \rightarrow q\bar{q}q\bar{q}$ are in agreement with Monte Carlo both without colour reconnection effects and with "realistic " colour reconnection scenarios. In other words effects are small. That is why the ARIADNE Monte Carlo, that predicts relatively large effects, does not agree with the data as well as the other LEP favoured Monte Carlo programs. Some models have even been rejected, for example, the Ellis-Geiger model and the Pythia "Toy Model".

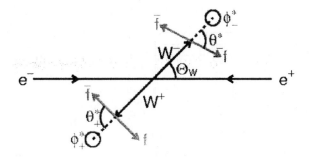

Figure 48: The angles determining the W-pair kinematic variables.

Bose-Einstein Correlations

Bose-Einstein Correlations (BEC) are well established in e^+e^-, hadron-hadron and nucleus-nucleus interactions. It is in fact quantum mechanical interference between identical bosons due to the symmetrization of the wave function, leading to enhanced production close in phase space. Given two pions with momenta k_1 and k_2 then one can define a correlation function:

$$C(k_1 k_2) = \frac{\rho(k_1, k_2)}{\rho_0(k_1, k_2)}, \tag{81}$$

where $\rho(k_1, k_2)$ is a two particle density function and $\rho_0(k_1, k_2)$ is the corresponding quantity for a reference sample without BEC.

The correlation function can be written as:

$$C(Q) = \frac{\rho(Q)}{\rho_0(Q)}, \tag{82}$$

where, $Q^2 = -(k_1 - k_2)^2 = M^2(\pi\pi) - 4m_\pi^2$. This allows us to define the function:

$$C(Q) \sim (1 + \lambda e^{-Q^2 R^2})[1 + \delta(Q)], \tag{83}$$

where λ is the correlation strength and R is the source radius.

An example of a search for BEC effects is the L3 analysis. The L3 collaboration uses a sample of $qqqq$ candidate events on which a jet finder is run in order to reconstruct W's. After the W's forming the final state have been identified track pairs are taken: within a single jet; between jets from a single W; and, between jets from different W's. Defining the double ratio:

$$C^* = \frac{\rho^{\pm\pm}}{\rho^{+-}} / \left(\frac{\rho^{\pm\pm}}{\rho^{+-}} \right)_{MC,noBE} \tag{84}$$

272

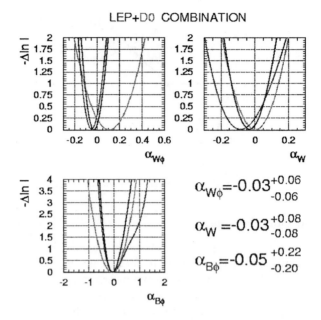

LEP+D0 COMBINATION

Figure 49: Log-likelihood curves for the experimental determination of the anomalous coupling constants from D0 and each of the LEP experiments. The combination of results is shown by the solid black curves. The dark grey lines are for LEP only results and the light grey lines represent the results from D0.

where the ρ parameters are defined between pairs of pions with the same and with different charges. The denominator is a Monte Carlo estimate without BECs. The double ratios obtained by the L3 collaboration, both for particle pairs within the same jet and for particles pairs defined between different W's, should, of course, be unity if the effect of BEC's is not measurable. There could be some evidence for small BEC effects at low values of Q (less than ~ 0.2 GeV/c^2) in the L3 analysis, as well as in the corresponding analyses of ALEPH, DELPHI and OPAL[62].

Overall the ALEPH and DELPHI analyses prefer no Bose-Einstein correlations between different W's. The analyses of OPAL and L3 are inconclusive although consistent with the latest BEC models. It is interesting to compare M_W measurements made with $qqqq$ and $qql\nu$ final states. If BEC's are evident one might expect a difference between these two mass determinations. The LEP results are shown in Fig. 47. The difference between the two mass measurements is consistent with zero: $M_W^{qqqq} - M_W^{qql\nu} = 0.08 \pm 0.16$GeV.

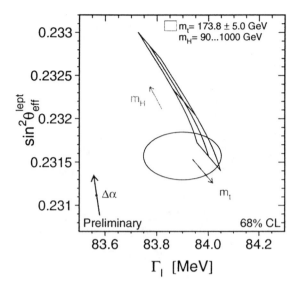

Figure 50: LEP-1 plus SLD measurements of $sin^2\theta^l_{eff}$ and Γ_l and the Standard Model prediction. The point shows the predictions if, among the electroweak variables, only the photon vacuum polarization correction is included. The corresponding arrow shows the variation of this prediction as $\alpha(M_Z^2)$ is changed by one Standard Deviation.

13.5 Trilinear Gauge Couplings

The most general Lagrangian that describes the triple gauge boson vertex (γWW or ZWW) depends linearly on seven terms [63], resulting in a total of 14 unmeasured parameters. Of these 14 parameters two violate C and P separately and 6 violate CP. As LEP-2 statistics are too meagre to determine all 14 couplings simultaneously additional theoretical restrictions must be applied. Invoking electromagnetic gauge invariance and using only terms corresponding to operators with dimension smaller than six which do not violate C, P or CP, we are left with five terms. The Standard Model predicts the values of these five terms to be, $\kappa_\gamma = \kappa_Z = g_1^Z = 1$ and $\lambda_\gamma = \lambda_Z = 0$.

These couplings can be related to a multipole expansion of the electric and magnetic charge of the W boson. Specifically, the magnetic dipole moment $\mu_W = (e/2M_W)(1 + \kappa_\gamma + \lambda_\gamma)$; and, the electric quadrupole moment $q_W = -(e/m_W)(\kappa_\gamma - \lambda_\gamma)$. If constraints from lower energy measurements are taken into account (which are mainly the oblique corrections to the Z propagator), the set of anomalous couplings to be investigated can be restricted to three

274

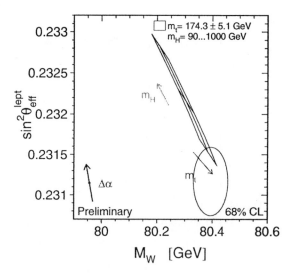

Figure 51: The 68% confidence level contour of of M_W versus $sin^2\theta^l_{eff}$ compared with the predictions of the Standard Model. The arrow in the bottom left of the plot shows the Born "prediction" where only the running of α is included (with one standard deviation uncertainty).

parameters:

$$\alpha_{W\phi} = \Delta g_1^Z cos^2\theta_W; \quad \alpha_{B\phi} = \Delta\kappa_\gamma - \Delta g_1^Z cos^2\theta_W; \quad \alpha_W = \lambda_\gamma. \quad (85)$$

where Δ refers to difference from the Standard Model value.

Anomalous couplings modify the WW production cross-section and lead to a distortion of the event kinematics. Experimentally the study of these couplings is performed by fitting both cross-sections and the distribution of W production and decay angles. These angles are defined in Fig. 48. Detector effects on the angular distributions are accounted for by using reweighting algorithms similar to those used in the M_W measurement.

As E_{cm} increases so too does the enhancement of the gauge boson pair cross-section due to anomalous couplings. Thus, in this area, we should expect the Tevatron experiments to have a distinct advantage the LEP experiments. However, backgrounds are larger at the Tevatron. It turns out that the sensitivity to gauge couplings at the Tevatron and at LEP is similar. The

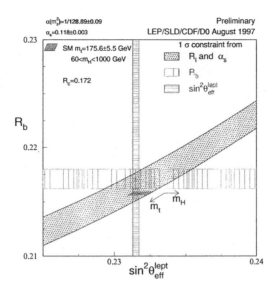

Figure 52: One standard-deviation bands of R_b, $sin^2\theta^l_{eff}$ and R_l measurements compared with Standard Model predictions.

log-likelihood curves for the experimental determination of the anomalous coupling constants from D0 and each of the LEP experiments are shown in Fig. 49 [64]. The combined measurements of the anomalous couplings are also shown in this figure.

14 Testing the Standard Model

The measurements of the electroweak observables discussed in previous sections can be analyzed within the framework of the Standard Model to ascertain whether the data enable us to discriminate the effects of electroweak radiative corrections. One can restrict this test to a subset of measurements: $sin^2\theta^l_{eff}$, Γ_l, M_W and R_b which are specifically sensitive to $\Delta\kappa$, $\Delta\rho$, Δr and the $Z^0 \to b\bar{b}$ vertex correction, respectively.

Fig. 50 [65] shows a comparison of the 68% probability contours obtained from measurements of $sin^2\theta^l_{eff}$ and Γ_l. The point shows that the (Born level) predictions – with only the photon vacuum polarization correction included – are not in agreement with the data. Also it can be seen that the data is in agreement with the Standard Model, at least for relatively low values of M_H.

The effect of electroweak corrections is even more apparent in the contour

276

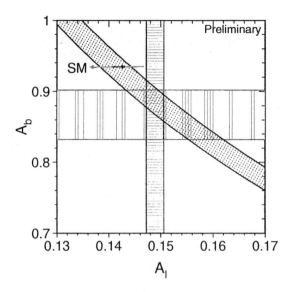

Figure 53: The measurements of the combined LEP+SLD value for A_l (vertical band), SLD value for A_b (horizontal band), and the LEP value for A_{FB}^b (diagonal band). The arrow pointing to the left shows the variation in the Standard Model expectations for M_H in the range 300^{+700}_{-210} GeV, and the arrow pointing to the right for M_t in the range 173.8 ± 5.0 GeV.

plot of M_W versus $sin^2\theta_{eff}^l$ [65] shown in Fig. 51. As before, the star shows the Born expectation where only the running of α is included. Again, the agreement of the data with the Standard Model is good. But, there is not agreement between the data and the naive Born expectation.

Fig. 52 show the intersection between the one standard deviation bands for R_b, R_l and $sin^2\theta_{eff}^l$ measurements at LEP-1. The important point here is the consistency of the intersection region with the Standard Model. It also provides a check of the direct and indirect determinations of R_b. As can be seen the intersection region is in good agreement with the Standard Model prediction. Note that this plot was made by the LEP Electroweak Working Group using the data available in the summer of 1997.

Currently no LEP measurements deviate from the Standard Model fit by more than two standard deviations. The largest deviation is seen for the b quark asymmetries A_{FB}^b and A_b. Using the measurement of A_l, A_b and A_c can be extracted from the LEP measurements of the b and c quark asymmetries. On the other hand the SLD experiment directly determines A_b and A_c. The LEP determinations of A_b and A_c are in good agreement with the SLD

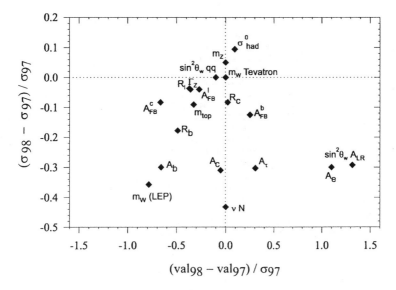

Figure 54: A synopsis of changes in the precision of electroweak measurements, utilized in the global electroweak fit made in the year preceding the Summer 1998 conferences. The vertical axis indicates the change in uncertainties relative to the uncertainties one year earlier (1997). The horizontal axis indicates the changes to the central values as a fraction of the uncertainties one year earlier (1997). A spread of the order of one standard deviation is observed for measurements with improved precision.

measurements and in reasonable agreement with the Standard Model expectation. However, the combination of LEP and SLD measurements of A_b is ~ 3.0 standard deviations below the Standard Model. This discrepancy is due to three independent results: the LEP measurement of $A_{FB}^{0,b}$ is low compared with Standard Model predictions; the combined LEP+SLD measurement of A_l is high compared with the Standard Model; and, the SLD measurement of A_b is low compared with the Standard Model. This can be seen summarized in Fig. 53[65].

In the year leading up to the 1998 summer conferences there was considerable improvement in the precision of a number of measurements that contribute to the electroweak fit. These improvements are summarized in Fig. 54[68]. The biggest impact of the improved precision on the electroweak fits arises from the following: the W mass determination from LEP data at an E_{cm} of 183 GeV; the improved precision of the tau polarization measurements (A_e and A_τ); the new A_{LR} measurement from SLD; and, the new neutrino scattering result from NuTeV[69].

	Measurement	Pull
m_Z (GeV)	91.1867 ± 0.0021	.09
Γ_Z (GeV)	2.4939 ± 0.0024	-.80
σ_{hadr}^0 (nb)	41.491 ± 0.058	.31
R_e	20.765 ± 0.026	.66
$A_{fb}^{0,e}$	0.01683 ± 0.00096	.73
A_e	0.1479 ± 0.0051	.25
A_τ	0.1431 ± 0.0045	-.79
$sin^2\theta_{eff}^{LEPT}$	0.2321 ± 0.0010	.53
m_W (GeV)	80.37 ± 0.09	-.01
R_b	0.21656 ± 0.00074	.90
R_c	0.1735 ± 0.0044	.29
$A_{fb}^{0,b}$	0.0990 ± 0.0021	-1.81
$A_{fb}^{0,c}$	0.0709 ± 0.0044	-.58
A_b	0.867 ± 0.035	-1.93
A_c	0.647 ± 0.040	-.52
$sin^2\theta_{eff}^{LEPT}$	0.23109 ± 0.00029	-1.65
$sin^2\theta_W$	0.2255 ± 0.0021	1.06
m_W (GeV)	80.41 ± 0.09	.43
m_t (GeV)	173.8 ± 5.0	.54
$1/\alpha$	128.878 ± 0.090	.00

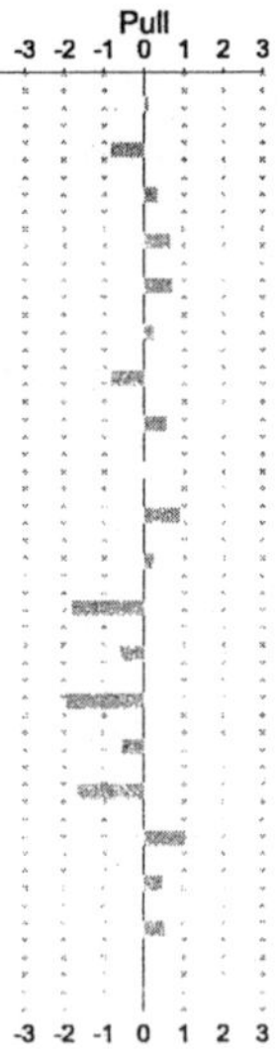

Figure 55: The precision measurements that were included in the global electroweak fit are shown together with the pulls. The overall χ^2 per degree of freedom is ~ 1.

A global electroweak fit was performed [65, 66] with the following input quantities: Z^0 measurement from LEP+SLD; W mass measurements from LEP-2, CDF, and D0; $sin^2\theta_W$ from the CCFR and NuTeV experiments; the top quark mass from CDF and D0; and $1/\alpha^{(5)}(M_Z^2) = 128.878 \pm 0.090$ (determined without top quark contribution [67]). The results of the global electroweak fit are given in Table 2 [68]. As can be seen the overall fit is reasonable for each of the three cases. The pulls of the individual measurements (defined as the difference between the measurement and the fit divided by the measurement error) for the fit – that uses all the data – are summarized in Fig. 55. The largest deviations are seen for the b quark asymmetries. The indirect measurements of M_W and M_t obtained from the global electoweak fit are compared with the direct measurements in Fig. 56.

14.1 Determining Limits on the Standard Model Higgs Boson Mass

The Direct Search for the Higgs Boson is continuing at LEP. The LEP lower limit on the Standard Model Higgs boson mass now stands at ~ 90 GeV at the 95% Confidence level [70]. Also, the indirect determination of the the Standard Model Higgs boson mass limit obtained from the global electroweak fit

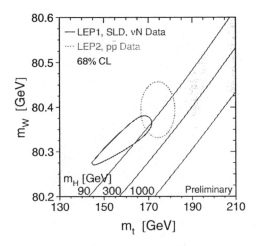

Figure 56: The 68% confidence level contours are shown for indirect and direct determinations of M_W and M_t. The diagonal band shows the Standard Model allowed region for three values of the Higgs boson mass.

is getting interesting. The direct measurements of M_W and M_t are compared with the indirect determination from the global fit in Fig. 56. The 68% confidence level contours show significant overlap and are consistent with Standard Model expectations for relatively light M_H. The sensitivity of the electroweak variables in the global fit to M_H is summarized in Fig. 57 and Fig. 58[65].

As has been shown in the section on radiative corrections the effect of M_H on the global electroweak fit is approximately logarithmic. Higgs boson hunting is also hampered by the strong correlation of $log_e(M_H)$ with M_t and $\alpha(M_Z)$. The strongest constraints on M_H come from the asymmetries but M_W, R_b and the lineshape variables are also significant. Fig. 59 shows the $\Delta\chi^2$ curve of the global electroweak fit as a function of Higgs boson mass. The shaded band that follows the curve is an estimate of the theoretical error due to missing higher order terms in the radiative corrections.

The global electroweak fit give an upper limit on the Higgs mass of $M_H <$ 262 GeV at the 95% confidence level, corresponding to the fitted value of, $M_H = 76^{+85}_{-47}$ GeV. Including the exclusion region defined by the direct search increases the 95% exclusion region by $\sim$ 30 GeV. The increase in χ^2 caused by fixing M_H to 1 TeV indicates that, as far as the global electroweak fit is

280

Preliminary

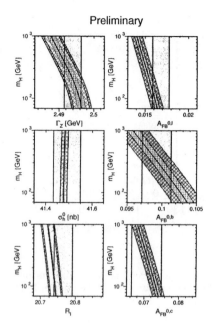

Figure 57: Comparison of LEP-1 measurements with the Standard Model prediction as a function of M_H. The measurement with its error is shown by the vertical band. the width of the Standard Model allowed band arises from the uncertainties on $\alpha(M_Z^2)$, $\alpha_s(M_Z^2)$, and M_t, The total width of the band is the linear sum of these effects.

concerned, a TeV scale Higgs boson is ruled out at the 5σ level.

Is the low value of the fitted value for M_H due to any particular measurements? The effect of individual inputs to the fit on the indirect determination of the Higgs mass is demonstrated by showing the results when each is removed, as shown in Fig. 60 [68].

The input measurements that are sensitive to Higgs boson mass can be divided into three categories : asymmetries; widths; and, the W mass. They test different components of the radiative corrections. One can further check the internal consistency of the global electroweak fit by using each of these different categories as the sole input to the fit, with the additional constraint that $\alpha_s = 0.119 \pm 0.002$ [71]. The results of this exercise are summarized in Table 3. As can be seen all the fits are consistent with a low Higgs boson mass. No particular set of measurements pull the Higgs boson mass down.

Although the Z asymmetry measurements are contributing most to the indirect Higgs mass determination it is clear from Table 3 that any future

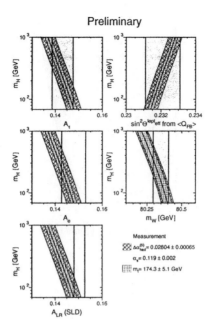

Figure 58: Comparison of LEP-1 measurements with the Standard Model prediction as a function of M_H. The measurement with its error is shown by the vertical band. The width of the Standard Model allowed band arises from the uncertainties on $\alpha(M_Z^2)$, $\alpha_s(M_Z^2)$, and M_t. The total width of the band is the linear sum of these effects. Also shown is the comparison of the SLD determination of A_{LR} with the Standard Model.

improvement will be limited by the uncertainty on $\alpha(M_Z^2)$. The precision with which M_W is known will improve by a significant factor in the near future. However, even if M_W is measured with a precision of 30 MeV, the error on $logM_H/GeV$ is going to be dominated by ΔM_t. Thus the error on $log(M_H/GeV)$ will not improve significantly until a precise measurement of M_t, for example to within 2 GeV, is obtained. In this case a precision on the determination of $log(M_H/GeV)$ of ~ 0.15 can be achieved.

The Higgs Boson – Mass Theory Meets experiment

There are theoretical limits to the Higgs boson mass arising from vacuum stability arguments, the absence of the Landau pole[72] and also from lattice gauge theories[73]. Explicit calculation of the decay widths for $H \to W^+W^-, Z^0Z^0$ including 2 loop diagrams, show that the 2-loop contribution exceeds the 1-loop term for $M_H > 930$ GeV[74].

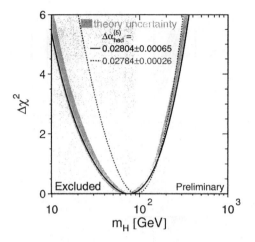

Figure 59: The $\Delta\chi^2 = \chi^2 - \chi^2_{min}$ versus M_H curve. The line is the result of a fit using all the data. The band represents an estimate of the theoretical error due to missing higher order corrections. The vertical band show the 95% confidence level exclusion limit on M_H from the direct search.

The behaviour of the quartic Higgs boson self coupling (λ) as a function of the energy scale (μ), is described by the renormalization groups equation. A lower bound on the Higgs boson mass can be obtained from the necessity to avoid unphysical quartic couplings from the negative top quark contribution. The requirement that the Higgs boson coupling remains finite and positive up to a scale Λ also constrains the Higgs boson mass. These bounds are summarized in Fig. 61 [75] as a function of the cut-off scale up to which the Standard Model Higgs sector can be extrapolated, when $M_t = 175 GeV$ and $\alpha_s = 0.018$. The allowed region is that area between the upper and lower curves. The black bands indicate the theoretical error due to the solution of the renormalization group equations [75]. It is somewhat depressing to note that the indirect determination of the allowed Standard Model Higgs mass range from the global electroweak fit is compatible with a value of M_H where λ can extend all the way to the Planck scale.

Table 2: Results of the global electroweak fits to LEP data alone, to all data except the direct W mass measurements, and to all data. Theoretical errors, due to uncalculated higher order corrections, are not included in the errors.

	LEP alone	all but m_W	all data
m_t (GeV)	160^{+13}_{-9}	171.0 ± 4.9	171.0 ± 4.9
m_H (GeV)	60^{+127}_{-35}	82^{+95}_{-51}	76^{+85}_{-47}
$\alpha_s(m_Z^2)$	0.121 ± 0.003	0.120 ± 0.003	0.119 ± 0.003
$\chi^2/$d.o.f.	4/9	15/13	15/15
$\sin^2 \theta_{eff}^{lept}$	0.23182 ± 0.00023	0.23159 ± 0.00020	0.23157 ± 0.00019
m_W (GeV)	80.314 ± 0.038	80.367 ± 0.029	80.371 ± 0.026

Table 3: The value of $log_e(M_H/GeV)$ for different types of measurements. The impact of the uncertainties on each input parameter are shown explicitly .

	$log\left(\frac{M_H}{GeV}\right)$	$[\Delta log\left(\frac{M_H}{GeV}\right)]^2 = \Delta_{exp}^2 + \Delta^2 M_t + \Delta^2(\alpha) + \Delta^2 \alpha_s$
Z asym's	$1.94^{+0.29}_{-0.31}$	$[0.29]^2 = [0.19]^2 + [0.12]^2 + [0.19]^2 + [0.01]^2$
Z widths	$2.21^{+0.36}_{-1.43}$	$[0.36]^2 = [0.31]^2 + [0.14]^2 + [0.08]^2 + [0.13]^2$
$M_W \& \nu N$	$2.04^{+0.45}_{-0.84}$	$[0.45]^2 = [0.41]^2 + [0.18]^2 + [0.08]^2 + [0.00]^2$

15 Using the Electroweak Data to Look Beyond the Standard Model

Currently, one of the most fashionable candidates for the theory that is expected to supercede the Standard Model is Minimal Supersymmetric Standard Model (MSSM) [72]. The MSSM is a renormalizable theory allowing a complete calculation of the electroweak precision variables in terms of: one Higgs mass (usually taken as the CP-odd 'pseudoscalar' mass, M_A); $tan\beta = v_2/v_1$, the ratio of the vacuum expectation values of the two Higgs doublets of the MSSM); a set of SUSY soft symmetry breaking parameters fixing the chargino/neutralino and scalar fermion sectors.

A global fit to the electroweak precision data [76], including the measurement of M_t from the Tevatron shows that the χ^2 per degree of freedom of the fit performed within the framework of the MSSM is not significantly better than in the case of the Standard Model, as shown in Fig. 62

15.1 Test for New Physics Using the ϵ Variables

As was discussed in the section on radiative corrections there are four important loop corrections contributing beyond tree level to predictions of elec-

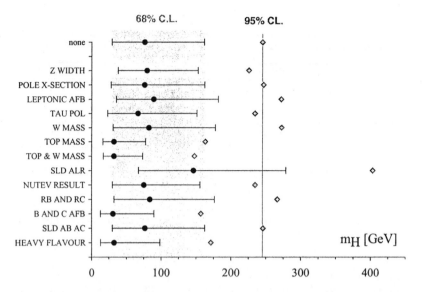

Figure 60: The vertical axis on the left show the particular observable removed from the global electroweak fit. The solid circle and error bar shows the central value and the 68% confidence interval, the diamonds indicate the 95% confidence interval. The very top point is the results with no measurements removed. The theoretical uncertainty is not included.

troweak observables. There are the three vacuum polarization corrections Δr, $\Delta \kappa$ and $\Delta \rho$ as well as the $Z \to b\bar{b}$ vertex correction. The presence of new physics could modify these corrections.

The ϵ parameters[77] are a set of four variables defined in terms of the electroweak observables (M_W, A_{FB}^0, Γ_l, and Γ_b) which quantify the deviations of these observables from the "Born" expectation (Born + QCD + QED effects). These variables were discussed in detail in R. Peccei's lectures at this meeting [72].

In the Standard Model the leading contributions to the four ϵ variables are:

$$\epsilon_1 = \frac{3G_F}{8\pi^2\sqrt{2}} M_t^2 - \frac{3G_f M_W^2}{4\pi^2\sqrt{2}} \frac{s_W^2}{c_W^2} ln\frac{M_H}{M_Z} + ... \qquad (86)$$

$$\epsilon_2 = \frac{G_f M_W^2}{2\pi^2\sqrt{2}} ln\frac{M_t}{M_Z} + ... \qquad (87)$$

$$\epsilon_3 = \frac{G_f M_W^2}{12\pi^2\sqrt{2}} ln\frac{M_H}{M_Z} + ... \qquad (88)$$

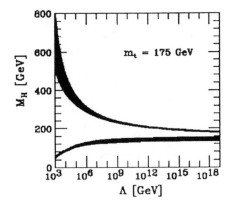

Figure 61: Theoretical limits on the Higgs boson mass derived from the absence of a Landau pole and from vacuum stability.

$$\epsilon_b = \frac{G_F}{4\pi^2 \sqrt{2}} M_t^2 + \dots \tag{89}$$

The variables ϵ_2 and ϵ_3 do not have contributions of order $G_F M_t^2$ and are thus more sensitive to possible new physics.

Searching for Technicolour and Supersymmetry Using the ϵ Variables

Two possible sources of new physics are considered here. First, the "heavy" MSSM model [78] in which all the SUSY particles are massive. In this case the MSSM predictions for the radiative corrections give essentially the same results as the Standard Model with a light Higgs boson ($M_H \leq 100$ GeV). The second class of models considered are Technicolour Models [79] where large positive corrections to ϵ_3 are hard to avoid.

Fig. 63 [15] compares the predictions of Technicolour and "heavy" MSSM models with the two dimensional plot of ϵ_1 against ϵ_3. The dotted lines show the expectation region for a "heavy" MSSM model. Not surprisingly the "heavy" MSSM is in good agreement with the data and with the predictions of the Standard Model. The dashed line corresponds to the expectation for simple Technicolour models [80]. As can be seen the fitted values of ϵ_1 and ϵ_3 are small enough to rule out a substantial class of Technicolour models [80]. The value of ϵ_b also tends to rule out a large class of technicolour models since in Extended Technicolour models the mechanism that gives rise to large M_t also generates large corrections to the $Z \to b\bar{b}$ vertex.

286

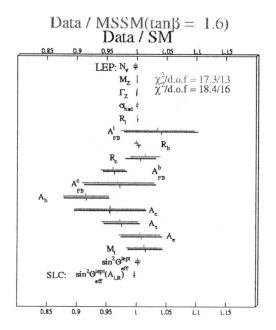

Figure 62: A global fit using the Electrwoweak data performed within the framework of the Standard Model and the MSSM, normalized to the data. The error bars are those arising from the data.

16 Conclusions

The LEP and SLC measurements have tested the electroweak aspects of the Standard Model to the level $\mathcal{O}(0.1\%)$. From Z boson observables the effects of purely weak radiative corrections are visible with a significance of over three standard deviations . However, the most impressive evidence for such corrections does not come from Z physics but from M_W and its relation to G_F, using the value of M_W measured at the TEVATRON and at LEP [81]. In this case there is a roughly seven standard deviation effect.

The top mass predicted by global electroweak fits ($M_t = 161^{+8.2}_{-7.1}$ GeV) is compatible, to within about 1.4σ, with the direct measurement from the Tevatron ($M_t = 173.8 \pm 5.0$ GeV). The W boson mass arising from the global electroweak fit ($M_W = 80.637 \pm 0.029$) is in good agreement with the direct measurement ($M_W = 80.39 \pm 0.06$ GeV). The mass of the Higgs boson obtained from the fit is low ($M_H = 76^{+85}_{-47}$ GeV) and $M_H < 262$ GeV at the 95% confidence level. At present the Standard Model and the MSSM are compatible

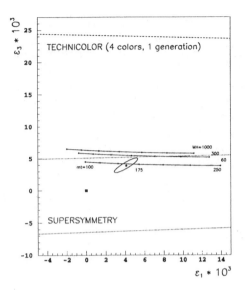

Figure 63: The 39% confidence level contour in the ϵ_1 versus ϵ_2 plane. The solid lines trace the predictions of the Standard Model for different values of M_H and M_t. the band defined by the dotted lines shows the expectation for the "heavy" MSSSM . The dashed line corresponds to the expectation for simple technicolour models. The " Born" expectation $\epsilon_i = 0$ is shown as a solid square.

with the data. However simple Technicolour models are largely ruled out by the data. Thus, at present, electroweak data obtained from e^+e^- colliders is completely consistent with the Standard Model.

17 Acknowledgements

The author wishes to thank Professors B. Campbell, F. C Khanna, and M. Vincter for a very stimulating meeting and a most enjoyable stay at Lake Louise. In addition, the author would like to thank the Conference Secretaries L. Grimard and A. Schaapman for their invaluable contribution to the success of the meeting. Behind the wealth of data and analysis so briefly sampled here lies the much appreciated work and dedication of many other fellow workers in the area of electroweak physics.

1. E. Lancon, contributed to *The IVth International Symposium on Radiative Corrections (RADCOR 98)*, Barcelona, Spain, September 8-12th, (1998).

2. K. MacFarland, contribution to *The XXXIIIth Renconres de Moriond,* Les Arcs, France, March 15-21, (1998).

3. R. Partridge, contributed to *The ICHEP98 Conference,* Vancouver, Canada, July 23-29, (1998).

4. OPAL Collaboration, *OPAL Technical Proposal,* CERN Technical Report CERN/LEPC/83-4, CERN, 1990; K. Ahmet, OPAL Collaboration, Nucl. Instr. Methods **A305**, 275, (1991).

5. D. Decamp *et al.,* ALEPH Collaboration, Nucl. Instrum,. Methods **A294**, 121 (1990).

6. P. Aarnio *et al.,* DELPHI Collaboration, Nucl. Instrum. Methods **A303**, 233, (1991).

7. B.Adeva *et al.,* L3 Collaboration, Nucl. Instrum. Methods **A239**, 35, (1990).

8. The SLD Collaboration, *SLD Design Report,* SLAC Report No. 273, (1984).

9. T. Maryana, E. Garwin, R. Prepost and G. Zapalac, Phys. Rev. **B46**, 4261, (1992).

10. L. Arnaudon *et al.,* Phys. Lett. **B284**, 431, (1992).

11. P. Melchior, *The Tides of the Earth,* Pergamon Press, (1983).

12. R. Assman *et al.,* Z. Phys. **C66**, 567, (1995).

13. L. Arnaudon, *et al.,* Nucl. Instrum. Methods **A357** 249, (1995).

14. P, Melchior, *The Tides of the Earth* , Pergamon Press, 1983.

15. M. Martinez, R. Miquel, L. Rolandi, R. Tenchini, Rev. Mod. Phys. **71**, 575, (1999).

16. B. W. Lynn, *Proceedings of the Workshop on Polarization at LEP,* CERN 88-06, ed. by G. Alexander, G. Altarelli, A. Blondel, G. Coignet, E. Keil, D. E. Plane and D. Treille, Vol. 1, 24, (1988).

17. W. Hollik, Fortschr. Phys. **38**, 165, (1990).

18. D. Bardin, W. Hollik, and T. Riemann, Z Phys. **C49**, 485, (1991).

19. R. Kleiss *et al,Proceedings of the Workshop on Z Physics at LEP 1,* CERN 89-08, eds G. Altarelli, R. Kleiss, and C. Verzegnassi. Vol.1 7, (1989).

20. H. Burkhardt and B. Pietrzyk, Phys. Lett. **B356**, p398, (1995). S. Eidelman and F. Jergerlehrner, Z. Phys. **C67**, 585, (1995).

21. G. Burgers, W. Holik and M. Martinez, *BHM Computer Code,* (1990).

22. G. Bonneau, and F. Martin, Nucl. Phys. **B27**, 381, (1971).

23. D. Bardin, W. Hollik, and G. Passarino, 1995, Eds. *Proceedings of the Working Group on Precision Calculations at the Z Resonance,* CERN 95-03, (1995).

24. D. Bardin, *et al.,* CERN-TH/6443-92, (1992).

25. M. Martinez *et al.*, Z. Phys. **C49**, 645, (1991).
26. S. Jadach and B. F. L. Ward, Phys. Lett. **B389**, 129, (1996).
27. D. Busilik *et al.*, ALEPH Collaboration,Z. Phys. **C62**, 539, (1994); P. Abreu *et al.*, DELPHI Collaboration, Nucl. Phys. **B418**, 403, (1994); M. Acciarri *et al.*, L3 Collaboration, Z. Phys. **C62**, 551, (1994); R. Akers *et al.*, OPAL Collaboration, Z. Phys. **C61**, 19, (1994).
28. D. Busilik *et al.*, ALEPH Collaboration, Z. Phys. **C62**, 539, (1994).
29. P. Abreu *et al.*, DELPHI Collaboration, Nucl. Phys. **B418**, 403, (1994).
30. For example, the ALEPH, DELPHI, L3 and OPAL Collaborations, The LEP Electroweak Working Group and the SLD Heavy Flavour Group, 1997, *A Combination of Preliminary Electroweak Measurements and Constraints on the Standard Model*, CERN-PPE/97-154, (1997).
31. T. Hebbeker *et al.*, Phys. Lett. **B331**, 165, (1994).
32. R. Akers *et al.*, OPAL Collaboration, Z Phys. **C65**, 17, (1995).
33. The JADE Collaboration, S. Bethke *et al.* Phys. Lett. **B213**, 235, 1988. Nucl. Instrum. & Methods,**A306**, 459, (1991).
34. For example: L. Bellantoni *et al.*, Nucl. Instrum. Methods, **A310**, 618, (1991); C. Bortlotto *et al.*, Nucl. Instrum. Methods, **A306**, 459 (1991).
35. For example, D. Buskulic *et al.*, Z Phys. **C62**, 1, (1994); DELPHI Collaboration, P. Abreu *et al.*, Phys. Lett. **B252**, 140, (1990); OPAL Collaboration, K. Ackerstaff *et al.*,Eur. Phys. J., **C1**, 439, (1997).
36. D. Busilik *et al.*, ALEPH Collaboration, Z. Phys. **C62**, 1, (1994).
37. ALEPH Collaboration, D. Buskulic *et al.*, Z. Phys. **C62**, 1, (1994).
38. For example, OPAL Collaboration, R. Akers *et al.*, Z. Phys. **C67**, 27, (1995).
39. OPAL Collaboration, R. Akers *et al.*, Phys. Lett. **B353**, 595, (1995).
40. M. H. Seymour, Nucl. Phys. **B435**, 136, (1995).
41. For example: SLD Collaboration, *A Measurement of R_b Using a Vertex Mass Tag*, Contribution to EPS EP 97, Jerusalem, EPS-118, (1997); DELPHI Collaboration, *Measurement of the Partial Decay Width $R_b^0 = \Gamma_{b\bar{b}}/\Gamma_{had}$ with the DELPHI Detector at LEP*, Contributed to EPS EP 97, Jerusalem, EPS-418, (1997); L3 Collaboration,*Measurement of R_b with the L3 Detector*, Contributed to EPS EP 97, Jerusalem, EP2-489, (1997); OPAL Collaboration, L. Ackerstaff *et al.*, Z. Phys. **C74**, 1, (1997).
42. For example, DELPHI Collaboration, P. Abreu *et al.*, Z. Phys. **C59**, 533, (1993).
43. For example: ALEPH Collaboration, *Measurement of R_c*, Contribution to ICHEP 96, Warsaw, PA-016, (1996).
44. DELPHI Collaboration, *Summary of R_C measurements at DELPHI*, Contributions to ICHEP 96, Warsaw, PA01-60, (1996).

45. For example: OPAL Collaboration, K. Ackerstaff *et al.*, Z. Phys. **C72**, 1, (1996).; ALEPH Collaboration *Study of Charmed Hadron Production in Z Decays* , Contribution to EPS EP 97, Jerusalem, EPS-623.

46. W. Hollik, *Precision Tests of the Standard Electroweak Model*, edited by Paul Langacker (World Scientific, Singapore), 37, (1993); W. Hollik, *Electroweak Theory*, hep-ph/9602380, (1996).

47. D. Busilik *et al.*, Z. Phys. **C62**, 179, (1994).

48. For example: ALEPH Collaboration, D. Buskilic *et al.*, Phys. Lett **B384**, 414, (1996); DELPHI Collaboration, P. Abreu *et al.*, Z. Phys. **C65**, 569, (1995). OPAL Collaboration, R. Akers *et al.*, Z. Phys. **C60**, 19, (1993).

49. ALEPH Collaboration, D. Buskilic *et al.*, Phys. Lett. **B352**, 479, (1995).

50. For example: DELPHI Collaboration, P. Abreu *et al.*, Z. Phys. **C66**, 341, (1995); OPAL Collaboration, G. Alexander *et al.*, Z. Phys. C73, 379, (1997).

51. OPAL Collaboration, R. Akers *et al.*, Z. Phys. **C65**, 17, (1995).

52. For example: A. Djouardi, B. Lampe, and P. M. Zerwas, Z. Phys. **C67**, 123, (1995).

53. ALEPH Collaboration, *WW Cross-section and W Branching Ratios at* $\sqrt{s} = 183$ *GeV*, ALEPH 90-019, Contribution to ICHEP 98, Vancouver, 23-29, July (1998).

54. N. Brown and W.J. Stirling, Phys. Lett. **B252**, 657, (1990); D.S.S. Bethke, Z. Kunst and W.J. Stirling, Nucl. Phys. **B370**, (1992); R.K. Ellis, W. J. Stirling and B. R. Webber, *QCD and Collider Physics*, Cambridge University Press, Cambridge, UK, (1996).

55. OPAL Collaboration, *Measurement of the W-pair Production Cross-section and Triple Gauge Boson Couplings at LEP*, OPAL Physics Note PN354, July 15th (1998).

56. A. Duperrin, *these proceedings*.

57. J. Ellis and K. Geiger, Phys. Lett **B404**, 230, (1992).

58. For example: T. Sjöstrand,*Pythia 5.7 and Jetset 7.4: Physics Manual*, LU-TP/95-20, CERN-TH.7112/93, Aug., (1995); T. Sjöstrand and M. Bentsson, Comp. Phys. Comm. **82**, 74, (1994).

59. G. Marchesini *et al.*, Comp. Phys. Comm., 465, (1992).

60. L. Lönnblad, Z. Phys. **C70**, 107, (1996).

61. For example: G. Abbiendi *et al.*, OPAL Collaboration, CERN-EP/98-196, (1998).

62. P. de Jong, *Reconnection and BE Effects in WW Decay*, contributed to the XXVIII International Symposium on Multiparticle Dynamics, Delph, Greece, Sept6-11th, (1998).

63. G. Altarelli, T. Sjostrand, and F. Zwirner, *Physics at LEP2*, CERN 96-

291

01, Vol.1, 525, (1996).

64. M. Verzocchi, *W Coupling Measurements at LEP*, contributed to the Second Latin Americam Symposium on High Energy Physics, San Juan (Puerto Rico), OPAL CR363, (1998).

65. The LEP Collbaorations & the SLD Heavy Flavour and Electrweak Groups, *A Combination of Preliminary Electrweak Measurements and Constraints on the Standard Model*, CERN-EP/99-15,(1999).

66. M. W. Grunewald, *Combined Analysis of Precision Electroweak Results*, 29th ICHEP Conference, Vancouver, Canada, July 23-29, (1998).

67. M. Steinhauser, Phys. Lett. **B429**, 158, (1998); S. Edelmann and F. Jegerlehner, Z Phys. **C67**, 585, (1995).

68. D. Karlen, *Experimental Status of the Standard Model*, 29th ICHEP Conference, Vancouver, Canada, July 23-29, (1998).

69. T. Bolton, *New Results from NuTeV*, 29th ICHEP Conference, Vancouver, Canada, July 23-29, (1998).

70. E. Gross, *these proceedings*

71. C. Caso *et al.*, The Particle Data Group, E. Phys. J. **C3**, 1, (1998).

72. R. Peccei, *these proceedings*.

73. For example: Kuti *et al.*, Phys. Rev. Lett. **61**, 678, (1988); P. Hasenfratz *et al.*, Nucl. Phys., **B317**, 81, (1989); M. Göckeler, H. Kastrup, T. Neuhas, and F. Zimmermann, Nucl. Phys. **B404**, 517, (1993).

74. A. Ghinculov, Nucl. Phys., **B455**, 21, (1995); A. Frink, B. Kniehl, and K. Riesselmann, Phys. Rev. **D54**, 4548, (1996).

75. T. Hambye and K. Riesselmann, Phys. Rev. **D55**, 7255, (1997).

76. W. de Boer, A. Dabelstein, W. Holli, W. Mösle, U. Schwickerath, Z. Phys. **C75**, 627, (1997).

77. G. Altarelli, R. Barbieri, and F. Caravvaglios, Phys. Lett., **B349**, 145, (1995).

78. G. Altarelli, Acta Phys. Pol. **B28**, 3, (1997).

79. S. Weinberg, Phys. Rev. **D13**, p974, (1976); E. Farhi and L. Susskind, Phys. Rept., **74**, 277, (1981).

80. J. Ellis, G. Fogli, and E. Lisi, Phys. Lett **B343**, 282, (1995).

81. F. Teubert, *Precision Test of the Standard Model from Z Physics*, Proceedings of the IVth International Symposium on Radiative Corrections, Barcelona, Spain, September 8-12th, (1998).

NONFACTORIZATION IN CABIBBO-FAVORED B DECAYS

F.M. Al-SHAMALI and A.N. KAMAL

Theoretical Physics Institute, Department of Physics, University of Alberta,
Edmonton, Alberta T6G 2J1, Canada.

We assume universal values for the color-singlet (ε_1) and color octet (ε_8) nonfactorization parameters in B Decays. Two sets of color-favored processes and one set of color-suppressed processes were used to give quantitative estimates of these parameters. It has been found (by calculating the branching ratios for a large number of Cabibbo-favored B Decays) that the values $\varepsilon_1(\mu_0) = -0.06 \pm 0.03$ and $\varepsilon_8(\mu_0) = 0.12 \pm 0.02$ improve significantly the predictions of the factorization model.

1 Introduction

The idea of nonfactorization in D and B decays was introduced in the last few years by several authors [1,2,3]. There are two equivalent ways of introducing nonfactorized contributions in a calculation: either, use the number of colors, N_c equal to 3 and explicitly add a nonfactorized contribution to each Lorentz scalar in the decay amplitude as in refs. [2,3], or introduce an effective number of colors, N_c^{eff}. The latter approach has been adopted in several papers [4,5,6] dealing with B decays into light mesons.

Though the nonfactorized amplitude remains incalculable, a few statements about it can be made. First, what is estimated to be the nonfactorized contribution depends on the model of form factors used to calculate the factorized contribution. Second, within a chosen model for the form factors, the nonfactorized contributions are process dependent. Third, if nonfactorization is characterized through an effective number of colors, N_c^{eff}, is it the same (as assumed in ref. [4] and ref. [5]) for the tree and the penguin generated processes or different (as in ref. [6])?

Unlike the papers listed in refs. [4,5,6], in this work we study B decays into a heavy and a light meson. As such they are all Cabibbo-favored $b \rightarrow c$ transitions. Following ref. [7], we use $N_c = 3$ and introduce nonfactorized contributions explicitly. To achieve simplicity of description, we assume that the nonfactorized effects associated with the three Lorentz-scalar structures in $B \rightarrow VV$ decays are the same. Using experimental data, we then calculate the average overall nonfactorization factors for color-favored $b \rightarrow c\bar{u}d$ and $b \rightarrow c\bar{c}s$ processes. We repeat this for color-suppressed $b \rightarrow c\bar{c}s$ processes. We relate these nonfactorization factors to the scale dependent parameters $\varepsilon_1(\mu)$ and $\varepsilon_8(\mu)$ of ref. [8] and determine them at $\mu_0 = 4.6$ GeV. The details are explained in the text of this paper. Finally, having isolated the parameters $\varepsilon_1(\mu_0)$ and $\varepsilon_8(\mu_0)$, we make predictions for the decay rates of Cabibbo-favored modes both measured and as yet unmeasured. The measured rates are shown to agree well with the predictions with few exceptions.

The paper is arranged as follows: Section II deals with the formalism and introduces notation. Sections III and IV deal with the calculation of nonfactorized amplitudes, the evaluation of $\varepsilon_1(\mu_0)$ and $\varepsilon_8(\mu_0)$, and the predictions of branching ratios. A discussion of the results and conclusions appear in Sec. V.

2 Formalism

2.1 Effective Hamiltonian

In the absence of strong interactions, the effective Hamiltonian for the process $b \to c\bar{u}d$ is given by

$$\mathcal{H}_{\text{eff}} = \frac{G_F}{\sqrt{2}} V_{cb} V_{ud}^* \, (\bar{c}b)_L \, (\bar{d}u)_L, \tag{1}$$

where $(\bar{c}b)_L = \bar{c}_i \gamma^\mu (1 - \gamma_5) b_i$, $(\bar{d}u)_L = \bar{d}_i \gamma^\mu (1 - \gamma_5) u_i$ and i is the color index. When QCD effects are included, the effective Hamiltonian is generalized to[9,10]

$$\mathcal{H}_{\text{eff}} = \frac{G_F}{\sqrt{2}} V_{cb} V_{ud}^* \left[C_1 \, (\bar{c}b)_L \, (\bar{d}u)_L + C_2 \, (\bar{c}u)_L \, (\bar{d}b)_L \right], \tag{2}$$

where $(\bar{c}b)_L \, (\bar{d}u)_L$ and $(\bar{c}u)_L \, (\bar{d}b)_L$ are current $\times$ current local operators. The Wilson coefficients, C_1 and C_2, include the short-distance QCD corrections. Their values depend on the renormalization scale μ. At the particular scale $(\mu_0 = 4.6 \text{ GeV} \sim m_b)$ we have $C_1 = 1.127 \pm 0.005$ and $C_2 = -0.286 \pm 0.008$. For the QCD scale we used the value $\Lambda_{\overline{\text{MS}}}^5 = 219 \pm 24$ MeV[11] which corresponds to $\alpha(m_Z) = 0.119 \pm 0.002$. The uncertainties in C_1 and C_2 are due to the uncertainties in $\Lambda_{\overline{\text{MS}}}^5$. Similarly, the other Cabibbo-favored process $(b \to c\bar{c}s)$ occurs through the Hamiltonian

$$\mathcal{H}_{\text{eff}} = \frac{G_F}{\sqrt{2}} V_{cb} V_{cs}^* \left[C_1 \, (\bar{c}b)_L \, (\bar{s}c)_L + C_2 \, (\bar{c}c)_L \, (\bar{s}b)_L \right], \tag{3}$$

with the same values for the Wilson coefficients mentioned above.

294

2.2 Factorization and Nonfactorization

Consider the color-favored decay $\overline{B}^0 \to D^+\pi^-$. The decay amplitude for this process is given by

$$
\mathcal{A}(\overline{B}^0 \to D^+\pi^-) = \frac{G_F}{\sqrt{2}} V_{cb} V_{ud}^* \, a_1 \left[\langle D^+|(\bar{c}b)|\overline{B}^0\rangle\langle\pi^-|(\bar{d}u)|0\rangle \right.
$$
$$
+ \langle D^+\pi^-|(\bar{c}b)\,(\bar{d}u)|\overline{B}^0\rangle^{nf}
$$
$$
\left. + \frac{C_2}{a_1} \langle D^+\pi^-|\frac{1}{2}\sum_a (\bar{c}\lambda^a b)\,(\bar{d}\lambda^a u)|\overline{B}^0\rangle^{nf} \right] \quad (4)
$$

where Fierz transformation and color-algebra have been used to rearrange the quark flavors. In (4)

$$
a_1(\mu) = \left(C_1(\mu) + \frac{C_2(\mu)}{N_c} \right), \quad (5)
$$

and λ^a are the Gell-Mann matrices. The second and third terms on the right hand side of (4) are nonfactorizable, while the first term is factorizable.

Following the conventions in ref. [8], we define the following color-singlet and color-octet nonfactorization parameters:

$$
\varepsilon_1^{(BD,\pi)} = \frac{\langle D^+\pi^-|(\bar{c}b)\,(\bar{d}u)|\overline{B}^0\rangle^{nf}}{\langle D^+|(\bar{c}b)|\overline{B}^0\rangle\langle\pi^-|(\bar{d}u)|0\rangle}, \quad (6)
$$

$$
\varepsilon_8^{(BD,\pi)} = \frac{\langle D^+\pi^-|\frac{1}{2}\sum_a (\bar{c}\lambda^a b)\,(\bar{d}\lambda^a u)|\overline{B}^0\rangle^{nf}}{\langle D^+|(\bar{c}b)|\overline{B}^0\rangle\langle\pi^-|(\bar{d}u)|0\rangle}. \quad (7)
$$

The decay amplitude then takes the form

$$
\mathcal{A}(\overline{B}^0 \to D^+\pi^-) = \frac{G_F}{\sqrt{2}} V_{cb} V_{ud}^* \, a_1 \xi_1^{(BD,\pi)} \langle D^+|(\bar{c}b)|\overline{B}^0\rangle\langle\pi^-|(\bar{d}u)|0\rangle, \quad (8)
$$

where

$$
\xi_1^{(BD,\pi)}(\mu) = \left(1 + \varepsilon_1^{(BD,\pi)}(\mu) + \frac{C_2}{a_1} \varepsilon_8^{(BD,\pi)}(\mu) \right). \quad (9)
$$

Similarily, the decay amplitude for the color-suppressed decay $\overline{B}^0 \to D^0\pi^0$, takes the form

$$
\mathcal{A}(\overline{B}^0 \to D^0\pi^0) = \frac{G_F}{\sqrt{2}} V_{cb} V_{ud}^* \, a_2 \, \xi_2^{(B\pi,D)} \langle\pi^0|(\bar{d}b)|\overline{B}^0\rangle\langle D^0|(\bar{c}u)|0\rangle, \quad (10)
$$

where

$$a_2(\mu) = \left(C_2(\mu) + \frac{C_1(\mu)}{N_c} \right) \tag{11}$$

and

$$\xi_2^{(B\pi,D)}(\mu) = \left(1 + \varepsilon_1^{(B\pi,D)}(\mu) + \frac{C_1}{a_2} \varepsilon_8^{(B\pi,D)}(\mu) \right). \tag{12}$$

In this work, we will assume that ε_1 and ε_8 are universal constants for all Cabibbo-favored B decays. For this reason, their superscripts will be dropped from now on. The decay amplitudes for the other processes of class I (color-favored) and processes of class II (color-suppressed) considered in this work, are derived from (8) and (10), respectively, by making appropriate replacements. However, for processes of class III, which receive contributions from both a_1 and a_2, we use a suitable combination of the above mentioned equations to derive their amplitude. For example, the decay amplitude for the process $B^- \to D^0 \pi^-$ is given by

$$\mathcal{A}(B^- \to D^0 \pi^-) = \frac{G_F}{\sqrt{2}} V_{cb} V_{ud}^* \left[a_1 \, \xi_1 \, \langle D^0 | (\bar{c}b) | B^- \rangle \langle \pi^- | (\bar{d}u) | 0 \rangle \right.$$
$$\left. + a_2 \, \xi_2 \, \langle \pi^- | (\bar{d}b) | B^- \rangle \langle D^0 | (\bar{c}u) | 0 \rangle \right]. \tag{13}$$

Eq's. (8) and (10) suggest the following definitions for the effective a_1 and a_2

$$a_1^{\it eff} = a_1 \, \xi_1 = a_1 \, [1 + \varepsilon_1] + C_2 \, \varepsilon_8 \tag{14}$$
$$a_2^{\it eff} = a_2 \, \xi_2 = a_2 \, [1 + \varepsilon_1] + C_1 \, \varepsilon_8. \tag{15}$$

From $1/N_c$ expansion [8,12,13] it is expected that $\varepsilon_1(\mu)/\varepsilon_8(\mu) = \mathcal{O}(1/N_c)$. A simple way to see this is to realize that whereas only one gluon exchange is needed to cause color-octet current to couple to color-singlet hadrons, two gluons are needed in the case of color-singlet currents.

2.3 Current Matrix Elements

The hadronic current matrix elements can be decomposed in terms of form factors and decay constants using Lorentz invariance. For this we use the conventions of ref. [14]. For the form factors, we use BSW model [14] to calculate their values at zero momentum transfer. Then, we extrapolate to the desired momentum using a monopole form for the form factors F_0 and A_1 and a dipole form for the form factors F_1, A_0, A_2 and V [15]. This is sometimes referred to as BSW II model. The two states $|\eta\rangle$ and $|\eta'\rangle$ are treated in the same way as in ref. [7] where the mixing angle and wavefunction normalizations are properly taken care of. Regarding the decay constants used see ref. [16].

3 Evaluation of Nonfactorization Contribution

The μ dependence of ε_1 and ε_8 is cancelled by the μ dependence of the Wilson coefficients such that the decay amplitude is μ independent. Our goal in this section is to estimate the values of $\varepsilon_1(\mu_0)$ and $\varepsilon_8(\mu_0)$ using available experimental data on color-favored and color-suppressed channels.

Consider, first, the color-favored processes of the type $b \to c\bar{u}d$. By calculating the branching ratios of these decays using the factorization assumption ($\varepsilon_1 = \varepsilon_8 = 0$) and then by comparing it with experimental measurements we calculate ξ_1^2 in each decay channel by dividing the experimental branching ratio by that calculated using factorization. We can see from the results that ξ_1^2 is almost channel independent except when the a_1 meson is present in the final state. By averaging the amount of nonfactorization in these channels we get the value, $\xi_1^2(\mu_0) = 0.83 \pm 0.05$. This corresponds to a 17% deviation from the factorization assumption. The second set of processes considered are the color-favored decays of the type $b \to c\bar{c}s$. The experimental value used for each decay mode was taken to be a weighted average over the charged and neutral decay channels. The weighted average of the amount of nonfactorization in these processes, $\xi_1^2(\mu_0) = 0.82 \pm 0.15$, is in good agreement with that found in the decay processes of the kind $b \to c\bar{u}d$. Finally, we considered the color-suppressed processes of the type $b \to c\bar{c}s$. These processes give predictions regarding the nonfactorization parameter $\xi_2^2(\mu_0)$. From the average, a value of $\xi_2^2(\mu_0) = 5.7 \pm 0.7$ is calculated.

Fig. 1, shows the regions in ε_1-ε_8 space that correspond to the average values of ξ_1^2 and ξ_2^2 predicted above. In the figure, we also show two intersecting solid lines corresponding to the equation $\varepsilon_1/\varepsilon_8 = \pm 1/N_c$. We realize that this relation need not be exact, nevertheless we use it to restrict our solutions.

As seen in Fig. 1, these regions intersect in four areas labeled 1, 2, 3 and 4. Areas 1 and 2 are excluded due to the severe violation of $\varepsilon_1/\varepsilon_8 = \pm 1/N_c$. Area 4 is also excluded because it produces negative a_2^{eff} which is not supported by the mainstream data on B decays. So, we are left with area 3. From the size and location of this area in the ε_1-ε_8 space, we get the following predictions for the color-singlet and color-octet nonfactorization parameters at μ_0:

$$\varepsilon_1(\mu_0) = -0.06 \pm 0.03,$$
$$\varepsilon_8(\mu_0) = \;\;\; 0.12 \pm 0.02 \; . \tag{16}$$

These values correspond to

$$\xi_1(\mu_0) = 0.91 \pm 0.03 \quad \longrightarrow \quad a_1^{eff} = 0.94 \pm 0.03 \, ,$$
$$\xi_2(\mu_0) = 2.39 \pm 0.23 \quad \longrightarrow \quad a_2^{eff} = 0.22 \pm 0.02. \tag{17}$$

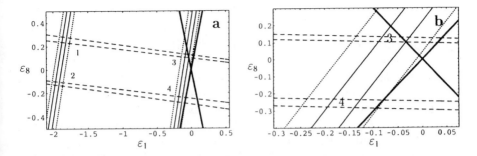

Figure 1: (a) The regions in ε_1-ε_8 space that correspond to the amount of nonfactorization estimated for ξ_1^2 and ξ_2^2. The two parallel regions bounded by thin solid lines correspond to the value $\xi_1^2 = 0.83 \pm 0.05$ calculated from the color-favored processes of the type $b \to c\bar{u}d$. The regions bounded by the dotted lines correspond to the value $\xi_1^2 = 0.82 \pm 0.15$ calculated from the color-favored processes of the type $b \to c\bar{c}s$. The regions bounded by the dashed lines correspond to the value $\xi_2^2 = 5.7 \pm 0.7$ calculated from the color-suppressed processes of the type $b \to c\bar{c}s$. The two intersecting solid lines correspond to the equation $\varepsilon_1/\varepsilon_8 = \pm 1/3$. (b) A magnification of the interesting region in ε_1-ε_8 space.

The estimated value of ε_8 is in agreement with that estimated in ref. [8] using a more limited set of processes.

4 Predictions of Branching Ratios

Assuming universality of the nonfactorization parameters, ε_1 and ε_8, we calculated the branching ratios for a large number of Cabibbo-favored decay channels which are shown in Table 1. The errors that appear in the theoretical calculations are due to the uncertainties in the Wilson coefficients, the decay constants, the B-meson lifetime and (where applicable) the mass of a_1 meson.

We start by studying six color-favored processes of the type $b \to c\bar{u}d$. From Table 1, it can be seen that the inclusion of nonfactorization improves the predictions of the factorization assumption and produces a much better fit to experimental measurements. Also, we show the predicted branching ratios for ten color-suppressed processes of the type $b \to c\bar{u}d$. As the results show, the inclusion of nonfactorization in these decays enhances their decay rates, pushing them closer to the present upper bounds. Table 1 also shows the predictions of the branching ratios for the color-favored and color-suppressed decays of the type $b \to c\bar{c}s$. By comparing these results with experimental data we find a significant improvement in the theoretical predictions, especially for the color-suppressed modes. The experimental value shown for each decay mode is a weighted average over the charged and neutral decay channels which differ in the flavor of their spectator quark. The average experimental values in ref. [11] for the three processes $B \to K\psi(2S)$, $B \to K^*J/\psi$ and $B \to K^*\psi(2S)$ are updated to include the latest measurements [17] by the CDF Collaboration. The four decay channels of B^- mesons receive contributions from two diagrams. As a result the decay amplitudes contain interference between a_1^{eff} and a_2^{eff}. Using the values of the nonfactorization parameters in (16) results in a constructive interference between the two diagrams and small modifications to the predictions of the factorization assumption. The predictions are in agreement with experimental results. However, if we use for the nonfactorization parameters the values $\varepsilon_1(\mu_0) = -0.17$ and $\varepsilon_8(\mu_0) = -0.26$, which are taken from area 4 in Fig. 1, the interference between the two diagrams becomes destructive. This results in lower values for the branching ratios and poor agreement with experiment supporting our choice of the solution in area 3 as discussed in the previous section.

Finally, we calculated the effect of nonfactorization contribution on the branching ratios of three sets of B_s decays. The results of these calculations are shown in Table 2. Even though, experimental data on B_s-decays are very limited, the measured branching ratio of $\overline{B}_s \to \phi J/\psi$ [11] shows encouraging agreement with our predictions. See ref. [16] for an extended study that includes polarization predictions.

5 Discussion and Conclusion

Strictly speaking, we know that the factorization assumption can not be correct. This is because it produces a scale dependent transition amplitude. However, quoting authors of ref. [8], "what one may hope for is that it provides a useful approximation if the Wilson coefficients (or equivalently the QCD coefficients a_1 and a_2) are evaluated at a suitable scale μ_f, the factorization point."

In B decays, the Wilson coefficients are usually evaluated at $(\mu \sim m_b)$. If this is the factorization scale, then we should expect the nonfactorization parameters $[\varepsilon_1(m_b)$ and $\varepsilon_8(m_b)]$ to vanish. However, since the predictions of the factorization assumption are generally not in agreement with the data, especially for color-suppressed decays, this could indicate that the nonfactorizable parts of the decay amplitude are not negligible at the B-mass scale.

In this work, we tried to answer the following question: is it possible via the introduction of a minimum number of new parameters to improve the predictions of the factorization assumption and explain the bulk of available experimental data on Cabibbo-favored B dB decays? In answering this question, we assumed universality of the color-singlet (ε_1) and the color-octet (ε_8)

Table 1: The branching ratios for a set of Cabibbo-favored processes, calculated in column 2 by taking $\varepsilon_1(\mu_0) = \varepsilon_8(\mu_0) = 0$ and, in column 3, by taking $\varepsilon_1(\mu_0) = -0.06 \pm 0.03$ and $\varepsilon_8(\mu_0) = 0.12 \pm 0.02$. The last column represents the available experimental measurements.

Process	Fac.	Nonfac.	Exp. [11]
	Branching Ratio $\times 10^{-3}$		
$\overline{B}^0 \to D^+\pi^-$	4.1 ± 0.1	3.4 ± 0.3	3.0 ± 0.4
$\overline{B}^0 \to D^+\rho^-$	9.9 ± 0.3	8.2 ± 0.6	7.9 ± 1.4
$\overline{B}^0 \to D^+a_1^-$	11.2 ± 1.0	9.3 ± 1.1	6.0 ± 3.3
$\overline{B}^0 \to D^{*+}\pi^-$	3.1 ± 0.1	2.6 ± 0.2	2.76 ± 0.21
$\overline{B}^0 \to D^{*+}\rho^-$	9.0 ± 0.3	7.4 ± 0.6	6.7 ± 3.3
$\overline{B}^0 \to D^{*+}a_1^-$	12.8 ± 1.2	10.6 ± 1.2	13.0 ± 2.7
	Branching Ratio $\times 10^{-4}$		
$\overline{B}^0 \to D^0\pi^0$	0.14 ± 0.04	0.79 ± 0.21	< 1.2
$\overline{B}^0 \to D^{*0}\pi^0$	0.19 ± 0.05	1.11 ± 0.30	< 4.4
$\overline{B}^0 \to D^0\eta$	0.08 ± 0.02	0.44 ± 0.12	< 1.3
$\overline{B}^0 \to D^{*0}\eta$	0.11 ± 0.03	0.61 ± 0.16	< 2.6
$\overline{B}^0 \to D^0\eta'$	0.02 ± 0.01	0.13 ± 0.04	< 9.4
$\overline{B}^0 \to D^{*0}\eta'$	0.03 ± 0.01	0.18 ± 0.05	< 14
$\overline{B}^0 \to D^0\rho^0$	0.09 ± 0.02	0.54 ± 0.14	< 3.9
$\overline{B}^0 \to D^{*0}\rho^0$	0.20 ± 0.05	1.16 ± 0.31	< 5.6
$\overline{B}^0 \to D^0\omega$	0.09 ± 0.02	0.53 ± 0.14	< 5.1
$\overline{B}^0 \to D^{*0}\omega$	0.20 ± 0.05	1.15 ± 0.31	< 7.4
	Branching Ratio $\times 10^{-3}$		
$B \to DD_s$	13.9 ± 3.0	11.5 ± 2.6	9.8 ± 2.4
$B \to DD_s^*$	12.2 ± 2.5	10.1 ± 2.2	9.4 ± 3.1
$B \to D^*D_s$	7.8 ± 1.7	6.5 ± 1.5	10.4 ± 2.8
$B \to D^*D_s^*$	25.9 ± 5.2	21.4 ± 4.6	22.3 ± 5.7
	Branching Ratio $\times 10^{-4}$		
$B \to KJ/\psi$	1.7 ± 0.3	9.6 ± 1.9	9.5 ± 0.8
$B \to K\psi(2S)$	0.9 ± 0.2	5.1 ± 1.0	5.8 ± 1.2
$B \to K^*J/\psi$	2.7 ± 0.5	15.5 ± 3.0	14.6 ± 1.4
$B \to K^*\psi(2S)$	1.6 ± 0.3	9.3 ± 1.9	9.6 ± 2.5
	Branching Ratio $\times 10^{-3}$		
$B^- \to D^0\pi^-$	5.1 ± 0.2	5.3 ± 0.4	5.3 ± 0.5
$B^- \to D^0\rho^-$	11.4 ± 0.3	10.7 ± 0.8	13.4 ± 1.8
$B^- \to D^{*0}\pi^-$	4.3 ± 0.1	4.5 ± 0.4	4.6 ± 0.4
$B^- \to D^{*0}\rho^-$	11.0 ± 0.3	11.3 ± 0.9	15.5 ± 3.1

300

Table 2: The predicted branching ratios for the color-favored B_s decays of the type $b \to c\bar{u}d$ calculated, in column 2, by taking $\varepsilon_1(\mu_0) = \varepsilon_8(\mu_0) = 0$ and, in column 3, by taking $\varepsilon_1(\mu_0) = -0.06 \pm 0.03$, $\varepsilon_8(\mu_0) = 0.12 \pm 0.02$. The last column represents the available experimental limits.

| Process | Branching Ratio $\times 10^{-3}$ | | |
	Fac.	Nonfac.	Exp. [11]
$\overline{B}_s \to D_s^+ \pi^-$	3.6 ± 0.2	3.0 ± 0.3	< 130
$\overline{B}_s \to D_s^+ \rho^-$	8.6 ± 0.4	7.2 ± 0.6	$-$
$\overline{B}_s \to D_s^+ a_1^-$	9.8 ± 1.0	8.1 ± 1.0	$-$
$\overline{B}_s \to D_s^{*+} \pi^-$	2.6 ± 0.1	2.2 ± 0.2	$-$
$\overline{B}_s \to D^{*+} \rho^-$	7.5 ± 0.4	6.2 ± 0.5	$-$
$\overline{B}_s \to D^{*+} a_1^-$	10.7 ± 1.1	8.9 ± 1.1	$-$
	Branching Ratio $\times 10^{-4}$		
$\overline{B}_s \to D^0 K^0$	0.18 ± 0.05	1.0 ± 0.3	$-$
$\overline{B}_s \to D^{*0} K^0$	0.25 ± 0.07	1.5 ± 0.4	$-$
$\overline{B}_s \to D^0 K^{*0}$	0.13 ± 0.03	0.7 ± 0.2	$-$
$\overline{B}_s \to D^{*0} K^{*0}$	0.27 ± 0.07	1.5 ± 0.4	$-$
	Branching Ratio $\times 10^{-4}$		
$\overline{B}_s \to \eta J/\psi$	0.44 ± 0.08	2.5 ± 0.5	< 38
$\overline{B}_s \to \eta \psi(2S)$	0.24 ± 0.05	1.4 ± 0.3	$-$
$\overline{B}_s \to \eta' J/\psi$	0.52 ± 0.09	2.9 ± 0.6	$-$
$\overline{B}_s \to \eta' \psi(2S)$	0.23 ± 0.04	1.3 ± 0.3	$-$
$\overline{B}_s \to \phi J/\psi$	1.83 ± 0.32	10.5 ± 2.1	9.3 ± 3.3
$\overline{B}_s \to \phi \psi(2S)$	1.25 ± 0.23	7.2 ± 1.5	$-$

nonfactorization parameters. Two sets of color-favored processes and one set of color-suppressed processes were used to give quantitative estimates of these parameters. It has been found (by calculating the branching ratios for a large number of Cabibbo-favored B decays) that the values $\varepsilon_1(\mu_0) = -0.06 \pm 0.03$ and $\varepsilon_8(\mu_0) = 0.12 \pm 0.02$ improve significantly the predictions of the factorization assumption even though the Wilson coefficients were evaluated at leading order and only the tree diagrams were considered. These results support the argument that nonfactorization plays an important role in B decays into two hadrons.

Acknowledgments

This research was partially supported by a grant to A.N.K. from the Natural Sciences and Engineering Research Council of Canada.

1. H.Y. Cheng, Phys. Lett. **B335**, 428, 1994; Z. Phys. **C69**, 647, 1996.
2. A.N. Kamal and A.B. Santra, Alberta Thy 31-94 (1994); Z. Phys. **C72**, 91, 1996.
3. J. Soares, Phys. Rev. **D51**, 3518, 1995.
4. A. Ali and C. Greub, Phys. Rev. **D57**, 2996, 1998.
5. A. Ali, G. Kramer and C.D. Lu, Phys. Rev. **D58**, 094009, 1998.
6. H.Y. Cheng and B. Tsang, Phys. Rev. **D58**, 094005, 1998.
7. F.M. Al-Shamali and A.N. Kamal, Eur. Phys. J. **C4**, 669, 1998.
8. M. Neubert and B. Stech, CERN-TH/97/99, hep-ph/9705292. To appear in the Second Edition of Heavy Flavours, Edited by A.J. Buras and M. Lindner (World Scientific, Singapore).
9. A.J. Buras, M. Jamin, M. Lautenbacher and P. Weisz, Nucl. Phys. **B370**, 69, 1992.
10. G. Buchalla, A. Buras and M. Lautenbacher, Rev. Mod. Phys. **68**, 1125, 1996.
11. C. Caso et al., (Particle Data Group), Eur. Phys. J. **C3**, 1, 1998.
12. E. Witten, Nucl. Phys. **B160**, 57, 1979.
13. A.J. Buras, J.M. Gérard and R. Rückl, Nucl. Phys. **B268**, 16, 1986.
14. M. Bauer, B. Stech and M. Wirbel, Z. Phys. **C34**, 103, 1987; M. Wirbel, B. Stech and M. Bauer, Z. Phys. **C29**, 637, 1985.
15. M. Neubert and V. Rieckert, Nucl. Phys. **B382**, 97, 1992.
16. F.M. Al-Shamali and A.N. Kamal, Phys. Rev. **D59**, 054020, 1999.
17. F. Abe et al. (CDF Collaboration), Phys. Rev. **D58**, 072001, 1998.

RECENT RESULTS ON RARE TAU DECAYS FROM CLEO

ANTON ANASTASSOV

Department of Physics, The Ohio State University,
174 W.18th Avenue, Columbus, OH 43210, U.S.A.

Recent results on rare decays of the tau lepton obtained by the CLEO Collaboration are reported. The results include the first observation of $\tau^- \to K^{*-}\eta\nu_\tau$, measurements of the branching fractions of three prong decays with charged kaons and study of the decay $\tau^- \to 2\pi^-\pi^+3\pi^0\nu_\tau$. The results are compared with Standard Model predictions.

1 Introduction

In this paper we present the results of three recent analyses of rare tau decays. The data were obtained by the CLEO II [1] detector at the Cornell Electron Storage Ring at center of mass energy of 10.6 GeV. The sample used in these analyses correspond to an integrated luminosity of $4.7fb^{-1}$ and contains 4.3 million $\tau^+\tau^-$ pairs.

2 Three prong tau decays with charged kaons

The large data sample accumulated at CLEO makes possible studies of tau decays to three prong final states with charged kaons. Here we present the results on measurements of four ratios of branching fractions:

$$B(\tau^- \to K^-h^+\pi^-(\pi^0)\nu_\tau)/B(\tau^- \to \pi^-\pi^+\pi^-(\pi^0)\nu_\tau) \text{ and}$$

$$B(\tau^- \to K^-K^+\pi^-(\pi^0)\nu_\tau)/B(\tau^- \to \pi^-\pi^+\pi^-(\pi^0)\nu_\tau),$$

where $h^{\pm}$ stands for either a charged kaon or pion [a]. Since we are interested in decays directly to three or four mesons we treat the decays $\tau^- \to K^-K_S(\pi^0)\nu_\tau$ as background.

The selected events are required to have 1 vs 3 topology (tag vs signal). The tag side must be consistent with one of the following decay modes: $\tau^- \to e^-\nu_\tau\bar{\nu}_e$, $\mu^-\nu_\tau\bar{\nu}_\mu$, $\pi^-\nu_\tau$, $\rho^-\nu_\tau$. The signal side must contain three charged tracks, and for the modes with a π^0, two isolated showers in the calorimeter with energy, and lateral profile consistent with the photon hypothesis. Two photon backgrounds are suppressed by cuts on the visible energy (E_{vis}) and

---

[a] Charge conjugate states are implied throughout this paper.

the transverse momentum (P_t): $2.5\ GeV < E_{vis} < 10\ GeV$, $P_t > 0.3\ GeV/C$. Hadronic backgrounds are rejected by requiring the invariant mass on the signal side to be less than the tau mass. Contributions from decays proceeding through an intermediate K_S with $K_S \to \pi^+\pi^-$ are suppressed by requirements on the impact parameters on the charged tracks.

The number of events with and without charged kaons is determined using a statistical approach based on the analysis of the specific ionization loss measured in the central drift chamber. The deviation δ_k of the specific ionization loss relative to that expected for kaons is calculated for tracks in the signal hemisphere. For true kaons this parameter is distributed as a zero-centered unit Gaussian, while for pions the distribution has Gaussian-like shape and is shifted from zero in a momentum dependent manner. We concentrate on tracks with $p > 1.5$ GeV/c. The K/π separation varies slowly in this region and the systematics are more traceable. The number of kaon and pion tracks in the signal hemisphere is found by fitting the δ_k distribution with the sum of a unit Gaussian and a Johnson distribution [2] representing the kaon and pion shapes, respectively. Since K/π separation depends on track momentum and the number of hits in the drift chamber, we perform individual δ_k fits for 36 bins in this two parameter space. This also makes possible the reconstruction of the momentum spectra of kaons and pions. Fig. 1 shows an example δ_k fit. For the $\tau^- \to K^- h^+ \pi^- (\pi^0)\nu_\tau$ analysis the two same sign tracks participate in

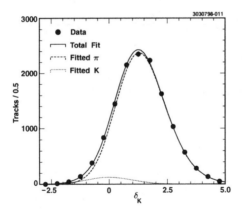

Figure 1: Fit to δ_k distribution for charged tracks in the signal hemisphere. The confidence level of the fit is 92%; C.L. drops to less that 10^{-3} if the kaon contribution is not included.

the fits, while for $\tau^- \to K^- K^+ \pi^- (\pi^0)\nu_\tau$ only tracks with charge opposite to the parent tau are considered. This allows us to relate the observed number of kaons and pions to the number of decays of interest using simple algebra. Here we ignore unphysical decays such as $\tau^- \to \pi^- K^+ \pi^- \nu_\tau$ and $\tau^- \to K^- \pi^+ K^- \nu_\tau$ as well as the phase space suppressed $\tau^- \to K^- K^+ K^- \nu_\tau$ (its decay rate is expected to be less than 1% relative to that of $\tau^- \to K^- \pi^+ \pi^- \nu_\tau$). For the decay modes with a π^0 the δ_k fits are performed separately for events with photon-pair invariant mass in the signal and sideband regions and sideband subtraction is performed to determine the number of K/π's.

The total number of kaons and pions is obtained by extrapolation of the measured yields to the region $p < 1.5$ GeV/c using the Monte Carlo shapes. Uncertainties in the modeling are included in the systematic errors. Fig. 2 shows the spectra for kaons and pions obtained in the study of the decay $\tau^- \to K^- h^+ \pi^- \nu_\tau$.

The numerical results for the ratios of branching fractions are summarized

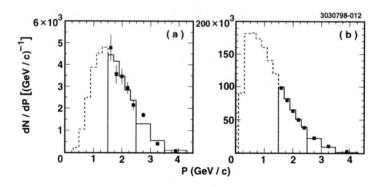

Figure 2: Reconstructed a) pion and b) kaon momentum spectra for $\tau^- \to K^- h^+ \pi^- \nu_\tau$ candidates. Solid squares with error bars are data points and the histogram is the MC shape.

below. The first error is statistical and the second is systematic.

$$B(\tau^- \to K^- h^+ \pi^- \nu_\tau)/B(\tau^- \to \pi^- \pi^+ \pi^- \nu_\tau) = (5.44 \pm 0.21 \pm 0.53) \times 10^{-2}$$
$$B(\tau^- \to K^- h^+ \pi^- \pi^0 \nu_\tau)/B(\tau^- \to \pi^- \pi^+ \pi^- \pi^0 \nu_\tau) = (2.61 \pm 0.45 \pm 0.42) \times 10^{-2}$$
$$B(\tau^- \to K^- K^+ \pi^- \nu_\tau)/B(\tau^- \to \pi^- \pi^+ \pi^- \nu_\tau) = (1.60 \pm 0.15 \pm 0.30) \times 10^{-2}$$
$$B(\tau^- \to K^- K^+ \pi^- \pi^0 \nu_\tau)/B(\tau^- \to \pi^- \pi^+ \pi^- \pi^0 \nu_\tau) < 0.0157 \ (95\% C.L.).$$

From these ratios one can also extract results for the exclusive decays $\tau^- \to K^- \pi^+ \pi^- \nu_\tau$ and $\tau^- \to K^- \pi^+ \pi^- \pi^0 \nu_\tau$. The decays of interest can be normalized to $\tau^- \to h^- h^+ h^- (\pi^0) \nu_\tau$ and using the CLEO results for the branching fractions for these decays [3] one can obtain the absolute branching fractions.

Table 1: Absolute branching fractions and comparison with ALEPH results and theoretical predictions

Decay mode	Measurement	Branching fraction, 10^{-2}
$\tau^- \to K^- \pi^+ \pi^- \nu_\tau$	ALEPH	$0.214 \pm 0.037 \pm 0.029$
	This analysis	$0.346 \pm 0.023 \pm 0.056$
	Theory [5]	0.77
	Theory [6]	0.18
$\tau^- \to K^- K^+ \pi^- \nu_\tau$	ALEPH	$0.163 \pm 0.021 \pm 0.017$
	This analysis	$0.145 \pm 0.013 \pm 0.028$
	Theory [5]	0.22
$\tau^- \to K^- \pi^+ \pi^- \pi^0 \nu_\tau$	ALEPH	$0.061 \pm 0.039 \pm 0.018$
	This analysis	$0.075 \pm 0.026 \pm 0.018$
$\tau^- \to K^- K^+ \pi^- \pi^0 \nu_\tau$	ALEPH	$0.075 \pm 0.029 \pm 0.015$
	This analysis	$0.033 \pm 0.018 \pm 0.007$

These results are compared with results from ALEPH [4] and theoretical predictions in Table 1.

3   First observation of the decay $\tau^- \to K^{*-} \eta \nu_\tau$

Tau decays involving η mesons are very rare. The Wess-Zumino-Witten anomaly plays an important role in many of these decays. CLEO has recently measured two such decays: $\tau^- \to K^- \eta \nu_\tau$ [7], $\tau^- \to 2h^- h^+ \eta \nu_\tau$ [8]. In both cases the theoretical predictions for the branching fractions available at the time were more than two orders of magnitudes lower than the experimental results. However, recent calculations by Li [6,9] are in good agreement with experiment. The decay $\tau^- \to K^{*-} \eta \nu_\tau$ is the next step in the continued studies of such decays by CLEO.

The candidate events are selected from samples with 1 vs 1 and 1 vs 3 topologies reflecting the reconstruction of K^{*-} from two decay chains: $K^{*-} \to K^- \pi^0$ and $K^{*-} \to K_S \pi^- \to [\pi^+ \pi^-] \pi^-$. The η is reconstructed using the $\gamma\gamma$ decay channel. As in the previous analysis, the tag side must be consistent with one of the the the decays $\tau^- \to e^- \nu_\tau \bar{\nu}_e$, $\mu^- \nu_\tau \bar{\nu}_\mu$, $\pi^- \nu_\tau$, $\rho^- \nu_\tau$. The hadronic background is suppressed by requiring the invariant mass in the signal hemisphere to be less than the tau mass. Two-photon, Bhabha, and additional hadronic background suppression is achieved by the requirement on the total visible energy, $0.25 < E_{tot}/E_{cm} < 0.85$, and the total transverse momentum of the event, $P_t > 0.3$ GeV/c.

Fig. 3 shows the invariant mass spectrum of $\gamma\gamma$ pairs accompanying K^{*-}

306

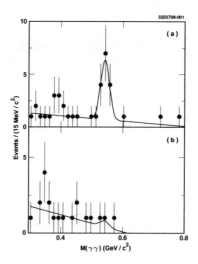

Figure 3: Invariant mass of the photon pairs in the signal hemisphere accompanying a π^- and reconstructed K_S. The $K_S\pi^-$ invariant mass is in (a) the K^{*-} signal and (b) sideband region, respectively. The curves show fits to the data.

candidates reconstructed from the decay chain $K^{*-} \to K_S\pi^- \to \pi^-\pi^+\pi^-$. There is a clear η signal in the K^{*-} signal region corresponding the first observation of the decay $\tau^- \to K^{*-}\eta\nu_\tau$. The second decay mode of K^{*-} used in the analysis $K^{*-} \to K^-\pi^0$ requires K/π separation. Particle identification is based on specific ionization and time of flight information.

The branching fractions obtained are listed below.

$$B(\tau \to K^{*-}\eta\nu_\tau) \times B(K^{*-} \to K_S\pi^-) = (1.18 \pm 0.38 \pm 0.12) \times 10^{-4}$$
$$B(\tau \to K^{*-}\eta\nu_\tau) \times B(K^{*-} \to K^-\pi^0) = (0.69 \pm 0.36 \pm 0.28) \times 10^{-4}.$$

Combining these results with the isospin requirement $B(K^{*-} \to K_S\pi^-) = B(K^{*-} \to K^-\pi^0) = \frac{1}{3}$ we obtain the absolute branching fraction:

$$B(\tau \to K^{*-}\eta\nu_\tau) = (2.90 \pm 0.80 \pm 0.42) \times 10^{-4}.$$

We have also measured the inclusive branching fractions without requiring a K^{*-}:

$$B(\tau \to K^-\pi^0\eta\nu_\tau) = (1.77 \pm 0.56 \pm 0.62) \times 10^{-4}$$
$$B(\tau \to K_S\pi^-\eta\nu_\tau) = (1.10 \pm 0.35 \pm 0.11) \times 10^{-4}.$$

The result for $B(\tau^- \to K^{*-}\eta\nu_\tau)$ is higher than the theoretical prediction by Li [6,9] $B(\tau^- \to K^{*-}\eta\nu_\tau) = 1.01 \times 10^{-4}$. The measured inclusive branching ratios are significantly higher than the predictions by Pich [10]: $B(\tau^- \to K^-\pi^0\eta\nu_\tau) \sim 8.8 \times 10^{-6}$, $B(\tau^- \to \pi^-\bar{K}^0\eta\nu_\tau) \sim 2.2 \times 10^{-5}$.

4 Study of the decay $\tau^- \to 2\pi^-\pi^+3\pi^0\nu_\tau$.

There are three six-pion tau decays: $\tau^- \to 2\pi^-\pi^+3\pi^0\nu_\tau$, $\tau^- \to 3\pi^-2\pi^+\pi^0\nu_\tau$, and $\tau^- \to \pi^-5\pi^0\nu_\tau$, which are related by isospin symmetry. The Conserved Vector Current (CVC) hypothesis relates the branching fractions of these decays to the e^+e^- annihilation cross-section to isovector six-pion states.

The decay $\tau^- \to \pi^-5\pi^0\nu_\tau$ is a major experimental challenge due to the presence of 10 photons in the final state leading to huge combinatorial background. This decay has not been observed yet. The other two decays have been observed but in the case of $\tau^- \to 2\pi^-\pi^+3\pi^0\nu_\tau$ the precision is not satisfactory. The decay $\tau^- \to 2\pi^-\pi^+3\pi^0\nu_\tau$ has been recently studied by the CLEO collaboration. We selected events with 1 vs 3 topology. The π^0 candidates are exclusively reconstructed from photon pairs in the signal hemisphere. The selection is based on the requirement on $S_{\gamma\gamma} = (m_{\gamma\gamma} - m_{\pi^0})/\sigma_{\gamma\gamma}$, where $\sigma_{\gamma\gamma}$ is the mass resolution. The hadronic background is suppressed by cuts on the invariant mass and the magnitude of the total momentum of the particles calculated in the tau rest frame. The measured branching ratio is

$$B(\tau^- \to 2\pi^-\pi^+3\pi^0\nu_\tau) = (2.85 \pm 0.56 \pm 0.51) \times 10^{-4}.$$

This result is considerably lower and has smaller errors than the one reported by ALEPH [11] $B(\tau^- \to 2\pi^-\pi^+ \geq 3\pi^0\nu_\tau) = (1.1 \pm 0.4 \pm 0.5) \times 10^{-3}$.

In order to compare the experimental results with the CVC prediction one should determine the vector current contribution to $\tau^- \to 2\pi^-\pi^+3\pi^0\nu_\tau$. This requires investigation of the substructure of the decay. Fig. 4 shows the invariant mass spectrum of the $\pi^+\pi^-\pi^0$ system. There is a clear signal in the ω mass region corresponding to the first observation of the decay $\tau^- \to \pi^-2\pi^0\omega\nu_\tau$. The branching fraction is measured to be

$$B(\tau^- \to \pi^-2\pi^0\omega\nu_\tau) = (1.89^{+0.74}_{-0.67} \pm 0.40) \times 10^{-4}.$$

The small η signal results from the decay $\tau^- \to \pi^-2\pi^0\eta\nu_\tau$ with $\eta \to \pi^+\pi^-\pi^0$ which was observed [12] in the decay channel $\eta \to \gamma\gamma$ with much higher statistics. In addition to these decay channels, the final state has a contribution from $\tau^- \to 2\pi^-\pi^+\eta$ with $\eta \to 3\pi^0$. The channels with ω and η mesons in the intermediate state saturate the investigated decay.

308

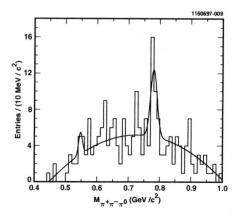

Figure 4: Invariant mass spectrum of $\pi^+\pi^-\pi^0$ combinations (6 entries/event).

We can test isospin symmetry by comparing the relative branching fractions of the observed decay modes with six pions in the final state:

$$f_{2\pi^-\pi^+3\pi^0} = \frac{B(\tau^- \to 2\pi^-\pi^+3\pi^0\nu_\tau)}{B(\tau^- \to (6\pi)^-\nu_\tau)}; \quad f_{3\pi^-2\pi^+\pi^0} = \frac{B(\tau^- \to 3\pi^-2\pi^+\pi^0\nu_\tau)}{B(\tau^- \to (6\pi)^-\nu_\tau)}.$$

The world average [13] of $B(\tau^- \to 3\pi^-2\pi^+\pi^0\nu_\tau)$ is used in the comparison. Before calculating the ratios we correct the branching fractions for η contributions using the CLEO results [12]. The results are plotted on Fig. 5 The triangle represents the region allowed by isospin symmetry. The isospin states are labeled and shown as filled circles. There are no data on the decay $\tau^- \to \pi^-5\pi^0\nu_\tau$ and the experimental result can only be represented as a line instead of a point. Since the branching fraction for this decay is expected to be very small it looks like the current result prefers the state $(\pi\rho\omega)$.

There are two results that can be compared with the CVC predictions. One approach is to take the contribution of $\tau^- \to \pi^-2\pi^0\omega\nu_\tau$ to represent the vector current contribution $B_{V_\omega}(\tau^- \to 2\pi^-\pi^+3\pi^0\nu_\tau) = (1.68 \pm 0.72) \times 10^{-4}$. Alternatively, one can correct the inclusive branching fraction for the contribution from the decays with η mesons in the intermediate states using the CLEO results [12] leading to $B_V(\tau^- \to 2\pi^-\pi^+3\pi^0\nu_\tau) = (1.41 \pm 0.76) \times 10^{-4}$. These values should be compared to the predictions by Sobie [14] and Eidelman [15] $B_V(\tau^- \to 2\pi^-\pi^+3\pi^0\nu_\tau) = (1.9-7.3) \times 10^{-4}$ and $(2.5 \pm 0.4) \times 10^{-4}$, respectively. Our result is somewhat lower than the expectation, indicating that there could be substantial $I = 0$ contribution to $e^+e^- \to 6\pi$ that has to be removed before calculating the CVC prediction.

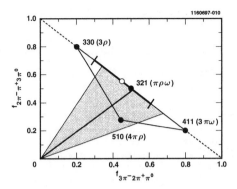

Figure 5: $f_{2\pi^-\pi^+3\pi^0}$ vs $f_{3\pi^-2\pi^+\pi^0}$. The solid line from the origin represents the experimental result when using the branching fractions corrected for η contributions. The shaded area indicates the one standard deviation region. The open circle with error bar uses B_{V_ω} to represent the vector current contribution to $\tau^- \to 2\pi^-\pi^+3\pi^0\nu_\tau$ together with the assumption
$$B(\tau^- \to \pi^-5\pi^0\nu_\tau) = 0.$$

References

1. Y. Kubota, Nucl. Instrum. Methods **A320**, 66, (1992).
2. W.P. Elderton and N.L. Johnson, *Systems of Frequency Curves*, (Wiley & Sons, New York, NY, 1994).
3. CLEO Collaboration R. Balest *et al*, Phys. Rev. Lett. **75**, 3809, (1995).
4. ALEPH Collaboration, R. Barate *et al.* CERN-PPE/97-69 (1997).
5. M. Finkemeier and E. Mirkes, Z. Phys. **C69**, 243, (1996).
6. B.A. Li, Phys. Rev. **55**, 1436, (1997).
7. CLEO Collaboration, J. Bartlet *et al*, Phys. Rev. Lett. **76**, 4119, (1996).
8. CLEO Collaboration, J. Bergfeld *et al*, Phys. Rev. Lett. **79**, 2406, (1997).
9. B.A. Li, Phys. Rev. **57**, 1790, (1998).
10. A. Pich, Phys. Lett. **196**, 561, (1987).
11. ALEPH Collaboration, D. Buskulic *et al*, Z. Phys. **70**, 579, (1996).
12. CLEO Collaboration, T. Begfeld *et al*, Phys. Rev. Lett. **79**, 2406, (1997).
13. Particle Data Group, C. Caso *et al*, *Review of Particle Properties*, Eur. Phys. J. **C3**, 1 (1998).
14. R.J. Sobie, Z. Phys. **69**, 99, (1995).
15. S.I. Eidelman and V.N. Ivanchenko, Nucl. Phys. **55**, 181, (1997).

D^0 MIXING AND CABIBBO SUPPRESSED DECAYS

D.M. ASNER

Department of Physics, University of California,
Santa Barbara 93106-9530, USA

The CLEO collaboration reported observation of the 'wrong sign' decay $D^0 \to K^+\pi^-$ in 1993. Upgrades have been made to the CLEO detector [1], including installation of a silicon vertex detector [2], which provide substantial improvements in sensitivity to $D^0 \to K^+\pi^-$. The vertex detector enables the reconstruction of the proper lifetime [3] of the D^0, and so provides sensitivity to D^0–$\overline{D}^0$ mixing. We will give preliminary results on the rate of 'wrong sign' decay and D^0–$\overline{D}^0$ mixing using data from the 9.1 fb^{-1} of integrated luminosity that has been accumulated with the upgrades in place. In addition, we will give sensitivity estimates of on-shell D^0–$\overline{D}^0$ mixing derived from measurement of the lifetime measured with decays of the D^0 to CP eigenstates such as K^+K^-, $\pi^+\pi^-$, and $K_S\phi$.

1 Introduction

Ground state mesons such as the K^0, D^0, and B^0, which are electrically neutral and contain a quark and antiquark of different flavor, can evolve into their respective antiparticles, the $\overline{K}^0$, $\overline{D}^0$, and $\overline{B}^0$. The rate measurements of $K^0 - \overline{K}^0$ mixing and B^0–$\overline{B}^0$ mixing have guided both the elucidation of the structure of the Standard Model and the determination of the parameters that populate it. These mixing measurements permit crude, but accurate, estimates of the masses of the charm and top quark masses prior to direct observation of those quarks at the high energy frontier.

Within the framework of the Standard Model the evolution of a D^0 into a $\overline{D}^0$ is expected to be infrequent, for two reasons. First, the overall D^0 decay amplitude is not Cabibbo suppressed, in distinction to the K^0 and B^0 cases. In all cases the mixing amplitude is (at least) double Cabibbo suppressed; consequently, the magnitudes of x and y, which are the ratios of the mixing amplitude via virtual and real intermediate states, respectively, to the mean decay amplitude, are not expected to exceed $\tan^2 \theta_c \approx 0.05$ for D^0–$\overline{D}^0$ mixing[4].

$$ x = \frac{\Delta M}{\overline{\Gamma}} \qquad y = \frac{\Delta \Gamma}{2\overline{\Gamma}} $$

Three out of four of the analogous ratios for the K^0 and B^0 systems have been measured and are all close to unity. Second, the near degeneracy in mass of the

d and s quarks relative to the W boson causes the Glashow-Illiopolous-Maini (GIM) cancellation to be particularly effective [5]. This drives the relative D^0 amplitudes down by a rather uncertain additional factor of 10 to 10^3. It was the *absence* of perfect GIM cancellation that permitted the inference of crude values of m_c and m_t from the various measurements of K^0 and B^0 mixing, prior to the direct observation of the c and t quarks.

The observation of a value of $|x|$ in the $D^0 - \overline{D}^0$ system in excess of about 5×10^{-3} might be evidence of incomplete GIM-type cancellations among new families of particles, such as supersymmetric partners of quarks [6]. The evidence would be most compelling if either the mixing amplitude exhibited a large CP violation, or if the Standard Model contributions could be decisively determined. It is possible that in the Standard Model that $|y| > |x|$, [7] and a determination of y allows the estimation of at least some of the long-distance Standard Model contributions to x.

The Standard Model predicts that $D^0 - \overline{D}^0$ mixing is likely to proceed through real intermediate states and will cause the decays to CP+ final states to have the shorter lifetime. This situation would cause *constructive* interference between mixing and decay in the process $D^0 \to K^+\pi^-$.

The study of Cabibbo suppressed decays of the D^0 to pairs of pseudo-scalars provides two avenues into the study of $D^0 - \overline{D}^0$ mixing. First, for single Cabibbo suppressed decays, the final states $\pi^+\pi^-$ and K^+K^- (Fig. 1a and 1b) are common to both the D^0 and $\overline{D}^0$, and so these final states provide innate sensitivity to mixing. Because these final states are also CP eigenstates, on-shell mixing of the D^0 with the $\overline{D}^0$ can change the exponential lifetime of the D^0 as measured exclusively with $\pi^+\pi^-$ and K^+K^-. The shift in lifetime as measured with $CP = +1$ final states, such as $\pi^+\pi^-$ and K^+K^- (Fig. 1d and 1e), should be equal in magnitude and opposite in sign to the lifetime shift as measured with $CP = -1$ final states, such as $\rho^0 K_S$ and ϕK_S (Fig. 1c).

The $D^{*\pm}$ tag, used to identify the flavor of the decaying D^0 or $\overline{D}^0$, opens up the second avenue to mixing. The tag is essential to distinguish the nominally double-Cabibbo suppressed decay (DCSD), $D^0 \to K^+\pi^-$, from the Cabibbo-favored $\overline{D}^0 \to K^+\pi^-$. The time-integrated rate of $D^0 \to K^+\pi^-$ can then be used to limit the mixing process $D^0 \to \overline{D}^0 \to K^+\pi^-$. The proper time distribution for this decay has three components - DCSD $\propto e^{-t}$, on-shell mixing $\propto te^{-t}$ and off-shell mixing $\propto t^2 e^{-t}$. The contribution of DCSD is important to measure because the smaller the DCSD contribution is, the *greater* the sensitivity to mixing.

312

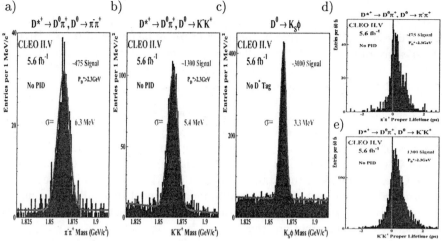

Figure 1: CP^+ Mass Distributions a) $D^0 \to \pi^+\pi^-$ b) $D^0 \to K^+K^-$ with $D^{*\pm}$ tag. CP^- Mass Distribution c) $D^0 \to \phi K_S$. D^0 decay time d) $D^0 \to \pi^+\pi^-$ e) $D^0 \to K^+K^-$.

2 Formalism

Wrong-sign hadronic decays occur via DCSD or mixing. In the limit of small mixing and no CP violation the decay time distribution depends on the rates, R_{DCSD} and R_{Mix}.

$$w(t) = (R_{\mathrm{DCSD}} + \sqrt{2R_{\mathrm{DCSD}}R_{\mathrm{Mix}}}(\cos\phi)\, t + \frac{1}{2}R_{\mathrm{Mix}}t^2)e^{-t} \qquad (1)$$

where, in terms of the other usual parameters,

$$R_{\mathrm{Mix}} = \frac{1}{2}(x^2 + y^2), \quad \phi = \tan^{-1}\left(-2\frac{\Delta M}{\Delta\Gamma}\right) + \delta_s = \tan^{-1}\left(-\frac{x}{y}\right) + \delta_s \qquad (2)$$

The strong phase between $D^0 \to K^+\pi^-$ and $\overline{D}^0 \to K^+\pi^-$ amplitudes, δ_s, is small by theoretical bias[9]. The time-integrated wrong-sign rate is,

$$R_{\mathrm{WS}} = R_{\mathrm{DCSD}} + \sqrt{2R_{\mathrm{DCSD}}R_{\mathrm{Mix}}}\cos\phi + R_{\mathrm{Mix}}, \qquad (3)$$

and the mean wrong-sign decay time is,

$$\langle t_{\mathrm{WS}}\rangle = \frac{R_{\mathrm{DCSD}} + 2\sqrt{2R_{\mathrm{Mix}}R_{\mathrm{DCSD}}}\cos\phi + 3R_{\mathrm{Mix}}}{R_{\mathrm{DCSD}} + \sqrt{2R_{\mathrm{Mix}}R_{\mathrm{DCSD}}}\cos\phi + R_{\mathrm{Mix}}} \qquad (4)$$

The behavior of $\langle t_{\mathrm{WS}}\rangle$ is shown as a function of $R_{\mathrm{Mix}}/(R_{\mathrm{DCSD}} + R_{\mathrm{Mix}})$ in Fig. 2a for the cases of $\cos\phi = \pm 1$ and $\cos\phi = 0$.

a)   b)

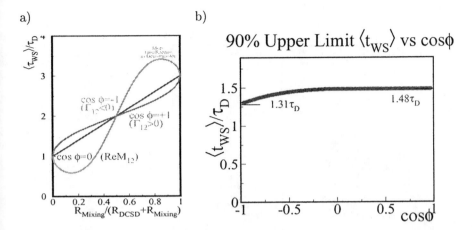

Figure 2: a) $\langle t_{WS} \rangle \tau_D$ vs $R_{Mix}/(R_{DCSD} + R_{Mix})$ b) $\langle t_{WS} \rangle \tau_D$ vs $\cos \phi$, 90% C.L. Upper Limit.

3   Wrong-Sign rate R_{ws} and Mean Decay time $\langle t_{ws} \rangle$

A binned maximum likelihood fit of the MC-generated background components to the two dimensional data on the $M_{K\pi\pi} - M_{K\pi}$ vs. $M_{K\pi}$ plane[10] determines R_{ws}.

$$R_{WS} = \frac{\Gamma(D^0 \to K^+\pi^-)}{\Gamma(D^0 \to K^-\pi^+)} = 0.0031 \pm .0009(stat) \pm .0007(syst) \qquad (5)$$

The fit also yields a breakdown of the background event content in Fig. 3a and 3b. The mean Wrong-sign decay time can be determined from Fig. 3c using the mean decay time for $\overline{D}^0$ and uds backgrounds of $\tau = 1$ and $\tau = 0$, respectively, combined with the background composition, we evaluate:

$$\langle t_{WS} \rangle = (0.65 \pm 0.40) \quad (\times \tau_{D^0}). \qquad (6)$$

Proper renormalization to the physical regions of t_{ws} (Fig. 2a) is required. The 90% C.L. Upper Limit on $\langle t_{WS} \rangle \tau_D$ vs $\cos \phi$ is shown in Fig. 2b. We obtain limits in the two dimensional space of R_{Mix} vs. R_{DCSD} from the rate of Wrong Sign decay, and the mean $\langle t_{WS} \rangle$.

4   Previous $D^0 - \overline{D}^0$ Mixing Limits

Three groups have reported non-zero measurements of R_{WS} all with analysis evaluated for the case $\cos \phi = 0$, and with neglect of CP violation:

314

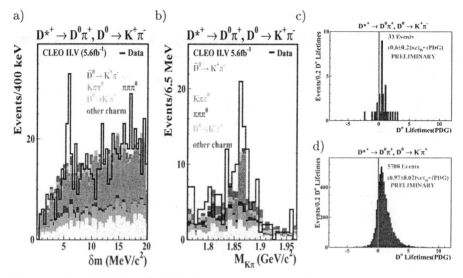

Figure 3: Binned likelihood fit of a) $M_{K\pi\pi} - M_{K\pi} - M_\pi$ vs b) $M_{K\pi}$ determines composition of background. t_{ws} for c) $D^0 \to K^+\pi^-$ d) $D^0 \to K^-\pi^+$.

- CLEO-II[11], equivalent to $R_{\mathrm{WS}} = R_{\mathrm{DCSD}} + R_{\mathrm{Mix}} = (0.77 \pm 0.35)\%$.

- E791 [12], where $R_{\mathrm{DCSD}} = (0.68 \pm 0.35)\%$, and $R_{\mathrm{Mix}} = (0.21 \pm 0.09)\%$, where, for R_{Mix}, $D^0 \to K^+\pi^-\pi^+\pi^-$ contribute in addition to $D^0 \to K^+\pi^-$; no report of a non-zero R_{Mix} was made.

- Aleph [13], where $R_{\mathrm{DCSD}} = (1.84 \pm 0.68)\%$, and an upper limit of $R_{\mathrm{Mix}} < 0.92\%$ is obtained, at 95% C.L.

Additionally, there are two other relevant limits on R_{WS}. The E691 collaboration [14] limited $R_{\mathrm{Mix}} < 0.37\%$, at 90% C.L., where again $D^0 \to K^+\pi^-\pi^+\pi^-$ contribute in addition to $D^0 \to K^+\pi^-$, and $R_{\mathrm{DCSD}} < 1.5\%$ at 90% C.L. The E791 [15] collaboration sought $D^0 \to K^+\ell^-\overline{\nu}_\ell$, and limited $R_{\mathrm{Mix}} < 0.5\%$. The regions allowed by the above work, in the R_{Mix} vs. R_{DSCD} plane, for $\cos\phi = 0$, are shown in Fig. 4a.

5   CLEO-II.V Charm Mixing Limits

The limits on $D^0 - \overline{D}^0$ determined from $D^0 \to K^+\pi^-$ with $5.6 fb^{-1}$ of CLEO-II.V data are shown in Fig. 4b-c and in column 2 of Table 1. Combining with $D^0 \to CP^+$ analysis from E791[16] improves limits on R_{Mix} and y (Table 1, column 3). The CLEO-II.V sensitivity ($9.1 fb^{-1}$) combining $D^0 \to$

a)   b)   c)

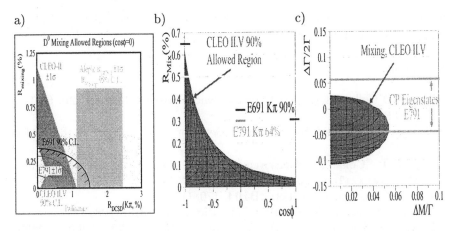

Figure 4: a) World mixing limits ($\cos \phi = 0$). CLEO-II.V 90% Mixing limits obtained from $D^0 \to K^+\pi^-$ b) R_{Mix} vs $\cos \phi$. c) x vs y.

$K^+\pi^-$, $K^+\pi^-\pi^0$, $K^+\pi^-\pi^+\pi^-$ and $D^0 \to CP$ analyses is listed in column 4. A factor of 2-5 (3-10) improvement in precision is obtained over the PDG[17] with $5.6fb^{-1}$ ($9.1fb^{-1}$). It is noteworthy that the CLEO II.V limit for x $\sim \tan^2 \theta_{\text{Cabibbo}}$, is more or less the largest level that $D^0 - \overline{D}^0$ mixing can be in the Standard Model.

Table 1: Current limit on $D^0 - \overline{D}^0$ Mixing Limits and projected CLEO-II.V sensitivity.

	CLEO-II.V ($5.6fb^{-1}$)	CLEO-II.V +E791	CLEO-II.V (Complete)	RPP98
x	$|x| < .054$	$|x| < .054$	$|x| < .03$	$|x| < .096$
y	$-.108 < y < .027$	$-.042 < y < .027$	$|y| < .01$	$|y| < .10$
R_{ws}	$.31 \pm .09\%$	$.31 \pm .09\%$	$\pm.05\%$	$.72 \pm .25\%$
R_{Mix}	$< 1.1\%$	$< 0.25\%$	$< 0.05\%$	$< 0.5\%$

Acknowledgments

We gratefully acknowledge the effort of the CESR staff in providing us with excellent luminosity and running conditions. This work was supported by the National Science Foundation, the U.S. Department of Energy, Research Corporation, the Natural Sciences and Engineering Research Council of Canada, the A.P. Sloan Foundation, the Swiss National Science Foundation, and the Alexander von Humboldt Stiftung.

References

1. CLEO Collab., CLEO-II Detector, Nucl. Instru. Meth. **A320**, 66-113, (1992).
2. T.S. Hill, CLEO-II Silicon Vertex Detector, Nucl. Instru. Meth. **A418**, 32-39, (1998).
3. CLEO Collab., Measurement of Charm Meson Lifetimes, http://xxx.lanl.gov/abs/hep-ex/9902011.
4. L. Wolfenstein, Phys. Lett. **B164**, 170, (1985).
5. A. Datta and D. Kumbhakar, Zeit. Phys. **C27**, 515, (1985).
6. A.G. Cohen *et al.* Phys. Rev. Lett. **78**, 2300, (1997), http://xxx.lanl.gov/abs/hep-ph/9610252.
7. E. Golowich and A. Petrov, Phys. Lett. **B427**, 172, (1998), http://xxx.lanl.gov/abs/hep-ph/9802291.
8. T. Liu, An Overview of $D^0 - \overline{D}^0$ Mixing Search Techniques, http://xxx.lanl.gov/abs/hep-ph/9508415.
9. T.E. Browder, S. Pakvasa, Phys. Lett. **B383**, 475-481, (1996), http://xxx.lanl.gov/abs/hep-ph/9508362.
10. J. Gronberg, D^0 Mixing at CLEO-II, http://xxx.lanl.gov/abs/hep-ph/9903368.
11. CLEO Collab., Observation of $D^0 \to K^+\pi^-$, Phys. Rev. Lett. **72**, (1994).
12. E791 Collab., Search for $D^0-\overline{D}^0$ Mixing and Doubly-Cabibbo-suppressed Decays of the D^0 in Hadronic Final States, Phys. Rev. **D57**, 133, (1998).
13. Aleph Collab., Study of $D^0-\overline{D}^0$ Mixing and D^0 Doubly Cabibbo Suppressed Decays, Phys. Lett. **B436**, 211-221, (1998).
14. E691 Collab., A Study of $D^0-\overline{D}^0$ Mixing, Phys. Rev. Lett. **60** (1988).
15. E791 Collab., A Search for $D^0-\overline{D}^0$ Mixing in Semileptonic Decay Modes, Phys. Rev. Lett. **77**, 2384-2387, (1996).
16. E791 Collab., Measurement of Lifetimes and a Limit on the Lifetime Difference in the Neutral D-Meson System, http://xxx.lanl.gov/abs/hep-ph/9903012.
17. C. Caso *et al.*, European Physical Journal **C3**, 1, (1998).

FIRST MEASUREMENT OF $\Xi^0 \to \Sigma^+ e^- \bar{\nu}$ FORM FACTORS

STEVE BRIGHT

for the KTeV Collaboration

The Enrico Fermi Institute and the Department of Physics,
The University of Chicago, Chicago, Illinois 60637

The KTeV experiment at the Fermi National Accelerator Laboratory has identified over 900 $\Xi^0 \to \Sigma^+ e^- \bar{\nu}$ events during the 1997 fixed target run. Using about 70% of our data, we present the first measurement of S_e, the polarization of the Σ^+ along the direction of the electron in the Σ^+ frame. By reconstructing the directions of the electron and the proton in the Σ^+ frame, we measure S_e to be $1.020 \pm 0.037_{(stat)} \pm 0.043_{(syst)}$. If we assume that the weak magnetism term f_2 has its predicted CVC value of 2.56, and that the second class current term g_2 is equal to zero, we can combine this result with the measured preliminary branching ratio of $\Gamma(\Xi^0 \to \Sigma^+ e^- \bar{\nu})/\Gamma(\Xi^0 \to All) = 2.54 \pm 0.11_{(stat)} \pm 0.16_{(syst)}$ to compare the measured f_1 and g_1 values with various theoretical predictions.

1 $SU(3)_f$ Symmetry

The most general transition amplitude for the semileptonic decay of spin 1/2 baryons $(B \to b\,e^- \bar{\nu})$ is:

$$\mathcal{M} = G_S \frac{\sqrt{2}}{2} \overline{u}_b (O_\alpha^V + O_\alpha^A) u_B \overline{u_e} \gamma^\alpha (1 + \gamma_5) v_\nu + H.c., \qquad (1)$$

where

$$O_\alpha^V = f_1 \gamma_\alpha + \frac{f_2}{M_B} \sigma_{\alpha\beta} q^\beta + \frac{f_3}{M_B} q_\alpha,$$

$$O_\alpha^A = (g_1 \gamma_\alpha + \frac{g_2}{M_B} \sigma_{\alpha\beta} q^\beta + \frac{g_3}{M_B} q_\alpha) \gamma_5,$$

$$q^\alpha = (p_e + p_\nu)^\alpha = (p_B - p_b)^\alpha, \quad \text{and} \qquad (2)$$

$$G_S = \begin{cases} G_F V_{us} \; for \mid \Delta S \mid = 1 \\ G_F V_{ud} \; for \;\; \Delta S = 0. \end{cases}$$

For the fundamental $SU(3)_f$ octet, in the limit of exact $SU(3)_f$ symmetry, any one of the form factors is given by:

$$f_i = C(B,b)_F * F_{f_i} + C(B,b)_D * D_{f_i}$$

$$g_i = C(B,b)_F * F_{g_i} + C(B,b)_D * D_{g_i} \qquad (3)$$

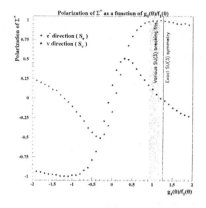
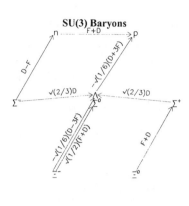

Figure 1: The polarization of the Σ^+ in the electron and neutrino directions (left) and a diagram of the octet baryons, with the F and D coefficients for the observed hyperon beta decays (right).

where the $C(B, b)$ are $SU(3)$ Clebsch-Gordan coefficients and the F and D's are symmetric and anti-symmetric coupling constants to be determined from experiment.

Thus, in this limit, the decay $\Xi^0 \to \Sigma^+ e^- \bar{\nu}$ should have the same form factors (f_i and g_i) as $n \to p e^- \bar{\nu}$. Deviations from this exact symmetry should arise from the mass and charge differences between the quarks. Details of $SU(3)_f$ breaking can be studied through the experimental determination of the form factors.

2   The Detector

An 800 GeV/c proton beam, with up to 5×10^{12} protons per 19 s Tevatron spill every minute, was targeted at a vertical angle of 4.8 mrad on a 1.1 interaction length (30 cm) BeO target. Photons were converted by 7.6 cm of lead 18 m downstream of the target. Charged particles were removed by a series dipole magnets located between 2 m and 90 m downstream of the target. Collimators defined two 0.25 μsr neutral beams that entered the KTeV apparatus

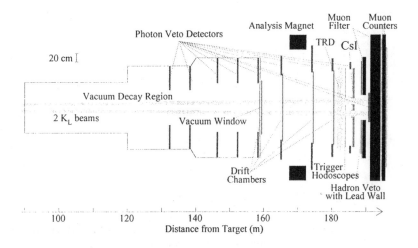

Figure 2: The KTeV Spectrometer as configured for E799 running.

(Fig. 2) 94 m downstream from the target. The 65 m vacuum ($\sim 10^{-6}$ Torr) decay region extended to the first drift chamber. The charged particle spectrometer consisted of a dipole magnet surrounded by four (1.28×1.28 m^2 to 1.77×1.77 m^2) drift chambers with ~ 100 μm position resolution in both horizontal and vertical views. To reduce multiple scattering, helium filled bags occupied the spaces between the drift chambers. The magnetic field imparted a ± 205 MeV/c horizontal momentum component to charged particles, yielding a momentum resolution of $\sigma(P)/P = 0.38\% \oplus 0.016\%$ P (GeV/c). The magnet polarity was reversed on a daily basis.

The electromagnetic calorimeter (CsI) consisted of 3100 pure CsI crystals. Each crystal was 50 cm long (27 radiation lengths, 1.4 interaction lengths). Crystals in the central region (1.2×1.2 m^2) had a cross-sectional area of 2.5×2.5 cm^2 while those in the outer region ($1.2 - 1.9 \times 1.2 - 1.9$ m^2) had a 5×5 cm^2 area. After calibration, the CsI energy resolution was 0.85% for the electron momentum spectrum in this analysis. The position resolution was ~ 1 mm.

Nine photon veto assemblies were used to detect particles leaving the fiducial volume. Two scintillator hodoscopes in front of the CsI were used to trigger

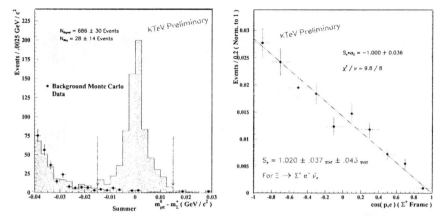

Figure 3: Reconstructed $\Sigma^+ \to p + \pi^0$ from Ξ^0 beta decays, (left) and acceptance corrected proton-electron correlation (right).

on charged particles. Another scintillator plane (hadron-anti), located behind both the CsI and a 10 cm lead wall, was used as a hadron shower veto. The hodoscopes and the CsI detectors had two holes (15×15 cm^2 at the CsI) and the hadron-anti had a single 64×34 cm^2 hole to let the neutral beams pass through without interaction. Charged particles passing through these holes were detected by 16×16 cm^2 scintillators (hole counters) located along each beam line in the hole region just downstream of the hadron-anti. There were also eight (2.1×2.1 m^2) MWPC transition radiation detectors (TRD) used to obtain further π / e rejection. Further description of the apparatus and the discovery of $\Xi^0 \to \Sigma^+ e^- \bar{\nu}$ at KTeV can be found in Affolder et al.[3]

3   Final State Polarization

From the transition amplitude in Eq. 1, we can determine the values of various observable quantities as functions of the form factors. The form factors f_3 and g_3 will always have contributions proportional to the electron mass, and may therefore be neglected. Values for S_e and the total rate for the process for different values of the form factors can be calculated from Bright et al.[1] and Winston and Watson [2].

Here we present expressions for the total rate of the decay and S_e, correct to order $\delta = (M_{\Xi^0} - M_{\Sigma^+})/M_{\Xi^0}$

$$R = R_0[(1 - \frac{3}{2}\delta)f_1^2 + (3 - \frac{9}{2}\delta)g_1^2 - (4\delta)g_1g_2] \qquad (4)$$

$$RS_e = R_0[(2 - \frac{10}{3}\delta)g_1^2 + (2 - \frac{7}{3}\delta)f_1g_1 - (\frac{1}{3}\delta)f_1^2] \qquad (5)$$

$$-(\frac{2}{3}\delta)f_1f_2 + (\frac{2}{3}\delta)f_2g_1 - (\frac{2}{3}\delta)f_1g_2 - (\frac{10}{3}\delta)g_1g_2]$$

$$R_0 = \frac{G_S^2(\delta M_{\Xi^0})^5}{60\pi^3}. \tag{6}$$

The polarization of the Σ^+ is observed via its two body decay $\Sigma^+ \to p\pi^0$ which has a large asymmetry $\alpha_\Sigma = -0.98$. The distribution of the cosine of the angle between the proton and the e^- in the Σ^+ frame is equal to

$$\frac{1}{\Gamma}\frac{d\Gamma}{d\Omega_p} = \frac{1}{4\pi}(1 + S_e\alpha_\Sigma\hat{p}\cdot\hat{e}). \tag{7}$$

4   The Data

The decay $\Xi^0 \to \Sigma^+ e^- \bar{\nu}$ is identified by: a positively charged high momentum track which travels down the beam hole, a negatively charged track which hits the calorimeter and deposits energy within 10% of the momentum, and two extra clusters in the calorimeter in time with the charged tracks. The Σ^+ is found from the point along the high momentum positive track where the invariant mass of the two photons is the π^0 mass. The Σ^+ momentum is then calculated and traced back to where it comes closest to the negative track. That point is identified as the Ξ^0 vertex.

Further kinematic cuts are applied to reduce the background from $\Xi^0 \to \Lambda\pi^0$ (with both $\Lambda \to p\pi^-$ and $\Lambda \to pe^-\bar{\nu}$), $K_L \to \pi^0\pi^+e^-\bar{\nu}$, $K_L \to \pi^+e^-\bar{\nu}\gamma$ with an accidental γ , $K_L \to \pi^+e^-\bar{\nu}$ with an accidental π^0 and $K_L \to \pi^+\pi^-\pi^0$. Also, the TRDs are used to obtain an additional factor of 10 π^- rejection.

After all event selection criteria are applied, $\Xi^0 \to \Sigma^+ e^- \bar{\nu}$ are identified by events with a proton-π^0 invariant mass within 15 MeV of the known Σ^+ mass ($1.18937\ GeV/c^2$). We find 714 candidate events with an estimated background of 28 ± 14 events (Fig. 3). The background is almost entirely $K_L \to \pi^+e^-\bar{\nu}\gamma$ with an accidental γ and $K_L \to \pi^+e^-\bar{\nu}$ with an accidental π^0 . The uncertainty in the number and measured proton-electron correlation of the background is the largest single contribution to the systematic error on S_e (0.03).

We subtract the predicted background events from the data and correct the proton-electron correlation distribution for acceptance. We then fit the p-e distribution to a straight line as seen in Fig. 3. We must divide by the $\Sigma^+ \to p\pi^0$ asymmetry of $-.98$ to obtain

$$S_e = 1.020 \pm 0.037_{(stat)} \pm 0.043_{(syst)}.$$

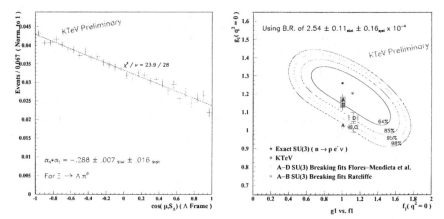

Figure 4: Acceptance corrected proton-lambda spin correlation for $\Xi^0 \to \Lambda \pi^0$ with $\Lambda \to p\pi^-$ (left) and confidence interval plot for f_1 and g_1 for $\Xi^0 \to \Sigma^+ e^- \bar{\nu}$ using measured S_e and branching ratio (right).

5 The Decay $\Xi^0 \to \Lambda \pi^0$ with $\Lambda \to p\pi^-$

In order to check the geometrical acceptance correction, a similar analysis was performed in a large sample of $\Xi^0 \to \Lambda \pi^0$ with $\Lambda \to p\pi^-$ decays.

The polarization of the Λ is equal to

$$\vec{P_\Lambda} = \alpha_{\Xi^0} \hat{\Lambda}. \tag{8}$$

For this decay, the distribution of the cosine of the angle between the proton and the polarization direction of the Λ is equal to:

$$\frac{1}{\Gamma} \frac{d\Gamma}{d\Omega_p} = \frac{1}{4\pi}(1 + \alpha_{\Xi^0}\alpha_\Lambda \hat{p} \cdot \vec{P_\Lambda}). \tag{9}$$

Previous measurements of this quantity give the value

$$\alpha_{\Xi^0}\alpha_\Lambda = -0.264 \pm .013 (PDG\,Value).$$

From a sample of $78,000$ such events, we measure:

$$\alpha_{\Xi^0}\alpha_\Lambda = -0.288 \pm .007(stat) \pm .016(syst).$$

6 Conclusions

The branching ratio for this decay was measured to be [4]:

$$\Gamma(\Xi^0 \to \Sigma^+ e^- \bar{\nu})/\Gamma(\Xi^0 \to All) = 2.54 \pm 0.11_{(stat)} \pm 0.16_{(syst)}$$

We can combine the results for the branching ratio and S_e to obtain f_1 and g_1. A contour plot of the confidence intervals for f_1 and g_1 (statistical and systematic errors were added in quadrature) is shown in Fig. 4. We see that our result ($f_1 = 1.10 \pm 0.35, g_1 = 1.20 \pm 0.13$) favors the $SU(3)_f$ prediction [5] and predictions in Ratcliffe[6] over those given by Flores-Mendieta et al.[7]. Analysis is underway to determine the final state polarization along the neutrino direction as well as the electron spectrum and electron neutrino correlation for the existing data. Also, we plan to collect 4 times more data in the upcoming 1999 fixed target run at Fermilab.

References

1. S. Bright et al., in preparation.
2. J.M. Watson and R. Winston, Phys. Rev. **181**, 1907, (1969). In order to make their phase space correct to second order, one must replace M_b/M_B with $(E_b + M_b)/2M_B$.
3. A. Affolder et al., Phys. Rev. Lett. **82**, 3751, (1999).
4. A. Alavi, (Los Alamos preprint hep-ex/9903031) .
5. Particle Data Group, C. Caso et al., Eur. Phys. J. **C3**, 1, (1998).
6. P.G. Ratcliffe, Phys. Rev. **D59**, 014038, (1999).
7. R. Flores-Mendieta et al., Phys. Rev. **D58**, 094028, (1998).

MASS MEASUREMENTS ON RADIOACTIVE ISOTOPES USING THE ISOLTRAP SPECTROMETER

J. DILLING, F. HERFURTH, H.-J. KLUGE, A. KOHL, E. LAMOUR, G.H. MARX, S.C. SCHWARZ

GSI, Planckstr.1, D-64291 Darmstadt, Germany

G. BOLLEN

CERN, CH-1211 Geneva 23, Switzerland

A.G. KELLERBAUER and R.B. MOORE

McGill University, 3600 University St. Montreal, QC, H3A 2T8, Canada

S. HENRY

CSNSM-IN2P3-CNRS, F-91405 Orsay-Campus, France

ISOLTRAP is a Penning trap mass spectrometer installed at the on-line isotope separator ISOLDE at CERN. Direct measurements of the masses of short-lived radio isotopes are performed using the existing triple-trap system. This consists of three electromagnetic traps in tandem: a Paul trap to accumulate and bunch the 60 keV dc beam, a Penning trap for cooling and isobar separation, and a precision Penning trap for the determination of the masses by cyclotron resonance. Measurements of masses of unknown mercury isotopes and in the vicinity of doubly magic ^{208}Pb are presented, all with an accuracy of $\delta m/m \approx 1 \times 10^{-7}$. Developments to replace the Paul trap by a radiofrequency quadrupole ion guide system to increase the collection efficiency are presently under way and the status is presented.

1   Introduction

The binding energy of the atomic nucleus is one of the fundamental properties of such a many-body system. Accurate mass data serve as a testing ground for nuclear models and stimulate their further improvement. Furthermore, systematic investigations of the binding energy as a function of proton and neutron number allows the direct observation of nuclear properties like pairing, shell and sub-shell closure, as well as deformation effects, and lead to a deeper understanding of nuclear structure. In addition, very precise mass differences are for example required in the context of weak interaction studies in nuclear beta decay. Therefore, large efforts are devoted to the application of classical as well as new spectrometric techniques, such as time-of-flight, Smith-RF or Schottky mass spectrometry, for the accurate mass determination of short-lived isotopes far from the valley of beta stability [1], [2].

Penning traps have been proven to be very accurate mass spectrometers [3]. A large variety of mass measurements with highest accuracy have been performed on stable, mostly light particles. ISOLTRAP is so far the only spectrometer operational for the investigation of short-lived radioactive isotopes. More than 100 mass values have been investigated with a typical accuracy of better than 20 keV. The more recent measurements have been carried out on rare earth isotopes in the vicinity of ^{164}Gd, on neutron-deficient mercury isotopes, and on isotopes with Z=82-85.

2   Principle and Experimental Setup

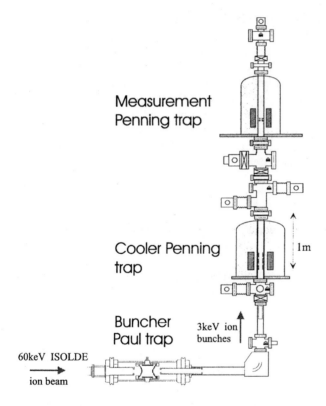

Figure 1: Experimental set-up of the ISOLTRAP spectrometer at ISOLDE/CERN.

The basic principle of mass measurements with ISOLTRAP is the determination of the cyclotron frequency $\omega_c = q/mB$ of ions with charge-to-mass ratio q/m stored in a Penning trap with known magnetic field B. Fig. 1 shows the present layout of the ISOLTRAP spectrometer[4]. The first section of the spec-

Table 1: Table of isotopes investigated with ISOLTRAP, which are discussed in the following.

Element	Mass Number
Hg	184,185g+m, 186 - 190, 191m, 192,193g+m,194-196,197g
Pb	196,198,208 (reference isotope)
Bi	197
Po	198
At	203

trometer has the task to stop the 60 keV ISOLDE beam and to prepare it for an efficient transfer into the cooler trap. In the past, a stopping/reionization technique was applied which limited the applicability of ISOLTRAP to surface ionizable elements. Recently, the system was considerably improved by the installation of an RFQ trap ion beam buncher [5], which allows to capture the continuous ISOLDE beam in flight. The lower Penning trap [4] has the task to accumulate, cool, and mass separate the ions delivered from the ion preparation section and to bunch them for an efficient delivery to a second Penning trap. This precision trap is the actual mass spectrometer where the cyclotron frequency of the captured ions is determined.

3   Mass Measurements

For the first time, mass measurements are carried out using the full capability of the ISOLTRAP spectrometer, including capture and bunching of the ISOLDE ion beam in flight and separation of isobars and isomers. Table 1 shows the isotopes investigated with the ISOLTRAP spectrometer which are discussed below.

3.1   Neutron-deficient mercury isotopes

The interest for nuclear structure investigations and mass measurements in the region of neutron-deficient mercury isotopes arises from the appearance of shape coexistence at low excitation energies in the region around the shell closure at Z=82. The onset of rotational bands built on low-lying 0+ states was found [9] in even-even Pt, Hg, Pb and Po isotopes, mid-shell between N = 82 and N = 126. A large staggering in the $\delta < r^2 >$ values determined from isotopic shift measurements was observed for A $\leq$ 185 for the ground-states of the light Hg isotopes, a jump from small to strong deformation in the neighboring Au isotopes at A $\leq$ 186 and a smooth transition in the Pt isotopes[6,7,8]. However, until recently no mass values were known in this mass region. Today, precise information is still lacking for A ≤ 185, where structural changes are very pronounced.

The neutron-deficient isotopes of elements around Z = 82 are all members of long α-decay chains with well-known Q-values. Therefore, an accurate determination the masses of such isotopes allows to fix these chains, making a large impact on a whole mass area starting at the upper part of the rare earth region and reaching to the border of known proton-rich isotopes.

With ISOLTRAP, a first series of mass measurements on the neutron-deficient mercury isotopes $^{185-197}$Hg was carried out after the installation of the RFQ trap ion beam buncher. In the case of the even isotopes, where no isomeric states exist, the evaluation was straightforward and an accuracy of $\delta m \approx 20$ keV can be assigned to all mass values. However, in the case of the odd isotopes long-lived isomers exist and are produced at ISOLDE. The excitation energies of typically 100 - 150 keV of these isotopes are very low. In those first measurements, the spectrometer was operated with a resolving power of R ≈ 500.000, which corresponds to a mass resolution of 300 keV in this mass range. Therefore it was not possible to resolve isomeric and ground state.

In a second run, the attempt was made to verify the production of the isomers and to resolve them and the corresponding ground states. For this purpose the spectrometer was operated with resolving powers up to R = 5 million corresponding to a mass resolution of $\delta m \approx 30$ keV. This way it was possible to resolve isomeric and ground state in the cases of ^{185}Hg (see Fig. 2) and ^{193}Hg. Furthermore, it was verified that ^{197}Hg is dominantly produced in the ground state, while for ^{195}Hg only the isomeric state has been seen, for which the excitation energy is known. For all these isotopes the ground state masses are now identified and determined with an accuracy of 20 keV. In addition, during this run it was possible to extend the measurements in the mercury chain to ^{184}Hg.

3.2 Isotopes with Z= 82 - 85

Using the Paul trap ion beam buncher, the investigation of a new region of isotopes with Z ≥ 82 was started very recently. In a first experiment isotopes were selected which are members of long alpha decay chains, those either not linked to an isotope with known mass or to one with a large mass uncertainty. Fig. 3 shows the decay chains and the isotopes investigated by ISOLTRAP. Due to the high accuracy of the ISOLTRAP data together with the known Q_α-values accurate information on nuclear binding energies is now available even for very heavy proton-rich isotopes like ^{210}Th,^{213}Pa,^{218}U, situated at the

328

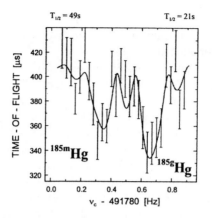

Figure 2: Time-of-flight spectrum of ^{185}Hg. Ground state and isomer are resolved. Excitation energy is $\Delta E = 118 \pm 5\,keV$.

borderline of known nuclei.

4 New Developments: Linear RFQ Cooler and Buncher

The success of the Paul trap system to capture the 60 keV ion beam in flight led to the development of a radiofrequency quadrupole rod structure [10]. Fig. 4 shows the principle of the new system. The RFQ 4 rod structure creates a quadrupole field for radial confinement of the ions.

Segmentation of the single rods allows to apply a longitudinal DC potential. Buffer gas cools the ions and they can be finally trapped in the end section of the structure. Static potentials are applied here to form a linear Paul trap where the ions are accumulated. They are then extracted as a bunch by switching the axial confining potential at the end segments. The complete system is operated on a 60kV platform to allow for electrostatic retardation of the ISOLDE beam. The Helium pressure in the structure is about $2 \times 10^{-2}\,mbar$ (center) to $4 \times 10^{-3}\,mbar$ (exit-region). Simulations of a typical ISOLDE ion beam show an acceptance of about 15% using Monte Carlo calculations including interactions with the buffer gas. First tests with the separator beam are performed showing an acceptance of the 60 keV beam into the deceleration electrodes and the structure of about 13%. The testing and optimization of the operation parameters with a 60 keV off-line source are under way.

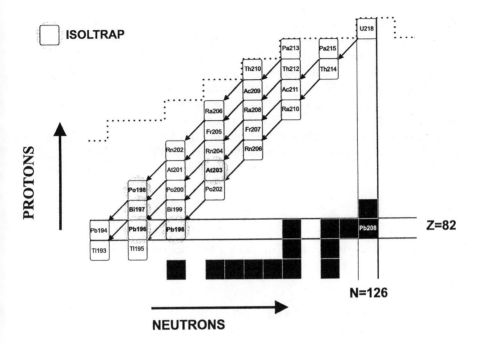

Figure 3: Q_α-value decay chains with isotopes investigated by ISOLTRAP (shaded). The dashed line indicates the borderline of known nuclei.

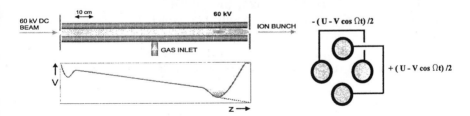

Figure 4: Basic principle of the linear RFQ structure. Left: side view of the system also shown is the potential that can be applied because of the segmentation. Right: the four rods generate a quadrupole field.

330

5 Conclusions

Penning trap mass spectrometry offers the possibility to investigate masses of radio nuclides far from the valley of stability. Here we presented experimental results from the ISOLTRAP facility. Masses of the mercury chain and in the region of the doubly magic lead were measured for the first time with a mass resolution of $\delta m \approx 30\,keV$. These measurements give important information on nuclear binding energies and two-neutron-separation energies. Hints were found for shape coexistence. With the new linear RFQ buncher and cooler it looks feasible to increase the efficiency of ISOLTRAP that will allow an extension of this studies to regions even further from stability and the investigation of new masses. The completion of this apparatus is planed for the summer of 1999.

Acknowledgments

This work is supported by the European TMR Network EXOTRAPS.

References

1. G. Bollen, Nucl. Phys. **A626**, 197c, (1997).
2. W. Wittig, *et al.*, Annul. Rev. Nucl. Sci. **47**, 22, (1997).
3. Proc. of The Intl. Conf. On Trapped Charged Particles and Fundamental Physics, Asilomar, California, USA (1998), AIP Conference Proceedings 457.
4. H. Raimbaut-Hartmann, *et al.*, Nucl. Instr. Meth. **B126**, 374, (1997).
5. S.C. Schwarz, PhD thesis University of Mainz (1998) and to be published.
6. J.L. Ulm, *et al.*, Z. Phys. **A325**, 409, (1986).
7. G. Passler, *et al.*, Nuc. Phys. **A580**, 173, (1994).
8. T. Hilberath, *et al.*, Z. Phys. **A342**, 1, (1992).
9. J.L. Wood, *et al.*, Phys. Rep **215**, 101-201, (1992).
10. T. Kim, PhD thesis McGill University, Canada (1998).

W^+W^- CROSS SECTION AND W MASS MEASUREMENT WITH THE DELPHI DETECTOR AT LEP

A. DUPERRIN

Université Claude Bernard de Lyon, IPNL, IN2P3-CNRS,
F-69622 Villeurbanne Cedex, France
E-mail: duperrin@in2p3.fr

This document reviews the methods and the results for the W^+W^- cross section and W mass measurements using the data collected up to the year 1999 with the DELPHI detector.

1 Introduction

The LEP2 programme started in the second half of 1996. The energy was first set at 161 GeV, the most favorable energy for the measurement of m_W from the cross-section for $e^+e^- \to W^+W^-$ at threshold. The energy was then gradually brought up to 172, 183, and 189 GeV leading to an integrated luminosity of about 230 pb^{-1} in total for DELPHI which is one of the four detectors at LEP. The luminosity measurement is monitored by the Small-Angle Bhabha Scattering $e^+e^- \to e^+e^-$, a process with large statistics and small theoretical uncertainties. The energy will be increased up to a maximum of about 200 GeV to be reached in mid '99, and LEP2 has so far been approved to run until the end of 2000, before the shutdown for the installation of the LHC. The main goals of LEP2 are the search for the Higgs and for new particles, the measurement of m_W, and the investigation of the triple gauge vertices WWZ and $WW\gamma$. A complete updated survey of the LEP2 physics is collected in ref.[1].

The possibility of performing precision tests is based on the formulation of the Standard Model as a renormalizable quantum field theory preserving its predictive power beyond tree level calculations. When the experimental accuracy is sensitive to the loop effects, the Higgs sector of the Standard Model can be probed. This document reports on the W^+W^- cross section and W mass measurement which have been performed with the DELPHI detector up to the year 1999.

2 W properties at LEP2

In the Born approximation three doubly resonant Feynman diagrams contribute to the $e^+e^- \to W^+W^-$ process, as shown in Fig. 1. These mini-

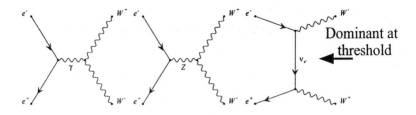

Figure 1: Lowest order Feynman diagrams for $e^+e^- \to W^+W^-$.

mal tree diagrams involve s-channel γ and Z exchange and t-channel ν exchange, the third one being dominant at threshold. Depending on the decay mode of each W, fully hadronic $WW \to q\bar{q}q\bar{q}$, mixed hadronic-leptonic ("semi-leptonic") $WW \to q\bar{q}l\nu$ or fully leptonic final states $WW \to l\nu l\nu$ are obtained, and the Standard Model branching fractions are 45.9%, 43.7% and 10.4%, respectively. About 3500 W^+W^- pairs were produced in each experiment at LEP2 so far. In addition to their production via the minimal diagrams, the four-fermion final states corresponding to these decay modes may be produced via other diagrams involving either zero, one, or two massive vector bosons. The interference between the three previous diagrams and the additional diagrams are generally expected to be negligible at current energies, except for final states with electrons or positrons. For the W mass reconstruction, the event generator EXCALIBUR [2] was used for the simulation of all four-fermion final states while the minimal Born processes (including radiatives corrections) were generated with the PYTHIA [3] event generator for the cross-section measurement.

This in turn implies that, besides the processes concerning the production and decay of a W-boson pair, one has to consider also those processes that lead to the same final state, but via different intermediate states, called background processes. The main background processes (ZZ, $q\bar{q}(\gamma)$, Zee, Weν ...) will be described below in each section dedicated to the three possible WW final states. All final states can be included in the cross section measurement, while for the mass measurement, only the fully hadronic channel and the three semi-leptonic final states involving an e, a μ or a τ provide enough information. In the fully leptonic final state, the only mass information is given by the charged lepton energy spectrum. As the distribution is considerably smeared, this technique gives a marginal improvement on the mass measurement and has not been implemented.

3   Event selection and cross-sections

The selection of WW events uses the criteria described in detail in ref. [4,5] for the cross-section measurements at $\sqrt{s} = 161$ and 172 GeV. In the fully-hadronic final state, however, a neural network analysis has been used for 189 GeV data. These criteria are briefly reviewed in the following sections. The selection for the W mass measurement is similar with slight differences.

The cross-sections are determined for the minimal diagrams. Corrections which account for the interference between the minimal diagrams and the additional diagrams were found to be consistent with unity within the statistical precision of ±2%.

3.1   Fully hadronic final state

The event selection criteria were optimized with sequential cuts at $\sqrt{s} = 161$ GeV, 172 GeV, and 183 GeV in order to ensure that the final state was purely hadronic and in order to reduce the background dominated by electron-positron annihilation into $q\bar{q}(\gamma)$ and ZZ, with a small contamination from $WW \rightarrow q\bar{q}l\nu$. These background processes are generated with the PYTHIA [3] event generator. At 189 GeV, a feed forward neural network is used to improve the efficiency × purity of the selection quality by about 5 %. Input variables are different jet or event observables to characterise $WW \rightarrow q\bar{q}q\bar{q}$ events by high multiplicity, high visible energy, and a four-jet structure.

The overall selection efficiency was 90.2 ± 0.9 % and the cross-section for the expected total background was estimated to be 2.06 ± 0.10 pb. A binned likelihood fit to the distribution of the neural network output variable (Fig. 2), corresponding to 1340 selected events at 189 GeV, leads to the cross-section measurement : $\sigma_{WW}^{q\bar{q}q\bar{q}} = 7.37 \pm 0.26$ (stat.) ± 0.24 (syst.) pb. The dominant contribution to the systematic error comes from the uncertainty on the background and a conservative contribution of 0.20 pb has been added to the systematic error to account for differences with respect to the result of alternative analyses that were performed as a cross-check.

3.2   Semi-leptonic final state

For $WW \rightarrow q\bar{q}l\nu$ events, sequential cuts were applied to keep events with an isolated high energetic lepton in the case where the lepton is a μ or an e, or events with low multiplicity jets if the lepton is a τ. These events are also characterized by hadronic activity in the detector and missing energy.

Fig. 3 shows the distribution of the momentum of the selected leptons at 189 GeV. At this energy, the total event selection efficiency was estimated

334

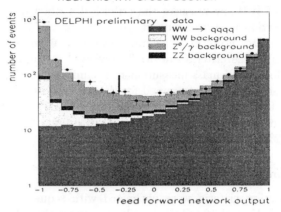

Figure 2: Neural Net output for selection of WW → qq̄qq̄ events at 189 GeV for the cross-section measurement. The points show the data and the histograms are the predicted distributions for signal (right side) and background (left side). The arrow indicates the cut value.

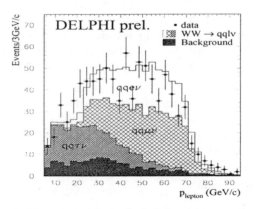

Figure 3: Distribution of the lepton momentum for semi-leptonic events at 189 GeV. The points show the data and the histograms are the predicted distributions for signal and background (dark area).

to be 75.4 ± 1.4 %, for a background contamination of 0.81 ± 0.06 pb. The errors include all systematic uncertainties from efficiency and background determination, the minimal Born diagram correction, and measurement of the luminosity. The cross-section result from 906 selected events is : $\sigma_{WW}^{q\bar{q}l\nu} = 6.72 \pm 0.26$ (stat.) ± 0.15 (syst.) pb.

3.3 Fully leptonic final state

These events are characterized by low track multiplicity with a clean two-jet topology with two energetic, acolinear and acoplanar leptons of opposite charge and by large missing momentum and energy. A total of 188 events was selected at 189 GeV with an overall efficiency of 63.4 ± 2.3 % and a residual background of 0.155 ± 0.016 pb. The cross-section value from a likelihood fit yields : $\sigma_{WW}^{l\nu l\nu} = 1.68 \pm 0.14$ (stat.) ± 0.07 (syst.) pb at 189 GeV. The systematic uncertainty includes track reconstruction efficiency.

4 Determination of total cross-section and branching fractions

The total cross-sections for WW production, with the assumption of Standard Model values for branching fractions are shown in Table 1 and represented in Fig. 4. From all final states combined, the branching fractions have been measured. Combining the results at all energies, the results are consistent with lepton universality. Assuming lepton universality, one can derive the leptonic and the hadronic branching fraction. They are shown in Table 2. The hadronic branching fraction is in agreement with the Standard Model prediction of 0.677.

Table 1: WW cross-sections at different centre-of-mass energies. The uncertainy (bkg.) from the QCD background (column 6) is included in the systematic error (column 5). The result at 188.63 GeV is preliminary.

Energy (GeV)	luminosity (pb^{-1})	WW cross-section (pb)	stat. error (pb)	syst. error (pb)	(bkg.) (pb)
161.31	10.07	3.61	0.90	0.19	0.13
172.14	10.12	11.37	1.37	0.32	0.13
182.65	52.52	15.86	0.69	0.26	0.10
188.63	154.	15.79	0.38	0.31	0.09

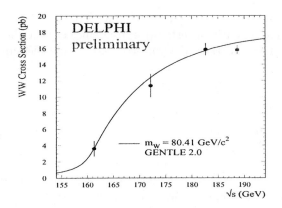

Figure 4: Measurements of the W^+W^- cross-section compared with the Standard Model prediction using $M_W = 80.41$ GeV/c^2.

Table 2: W branching fractions from the combined 161, 172, 183 and 189 GeV data. The uncertainy from the QCD background (col. 5) is included in the systematic error (col. 4).

channel	B.R.	assuming lepton universality		
		stat. error	syst. error	(bkg.)
$W \to \ell\nu$	0.1068	0.0024	0.0019	0.0006
$W \to$ hadrons	0.6796	0.0073	0.0058	0.0018

5 W mass reconstruction measurement

The event generator EXCALIBUR [2] was used for the mass measurement. The main points of the analyses both for the WW $\to$ q$\bar{\text{q}}$lν and the WW $\to$ q$\bar{\text{q}}$q$\bar{\text{q}}$ channels are described in ref. [5,6] for 172-183 GeV data. The results given in the following for 189 GeV are still very preliminary.

5.1 Reconstruction of the mass in the semi-leptonic and fully hadronic decay channels

- The WW $\to$ q$\bar{\text{q}}$lν channel :

Events were selected using sequential cuts from the data sample recorded while all detectors essential to this measurement were fully efficient. The criteria used to tag the WW → qq̄lν are similar to those described above for the cross-section measurement. They rely on lepton identification and isolation cuts. After the selection, 88 electron and 109 muon candidates remained in the data at 183 GeV. The number of expected events from simulation is 80.2 with a purity of 91.0% in the electron channel and 100.7 with a purity of 94.5% in the muon channel.

A kinematic reconstruction is applied to obtain an optimal resolution on the 4-momenta of the jets for the mass reconstruction. Energy and equality of the two reconstructed masses are required leading to a two-constraint fit. The distribution of the reconstructed masses is shown in Fig. 5 for real and simulated data in the electron and muon channels.

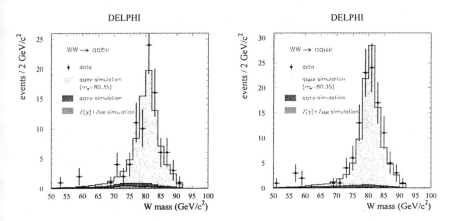

Figure 5: The distributions of the reconstructed masses for the electron and muon channels at 183 GeV.

The W mass was extracted from the reconstructed mass distribution using the same method for 172, 183 and 189 GeV : an event-by-event maximum likelihood fit to a relativistic Breit-Wigner convoluted with a Gaussian resolution function plus a background shape. The background distribution was taken from the simulation. The error on the reconstructed mass from the constrained kinematic fit was used as the width of the Gaussian for the corresponding event. Only events in the mass range between 69 and 91 GeV/c were used in the fit. The events with good resolution have much impact using this method. However, the parametrisation adopted is approximate since detector and physics effects, such as initial state radiation, are not all taken into account. A calibration curve is needed to correct the bias which is estimated with

338

various sample of known generated masses. The same procedure is necessary for the fully hadronic channel.

- The WW → qq̄qq̄ channel :

In the fully hadronic channel, the main background contamination comes from the Z → qq̄(γ) events. The priority for the event selection was a high efficiency, rather than a high signal to background ratio, in order to suppress possible biases on the mass determination from a tighter event selection as for the cross-section measurement. With the sequential cuts, the efficiency and purity of this selection were estimated to be 88% and 66% respectively at 183 GeV. A total of 540 events were selected from the data (see Fig. 6). The number of events expected from simulation was 518. However, an independent analysis cross-checked the W mass result using an neural network tagging.

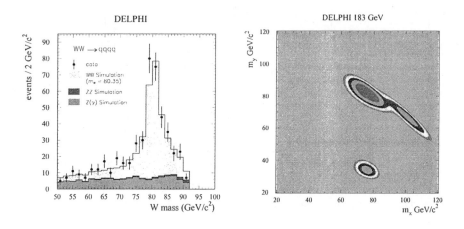

Figure 6: Mass plot (left) for the selected $W^+W^- \to q\bar{q}q\bar{q}$ candidates at 183 GeV showing only one reconstructed mass per event (that with the best χ^2), and using an equal-mass constraint (not used in the determination of M_W, see text). Examples of probability density function (right) for a 4-jet hadronic event. The first 5 sigma contours are shown. The three possible pairings are clearly distiguishable and the procedure $m_x \simeq 80$ GeV/c^2 , $m_y \simeq 80$ GeV/c^2 has more weight than the two others.

5.2 Event by event likelihood in the WW → qq̄qq̄ channel

A constrained fit[5] was used to improve the precision on the 4-momenta of the jets, requiring energy and momentum conservation. A specific difficulty of the fully hadronic channel (compared to the semi-leptonic channel) comes from the different possible attribution of the jets to each of the two W's. For a four jets event, there are three possible pairing while for a five jets event, this number increases to ten possible associations (two jets for one W and three jets for the opposite W).

To avoid the difficulty of choosing one solution among the different possibilities, the solution adopted is to use all the three and the ten jets solutions. Each solution is then weighted by a probability density function (PDF). The probability density function $p_i(m_x, m_y)$ that this attribution corresponds to two objects with masses m_x and m_y was computed as follows : a 6C-fit with constraints from energy and momentum conservation was performed, fixing the two masses to m_x and m_y, and the probability p_i was derived from the resulting χ^2 of the constrained fit as $p_i(m_x, m_y) \propto \exp(-\frac{1}{2}\chi_i^2(m_x, m_y))$. An example of PDF is shown in Fig. 6 for a 4-jet event. Further improvements in the jet association are implemented which take into account angular distribution between the jets, gluon radiation in five jets events, and results obtained with different jets algorithms.

The W mass result in the fully hadronic channel is cross-checked with an independent analysis at 183 GeV. For this second analysis, a neural network performs both the event selection and the jet assignment. The W boson mass was extracted in this method from a likelihood fit to the two-dimensional plot formed by the average and the difference of the two W-masses obtained with a fast kinematic fit, using the distribution predicted by the full simulation. In order to obtain the Monte Carlo spectrum for arbitrary values of M_W, a Monte Carlo reweighting technique was used. The mass found is in full agreement with the result of the event by event likelihood analysis which is given later.

5.3 Systematic errors

Any error in the simulation will cause a systematic error on the mass. The different sources of systematic errors are discussed in detail in ref. [5]. The list of the largest ones is presented in Table 3.

In the semi-leptonic channel, the dominant systematic effects are due to the uncertainty on the absolute energy calibrations. Bhabha and Compton scattering events showed an uncertainty on the electron energy of 1%, while the systematic uncertainty on the muon momentum was estimated from $Z^0 \to \mu^+\mu^-$ events to be 0.5%. The uncertainty on the jet energy was estimated to be 2% from $Z^0 \to q\bar{q}$ events. In the fully hadronic channel, the present dominant systematic uncertainty comes from interactions among the W decay products (FSI). Since the two W's under LEP2 conditions decay much closer to each other than the typical hadronization scale of 0.5 – 1.0 fm, cross-talk between the two W may lead to systematic error because this is not taken into account in the simulation. Two possible sources of such effects have been identified: colour reconnection (CR) among partons from the two different colour singlet systems and Bose-Einstein correlations (BEC) among identical bosons in the final state. The uncertainty on the LEP beam energy cause an uncertainty on the mass of the W which should decrease to about 10 MeV/c^2 for the forthcoming data taking.

Table 3: Contributions to the systematic error on the W boson mass measurement at 183 GeV. The error sources have been separated into those uncorrelated (uncor.) and correlated (cor.) between the different LEP experiments.

systematic error (MeV/c^2)	Electrons	Muons	Leptons	Hadrons	Combined
Stat. calibrat.	23	18	14	9	8
Lepton energy	40	35	26	-	9
Jet energy	50	30	38	20	26
Background level	5	-	2	5	3
Background shape	-	-	-	5	3
Lepton isolat.	20	-	8	-	3
Total uncor.	**71**	**50**	**49**	**23**	**29**
Fragmentation	10	10	10	20	17
I.S.R.	10	10	10	10	10
Total cor.	**14**	**14**	**14**	**22**	**20**
LEP energy	**21**	**21**	**21**	**21**	**21**
CR	-	-	-	50	33
BEC	-	-	-	20	13
Tot. FSI	-	-	-	**54**	**35**

6   W mass combined results

The masses measured in the mixed and hadronic decays analysis are in good agreement within statistics and are displayed in Fig. 7. Combining results at 161, 172, 183, and 189 GeV yields as preliminary W mass measurement :

$$M_W = 80.342 \pm 0.078 \text{ (stat.)} \pm 0.051 \text{ (syst.) GeV/c}^2$$

where (syst.) includes all the systematic uncertainties which have been evaluated so far.

DELPHI W MASS (161+172+183+189 GeV)

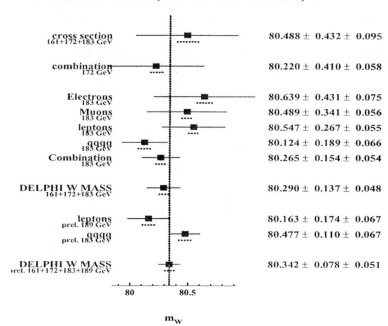

Figure 7: Summary of the DELPHI W mass measurements both in the semi-leptonic channel and the fully hadronic channel for the year 1996-1998.

7 Summary and Outlook

The cross-section measurements at 161, 172, 183, and 189 GeV show the clear evidence for the existence of the non-abelian coupling. The measurement of the DELPHI cross-section at 189 GeV is still preliminary : $\sigma_{WW}^{tot} = 15.79 \pm 0.38$ (stat.) ± 0.31 (syst.) pb. This value is 1.5 standard deviations lower than the Standard Model prediction using GENTLE [7] and assuming a theoretical error of 2 %. The present W boson mass measurement in DELPHI is :

$M_W = 80.342\pm0.078$ (stat.)±0.051 (syst.) GeV/c². This measurement is compatible with the present world average of the direct mass measurement including TEVATRON and LEP2 [8] : $M_W = 80.410\pm0.044$ GeV/c², but also with indirect mass value from a global electroweak fit : $M_W = 80.364\pm0.029$ GeV/c². The statistical error on the mass is now at the level of the quoted systematics. We therefore need to improve our understanding of the systematic error both for detectors and physics effects. Future new measurements of the W mass with more than $\mathcal{L} \simeq 300$ pb^{-1} are expected soon and we should achieve a final LEP2 statistical error below 25 MeV/c² combining results from the DELPHI detector with those of ALEPH, L3, and OPAL experiments.

References

1. G. Altarelli, T. Sjöstrand and F. Zwirner, *Physics at LEP2*, CERN 96-01 (1996) Vol.1.
2. F.A. Berends, R. Pittau and R. Kleiss, Comp. Phys. Comm. **85**, 437, (1995).
3. T. Sjöstrand, Comp. Phys. Comm. **82**, 74, (1994).
4. DELPHI Collaboration, P. Abreu *et al.*, Phys. Lett. **B397**, 158, (1997).
5. DELPHI Collaboration, P. Abreu *et al.*, E. Phys. J. **C2**, 581, (1998).
6. DELPHI Collaboration, P. Abreu *et al.*, 99-41 MORIO CONF 240, *Measurement of the mass of the W boson using direct reconstruction*, (To be submitted to Phys. Lett. **B**).
7. D. Bardin *et al.*, DESY 95-167 (1995).
8. XXXIVth Rencontres de Moriond, Les Arcs, France, 15-21 March, 1999. http://www.cern.ch/LEPEWWG/, http://www.lal.in2p3.fr/CONF/Moriond/ElectroWeak/electroweak.html

A LARGE NEUTRINO DETECTOR FACILITY AT THE SPALLATION NEUTRON SOURCE AT OAK RIDGE NATIONAL LABORATORY

YU. V. EFREMENKO

(on behalf of ORLaND collaboration)

Oak Ridge National Laboratory, MS-6372, Oak Ridge,
TN 37831-6372, USA
E-mail: efremenk@utkux1.utk.edu

The ORLaND (Oak Ridge Large Neutrino Detector) collaboration proposes to construct a large neutrino detector in an underground experimental hall adjacent to the first target station of the Spallation Neutron Source (SNS) at the Oak Ridge National Laboratory. The main mission of a large (2000 ton) Scintillation-Cherenkov detector is to measure $\bar{\nu}_\mu \to \bar{\nu}_e$ neutrino oscillation parameters more accurately than they can be determined in other experiments, or significantly extending the currently measured parameter space ($\sin^2 2\theta \leq 10^{-4}$). In addition to the neutrino oscillation measurements, ORLaND would be capable of making precise measurements of $sin^2\theta_W$, search for the magnetic moment of the muon neutrino, and investigate the anomaly in the KARMEN time spectrum, which has been attributed to a new neutral particle. With the same facility an extensive program of measurements of neutrino nucleus cross sections is also planned to support nuclear astrophysics.

1 The Spallation Neutron Source SNS and ORLaND

The Spallation Neutron Source (SNS) under construction at the Oak Ridge National Laboratory will produce the most intense pulsed-proton beam in the world. It consists of a 493-m long linac, an accumulator ring with a radius of just over 35 meters, and a target station. A second accumulator ring and target station is planned to be built in the future. The SNS will be used primarily for material science research. However, it also will be the world's most powerful neutrino factory. The high intensity and the pulsed nature of the accumulator ring provide an ideal laboratory for neutrino physics research at medium energies. The short-pulsed beam ($\leq 0.55\mu sec, 60Hz$) provides a virtually cosmic-ray-free measurement of various neutrino interactions. It also permits the separation of neutrinos from pion decay and muon decay. A plan view of the completed SNS facility is shown in Fig 1.

The existence of a second accumulator ring and target station would make SNS a unique neutrino facility with two simultaneously operating sources with known time structures and different flight paths. The final planned beam current (4 mA) and proton energy (1 GeV) will produce neutrino fluxes 10

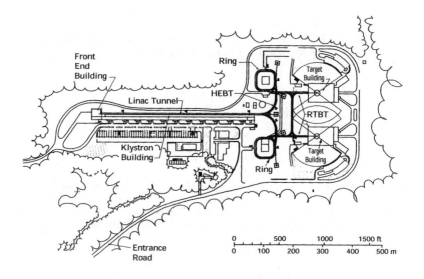

Figure 1: Plan view of the completed SNS facility.

times as intense as those at LANCE[1], but with a pulse time structure similar
to that used successfully by KARMEN at the ISIS neutron source[2].

Table 1 shows the design parameters of the SNS.

The ORLaND detector will be placed in a bunker located immediately
adjacent to the SNS first target building. It will be totally underground, with
the top of the concrete cap level with the finished grade of the site. The
distance between the center of the neutrino detector and the center of the first
SNS target is approximately 139 feet (42.3 meters).

2 Neutrino Oscillation Search

2.1 Search for $\bar{\nu}_\mu \to \bar{\nu}_e$ oscillations

The ORLaND neutrino oscillation search, similar to LSND and KARMEN is
based on a unique feature of neutrino beams from spallation sources. The
enormous suppression of $\bar{\nu}_e$ production allows the attribution of $\bar{\nu}_e$ events to
$\bar{\nu}_\mu \to \bar{\nu}_e$ oscillation.

This feature has the following explanation. In the SNS, pions are copiously
produced from the interaction of 1.0 GeV protons with the mercury target. The
production rate is approximately   0.068 π^+ and 0.049 π^- per proton.

Table 1: SNS Design Parameters

	baseline	after upgrade
beam power on the target	1 MW	4 MW
beam energy on the target	1 GeV	
average beam current	1 mA	4 mA
repetition rate	60 Hz	10 Hz, 60 Hz
ion type, source-linac	H^-	
particles stored in ring (ppp)	1×10^{14}	2×10^{14}
extracted pulse length	550 nsec	
peak current on target	30 A	120 A
number of target stations	1	2
target material	mercury	
beam spot on target	7×21 cm	
neutron moderators	4	8
neutron beam ports	18	36

Absorption of π^+ and its decay μ^+ are negligible. However most π^- are quickly stopped and captured in the target materials. Detailed simulations show that about 99.6% of the negative pions produced in the Hg target are captured before they have a chance to decay. Moreover, a significant number of μ^- (94%) are captured by heavy elements in the target and in the surrounding lead reflector. This results in a large asymmetry between $\bar{\nu}_e$ and $\bar{\nu}_\mu$ production. In Fig. 2, and Fig. 3 the calculated time and energy spectra for four types of neutrino are shown. Several conclusions can be drawn from these distributions:

1. There is a large imbalance between the two antineutrino flavors in favor of $\bar{\nu}_\mu$.

2. Most of the $\bar{\nu}_\mu$ are produced from decay at rest (DAR) due to the large stopping power in heavy and dense target materials. Production of high-energy $\bar{\nu}_\mu$ (above 52.83 MeV kinetic energy) is due to the decays in flight (DIF). This production is suppressed by the design of the spallation target station, the large size of the mercury target, and the lead reflector.

3. There is a strong time dependence of the $\bar{\nu}_e$ to $\bar{\nu}_\mu$ ratio. This is because of the fast capture rate of μ^- in the high Z materials (mercury in the target and lead in the neutron reflector). This time dependence allows us to experimentally measure background from $\bar{\nu}_e$, which is the largest beam-related irreducible background.

346

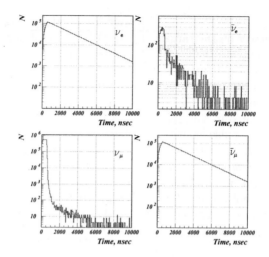

Figure 2: Time distributions for four neutrino types produced at SNS.

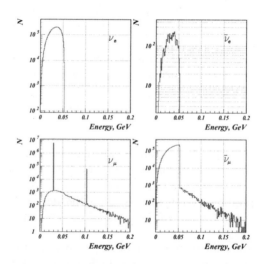

Figure 3: Energy distributions for four neutrino types produced at SNS. The monoenergetic line for the muon neutrino near 100 MeV does not include broadening due to nuclear effects.

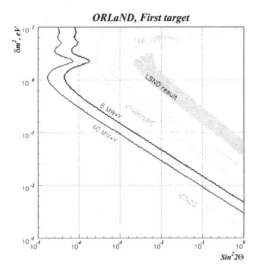

Figure 4: Limits for mixing parameters for neutrino oscillations reachable at ORLaND compared with that of Mini-BooNE, MINOS and I-216 proposals

Based on detailed simulations for the SNS target station, we concluded that the ratio of $\bar{\nu}_e$ to $\bar{\nu}_\mu$ in ORLaND will be $2.4 \cdot 10^{-4}$, three times smaller than the ratio calculated for the LSND experimental setup.

In Fig. 4 the possible limits for neutrino oscillation parameters for OR-LaND are compared with that of LSND, Mini-BooNE, MINOS, and CERN I-216 proposals. For ORLaND two limits with the first target station are shown. One results from the first three years of operation with integrated intensity of 6 MW*Y. The second one corresponds to the limits achievable over the life time of facility - 40 MW*Y. The ORLaND will not only cover the entire region where LSND indicates an oscillation effect but will probe a significantly larger area of possible mixing parameters. Compared with Mini-BooNE, ORLaND will have almost an order of magnitude better sensitivity in both mixing angle and δm^2 and with much less background.

The very intense SNS-based neutrino source together with the massive size of the ORLaND detector provides an unprecedented opportunity for a high-statistics neutrino oscillation search. In one week the ORLaND detector will be able to accumulate the same statistics as LSND did in three years. If LSND results are confirmed by early experiments (KARMEN, Mini-BooNE), ORLaND would be uniquely positioned to measure precisely the neutrino mix-

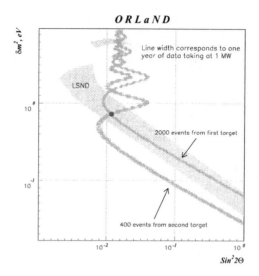

Figure 5: The precision of the measurment of the mixing parameters in the LSND region. Narrow bands correspond to regions of allowed parameters, using the number of events from the first and second target stations for one year of data collection with a spallation source power of 1 MW per target.

ing parameters. An example of such a measurement is shown in Fig. 5. The mixing parameters $sin^2 2\theta \approx 1.5 \cdot 10^{-2}$ and $\delta m^2 \approx 0.7$ eV were assumed in this figure. For one year of data taking with spallation power of 1 MW per target ORLaND will detect 2,000 events from the first and 400 events from the second spallation target. Corresponding regions of allowed parameters are shown by the two narrow bands. The intersection of these bands corresponds to the accuracy of measurement of the mixing parameters.

2.2 Search for $\nu_\mu \to \nu_e$ transitions with monoenergetic neutrinos

The time structure and monoenergetic character of the spectrum of ν_μ due to the decay $\pi^+ \to \mu^+ + \nu_\mu$, shown in Fig. 2 and Fig. 3, can be used in ORLaND to search for $\nu_\mu \to \nu_e$ oscillations. These neutrinos will oscillate to ν_e with an observable probability for some values of $sin^2 2\theta$ and δm^2 in the range reported by the LSND collaboration. The reaction $^{12}C(\nu_e, e^-)\, ^{12}N_{g.s.}$ results in a Cerenkov ring, followed by the decay $^{12}N_{gs} \to^{12} C + e^+ + \nu_e$. There will, of course, be a background of ν_e from the broad spectrum due to the decay of μ^+; however, it will be suppressed by making use of the time structure of the neutrinos shown in Fig 2. The events that occur due to $\nu_\mu \to \nu_e$ oscillations would cause a significant distortion of the background spectrum as shown in Fig. 6.

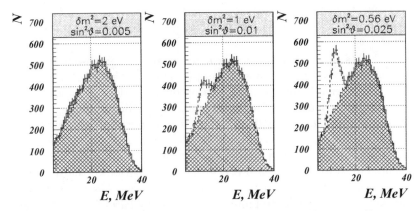

Figure 6: The energy spectrum of electrons from the reaction $^{12}C(\nu_e, e^-)^{12}N_{gs}$. The cross-hatched region contains events due to ν_e from the decay of μ^+. The distributions at the lower energy edge are those predicted for ν_μ to ν_e oscillations for 3 years of data acquisition with ORLaND with a proton beam of 2 mA.

Detection of neutrino oscillations for both neutrinos and antineutrinos, will give the possibility to test CP invariance in the neutrino sector.

3   Other Neutrino Physics at ORLaND

3.1   Measurement of Neutrino-Nucleus Cross Sections

The similarity of the neutrino spectra from π^+ and μ^+ decay at rest at the neutron spallation source, and those produced during supernovae core collapse is striking. This coupled with the intensity and pulse structure at the SNS makes the proposed ORLaND facility ideal for making measurements that directly support theoretical nuclear astrophysics.

Supernovae involve such a large number of neutrino-nucleus reactions, some involving radioactive nuclei, that laboratory measurements of all of them are impossible. Accordingly, random phase approximation models are used [3], which themselves have parameters. It is important to test and tune these models in a few important cases which can be measured in the small segmented detector planned for the ORLaND facility. In this detector, materials foreign to the scintillator can be introduced in a variety of ways.

3.2 Precision $sin^2\theta_w$ Measurement Using $\nu - e$ Elastic Scattering

The high intensity of the SNS neutrino beam in combination with its unique time structure will let ORLaND measure $sin^2\theta_w$ with a statistical accuracy approaching 1%.

Measurements of the masses of the intermediate vector bosons made in the past few years provide the most accurate determination of the parameter $sin^2\theta_w$. However two independent measurements performed at LEP and SLAC differ by more than 1.8σ. This contradicts the prediction of the SM that $sin^2\theta_w$ is the sole parameter in this theory. It is important that $sin^2\theta_w$ be tested with neutrinos as a probe with higher accuracy. Deep-inelastic scattering on nucleons has proved to be the most accurate method so far but it has the problem common to hadronic processes that calculation of the result involves theoretical uncertainties at the level of a few percent. Neutrino-electron scattering experiments have been limited in the past by lack of statistics, subtraction of background, and difficulties in the neutrino flux calculations. The most precise neutrino-electron measurements have been of the ratio of neutrino to antineutrino scattering. A substantial contribution to the systematic error in these experiments arises from the uncertainty in the characteristics of the two neutrino beams. We propose to measure the ratio,

$$R = \frac{\sigma(\nu_\mu e)}{\sigma(\nu_e e) + \sigma(\bar{\nu}_\mu e)} = \frac{3}{4}\frac{1 - 4s^2 + 16/3s^4}{1 + 2s^2 + 8s^4} \tag{1}$$

where $s^2 = sin^2\theta_W$. If we assume $sin^2\theta_w = 0.23$, then R = 0.13.

ORLaND proposes to make a precision measurement of the ratio of neutrino electron scattering with neutrinos from pion decay and muon decay simultaneously.

We have estimated errors on the basis of a two-kiloton detector located 42 meters from the beam-stop. For three years of data taking at SNS intensity of 2 MW there will be about 50 000 ($\nu_e e$) and 9 000 ($\nu_\mu e$) and ($\bar{\nu}_\mu e$) interactions in the detector. These statistics will let us measure $sin^2\theta_w$ with an accuracy close to 1%. Of course intensive studies will be required to evaluate the magnitude of systematic errors.

3.3 KARMEN's Anomalous Bump on the Time Decay Curve

In 1995 the KARMEN collaboration reported an anomaly in the time distribution[4] of neutrinos from the pulsed neutron spallation source ISIS. The anomaly is a peak with a width of 1 μs at 3.6 μs after the beam pulse, on top of the $\tau = 2.2\mu$s time spectrum of neutrinos from muon decays at rest. The anomaly has been observed only in *single pronged* events below 35 MeV. Above this energy the time spectrum is flat. The time distribution of the sequential charged current reaction $^{12}C(\nu_e, e^-)^{12}N_{gs}$ showed the expected 2.2 μs decay time, thus indicating that the anomaly is not caused by a variation in muon production or decay. After investigating and excluding several possible sources for this effect, like albedo neutrons from a hill behind the experiment, electronics, and others, KARMEN proposed as a working hypothesis a heavy weakly interacting neutral particle X produced in pion decay: $\pi^+ \rightarrow \mu^+ + X$ with velocity $\frac{1}{60}c$ (17.6m/3.6μs), having a very small kinetic energy, which leaves 33.9 MeV for the rest mass of the particle. This particle would not be observed directly in the detector, but would decay for example into $\gamma + \nu$ or $e^+ + e^- + \nu$. The detectable decay products would then give a broad energy spectrum between 0 and 33.9 MeV. Since this is the same energy range as for CC and NC neutrino reactions with ^{12}C, the small number of events can not be separated from the normal neutrino background. The excess events have been observed over a time period of 7 years. After a major upgrade of the KARMEN Veto system in 1996, which also involved significant changes in the passive shielding of the detector, the number of events is still increasing with the same rate[5]. However with the low statistics of this effect it will take another two years to verify with 5 σ that it is not a statistical fluctuation.

ORLaND would be able to look for this effect with much higher statistics. Since ORLaND is farther away from the target, the X particle would be observed in a time window between 8.1 and 9.4 μs, outside the normal neutrino background. Fig. 7 shows the lower limit of the parameter range and expected events rate in ORLaND for lifetimes of X from 10^{-7} to 10^3 sec.

3.4 Search for Effects of the Magnetic Moment of the muon neutrino

The existence of neutrino mass leads to the possibility for Dirac type neutrinos to have a magnetic dipole moment or non-diagonal magnetic transition moments for Majorana neutrinos. In that case neutrino interactions include an electromagnetic part in addition to the standard weak interaction. The electromagnetic part has a characteristic energy behavior with an increase of the differential cross section, while the recoil energy of the scattering target decreases. The simplest way to measure the neutrino magnetic moment would

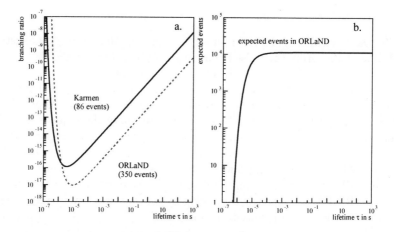

Figure 7: (a) Branching ratio of the Pion decay versus lifetime of the X particle. The dashed curve shows the sensitivity of ORLaND for 350 events. (b) Expected events for X particle in ORLaND, based on the KARMEN result for 4 MW Year of data taking.

be to measure the excess of neutrino differential cross section in the low energy of the recoil particle region. From this point of view the ORLaND neutrino facility is not competitive to measure the electron neutrino magnetic moment as compared to reactor neutrino experiments. However, for muon neutrinos, there might be a unique opportunity to improve the existing restriction on the magnetic moment, $\mu < 7.4 \cdot 10^{-10}\mu_B$, by about an order of magnitude.

4   Acknowledgements

This work was supported in part by U.S. Department of Energy under contract DE-AC05-960R22464 with Lockheed Martin Energy Research Corporation.

References

1. C. Athanassopoulos *et. al.*, PRL **81**, 1774, (1998).
2. B. Armbruster *et. al.*, PLB **348**, 19, (1995), (the KARMEN collaboration).
3. E. Kolb, K.-H. Langanke, S. Krewald, and f.-K. Thielemann, NBA **A540**, 599, (1992).
4. Armbruster *et al.*, PLB **348**, 19, (1995).
5. Karmen collaboration, private communication 1998.

UPDATE OF THE KARMEN2 $\bar{\nu}_\mu \to \bar{\nu}_e$ OSCILLATION SEARCH

K. EITEL [1]

Institut für Kernphysik I, Forschungszentrum Karlsruhe
Postfach 3640, D-76021 Karlsruhe, Germany
E-mail: klaus@ik1.fzk.de
for the KARMEN collaboration

The neutrino experiment KARMEN situated at the beam stop neutrino source ISIS is taking data (since February 1997) with its upgraded veto shield. The results of the oscillation search $\bar{\nu}_\mu \to \bar{\nu}_e$ in the appearance mode shown here are based on data taken until November 1998. Applying stringent cuts to delayed coincidences from the detection reaction $p(\bar{\nu}_e, e^+)n$, 7 sequences are extracted with an expected background of 9.3 ± 0.7 events. A maximum likelihood analysis of these sequences shows no oscillation signal thus leading to an upper limit for the mixing angle of $\sin^2(2\Theta) < 1.9 \cdot 10^{-3}$ (90% CL) at large Δm^2. The corresponding exclusion curve in $(\sin^2(2\Theta), \Delta m^2)$ is discussed and compared with the favored area given by the 1993-98 data of the LSND experiment.

1 Introduction

The search for neutrino oscillations and hence massive neutrinos is one of the most fascinating fields of modern particle physics with more and more evidence especially for the oscillation of atmospheric neutrinos by the SuperKamiokande experiment. The **K**arlsruhe **R**utherford **M**edium **E**nergy **N**eutrino experiment KARMEN searches for neutrino oscillations in different appearance ($\nu_\mu \to \nu_e$ and $\bar{\nu}_\mu \to \bar{\nu}_e$) and disappearance modes ($\nu_e \to \nu_x$) [2]. The physics program of KARMEN also includes the investigation of ν–nucleus interactions [3] as well as the search for lepton number violating decays of pions and muons and the test of the V–A structure of μ^+ decay [4].

Intermediate results are presented here of the oscillation search in the appearance channel $\bar{\nu}_\mu \to \bar{\nu}_e$ on the basis of data taken from February 1997 until November 1998 with the upgraded experimental configuration (KARMEN2) corresponding to about 40% (or 3700C accumulated proton charge on target) of the envisaged accumulated neutrino flux from ISIS through 2001.

2 Neutrino Production and Experiment Setup

The KARMEN experiment is performed at the neutron spallation facility ISIS of the Rutherford Appleton Laboratory, Chilton, UK. The neutrinos are produced by stopping 800 MeV protons in a beam stop target of Ta-D_2O. In addition to spallation neutrons, there is the production of charged pions. The π^-

are absorbed by the target nuclei whereas the π^+ decay at rest. Muon neutrinos ν_μ therefore emerge from the decay $\pi^+ \rightarrow \mu^+ + \nu_\mu$. The produced μ^+ are also stopped within the massive target and decay via $\mu^+ \rightarrow e^+ + \nu_e + \bar{\nu}_\mu$. Because of this π^+-μ^+-decay chain produced at rest, ISIS represents a ν-source with identical intensities for ν_μ, ν_e and $\bar{\nu}_\mu$ emitted isotropically ($\Phi_\nu = 6.37 \cdot 10^{13} \, \nu/s$ per flavor for a proton beam current $I_p = 200 \, \mu A$). There is a minor fraction of π^- decaying in flight (DIF) with the following μ^- DAR in the target station which again is suppressed by muon capture of the high Z material of the spallation target. This decay chain leads to a very small contamination [5] of $\bar{\nu}_e/\nu_e < 6.2 \cdot 10^{-4}$, which is even further reduced by the final evaluation cuts.

The energy spectra of the ν's are well defined due to the DAR of both the π^+ and μ^+ (Fig. 1a). The ν_μ's from π^+–decay are monoenergetic with $E(\nu_\mu)=29.8\,\mathrm{MeV}$, the continuous energy distributions of ν_e, $\bar{\nu}_\mu$ up to 52.8 MeV can be calculated using the V–A theory and show the typical Michel shape. Two parabolic proton pulses of 100 ns base width and a gap of 225 ns are

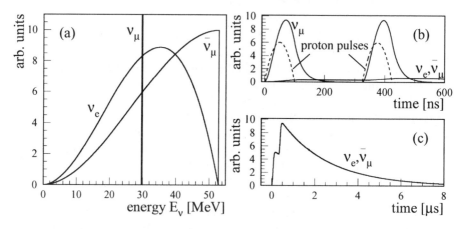

Figure 1: Neutrino energy spectra (a) and production times of ν_μ (b) and $\nu_e, \bar{\nu}_\mu$ (c) at ISIS.

produced with a repetition frequency of 50 Hz. The different lifetimes of pions ($\tau = 26\,\mathrm{ns}$) and muons ($\tau = 2.2\,\mu s$) allow a clear separation in time of the ν_μ-burst (Fig. 1b) from the following ν_e's and $\bar{\nu}_\mu$'s (Fig. 1c). Furthermore the accelerator's duty cycle of 10^{-5} allows effective suppression of any beam uncorrelated background.

The neutrinos are detected in a rectangular tank filled with 56 t of a liquid scintillator. This central scintillation calorimeter is segmented by double acrylic walls with an air gap allowing efficient light transport via total internal reflection of the scintillation light at the module walls. The event position is determined by the individual module and the time difference of the PM signals at each end of this module. Due to the optimized optical properties of the organic liquid scintillator and an active volume of 96% for the calorimeter, an energy resolution of $\sigma_E = 11.5\%/\sqrt{E[MeV]}$ is achieved. In addition, Gd_2O_3 coated paper within the module walls provides efficient detection of thermal neutrons due to the very high capture cross section of the $Gd(n,\gamma)$ reaction.

A massive blockhouse of 7000 t of steel in combination with a system of two layers of active veto counters provides shielding against beam correlated spallation neutron background, suppression of the hadronic component of cosmic radiation as well as reduction of the flux of cosmic muons. On the other hand, this shielding is a source of energetic muon induced neutrons produced by deep inelastic muon nucleon scattering and the nuclear capture of μ^-. These neutrons produced in the steel can penetrate the anti counter systems undetected and simulate a $\bar{\nu}_e$ detection sequence. The prompt signal is caused e.g. by a $n - p$ scattering followed by the delayed capture of the thermalized neutron. These neutrons were the major background source in the KARMEN1 experiment. In 1996 an additional third anti-counter system with a total area of $300\,\mathrm{m}^2$ was installed within the 3 m thick roof and the 2–3 m thick walls of the iron shielding [6]. By detecting the muons in the steel at a distance of 1 m from the main detector and vetoing the successive events, this background has been reduced by a factor 40 compared to the KARMEN1 data.

3 Search for $\bar{\nu}_\mu \rightarrow \bar{\nu}_e$ Oscillations

The probability for ν–oscillations $\bar{\nu}_\mu \rightarrow \bar{\nu}_e$ can be written in a simplified 2 flavor description as

$$P(\bar{\nu}_\mu \rightarrow \bar{\nu}_e) = \sin^2(2\Theta) \cdot \sin^2(1.27\frac{\Delta m^2 L}{E_\nu}) \tag{1}$$

where L and E_ν are given in meters and MeV, Δm^2 denotes the difference of the squared mass eigenvalues $\Delta m^2 = |m_1^2 - m_2^2|$ in eV^2/c^4. With $\langle L \rangle = 17.7\,\mathrm{m}$ and $E_\nu < 52.8\,\mathrm{MeV}$, KARMEN is sensitive to small mixings $\sin^2(2\Theta)$ for oscillation parameters $\Delta m^2 \gtrsim 0.1\,eV^2/c^4$.

The signature for the detection of $\bar{\nu}_e$'s is a spatially correlated delayed coincidence of positrons from $p(\bar{\nu}_e, e^+)n$ with energies up to $E_{e+} = E_{\bar{\nu}_e} - Q = 52.8 - 1.8 = 51.0\,\mathrm{MeV}$ (Fig. 2a) and γ emission of either of the two neutron capture processes $p(n,\gamma)d$ with one γ of $E(\gamma) = 2.2\,\mathrm{MeV}$ or $Gd(n,\gamma)Gd$ with 3 γ–quanta in average and a sum energy of $\sum E(\gamma) = 8\,\mathrm{MeV}$ (Fig. 2c). The positrons are expected in a time window of several μs after beam–on–target

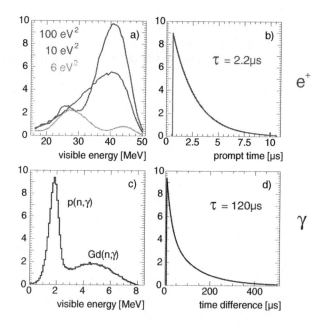

Figure 2: Expected signature in the KARMEN calorimeter for $\bar{\nu}_\mu \to \bar{\nu}_e$ oscillation (simulation including all detector response functions): a) visible energy of prompt positron for three different values of the oscillation parameter Δm^2; b) time of prompt event relative to ISIS beam–on–target reflecting the $2.2\mu s$ lifetime of μ^+ in the spallation target; c) visible energy of delayed γ's from neutron capture on Gadolinium in the module walls or free protons of the scintillator; d) time difference between prompt e^+ and delayed capture γ's defined by the neutron thermalisation and diffusion

(Fig. 2b) with a $2.2\,\mu s$ exponential decrease due to the μ^+ decay. The time difference between the e^+ and the capture γ is given by the thermalization, diffusion and capture of neutrons (Fig. 2d).

The raw data investigated for this oscillation search were recorded in the data-taking period of February 1997 to November 1998 which corresponds to $3731\,C$ protons on target or $1.1 \cdot 10^{21}$ $\bar{\nu}_\mu$ produced in the ISIS target. A positron candidate is accepted only if there is no activity in the central detector, the inner anti, and outer shield up to $24\,\mu s$ prior. If only the outermost veto counter was hit, a dead time of $14\,\mu s$ is applied. These conditions reduce significantly background induced by cosmic muons: penetrating μ, decay products of stopped muons and neutrons from deep inelastic muon scattering. The required cuts in energy and time are: $0.6 \le t_p \le 10.6\,\mu s$, $16.0 \le E_p \le 50.0\,\text{MeV}$. The cuts on the delayed expected neutron event are

Table 1: Background contributions to the $\bar{\nu}_\mu \to \bar{\nu}_e$ search and their determination methods.

background	expectation	determination
induced by cosmic μ	1.9 ± 0.1	pre-beam measurement
^{12}C$(\nu_e, e^-)^{12}$N$_{\text{g.s.}}$ sequences	3.4 ± 0.4	measurement in diff. E/t-window
ν-induced accidental coinc.	3.0 ± 0.6	measurement in diff. E/t-window
ISIS source contamination	0.9 ± 0.1	MC simulation
total background	9.3 ± 0.7	

as follows: $5.0 \leq t_d - t_p \leq 500\,\mu$s, $E_d \leq 8\,$MeV and a volume of $1.3\,$m^3 for the spatial coincidence ^{a}. Applying all cuts, the background expectation amounts to 9.3 ± 0.7 sequences. The individual background sources are listed in Table 1 together with their evaluation method. All background contributions but the small intrinsic $\bar{\nu}_e$ contamination are measured online with high statistical precision. Analysing the data results in 7 sequential events which satisfy all conditions.

4 Results and Comparison with LSND

To extract a possible $\bar{\nu}_\mu \to \bar{\nu}_e$ signal a maximum likelihood analysis of these 7 sequences is applied making use of the precise knowledge of all background sources and a detailed MC description of the oscillation signature in the detector (see Fig. 2). As this analysis shows no evidence for an oscillation signal, an upper limit of 3.0(2.6) oscillation events for $\Delta m^2 < 1\,$eV^2/c^4($\Delta m^2 > 20\,$eV^2/c^4) can be extracted ^{b} at 90% CL. Assuming maximal mixing ($\sin^2(2\Theta) = 1$), 1384 ± 152 $(e^+,$n$)$ sequences from oscillations with large Δm^2 were expected leading to an upper limit on the mixing amplitude of $\sin^2(2\Theta) < 1.9 \times 10^{-3}$. The complete exclusion curve in $(\Delta m^2, \sin^2(2\Theta))$ can be seen in Fig. 3 together with limits from other experiments [8,9,10]. Also shown are the updated favored regions from LSND analysing the entire 1993-98 data set [11]. The KARMEN2 result excludes the complete parameter region of LSND for $\Delta m^2 > 10\,$eV^2/c^4 whereas for small Δm^2 there is no conclusive result yet.

---

^{a}These cuts are less stringent than the first cuts on the KARMEN2 data reported earlier [7]. With increasing statistics, a method of maximum likelihood on data with looser cuts is much more sensitive than a simple counting experiment, since the individual event parameters, i.e. energy, position and time, are measured with high accuracy.

^{b}Since the positron energy used in this analysis is affected by the oscillation parameter Δm^2 the result also depends on Δm^2(see Fig. 2a).

358

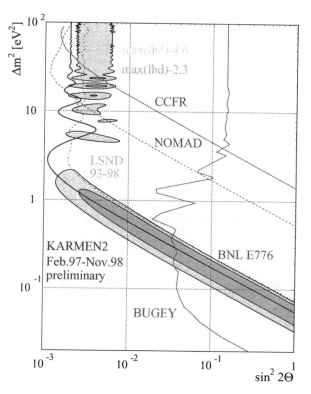

Figure 3: KARMEN 90% CL exclusion curve for the oscillation parameters Δm^2,$\sin^2(2\Theta)$.
Also shown are limits from other experiments and the favored region for $\bar{\nu}_\mu \to \bar{\nu}_e$ from the
4-dimensional maximum likelihood analysis of the LSND 1993-98 data.

Table 2 shows a comparison of some input parameters for the maximum
likelihood analysis of both the KARMEN and LSND experiments. Also shown
are the results of the maximum likelihood analyses with the appropriate parameters for an example of $\Delta m^2 = 0.5 \, \mathrm{eV}^2/c^4$. The mixing which fits best
the LSND data ($\sin^2(2\Theta) = 0.013$) corresponds to an expected event number of $N_{exp} = 2.8$ in the KARMEN detector. The errors thereby correspond
to the contour line 2.3 units below the absolute maximum likelihood value.
The slightly lower oscillation efficiency $\epsilon = 0.149$ of the KARMEN experiment
reflects the smaller distance $L = 17.7 \, \mathrm{m}$ to the neutrino source compared to
LSND ($L = 30 \, \mathrm{m}$)[c]. The smaller oscillation expectation for KARMEN2 is al-

---

[c]The efficiency ϵ is in fact the integral over the neutrino energy spectrum and the detector
volume of the oscillation probability $P(\bar{\nu}_\mu \to \bar{\nu}_e)$ (equ. 1) including detection thresholds and

Table 2: Comparison of the input of the maximum likelihood analysis of KARMEN and LSND to investigate the $(\sin^2(2\Theta), \Delta m^2)$ parameter plane and results for an oscillation parameter of $\Delta m^2 = 0.5\,\mathrm{eV}^2/\mathrm{c}^4$.

	LSND	KARMEN
selection cuts	$20 \leq E \leq 60\,\mathrm{MeV}$	$16 \leq E_{pr} \leq 50\,\mathrm{MeV}$
	no R–cut	$0.6 \leq t_{pr} \leq 10.6\,\mu\mathrm{s}$
extracted events	3049	7
beam related bg	1349 ± 148	7.4 ± 0.7
beam unrelated bg	1467 ± 38	1.9 ± 0.1
total background expectation	2816 ± 153	9.3 ± 0.7
oscillation expectation for $\sin^2(2\Theta) = 1, \Delta m^2 = 100\,\mathrm{eV}^2/\mathrm{c}^4$	13950 ± 1530	1384 ± 152
max. likelihood parameters	$(E, R, L, \cos\theta)$	$(E_{pr}, t_{pr}, E_{del}, t_{del})$
$\Delta m^2 = 0.5\,\mathrm{eV}^2/\mathrm{c}^4$	$\epsilon = 0.188$ $\sin^2(2\Theta) = 0.013^{+0.007}_{-0.005}$ $N_{osc} = 68^{+37}_{-26}$	$\epsilon = 0.149$ $N_{exp} = 2.8^{+1.4}_{-1.0}$ $N_{osc} < 3.0\ (90\%\,\mathrm{CL})$

most completely due to the lower accumulated neutrino flux at ISIS compared to LANSCE [d].

5 Conclusion and Outlook

The present results of the $\bar{\nu}_\mu \to \bar{\nu}_e$ oscillation search obtained by KARMEN are the most stringent limits for $\nu_e \leftrightarrow \nu_\mu$ mixing from accelerator experiments in the range of $\Delta m^2 > 0.1\,\mathrm{eV}^2/\mathrm{c}^4$. Beyond that, KARMEN2 continues to take data and envisages an accumulated proton charge on the ISIS spallation target of about $9200\,\mathrm{C}$ therefore more than doubling the actual statistical basis of the oscillation search in spring 2001. The anticipated sensitivity will then reach $\sin^2(2\Theta) \simeq 1 \times 10^{-3}$ for large Δm^2 . A detailed statistical analysis of the LSND oscillation parameter region based on a recently proposed frequentist's approach [12] combined with the more sensitive KARMEN2 result is then expected to determine a correct confidence region in $(\Delta m^2, \sin^2(2\Theta))$.

---

resolution functions. Therefore, assuming values of $\Delta m^2 > 100\,\mathrm{eV}^2/\mathrm{c}^4$ for both experiments $\epsilon = 0.5$.

[d] Applying a stringent coincidence cut of $R \geq 30$ which corresponds to the KARMEN coincidence requirements for the 7 extracted events, the LSND sample reduces from 3049 to 70 events, with the oscillation expectation decreasing to 36% of the original 13950 signal events for $\sin^2(2\Theta) = 1, \Delta m^2 = 100\,\mathrm{eV}^2/\mathrm{c}^4$.

References

1. present address: *Los Alamos National Laboratory, Los Alamos, New Mexico 87545, USA, E-mail: eitel@lanl.gov*
2. B. Armbruster *et al*, Phys. Review **C57,** 3414, 1998.
3. R. Maschuw *et al*, Prog. Part. Nucl. Physics **40,** 183, 1998.
4. B. Armbruster *et al*, Phys. Rev. Lett. **81,** 520, 1998.
5. R. Burman *et al*, Nucl. Instrum. Methods **368,** 416, 1996.
6. G. Drexlin *et al*, Prog. Part. Nucl. Physics **40,** 193, 1998.
7. K. Eitel and B. Zeitnitz, hep-ex/9809007, in *Proceed. XVIII Int. Conf. on Neutrino Physics and Astrophysics* (Neutrino'98), Takayama, Japan
8. L. Borodovsky *et al*, Phys. Rev. Lett. **68,** 274, 1992.
9. A. Romosan *et al*, Phys. Rev. Lett. **78,** 2912, 1997.
10. B. Achkar *et al*, Nucl. Phys. **B434,** 503, 1995.
11. R. Tayloe for the LSND collaboration, *these proceedings*
12. G.J. Feldman and R.D. Cousins, Phys. Rev. **D57,** 3873, 1998.

HELICITY AND PARTIAL WAVE AMPLITUDE ANALYSIS OF $D \to V_1 V_2$ DECAY

E. EL AAOUD AND A.N. KAMAL[a]

Theoretical Physics Institute and Department of Physics,
University of Alberta, Edmonton, Alberta T6G 2J1, Canada.

We have carried out an analysis of helicity and partial-wave amplitudes for the process $D \to V_1 V_2$. In particular we have studied the Cabibbo-favored decays $D_s^+ \to \rho\phi$ and $D \to K^*\rho$ in the factorization approximation using several models for the form factors. All the models, with the exception of one, generate partial-wave amplitudes with the hierarchy $\mid S \mid > \mid P \mid > \mid D \mid$. Even though in most models the D-wave amplitude is an order of magnitude smaller than the S-wave amplitude, its effect on the longitudinal polarization could be as large as 30%. Due to a misidentification of the partial-wave amplitudes in terms of the Lorentz structures in the relevant literature, we cast doubt on the veracity of the listed data for the decay $D \to K^*\rho$, particularly the partial-wave branching ratios.

1 Motivation

One peculiarity of a pseudoscalar meson, P, decaying into two vector mesons is that the final-state particles are produced in different correlated polarization states. The hadronic matrix element, $A = \langle V_1 V_2 \mid H_{weak} \mid P \rangle$ can be expressed in terms of three different, but equivalent, bases; the helicity basis $\mid + + \rangle, \mid - - \rangle, \mid 00 \rangle$, the partial-wave basis (or the LS-basis) $\mid S \rangle, \mid P \rangle, \mid D \rangle$ and the transversity basis $\mid 0 \rangle, \mid \parallel \rangle, \mid \perp \rangle$. As a results of the completeness of each one of these bases, the decay rate $\Gamma(D \to V_1 V_2)$ is expressed as an incoherent sum of the helicity, or the partial-wave amplitudes [1]. This imposes some constraints on the helicity and the partial-wave branching fractions B as they should add up to the total branching fraction for the mode $D \to V_1 V_2$ as follows: $B_{++} + B_{--} + B_{00} = B_S + B_P + B_D = B_{V_1 V_2}$. An obvious problem with the $D^0 \to K^{*0}\rho^0$ data [2] is that this constraint is violated: the sum of the branching fractions into S and D states exceeds the total branching fraction. The fact that this sum also exceeds the transverse branching fraction is, by itself, not a problem due to the interference between the S and D waves. However, the problem with the data [2] is that the transverse branching fraction saturates the total branching fraction. There is, therefore, an internal inconsistency in the data: all the data listings can not be correct.

The branching ratio and the longitudinal polarization in $D_s^+ \to \phi\rho^+$ have now been measured [2,3]:

---

[a]talk presented by E.E.

$$B(D_s^+ \to \phi\rho^+) = (6.7 \pm 2.3)$$
$$P_L(D_s^+ \to \phi\rho^+) \equiv \Gamma_L/\Gamma = (0.370 \pm 0.035 \pm 0.038). \tag{1}$$

Theoretically, Gourdin *et al.* [1] studied the ratio

$$R_h \equiv B(D_s^+ \to \phi\rho^+)/B(D_s^+ \to \phi\pi^+) = 1.86 \pm 0.26 \pm \begin{smallmatrix} 0.29 \\ 0.40 \end{smallmatrix} \tag{2}$$

as a function of the ratio $x = A_2(0)/A_1(0)$ and $y = V(0)/A_1(0)$ in three different scenarios for the q^2 dependence of the form factors.

An important point to be made is that though the decay rate $\Gamma(P \to V_1 V_2)$ does not depend on the amplitudes phases, the longitudinal polarization does depend on the phase difference $\delta_S - \delta_D$. We have studied the data shown in (1) within the context of factorization, we also include the effect of final state interaction by allowing for nonzero $S-$, $P-$ and $D-$wave phases.

2   Details of Calculation

The decays $D \to K^*\rho$ and $D_s^+ \to \phi\rho^+$ are Cabibbo-favored and are induced by the effective weak Hamiltonian which can be reduced to the following color-favored (CF) and color-suppressed (CS) forms [4]:

$$H_{CF} = \frac{G_F}{\sqrt{2}} V_{cs} V_{ud}^* [a_1 (\bar{u}d)(\bar{s}c) + c_2 O_8], \tag{3}$$

$$H_{CS} = \frac{G_F}{\sqrt{2}} V_{cs} V_{ud}^* [a_2 (\bar{u}c)(\bar{s}d) + c_1 \tilde{O}_8], \tag{4}$$

where $V_{qq'}$ are the CKM matrix elements. The brackets $(\bar{u}d)$ represent $(V - A)$ color-singlet Dirac bilinears. O_8 and $\tilde{O}_8$ are products of color octet-currents, a_1 and a_2 are the Wilson coefficients.

We consider the following four decays: (i) $D_s^+ \to \phi\rho^+$ (ii) $D^0 \to K^{-*}\rho^+$, (iii) $D^0 \to \bar{K}^{0*}\rho^0$, and (iv) $D^+ \to \bar{K}^{0*}\rho^+$. In the factorization approximation the decay amplitudes are given by:

$$A(D_s^+ \longrightarrow \rho^+\phi) = \frac{G_F}{\sqrt{2}} V_{cs} V_{ud}^* a_1 \langle \phi \mid \bar{s}c \mid D_s^+ \rangle \langle \rho^+ \mid \bar{u}d \mid 0 \rangle , \tag{5}$$

$$A(D^0 \to K^{-*}\rho^+) = \frac{G_F}{\sqrt{2}} V_{cs} V_{ud}^* a_1 \langle \rho^+ \mid \bar{u}d \mid 0 \rangle \langle K^{-*} \mid \bar{s}c \mid D^0 \rangle , \tag{6}$$

$$A(D^0 \to \bar{K}^{0*}\rho^0) = \frac{G_F}{\sqrt{2}} V_{cs} V_{ud}^* \frac{a_2}{\sqrt{2}} \langle \bar{K}^{0*} \mid \bar{s}d \mid 0 \rangle \langle \rho^0 \mid \bar{u}c \mid D^0 \rangle , \tag{7}$$

$$A(D^+ \to \bar{K}^{0*}\rho^+) = \frac{G_F}{\sqrt{2}} V_{cs} V_{ud}^* \left\{ a_1 \langle \rho \mid \bar{u}d \mid 0 \rangle \langle \bar{K}^{*0} \mid \bar{s}c \mid D^+ \rangle \right.$$

$$\left. + a_2 \langle \bar{K}^{0*} \mid \bar{s}d \mid 0 \rangle \langle \rho^+ \mid \bar{u}c \mid D^+ \rangle \right\}$$

$$= A(D^0 \to K^{-*}\rho^+) + \sqrt{2} A(D^0 \to \bar{K}^{0*}\rho^0). \qquad (8)$$

In terms of the helicity amplitudes the decay rate is given by

$$\Gamma(D \to V_1 V_2) = \frac{p}{8\pi m_D^2} \left\{ |H_{00}|^2 + |H_{++}|^2 + |H_{--}|^2 \right\} \qquad (9)$$

where p is the center of mass momentum in the final state. H_{00}, H_{++} and H_{--} are the longitudinal and the two transverse helicity amplitudes, respectively, which for the decay $D^0 \to K^{-*}\rho^+$ they are given by:

$$H_{00}(D^0 \to K^{-*}\rho^+) = -i \frac{G_F}{\sqrt{2}} V_{cs} V_{ud}^* m_\rho f_\rho (m_D + m_{K^{-*}}) a_1 \left\{ \alpha A_1^{DK^*}(m_\rho^2) - \right.$$

$$\left. \beta A_2^{DK^*}(m_\rho^2) \right\} \qquad (10)$$

$$H_{\pm\pm}(D^0 \to K^{*-}\rho^+) = i \frac{G_F}{\sqrt{2}} V_{cs} V_{ud}^* m_\rho f_\rho (m_D + m_{K^{*-}}) a_1 \left\{ A_1^{DK^*}(m_\rho^2) \mp \right.$$

$$\left. \gamma V(m_\rho^2)^{DK^*} \right\}. \qquad (11)$$

$A_i(q^2), (i = 1, 2, 3)$ and $V(q^2)$ are invariant form factors defined in ref. [5] and α, β, γ are defined in ref. [6]: Equivalently one can work with the partial wave amplitudes which are related to the helicity amplitudes by [7],

$$H_{00} = -\frac{1}{\sqrt{3}}S + \sqrt{\frac{2}{3}}D, \qquad H_{\pm\pm} = \frac{1}{\sqrt{3}}S \pm \frac{1}{\sqrt{2}}P + \frac{1}{\sqrt{6}}D. \qquad (12)$$

For completeness, we introduce here the transversity basis, A_0, $A_\parallel$ and $A_\perp$, through

$$A_0 = H_{00} = -\sqrt{\frac{1}{3}}S + \sqrt{\frac{2}{3}}D$$

$$A_\parallel = \sqrt{\frac{1}{2}}(H_{++} + H_{--}) = \sqrt{\frac{2}{3}}S + \sqrt{\frac{1}{3}}D$$

$$A_\perp = \sqrt{\frac{1}{2}}(H_{++} - H_{--}) = P. \qquad (13)$$

The partial waves can be expressed in terms of their phases as follows

$$S = \mid S \mid \exp i\delta_S, \quad P = \mid P \mid \exp i\delta_P, \quad D = \mid D \mid \exp i\delta_D \qquad (14)$$

and the longitudinal polarization is defined by:

$$P_L = \frac{\Gamma_{00}}{\Gamma} = \frac{\mid H_{00} \mid^2}{\mid H_{++} \mid^2 + \mid H_{--} \mid^2 + \mid H_{00} \mid^2}. \tag{15}$$

Using equations (10 - 12) to solve for S, P and D in terms of form factors, we obtain (we drop a common factor of $i\frac{G_F}{\sqrt{2}}V_{cs}V_{ud}^* m_\rho f_\rho (m_D + m_{K^{-*}})a_1$):

$$S = \frac{1}{\sqrt{3}}\left\{(2+\alpha)A_1(q^2) - \beta A_2(q^2)\right\}, \quad P = -\sqrt{2}\gamma V(q^2) \text{ and}$$

$$D = \sqrt{\frac{2}{3}}\left\{(1-\alpha)A_1(q^2) + \beta A_2(q^2)\right\}. \tag{16}$$

The decay rate given by an incoherent sum,

$$\Gamma \propto \left(\mid H_{++} \mid^2 + \mid H_{--} \mid^2 + \mid H_{00} \mid^2\right) = \left(\mid S \mid^2 + \mid P \mid^2 + \mid D \mid^2\right),$$

is independent of the partial-wave phases. However, the polarization does depend on the phase difference, $\delta_{SD} = \delta_S - \delta_D$:

$$P_L = \frac{1}{3}\frac{\mid S \mid^2 + 2\mid D \mid^2 - 2\sqrt{2}\mid S \mid\mid D \mid \cos\delta_{SD}}{\mid S \mid^2 + \mid P \mid^2 + \mid D \mid^2}. \tag{17}$$

In our numerical calculation [3,6] we have used the following model for the form factors: i) Bauer, Stech and Wirbel (BSWI) [5], where an infinite momentum frame is used to calculate the form factors at $q^2 = 0$, and a monopole form for q^2 dependence is assumed to extrapolate all the form factors to the desired value of q^2; ii) BSWII [8] is a modification of BSWI; iii) Altomari and Wolfenstein (AW) model [9], where the form factors are evaluated in the limit of zero recoil, and a monopole form is used to extrapolate to the desired value of q^2; iv) Casalbuoni, Deandrea, Di Bartolomeo, Feruglio, Gatto and Nardulli (CDDGFN) model [10], where the form factors are evaluated at $q^2 = 0$ in an effective Lagrangian satisfying heavy quark spin-flavor symmetry in which light vector particles are introduced as gauge particles in a broken chiral symmetry; v) Isgur, Scora, Grinstein and Wise (ISGW) model [11], where a non-relativistic quark model is used to calculate the form factors at zero recoil and an exponential q^2 dependence is used to extrapolate them to the desired q^2; vi) Bajc, Fajfer and Oakes (BFO) model [12], where the form factors $A_1(q^2)$ and $A_2(q^2)$ are assumed to be flat, and a monopole behavior is assumed for $V(q^2)$; and finally (vii), a parametrization that uses experimental values (Exp.FF) [13] of the form factors at $q^2 = 0$ and extrapolates them using monopole forms.

3 Results and Discussion

3.1 $D_s^+ \to \phi\rho^+$ Decay

The results are summarized in Table 1. For the entries in the last column of Table 1 we have used the experimental values of the form factors at $q^2 = 0$: $A_1^{Ds\phi}(0) = 0.62 \pm 0.06, A_2^{Ds\phi}(0) = 1.0 \pm 0.3, V^{Ds\phi}(0) = 0.9 \pm 0.3$ [13] and extrapolated them with monopole forms. First, we note from the Table 1 that all models, except the CDDGFN and the one where experimentally measured form factors are used, overestimate the decay rate. This fact arises from an overestimate of the form factor A_1. Ref. [14] has noted this fact and attributes it to the imposing of chiral symmetry. Further, as ref. [15] has argued, more theoretical as well as experimental studies are needed for a better understanding of the q^2 dependence of form factors. Second, we observe that all the six sources of form factors allow a range for the polarization which overlaps with experiment with $\delta_{SD} \neq 0$. Note that the polarization is independent of the normalization of A_1. It is also found that most of the final state in the decay $D_s^+ \to \rho^+\phi$ is in the S wave. It is also seen from the Table 1 that the hierarchy of the partial wave amplitudes is: $| S | > | P | > | D |$. If we consider the final state to get contribution only from S wave the decay rates would only be reduced by (5 to 12) %, while the polarization would be $P_L = 0.33$. The hierarchy of the sizes of the partial wave amplitudes is in accordance with intuitive expectations based on threshold arguments. It is the S-wave dominance which makes an accurate determination of δ_{SD} difficult (the errors in δ_{SD} are large despite the fact that the errors in P_L are small) since the D-wave is an order of magnitude smaller than the S-wave.

3.2 $D^0 \to K^{*-}\rho^+$

We calculate the experimental value of polarization from the listing of Ref. [2]:

$$P_L = \frac{\Gamma(D^0 \to \rho^+\bar{K}^{*-}_{longitudinal})}{\Gamma(D^0 \to \rho^+\bar{K}^{*-})} = \frac{2.9 \pm 1.2}{6.1 \pm 2.4} = 0.475 \pm 0.271 \qquad (18)$$

In Table 2 we have summarized the results for the decay rates Γ, longitudinal polarization P_L, and partial-wave ratios $\frac{|S|}{|P|}$ and $\frac{|S|}{|D|}$ in different models. We note from Table 2 that the models CDDGFN, BFO, and the scheme that uses experimentally measured form factors, predict a decay rate within a standard deviation of the central measured value. All other models overestimate the rate by several standard deviations. As for the longitudinal polarization, given the freedom of the unknown $\cos\delta_{SD}$, all models are able to fit the data. In particular, all models except BFO are able to predict the polarization correctly for $\delta_{SD} = 0$; in the BFO model for $\delta_{SD} = 0$, $D^0 \to K^{*-}\rho^+$ becomes totally transversely polarized. This circumstance arises from the fact that BFO model predicts a large D-wave contribution, $\frac{|S|}{|D|} \approx \sqrt{2}$. It then becomes evident from Eq. (12) that H_{00} vanishes. All models except BFO also display the partial-wave-amplitude hierarchy: $| S | > | P | > | D |$; BFO model on the other hand predicts $| S | > | D | > | P |$, which we believe is less likely. The reasoning goes as follows: for decays close to threshold, one anticipates the Lth partial-wave amplitude to behave like $(p/\Lambda)^L$, where p is the center of mass momentum and Λ a mass scale. For $p \sim 0.4~GeV$ and $\Lambda \sim m_D$, one expects the hierarchy $| S | > | P | > | D |$.

3.3 $D^+ \to \bar{K}^{0*}\rho^+$

In contrast to the decay mode $D^0 \to \bar{K}^{-*}\rho^+$, here the data listing[2] is at best confusing. First, since the longitudinal and/or transverse branching ratios are not listed, it is not possible to calculate the longitudinal polarization. Second, though in refs. [16,17] the identification of the transversity amplitudes, (A_T, A_L and $A_{l=1}$ in the notation of ref. [16]) in terms of the partial-wave amplitudes is correct (see Eqns. (20) - (26) of ref. [16]), their identification of the partial-wave amplitudes S and D in terms of the Lorentz structure of the decay amplitude is incorrect. In Table II, and more succinctly in Eqns. (32) and (34) of ref. [16], S-wave amplitude is identified with the Lorentz structure that goes with the form factor A_1, and D-wave amplitude with that of A_2. In fact, the correct identification of the S- and D-wave amplitudes given in Eq. (16) shows that they are both linear superpositions of A_1 and A_2.

With the caveat that the identification of the partial waves in refs. [16,17] is incorrect, we take the S-, P- and D-wave branching ratios at their face value and calculate the 'experimental' ratios $\frac{|S|}{|P|}$ and $\frac{|S|}{|D|}$.

In Table 2, we have shown the calculated decay rate, the longitudinal polarization and the ratios of the partial-wave amplitudes in different models and compared them with the data. The BFO model is the only one that reproduces the total rate correctly. This model also generates a large D-wave amplitude, with the partial-wave hierarchy $| S | > | D | > | P |$. This feature of the BFO model is due to the exceptionally large value of the form factor A_2, which is in contradiction with the experimental determination of the form factor as shown in Table 3.

3.4 $D^0 \rightarrow \bar{K}^{0*}\rho^0$

Ref. [2] lists the branching ratio and the transverse branching ratio. This enables us to calculate the longitudinal polarization from

$$P_L = 1 - P_T = 1 - \frac{B(D^0 \rightarrow \bar{K}^{0*}\rho^0{}_{transverse})}{B(D^0 \rightarrow \bar{K}^{0*}\rho^0)} = 0.0 \pm {}^{0.4}_{0.0}. \quad (19)$$

Ref. [2] also lists the S- and D-wave branching ratios. However, our criticism of these numbers in the previous subsection applies also to $D^0 \rightarrow \bar{K}^{0*}\rho^0$ decay. With this caution, we have taken their ref. [2] numbers at face value and calculated the experimental and theoretical ratios of the partial wave amplitudes and listed them in Table 2. We note from Table 2 that the rate in the BFO model is too low by three standard deviations; the rates predicted in BSWI and BSWII models are 1.5 standard deviations too high, while all other models fit the rate within one standard deviation. As for the longitudinal polarization, all models predict a value consistent with the data. All models also satisfy the $\frac{|S|}{|P|}$ bound, but only the BFO model fits the $\frac{|S|}{|D|}$ ratio. This is because the BFO model generates a large D-wave amplitude.

4 S-wave- and $A_1(q^2)$-dominance

Since S-wave and D-wave amplitudes are linear superpositions of the form factors A_1 and A_2, see Eq. (16), the concept of S-wave-dominance is different from that of A_1-dominance. All the models we have discussed, with the exception of BFO model, predict that the S-wave amplitude is the dominant partial wave amplitude. Further, since ref. [16] identifies $S \sim A_1$ and $D \sim A_2$, we need to look at what is meant by S-wave-dominance and contrast it with A_1-dominance.

Consider first the concept of S-wave dominance. We see from Eq. (9, 12) that in this approximation, $\Gamma \propto |S|^2$, and $|H_{00}| = |H_{++}| = |H_{--}| = |\frac{S}{\sqrt{3}}|$. In practice, most of the models predict the S-wave amplitude to be roughly an order of magnitude larger than the D-wave amplitude. Consequently, D wave would contribute only 1% to the rate relative to the S-wave. However, it could influence the longitudinal polarization considerably through its interference with the S wave. Depending on the value of δ_{SD} the interference term could amount to a 30% correction to P_L (see also ref. [18]). However, regardless of the exact size of the D-wave amplitude, S-wave dominance would lead to $P_L \rightarrow \frac{1}{3}$, for $\delta_{SD} = \frac{\pi}{2}$.

Consider now the concept of A_1-dominance. From Eqns. (10) and (11), we see that $H_{00} \propto \alpha A_1$ and $H_{++} = H_{--} \propto A_1$. With $\alpha = 1.52$, the longitudinal helicity amplitude is the largest, and the longitudinal polarization becomes

$$P_L = \frac{\alpha^2}{2 + \alpha^2} = 0.54, \tag{20}$$

in contrast to a value $1/3$ (with an error from $S - D$ interference) for S-wave dominance. Further, from Eq. (16), we note that in A_1-dominance,

$$S \propto \frac{2 + \alpha}{\sqrt{3}} A_1(q^2), \quad D \propto \sqrt{\frac{2}{3}}(1 - \alpha)A_1(q^2), \tag{21}$$

which makes the S-wave amplitude five times larger than the D-wave amplitude - not quite what one would term "S-wave dominance."

Table 1: Decay rate and longitudinal polarization for $D_s^+ \longrightarrow \rho^+\phi$. The values of Γ must be multiplied by $10^{12}s^{-1}$. $\delta_{SD} = \delta_S - \delta_D$ is the value needed to get agreement with P_L data to one STD. The last column uses experimentally measured form factors. 'Expt.FF' stands for 'Experimental form factors' .

	$BSWI$	$BSWII$	AW	$CDDGFN$	$ISGW$	$Expt.FF$
Γ	0.32	0.32	0.35	0.15	0.37	0.18 ± 0.04 [2]
δ_{SD}	135 ± 45	138 ± 43	122 ± 32	140 ± 40	120 ± 35	134 ± 46
$\frac{\|S\|}{\|P\|}$	4.3	3.7	3.8	2.8	5.5	4.7
$\frac{\|S\|}{\|D\|}$	11.9	13.5	7.4	8.2	8.6	16.5
Experimental values of Γ and P_L			$\Gamma = 0.14 \pm 0.05$ [2], $P_L = 0.370 \pm 0.052$			

Acknowledgments

This research was partially funded by the Natural Sciences and Engineering Research Council of Canada through a grant to A.N.K.

References

1. A. Ali, J.G. Körner and G. Kramer, Z. Phys. **C1**, 269, (1979). A.S. Dighe, I. Dunietz, H.J. Lipkin and J.L. Rosner, Phys. Lett. **B369**, 144, (1996). M. Gourdin, A.N. Kamal, Y.Y. Keum, X.Y. Pham, Phys. Lett. **B339**, 173, (1994)
2. Particle Data Group, C. Caso et al., Eur. Phys. J. **C3**, 1, (1998).
3. El hassan El aaoud, Phys. Rev. **D58**, 037502, (1998).
4. A.N. Kamal, A.B. Santra, T. Uppal, R.C. Verma, Phys. Rev. **D53**, 2506, (1996).

Table 2: Decay rates for $D^{+,0} \longrightarrow \bar{K}^{0*}\rho^{+,0}$. The values of Γ must be multiplied by $10^{11}s^{-1}$. The parameter $z = \cos\delta_{SD}$. The experimental values of P_L are listed only if measurements of longitudinal or transverse branching ratios are available [2].

		BSI	BSII	AW	CDD	ISG	BFO	Exp.FF	Expt.
D^0	Γ	4.99	4.96	4.63	2.20	4.56	1.03	2.20	1.47± 0.58
↓	P_L	0.319 - 0.084z	.313 - .071z	.316 - .122z	.315 - .127z	.324 - .108z	.418 - .417z	.315 - .127z	.475± 0.271
K^{-*}	$\frac{|S|}{|P|}$	4.3	3.7	3.6	3.5	4.7	2.9	3.5	
ρ^+	$\frac{|S|}{|D|}$	10.6	12.3	7.0	6.7	8.3	1.4	6.7	
D^+	Γ	1.56	1.54	1.50	0.409	1.69	0.268	0.559	0.20± 0.12
↓	P_L	0.326 - 0.086z	.325 - .079z	.319 - .141z	.318 - .128z	.333 - .129z	.416 - .416z	.321 - 0.134z	
$\bar{K}^{0*}$	$\frac{|S|}{|P|}$	5.5	5.3	3.6	3.7	6.9	2.8	4.0	$> 2^{21}$
ρ^+	$\frac{|S|}{|D|}$	10.6	11.5	6.1	6.7	7.0	1.4	6.5	1.3± 0.8^{21}
D^0	Γ	0.481	.488	.426	.353	.351	0.124	0.267	.354± 0.080
↓	P_L	0.309 - 0.080z	.294 - .060z	.314 - .097z	.313 - .125z	.379 - .074z	.420 - .419z	.307 - .119z	$0.0^{+0.4}_{-0}$
$\bar{K}^{0*}$	$\frac{|S|}{|P|}$	3.4	2.7	3.6	3.3	3.1	3.0	3.0	$> 2.8^{\,21}$
ρ^0	$\frac{|S|}{|D|}$	10.7	13.7	8.9	6.8	11.5	1.4	7.1	1.21± 0.23^{21}

5. M. Wirbel, B. Stech and M. Bauer, Z. Phys. **C29**, 637, (1985); M. Bauer, M. Wirbel, Z. Phys. **C42**, 671, (1989).

6. El hassan El aaoud and A.N. Kamal, to be published in Phys. Rev. **D** (June 1999).

7. N. Sinha, Ph.D thesis, University of Alberta (1989). A.S. Dighe et al., in Ref. [1].

8. M. Neubert, V. Rieckert, B. Stech and Q.P. Xu, in *Heavy Flavours*, eds. A.J. Buras and M. Lindner (World Scientific, Singapore, 1992) p. 286.

9. T. Altomari and L. Wolfenstein, Phys. Rev. **D37**, 681, (1988).

10. R. Casalbuoni, A. Deandrea, N. Di Bartolomeo, R. Gatto, F. Feruglio, and G. Nardulli, Phys. Lett. **B299**, 139, (1993). S.M. Ryan and J.N. Simone, Phys. Rev. **D58**, 014506, (1998).

11. N. Isgur, D. Scora, B. Grinstein and M.B. Wise, Phys. Rev. **D39**, 799, (1989).

12. B. Bajc, Fajfer, R.J. Oakes, Phys. Rev. **D53**, 4957, (1996); B. Bajc, S. Fajfer, R.J. Oakes, S. Prelovšek, Phys. Rev. **D56**, 7207, (1997).

Table 3: Model and experimental predictions for the form factors : $A_{1,2}(q^2), V(q^2)$ and the ratios $x = \frac{A_2(0)}{A_1(0)}, y = \frac{V(0)}{A_1(0)}$ for the process $D \longrightarrow K^* \rho$.

	BSI	BSII	AW	CDD	ISG	BFO	Exp.FF
$A_1^{DK^*}(m_\rho^2)$	0.969	0.969	0.887	0.606	0.909	0.578	0.606
$A_2^{DK^*}(m_\rho^2)$	1.264	1.392	0.707	0.441	0.929	3.747	0.441
$V^{DK^*}(m_\rho^2)$	1.414	1.630	1.602	1.153	1.25	0.773	1.153
$x = \frac{A_2(0)}{A_1(0)}$	1.30	1.30	0.80	0.73	1.02	6.5	0.73 [20]
$y = \frac{V(0)}{A_1(0)}$	1.39	1.39	1.73	1.82	1.38	1.16	1.87 [20]
$A_1^{D\rho}(m_{K^*}^2)$	0.898	0.898	0.835	0.732	0.766	0.605	0.637
$A_2^{D\rho}(m_{K^*}^2)$	1.070	1.240	0.846	0.487	0.958	3.574	0.464
$V^{D\rho}(m_{K^*}^2)$	1.529	1.908	1.343	1.326	1.41	0.713	1.248

13. Particle Data Group, R.M. Barnett et al., Phys. Rev. **D54**, 1, (1996).

14. F. Buccella, M. Lusignoli, G. Miele, A. Pugliese and P. Santorelli, Phys. Rev. **D51**, 3478, (1995).

15. R. Casalbuoni, A. Deandrea, N. Di Bartolomeo, R. Gatto, F. Feruglio, and G. Nardulli, Phys. Rep. 281, 145, (1997).

16. MARK III Collaboration, D. Coffman et. al., Phys. Rev. **D45**, 2196, (1992).

17. D.F. DeJongh, Ph.D. thesis, California Institute of Thechnology, 1991.

18. H. Arenhövel, W. Leidemann and E.L. Tomusiak, Nucl. Phys. **A641**, 517, (1998).

19. The value of experimental form factors (Exp. FF) are calculated using the nearest pole approximation: $F^{DK^*}(m^2) = \frac{F^{DK^*}(0)}{1-\frac{m^2}{\Lambda^2}}$ where the values of $F^{DK^*}(0)$ are taken from [13] and the pole mass Λ are from [5]. In calculating $F^{D\rho^+}(m_{K^*}^2)$ we have used the approximation $F^{D\rho^+}(0) \approx F^{DK^*}(0)$.

20. E791 Collaboration, E.M. Aitala et al., hep-ex/9809026.

21. These values represent numbers extracted from ref. [2]. See, however, our criticism of the data.

STUDY OF ZZ PRODUCTION IN e⁺e⁻ COLLISIONS AT $\sqrt{s} = 189$ GeV WITH THE L3 DETECTOR AT LEP

M.A. FALAGÁN

CIEMAT, Edif. 2, Avda. Complutense 22,
28040 MADRID, SPAIN
E-mail: falagan@ae.ciemat.es

L3 COLLABORATION
CERN,
CH-1211, Geneva 23, Switzerland

A study of ZZ production using the 176 pb⁻¹ of integrated luminosity collected by the L3 detector at LEP at the centre-of-mass energy of 188.66 GeV is performed. The ZZ production cross section is measured to be $\sigma_{ZZ} = 0.54^{+0.16}_{-0.14}$ pb, in agreement with the Standard Model expectation. No evidence for the existence of anomalous triple gauge boson ZZZ and ZZγ couplings is found and limits on these couplings are set. All the presented results are preliminary.

1 Introduction

In e⁺e⁻ collisions at centre-of-mass energies above $2 * M_Z \simeq 182$ GeV, the production of a pair of on-shell Z bosons as intermediate state contributes to four-fermion final states. The study of these events is a new test of the Standard Model, as cross sections and distributions different from its predictions could signal the existence of new physics. Moreover, ZZ events are an important background for Higgs searches, and therefore must be well understood.

Although the ZZ production rate is low and there are large backgrounds, it is being measured at LEP since 1997, and deviations from the Standard Model are searched for.

Some of the Feynman Diagrams involved in four-fermion production are shown in Fig. 1. WW production is one of the dominant processes at LEP2 energies; its cross section is 25 times greater than the ZZ cross section, and gives rise to a large and partially irreducible background for some final states. On-shell ZZ production proceeds through the ZZ conversion diagram shown in Fig. 1. Multiperipheral diagrams give a large contribution to final states with electrons at low angle. Many other diagrams contribute to four-fermion final states [1].

In order to be highly sensitive to the on-shell ZZ production diagram and have a small contribution from the others, we define our ZZ signal in four-fermion final states, in terms of the following criteria:

- Two fermion-antifermion pairs with invariant masses in the range $(70, 105)$ GeV

- In final states with electrons, these electrons must verify $|\cos\theta_e| < 0.95$, θ_e being their angle with respect to the beam axis

- In final states with WW contribution, namely $u\bar{u}d\bar{d}$, $c\bar{c}s\bar{s}$ and $l^+l^-\nu\bar{\nu}$, the masses of the fermion pairs susceptible to come from W decay must be outside $(75,85)$ GeV

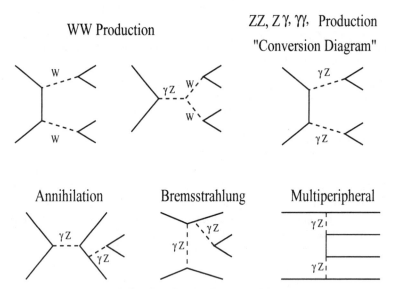

Figure 1: A few four-fermion Feynman Diagrams.

The ZZ cross section as calculated with our signal definition is shown in Fig. 2 as a function of the centre-of-mass energy. The EXCALIBUR [2] four-fermion Monte Carlo generator has been used for this calculation. There is a threshold at $\sqrt{s} \sim 2 * M_Z$ and the maximum is reached slightly above the LEP2 energies.

In Table 1 the cross sections for the different ZZ final states at $\sqrt{s} = 188.66$ GeV are shown. The small differences between the actual fractions and those calculated naively from the Z branching ratios are due to the contribution of non-resonant Feynman Diagrams to the signal, and to the cuts applied in the signal definition to minimize this non-resonant contribution.

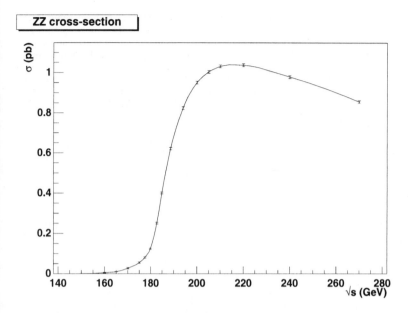

Figure 2: ZZ cross-section calculated with EXCALIBUR.

2 Data and Monte Carlo samples

The analysis presented here corresponds to a luminosity of 176 pb^{-1} collected by the L3 experiment at $\sqrt{s} \simeq 188.66$ GeV during 1998.

The EXCALIBUR Monte Carlo is used to simulate the four-fermion events. KORALW 1.21 [3] is also used for on–shell W^+W^- production. Backgrounds from fermion–pair production are simulated with PYTHIA 5.72 [4] ($e^+e^- \rightarrow q\bar{q}(\gamma)$), KORALZ 4.02 [5] ($e^+e^- \rightarrow \mu^+\mu^-(\gamma)$ and $e^+e^- \rightarrow \tau^+\tau^-(\gamma)$) and BH-WIDE 3 [6] ($e^+e^- \rightarrow e^+e^-(\gamma)$).

The L3 detector response is simulated using the GEANT 3.15 program [7], which takes into account the effects of energy loss, multiple scattering and showering in the detector. The GEISHA program [8] is used to simulate hadronic interactions in the detector.

3 ZZ Selection

ZZ events are characterized experimentally by two pairs of visible fermions, (leptons or hadronic jets), with high invariant masses, or a single pair of fermions with large visible and recoil masses.

Table 1: ZZ cross sections calculated with EXCALIBUR. The first column refers to the ZZ decay channel. The second column contains the cross sections. The third column gives the fraction of the total that the channel represents. The fourth column are the fractions calculated from the branching ratio of the Z boson, taken from the Particle Data Group.

ZZ →	σ (pb)	Fraction	'Naive' (B.R.)
$q_1\bar{q}_1q_2\bar{q}_2$	$0.2895 \pm 0.2\%$	46.8 %	48.9 %
$q\bar{q}\nu\bar{\nu}$	$0.1702 \pm 0.2\%$	27.4 %	28.0 %
$q\bar{q}l^+l^-$	$0.0973 \pm 0.2\%$	15.6 %	14.1 %
$l^+l^-\nu\bar{\nu}$	$0.0271 \pm 0.3\%$	4.4 %	4.0 %
$l_1^+l_1^-l_2^+l_2^-$	$0.0111 \pm 2.2\%$	1.8 %	1.0 %
$\nu_1\bar{\nu}_1\nu_2\bar{\nu}_2$	$0.0248 \pm 0.2\%$	4.0 %	4.0 %
ZZ	$0.6200 \pm 0.1\%$	100.0 %	100.0 %

Different cut-based selections as well as artificial neural networks are used to select the ZZ events, rejecting a large amount of the background. The results are sumarized in Table 2.

Table 2: ZZ selection results. The first column refers to the ZZ decay channel. The second one contains the experimental signatures. In the third column are the number of observed events in the data. In the fourth one shows the number of expected events from background. In the fifth column, the signal efficiencies are shown.

ZZ →	Experimental Signature	DATA	BKG	Eff.
$q_1\bar{q}_1q_2\bar{q}_2$	4 hadronic jets	211	207.5 ± 9.0	34 %
$q\bar{q}\nu\bar{\nu}$	2 jets + missing energy	36	22.7 ± 0.1	35 %
$q\bar{q}l^+l^-$	2 jets + 2 leptons	12	3.8 ± 0.5	77 % (qqee) 61 % (qqμμ) 32 % (qqττ)
$\nu_1\bar{\nu}_1\nu_2\bar{\nu}_2$	invisible!	0	0.0	0 %

The qqqq selection is based on the following criteria:

- A preselection of events is done requiring high multiplicity, large visible energy, small missing energy, no energetic leptons and at least four jets ($Y_{34} > .007$)

- A neural network is used to distinguish among ZZ, WW and qq. The inputs are variables related to the multiplicity, energy and direction of the jets, and kinematic variables such as invariant masses of pairs of jets

In Fig. 3(a) the output of the neural network is shown. The qqnn selection is based on the following criteria:

- High multiplicity events

- $\sin\theta_{\text{Pmiss}} > 0.7$

- 2 acoplanar, ($> 2.38^{o}$), and acollinear, ($> 2.35^{o}$), hadronic jets.

- $P_T/\sqrt{s} > 0.06$

- Visible Energy $\in (0.42, 0.56)\sqrt{s}$

- Large visible and recoil masses

In Fig. 3(b) the visible invariant mass is shown. For the qqll selection three selections, qqee, qq$\mu\mu$ and qq$\tau\tau$, are combined with a logical 'or'. The criteria are:

- qqee selection: 2 electrons and 2 jets with invariant masses above 50 GeV. Opening angle of the Z bosons above 130°. Large visible energy and small missing momentum.

- qq$\mu\mu$ selection: 2 muons and 2 jets with invariant masses above 50 GeV. Opening angle of the Z bosons above 130°.

- qq$\tau\tau$ selection: 2 low multiplicity jets and 2 high multiplicity jets with invariant masses in the range (70,120) GeV. Visible mass slightly below $\sqrt{s}$.

In Fig. 3(c) the invariant mass of the two leptons is shown.

4 Cross Section results

Combining the results from the qqqq, qq$\nu\nu$ and qqll selections, the ZZ cross section is measured. From the three experimental distributions shown previously, a negative Log Likelihood with one free parameter, the ZZ cross section, is minimized, as shown in Fig. 3(d). The measured cross section is $0.54^{+0.16}_{-0.14}$ pb which can be compared to the Standard Model prediction, 0.62 pb. The error is statistical only. Our previous published result at $\sqrt{s} = 183$ GeV [9], $0.30^{+0.22}_{-0.16}\,^{+0.07}_{-0.03}$ pb, was also compatible with the Standard Model prediction 0.25 pb.

376

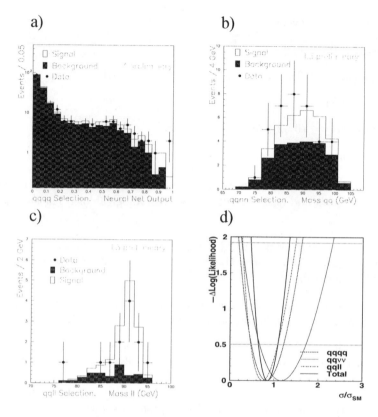

Figure 3: A few experimental distributions for the selected ZZ events in three different selections: NN output for the qqqq selected events (a), visible invariant mass for the qqnn selected events (b), invariant mass of leptons for the qqll selected events (c), minus log-likelihood of the distributions as a function of the ratio of the measured cross section to the Standard Model prediction (d).

5   ZZZ and ZZγ Anomalous Couplings

The study of ZZ production allows us to look for deviations from the SM in triple gauge boson couplings. Standard ZZ production is mainly due to the conversion diagram (see Fig. 1). The existence of ZZZ or ZZγ anomalous couplings would give rise to the new diagrams shown in Fig. 4.

Four possible couplings[10] are associated to this diagram: f_4^Z, f_4^γ, f_5^Z and f_5^γ, all of them vanish in the Standard Model at tree level. f_4^Z and f_4^γ are CP-violating whereas f_5^Z and f_5^γ are P-violating.

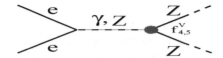

Figure 4: ZZ production diagrams with Anomalous Couplings.

The signatures of these anomalous couplings would be:

- Change in the total cross section $e^+e^- \rightarrow ZZ$

- Modification of the angular distributions of the Z bosons

- Change in the average polarization of the Z's, and therefore change in the angular distributions of the decay fermions from the Z's.

The amplitude for the anomalous coupling diagram has been calculated analytically, and the result has been added to the full set of amplitudes used by the EXCALIBUR program[11].

Therefore, for any four-fermion Monte Carlo event, knowing the four four-momenta, the corresponding amplitude squared is calculated as a function of the anomalous couplings. Then to obtain distributions with anomalous couplings, the SM distributions are reweighted, event by event, according to the ratio of matrix elements represented in Fig. 5.

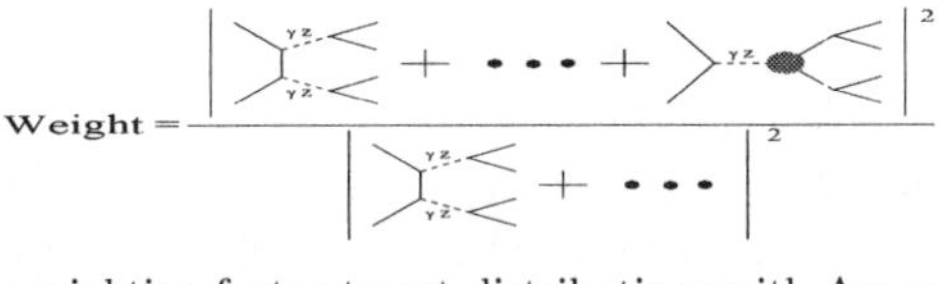

$$\text{Weight} = \frac{\left| \text{[diagram]} + \cdots + \text{[diagram]} \right|^2}{\left| \text{[diagram]} + \cdots \right|^2}$$

Figure 5: Reweighting factor to get distributions with Anomalous Couplings.

The histograms of the experimental variables shown previously are used to compare the measured data with the Monte Carlo expectations, (which include background and ZZ signal). The values of anomalous couplings are varied, reweighting in the Monte Carlo histogram,the four-fermion events susceptible to come from the decay of two Z's, in order to maximize a binned likelihood:

$$\mathcal{L} = \prod_{i=1}^{N_{BIN}} \left[\frac{N_{MC}(i)^{N_{DATA}(i)}}{N_{DATA}(i)!} e^{-N_{MC}(i)} \right]$$

where

$$N_{MC}(f_4^V, f_5^V) = N_{BKG} + N_{SIGNAL}(f_4^V, f_5^V)$$

From the qqll selection at $\sqrt{s} \simeq 189$ GeV, the measured values of the four possible anomalous couplings are compatible with zero, the Standard Model prediction. Limits at the 95 % confidence level are shown in Table 3. These new limits improve notably our previous published result at $\sqrt{s} = 183$ GeV[9].

Table 3: 95 % confidence level limits to ZZZ and ZZγ anomalous couplings, from the qqll selection at 189 GeV (first column), compared to our previous published result, from the combination of all the selections performed at 183 GeV (second column).

Limits with $\sqrt{s} \simeq 189$ GeV Data	Limits with $\sqrt{s} \simeq 183$ GeV Data
$-1.8 < f_4^Z < 1.8$	$-3.6 < f_4^Z < 3.4$
$-5.5 < f_5^Z < 5.4$	$-8.4 < f_5^Z < 7.9$
$-1.1 < f_4^\gamma < 1.0$	$-2.1 < f_4^\gamma < 2.1$
$-3.3 < f_5^\gamma < 3.2$	$-4.9 < f_5^\gamma < 4.8$

References

1. "Physics at LEP2", Eds. G. Altarelli, T. Sjöstrand and F. Zwirner, CERN 96-01 (1996).
2. F.A. Berends, R. Kleiss and R. Pittau, Nucl. Phys. **B424**, 308, (1994); Nucl. Phys. **B426**, 344, (1994); Nucl. Phys. (Proc. Suppl.) **B37**, 163, (1994); Phys. Lett. **B335**, 490, (1994); R. Kleiss and R. Pittau, Comp. Phys. Comm. **83**, 14, (1994).
3. M. Skrzypek et al., Comp. Phys. Comm. **94**, 216, (1996) ; M. Skrzypek et al., Phys. Lett. **B372**, 289, (1996).
4. T. Sjöstrand, CERN-TH/7112/93 (1993), revised August 1995; T. Sjöstrand, Comp. Phys. Comm. **82**, 74, (1994); T. Sjöstrand, CERN-TH 7112/93 (1993, revised August 1994).
5. S. Jadach, B.F.L. Ward and Z. Wąs, Comp. Phys. Comm. **79**, 503, (1994).
6. S. Jadach, W. Paczek and B.F.L. Ward, Phys. Lett. **B390**, 298, (1997).
7. R. Brun et al., preprint CERN DD/EE/84-1 (Revised 1987).
8. H. Fesefeldt, RWTH Aachen Report PITHA 85/02 (1985).
9. "Study of neutral current four-fermion and ZZ production in e^+e^- collisions at $\sqrt{s} = 183$ GeV". L3 Collab. M. Acciarri et al., Phys. Lett. **B**, accepted.
10. K. Hagiwara et al, Nucl. Phys. **B282**, 253, (1987).
11. J. Alcaraz et al., CIEMAT-870 Report, December 1998. http://xxx.lanl.gov/hep-ph/9812435

$B - \overline{B}$ MIXING AND MIXING-INDUCED CP VIOLATION AT CLEO

A.D. FOLAND

Wilson Synchrotron Laboratory
Cornell University
Ithaca, NY 14850, USA
E-mail: adf4@cornell.edu

I describe a new technique for measuring mixing-induced CP violation, integrating over the decay time of one B. The technique does not require the observation of the time difference or time-ordering of the two B decays. A measurement of the mixing parameter x_d, which is a necessary input for the extraction of $\sin 2\beta$ is discussed. I describe the experimental requirements of a CP violation measurement, and describe the extent to which CLEO has already met them with 9 fb^{-1} of data taken at the $\Upsilon(4S)$.

Observed CP violation in the neutral K system[1] is explained in the Standard Model by the Cabbibo-Kobayashi-Maskawa (CKM) mechanism[2,3]. Despite 35 years of study, no other system has been found to display CP violation, profoundly limiting the ability to test the CKM model. One of the most important proving grounds for this test is the neutral B system, which is expected to exhibit large CP asymmetries due to the interference of mixed and direct decays to CP eigenstates (henceforth referred to as mixing-induced CP violation). Upcoming asymmetric B factories, operating at the $\Upsilon(4S)$, propose to measure CP asymmetries. I examine a new method for use at the $\Upsilon(4S)$, in which the decay time of only one B is observed.

Extraction of a mixing-induced CP-violation parameters requires experimental input for the $B - \overline{B}$ mixing parameter $x_d \equiv \frac{\Delta m}{\Gamma}$. I describe a preliminary measurement of x_d at CLEO.

An experiment to measure CP violation consists of the following pieces: reconstruction of CP eigenstates; measurement of the CP eigenstate decay time; tagging of flavor of the other B in the event; and an unbinned maximum likelihood fitter to extract the value of $\sin 2\beta$ from the data. Finally, I discuss each of these aspects in relation to the current CLEO data set of 9 fb^{-1}.

1 CP Violation in B Mesons

Once the $\Upsilon(4S)$ has decayed to a pair of neutral B mesons, the $B^0\overline{B}^0$ exhibit coherent flavor oscillations until the time of the first meson's decay. Afterwards the second B meson oscillates freely, with period Δm, from its state at the time of the first B decay, until its own decay. When one B decays to a CP eigenstate, CP violation can be established from the other B meson via a flavor asymmetry in flavor-specific decays. Because the meson pair is in a $C = -1$ state, the excess vanishes if no time information is used. This leads one naturally to consider $\Upsilon(4S)$ production with asymmetric colliders, where the large boost allows time information to be easily observed.

In the existing literature on $\Upsilon(4S)$ measurements, observation of the time difference Δt between the two B decay times is often stated as a necessity. It is not uncommon to find such statements as "to this end [measuring CP violation], one needs to determine the time interval between the two B-meson decays" [4], "only if we can observe the decay time difference $T = t - t'$ [Δt] can we measure a nonzero asymmetry in the decay of a C=-1 $B\overline{B}$ state" [5], or "a determination of Δt is required for the observation of a CP asymmetry in experiments at the $\Upsilon(4S)$" [6]. Several methods have been proposed to measure CP violation using the observed time difference [7,8]. In this note, I show that for mixing-induced CP violation (such as the "golden mode" $B \to \psi K_S$), neither time ordering nor time-differences are required in order to be sensitive to CP violation at the $\Upsilon(4S)$.

1.1 Formalism for CP Violation

One may begin with the expression for decay rate of the correlated $B^0\overline{B^0}$ state into any pair of final states f_1 at time t_1 and f_2 at time t_2:

$$
\begin{aligned}
R(t_1, t_2) \propto e^{-\Gamma(t_1+t_2)} \{ & (|A_1|^2 + |\overline{A}_1|^2)(|A_2|^2 + |\overline{A}_2|^2) - 4R(\frac{q}{p}A_1^*\overline{A}_1)R(\frac{q}{p}A_2^*\overline{A}_2) \\
& - \cos(\Delta m(t_1 - t_2))[(|A_1|^2 - |\overline{A}_1|^2)(|A_2|^2 - |\overline{A}_2|^2) - 4I(\frac{q}{p}A_1^*\overline{A}_1)I(\frac{q}{p}A_2^*\overline{A}_2)] \\
& + 2\sin(\Delta m(t_1 - t_2))[I(\frac{q}{p}A_1^*\overline{A}_1)(|A_2|^2 - |\overline{A}_2|^2) - (|A_1|^2 - |\overline{A}_1|^2)I(\frac{q}{p}A_2^*\overline{A}_2)] \},
\end{aligned}
$$

$$(1)$$

where I use the notation found in ref. [9], with the standard [10] meanings for λ, p, and q: p and q are the coefficients for the quantum-mechanical admixture of B^0 and $\overline{B}^0$ which comprises the B mass eigenstates; $\lambda = \frac{q}{p}\frac{\overline{A_{f_{CP}}}}{A_{f_{CP}}}$. For the CP eigenstate ψK_s, $Im(\lambda_{CP})$ is equal to $\sin 2\beta$, where β is the usual angle of the unitarity triangle [9], and is nearly equal to $\arg(V_{td})$. A_i is the amplitude for a B^0 to decay to f_i, and $\overline{A}_i$ is the amplitude for $\overline{B^0}$ to decay to the same state f_i. As the B and $\overline{B}$ have few decay final states in common, $\frac{\Delta\Gamma}{\Gamma}$ is expected to be small, and has been set to 0 throughout this communication.

To measure CP violation, one observes a decay at time t_1 to a CP eigenstate f_{CP} and a decay at time t_2 for a flavor-tagging state f_{tag}, which may have sign $+$ or $-$ for $B^0(\overline{B}^0)$ decays. In the case of the flavor-tagging decay of a B^0, $A_2 = A_{tag}$ and $\overline{A}_2 = 0$. Integrating over time t_2 and taking $|\lambda_{CP}| = 1$ (i.e., neglecting direct CP violation), we obtain [11] for the expression for the distribution of decay times to a CP eigenstate:

$$
\begin{aligned}
R_{\pm}(t_{CP}) &= \frac{2C|\overline{A}_{tag}|^2 \||A_{CP}|^2}{\Gamma(1 + x^2)} e^{-\Gamma t_{CP}} \{(1 + x^2) \\
&\quad \pm Im(\lambda_{CP})[x\cos(\Delta m t_{CP}) - \sin(\Delta m t_{CP})]\} \\
&= \frac{2C|\overline{A}_{tag}|^2 \||A_{CP}|^2}{\Gamma} e^{-\Gamma t_{CP}} \{1 \mp \frac{Im(\lambda_{CP})}{\sqrt{1 + x^2}} [\sin(x(\frac{t_{CP}}{\tau} - \frac{\tan^{-1} x}{x}))]\}
\end{aligned}
$$

(2)

This is the primary result of this section.

Despite having integrated over t_{tag}, CP asymmetry information is still encoded in the distribution $R_{\pm}(t_{CP})$. (Note that the integral $\int_0^\infty dt_{CP} R_{\pm}(t_{CP})$ does equal 0, as expected). Hence even though the decay ordering and time difference are not observed, the decay distributions are different for events that are tagged as B^0 or $\overline{B}^0$.

Even in the limit of poor resolution, where the oscillation modulations are not directly observable, one may simply measure the mean decay time of the B CP eigenstate. This mean decay time will be different for events in which the other B is tagged as a $\overline{B}^0$ or B^0 decay, and the difference in the mean times is proportional to $\sin 2\beta$.

1.2 Measuring $\sin 2\alpha$ and Penguin Pollution

The technique of integrating over one decay time may also be applied to the CP side, by observing the flavor-tagging B's decay time while integrating over the CP tag decay time. This makes possible the observation of mixing-induced CP violation in all-neutral final states; it is often assumed that these modes can be useful only for measuring decay rates and direct CP violation. Such modes include $\pi^0\pi^0$, $K_S\pi^0$, or $K_S K_S$. Though expected to be very small, these modes may help us understand penguin contributions to $\sin 2\alpha$ measurements.

2 Preliminary Measurement of x_d

As noted earlier, one must first measure x_d in order to extract a value for $\sin 2\beta$. An initially created B^0 may decay as a $\overline{B}^0$, or *vice versa*; these are

382

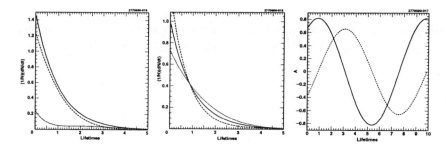

Figure 1: (a)Effect of $B - \overline{B}$ mixing on decay time distributions on a flavor-tagging final state. The dotted(dashed) line shows the distribution of the decay times when the opposite B is tagged as a mixed(unmixed), for $\Delta m{=}0.7$. The large average decay time of mixed events is readily apparent from the plot. (b) Effect of CP violation on decay time distributions of ΨK_S final state. The solid line shows the case $\sin 2\beta{=}0$; the dotted(dashed) line shows the distribution of the decay times when the opposite B is tagged as a $B^0(\overline{B^0})$, for $\sin 2\beta{=}0.7$. (c)The asymmetries for A_{CP} (dashed) and $A_{mixed/unmixed}$ (solid) as a function of the measured decay time. The plots are generated setting $x{=}0.7$, $\sin 2\beta{=}0.7$.

referred to as "mixed decays". We may integrate over *both* decay times to observe the time-integrated probability for mixing, denoted χ_d:

$$\chi_d = \frac{x^2 + y^2}{2(1 + x^2)(1 - y^2)} \ . \tag{3}$$

Since at the $\Upsilon(4S)$ the initial state contains both a B and $\overline{B}$, if we find two B's, or two $\overline{B}$'s, we know that mixing has occurred. By measuring the fraction of events that appear as mixed and unmixed, we measure χ_d, and therefore x_d and Δm.

In order to do this, we take advantage of the unique two-body kinematics of the decay chains $\Upsilon(4S) \rightarrow B\overline{B}$, $B^0 \rightarrow D^{*-}\pi^+$, $D^{*-} \rightarrow \overline{D}^0\pi^-$; and $\Upsilon(4S) \rightarrow B\overline{B}$, $B^0 \rightarrow D^{*-}\rho^+$, $D^{*-} \rightarrow \overline{D}^0\pi^-$.

The π or ρ (henceforth, h_W) from the decay of one B tags the flavor of that B. The decay flavor of the other B in the event is tagged by the lepton. The ratio χ_d, in the absence of backgrounds and mistags, is given by

$$\chi_d = \frac{h_W^+ \ell^+ + h_W^- \ell^-}{h_W^+ \ell^+ + h_W^- \ell^- + h_W^+ \ell^- + h_W^- \ell^+} \ . \tag{4}$$

As described in ref. [13], the B decay chain can be reconstructed without observing the D^0, which leads to high reconstruction efficiency. We select events which are consistent with originating from this decay chain, and which, in addition, have a high momentum lepton (e or μ) with momentum above 1.4 GeV.

Backgrounds to the measurement are low, because of the unique kinematics. In addition, most backgrounds are related to the signal mode either by isospin or by a missing soft track. Such backgrounds are >98% accurate at correctly determining the B flavor. Five percent of the events selected are random combinations of an h_W and π_s. We expect the total mistag rate for the hadronic tag to be 3.0±1.4%.

For the lepton tag, we must consider mistags due to secondary (charm) decays, ψ decays, hadronic fakes, π^0 dalitz decays, γ conversions, δ-rays The first three contributions, which constitute 99% of the mistag rate, are measured with experimental data. We find the mistag rate for the lepton tag to be 4.1±0.5%.

Finally, some B^+B^- events enter the sample, for instance due to two-body charged B decays to $D^{*0}h_W$, with a random π_s. We measure this contribution by examining the reconstructed decay angle distribution, which has a distinctive shape for random π_s. We find the charged B contribution to be 17.1±2.1%.

We use a part of the CLEO II dataset, for a total of 4 fb^{-1}. After subtraction of continuum ($q\bar{q}$) backgrounds, we find 148.6 mixed and 544.6 unmixed events. We correct for hadronic and lepton mistags and find $\chi_d = 0.195 \pm 0.026 \pm 0.016$, where the first uncertainty is statistical and the second systematic. With the full 9fb^{-1} dataset, we expect to find uncertainties of ±0.016 ± 0.009. Such a measurement would be among the best single measurements in the world, with systematic uncertainties very different from time-dependent measurements.

3 Experimental Aspects of a CP Violation Experiment

The experimental measurement method proceeds as follows. The experimenter selects decays in the $B \to \psi K_S$ mode. The $\psi \to \ell^+\ell^-$ vertex provides good decay time information when the beam position is well-known. The information is preserved, though diluted, even when only one dimension of the beam is well-determined, as is generally the case in the y dimension of an e^+e^- collider. The other B decay, with decay time unobserved, provides a tag as either a B^0 or $\overline{B^0}$. The decay time distributions for the ψK_S candidates in the two cases are fit with an unbinned maximum likelihood fitter. [Fig. 1] In the following sections we discuss CLEO's success in testing each piece of this method in its current 9 fb^{-1} dataset.

Table 1: Preliminary CLEO results for experimentally reconstructed B^0 decay modes useful for measuring $\sin 2\beta$. All results are for 6.3 fb^{-1}.

B^0 Decay Mode	Reconstructed Events	Expected Background
$\psi K_S,\ K_S \to \pi^0 \pi^0$	15	0.7
$\psi K_S,\ K_S \to \pi^+ \pi^-$	75	0.1
ψK_L	66	31
$\chi_{c1} K_S$	6	0.4
$\psi' K_S$	15	0.6
$\psi \pi^0$	7	0.6

3.1 Reconstruction of CP eigenstates

With 6 fb^{-1} of our 9 fb^{-1} dataset, CLEO has already reconstructed CP eigenstate decays in many decay modes [14]. The number of reconstructed events is shown in Table 1.

3.2 Flavor-Tagging

As we showed with the mixing measurements, there are many ways to tag the flavor of a decaying B. We can use soft pions from D^{*+} decays, hard pions from two-body and pseudo-two-body decays, leptons, or $K^{\pm}$ from the decay chain $b \to c \to s$ when particle ID is available. One may build generic flavor taggers combining all of these tags; often the package is implemented using a neural net. Monte Carlo studies at Babar and Belle indicate that effective separation εD^2 of 0.23-0.35 can be achieved. At CLEO we are currently studying a flavor-tagging package using our accumulated data; Our Monte Carlo studies indicate tagging separation of 0.28 with the CLEO II.V detector.

3.3 Measurement of CP eigenstate decay times and fits

Finally, one must measure the decay time of the CP eigenstate and fit the resulting distribution. The decay time t_{dec} is found by relating the observed decay distance l_{dec} to the decay time, $t_{dec} = \frac{l_{dec}}{\gamma \beta c}$. The decay distance l_{dec} is reconstructed by forming a vertex from the two leptons in the decay $\psi \to \ell^+ \ell^-$.

We use a maximum likelihood method, incorporating the result of the flavor tagging procedure, and known detector vertex resolutions. Other inputs to the fit include the B lifetime and x_d. To parameterize the detector resolution we use the results of examining $\gamma\gamma \to \mu\mu$ and $D \to K\pi$ data, which indicate the reported uncertainties underestimate the true uncertainties by about 10%, and that there is a 2% tail in a badly-misreconstructed tail.

4 CLEO Sensitivity

In order to estimate the experimental sensitivity to the CP violation parameter $\sin 2\beta$ at a symmetric B factory, one must make assumptions about decay resolutions, reconstruction efficiencies, and effective flavor tagging efficiency. Expected decay lengths for $B^0(\overline{B^0})$ tagged $B \to \psi K_S$ decays are $34(22)\mu$m at a symmetric $\Upsilon(4S)$ machine, for $\sin 2\beta = 0.7$ and $c\tau_B = 468\mu$m. The average y projections of these decay lengths are $19(12)$ μm. With an integrated luminosity of 30 fb^{-1}, effective flavor-tagging efficiency of 0.35, and average y-vertex resolution of 25 μm, Monte Carlo studies indicate that the ψK_S mode provides a measurement of $\sin 2\beta$ with statistical uncertainty ± 0.37. The CLEO II.V detector, in measurements of the D lifetimes [12], has demonstrated control of systematic uncertainties to a few μm. Due to the small decay lengths, sensitivity is nearly linear in the vertex resolution, in contrast to an asymmetric B factory. Of course, luminosity estimates given above are correspondingly sensitive to the square of the achievable resolution.

Acknowledgments

I would like to thank the conference organizers for their hospitality. I would like to thank my CLEO collaborators. Lawrence Gibbons, Frank Wuerthwein, and Alexey Ershov made important contributions to this paper. This work was supported by the National Science Foundation.

References

1. J.E. Christenson *et al.*, Phys. Rev. Lett. **13**, 138, (1964).
2. N. Cabbibo, Phys. Rev. Lett. **10**, 531, (1963).
3. M. Kobayashi and T. Maskawa, Prog. Theor. Phys. **49**, 652, (1973).
4. *Babar Technical Design Report*, SLAC-R-95-457, Stanford, CA (1995).
5. *Physics Rationale for a B Factory*, Lingel, *et al* (CESR B Working Group). Ithaca, NY (1993).
6. *Belle Technical Design Report*, KEK Report 95-1, Tsukuba (1995).
7. R. Aleksan, J. Bartelt, P.R. Burchat, and A. Seiden, Phys. Rev. **D39**, 1283, (1989).
8. K. Berkelman, Mod. Phys. Lett. **A10**, 165, (1995).
9. P.F. Harrison, H.R Quinn, ed. *The Babar Physics Book*, 1998.
10. A.B. Carter and A.I. Sanda, Phys. Rev. Lett. **45**, 952, (1980).
11. A.D. Foland, CLNS 99/1617, Ithaca, NY (1999).

12. Bonvicini, *et al. Measurement of Charm Meson Lifetimes* Submitted to Phys. Rev. Lett.
13. Brandenberg, *et. al.*, Phys. Rev. Lett. **80**, 5290, (1996).
14. Sasha Kopp, hep-ex/9904009.

NEW PARTICLE SEARCHES AT OPAL

D.I. FUTYAN

Department of Physics and Astronomy,
The University of Manchester,
Manchester M13 9PL, UK
E-mail: david.futyan@cern.ch

A number of searches for new physics have been performed using total of 265 pb^{-1} of data taken by the OPAL detector between $\sqrt{s}=2M_W$ and $\sqrt{s}=189$ GeV. A wide range topologies have been studied, each of which are signatures for a variety of new physics processes. No evidence for new particle production has been observed and limits have been set on the production cross-sections and masses of particles within the framework of theoretical models such as the MSSM.

1 Introduction

Since November 1995, the LEP e^+e^- collider has been operated at centre-of-mass energies well above the Z^0 peak. In 1997 and 1998, the OPAL detector collected integrated luminosities of 57.3 pb^{-1} at $\sqrt{s}$ =183 GeV and 187.2 pb^{-1} at $\sqrt{s}$ =189 GeV. This provided an opportunity to search for the existence of new particles, since the increased centre-of-mass energies has extended the range of new particle masses kinematically accessible. The increased integrated luminosity also gives greater sensitivity to processes which may have low cross-sections.

A number of searches for new particles have been performed at OPAL. Direct searches have been performed for particles predicted by supersymmetric[1] (SUSY) theories. In the framework of the Minimal Supersymmetric Standard Model[2] (MSSM), searches have been performed for scalar leptons, scalar top and bottom quarks, charginos and neutralinos. Searches have also been performed for SUSY particles with gravitino decay modes and for SUSY particles with R-parity violating decays. Other direct searches include the Standard Model and MSSM Higgs bosons, heavy and excited leptons and leptoquarks. In addition to these direct searches, topological searches have been performed which are sensitive to a range of new physics signals, such as a search for photonic events with missing energy and a search for long lived heavy charged particles.

This review concentrates on three of these searches: scalar leptons (in the context of a more general acoplanar lepton pair analysis)[3], charginos and neutralinos[4], and photonic events with missing energy[5].

2 Acoplanar Lepton Pairs

The acoplanar lepton pair topology consists of two charged leptons with significant missing transverse momentum. This is an experimental signature for the production of new particles that result in final states with two charged leptons accompanied by one or more invisible particles, such as neutrinos or the hypothesised lightest stable supersymmetric particle [2] (LSP), which may be the lightest neutralino, $\tilde{\chi}_1^0$, or the gravitino. Specifically, the following new particle decays are considered:

charged scalar leptons (sleptons): $\tilde{\ell}^\pm \rightarrow \ell^\pm \tilde{\chi}_1^0$, where $\tilde{\ell}^\pm$ may be a selectron ($\tilde{e}$), smuon ($\tilde{\mu}$) or stau ($\tilde{\tau}$) and $\ell^\pm$ is the corresponding charged lepton.

charged Higgs: $H^\pm \rightarrow \tau^\pm \nu_\tau$.

charginos: $\tilde{\chi}_1^\pm \rightarrow \ell^\pm \tilde{\nu}$ ("2-body" decays)   or $\tilde{\chi}_1^\pm \rightarrow \ell^\pm \nu \tilde{\chi}_1^0$ ("3-body" decays).

A number of Standard Model processes also lead to this experimental topology. By far the most common source is W^+W^- production in which both W's decay leptonically: $W^- \rightarrow \ell^- \bar{\nu}_\ell$. When the mass difference, Δm, between the parent particle and the invisible daughter particle is small, the background from two-photon processes, $e^+e^- \rightarrow e^+e^-\ell^+\ell^-$, is also important.

The event selection is performed in two stages. The first stage consists of a general selection of events containing two charged leptons and significant missing momentum in the transverse plane. The results of this selection are summarised in Table 1.

Table 1: Comparison between data and Monte Carlo of the number of events passing the general selection at each centre-of-mass energy. The total number of events predicted by the Standard Model is given, together with a breakdown into the contributions from individual processes. The Monte Carlo statistical errors are given.

$\sqrt{s}$ (GeV)	data	SM	$\ell^+\nu\ell^-\bar{\nu}$	$e^+e^-\ell^+\ell^-$	other
189	301	299.9±3.4	291.0±3.3	4.4±0.8	4.5±0.2
183	78	84.9±0.8	79.0±0.7	5.5±0.5	0.4±0.0

In the second stage of the event selection, a likelihood calculation combines the information from a number of discriminating variables to distinguish between new physics signals and the Standard Model background. Given an event for which the values of a set of variables x_i are known, the likelihood, L_S, of the event being consistent with the signal hypothesis is calculated as the product of the probabilities, $P_S(x_i)$, that the signal hypothesis would produce an event with variable i having value x_i:

$$L_S = \prod_i P_S(x_i).$$

The likelihood, L_B, of an event being consistent with the background hypothesis is calculated similarly. The quantity L_R is defined by:

$$L_R = \frac{L_S}{L_S + L_B}.$$

Event selection is performed by making a cut on the value of L_R rather than on the individual variables x_i.

Because the probability distributions $P_S(x_i)$, and therefore the value of L_R for a given event, vary considerably with the mass, m, of the parent particle (eg. smuon) and the mass difference, Δm, between the parent particle and the invisible daughter particle (e.g., lightest neutralino), L_R distributions are constructed using signal and background Monte Carlo for a range of points in the $(m, \Delta m)$ plane. The optimum cut value on L_R for each point is determined such that the *a priori* average value of the 95% CL upper limit on the cross-section for new physics is minimised.

No evidence for new physics is apparent, and the efficiencies, numbers of candidates and expected backgrounds are used to calculate 95% CL upper limits on the cross-section for all m and Δm. As an example, Fig. 1(a) shows the upper limits at 95% CL on the smuon pair cross-section at $\sqrt{s} = 189$ GeV times branching ratio squared for the decay $\tilde{\mu}^- \to \mu^- \tilde{\chi}_1^0$ as a function of smuon mass and lightest neutralino mass. These limits are compared with the predicted cross-sections and branching ratios to set limits on the masses of the particles searched for. Fig. 1(b) shows limits on right-handed smuons as a function of smuon mass and lightest neutralino mass for three assumed values of the branching ratio squared for $\tilde{\mu}_R^\pm \to \mu^\pm \tilde{\chi}_1^0$. For a branching ratio of 100% and for a smuon-neutralino mass difference exceeding 2 GeV, right-handed smuons are excluded at 95% CL for masses below 78 GeV.

3 Charginos and Neutralinos

Charginos, $\tilde{\chi}_j^\pm$, are the mass eigenstates formed by the mixing of the fields of the fermionic partners of the W and charged Higgs bosons. Neutralinos, $\tilde{\chi}_i^0$, are formed by the mixing of the fermionic partners of the γ, Z, and the neutral Higgs bosons. In the MSSM, there are two charginos and four neutralinos.

OPAL Preliminary

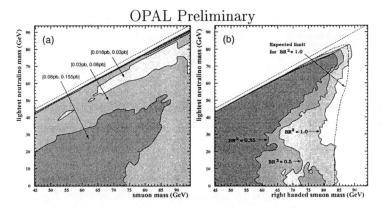

Figure 1: (a) Contours of the 95% CL upper limits on the smuon pair cross-section times $BR^2(\tilde{\mu} \to \mu\tilde{\chi}_1^0)$ at $\sqrt{s} = 189$ GeV based on combining the $\sqrt{s} = 183$ and 189 GeV data-sets. (b) 95% CL exclusion region for right-handed smuon pair production. The limits are calculated for the values of the branching ratio squared for $\tilde{\mu}_R^{\pm} \to \mu^{\pm}\tilde{\chi}_1^0$ indicated in the figure. Otherwise they have no supersymmetry model assumptions. The expected limit calculated from Monte Carlo alone is indicated by the dash-dotted line.

If charginos exist and are sufficiently light, they will be pair-produced in e^+e^- collisions at LEP through a γ or Z in the s-channel or through electron-sneutrino exchange in the t-channel. The production cross-section is large unless the sneutrino is light. The lightest chargino $\tilde{\chi}_1^+$ can decay into $\tilde{\chi}_1^0\ell^+\nu$, or $\tilde{\chi}_1^0 q\bar{q}'$, the decays occuring dominantly via a W^*. The two possible final states for each chargino result in three experimental topologies, listed in Table 2. The event sample is divided into three categories, motivated by the different topologies, and separate analyses are applied to the events in each category. Since the event topology depends on $\Delta M_+(\equiv m_{\tilde{\chi}_1^{\pm}} - m_{\tilde{\chi}_1^0})$, different selection criteria are applied to four ΔM_+ regions: (I) $\Delta M_+ \leq 10$ GeV ^{a}, (II) 10 GeV $< \Delta M_+ \leq m_{\tilde{\chi}_1^+}/2$, (III) $m_{\tilde{\chi}_1^+}/2 < \Delta M_+ \leq m_{\tilde{\chi}_1^+} - 20$ GeV and (IV) $m_{\tilde{\chi}_1^+} - 20$ GeV$< \Delta M_+ \leq m_{\tilde{\chi}_1^+}$. The number of background events expected and number of events observed are given in Table 2.

Neutralino pairs $(\tilde{\chi}_1^0\tilde{\chi}_2^0)$ can be produced through an s-channel virtual Z, or by t-channel $\tilde{e}$ exchange. The MSSM prediction for the cross-section is generally much lower than that for $\tilde{\chi}_1^-\tilde{\chi}_1^+$ production. The $\tilde{\chi}_2^0$ will decay, dominantly via a Z^*, into the final states $\tilde{\chi}_1^0\nu\bar{\nu}$, $\tilde{\chi}_1^0\ell^+\ell^-$ or $\tilde{\chi}_1^0q\bar{q}$. For the latter two cases this leads to an experimental signature consisting of an acoplanar

---

^{a}For the acoplanar lepton topology, region (I) is subdivided into (Ia) $\Delta M_+ \leq 5$ GeV and (Ib) 5 GeV $< \Delta M_+ \leq 10$ GeV.

Table 2: The numbers of total background events expected (with MC statistical error) and the observed number of events for the chargino search.

Topology	ΔM_+	I		II	III	IV
	Region	a	b			
jets $+ E_T^{miss}$	total bkg.	2.4±1.8		3.4±1.8	14.0±1.9	19.2±2.0
	observed	3		6	19	23
jets $+ 1\,\ell + E_T^{miss}$	total bkg.	3.6±2.2		5.7±2.2	3.3±0.3	6.9±0.5
	observed	3		8	7	10
2 acop. $\ell + E_T^{miss}$	total bkg.	8.0±1.3	6.3±1.0	12.6±1.1	68.5±1.7	126.9±2.1
	observed	8	8	16	68	119

pair of either leptons or jets. Separate analyses are carried out for the two topologies. Since the event shape of $\tilde{\chi}_1^0 \tilde{\chi}_2^0$ events depends on $\Delta M_0 (\equiv m_{\tilde{\chi}_2^0} - m_{\tilde{\chi}_1^0})$, the selection criteria are optimised for four ΔM_0 regions: (i) $\Delta M_0 \leq 10$ GeV, (ii) $10 < \Delta M_0 \leq 30$ GeV, (iii) $30 < \Delta M_0 \leq 80$ GeV and (iv) $\Delta M_0 > 80$ GeV. The number of background events expected and number of events observed are given in Table 3.

Table 3: The numbers of total background events expected (with MC statistical error) and the observed number of events for the neutralino search.

Topology	Region	i	ii	iii	iv
2 acop. $\ell + E_T^{miss}$	total bkg.	8.0±1.3	12.6±1.1	68.5±1.7	126.9±2.1
	observed	8	16	68	119
2 jets or monojet $+ E_T^{miss}$	total bkg.	3.0±2.1	6.1±2.2	12.1±2.2	39.8±3.2
	observed	4	10	19	37

Model-independent upper limits are obtained at 95% CL for the production cross-sections. This is done for $\tilde{\chi}_1^+ \tilde{\chi}_1^-$ assuming the specific decay mode $\tilde{\chi}_1^\pm \rightarrow \tilde{\chi}_1^0 W^{(*)\pm}$, and for $\tilde{\chi}_1^0 \tilde{\chi}_2^0$ assuming the $\tilde{\chi}_2^0 \rightarrow \tilde{\chi}_1^0 Z^{(*)}$ decay. The three analyses for charginos and two analyses for $\tilde{\chi}_1^0$ are combined using the likelihood ratio method [6], which assigns greater weight to the analysis with the greatest sensitivity. In the Constrained MSSM (CMSSM), chargino and neutralino phenomenology is determined by the four parameters M_2, μ, $\tan\beta$ and m_0. The cross-section limits are used to exclude regions of this parameter space at 95% CL. These exclusion regions are then transformed into mass limits on $\tilde{\chi}_1^\pm$, $\tilde{\chi}_1^0$ and $\tilde{\chi}_2^0$. Fig. 2(a) shows the gaugino mass limits as functions of M_2, for $\tan\beta$=1.5 and m_0=500 GeV. Fig. 2(b) shows the dependence of the $m_{\tilde{\chi}_{1,2}^0}$ limits on the value of $\tan\beta$. Of particular interest is the absolute lower limit, valid within the framework of the CMSSM, on the mass of the lightest neutralino of $m_{\tilde{\chi}_1^0} > 32.9$ GeV at 95% C.L. for $m_0 \geq 500$ GeV.

392

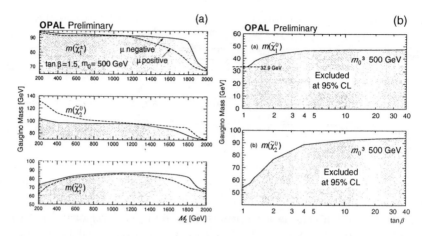

Figure 2: (a) The 95% C.L. mass limits for $\tilde{\chi}_1^+$, $\tilde{\chi}_2^0$, and $\tilde{\chi}_1^0$ for $\tan\beta = 1.5$ and $m_0 = 500$ GeV for slices of constant value of M_2. Limits are shown separately for $\mu < 0$ (solid lines) and $\mu > 0$ (dashed lines). Curves for larger values of $\tan\beta$ are in general between those shown for $\mu < 0$ and $\mu > 0$. (b) The 95% C.L. mass limits on $\tilde{\chi}_1^0$ and $\tilde{\chi}_2^0$ as a function of $\tan\beta$ for $m_0 \geq 500$ GeV. The mass limit on $\tilde{\chi}_2^0$ is for the additional requirement of $\Delta M_0 > 10$ GeV. The exclusion region for $m_0 \geq 500$ GeV is shown.

4 Photonic Events with Missing Energy

Experimental signatures consisting of photons and large missing transverse energy $\not{E}_T$ arise in a variety of models. Much of the recent interest in these final states has been motivated by supersymmetric models with gauge-mediated supersymmetry breaking in which decays of SUSY particles such as neutralinos into a photon and an invisible gravitino can dominate. Photon signatures can also be important in the MSSM scenario where the radiative decay $\tilde{\chi}_2^0 \to \tilde{\chi}_1^0 \gamma$ is possible and can be dominant in some regions of parameter space. Single and pair-production of excited neutrinos decaying as $\nu^* \to \nu\gamma$ can also lead to these final states. New-physics contributions can also arise from non-Standard Model $ZZ\gamma$ couplings, or large branching ratios of Higgs to $\gamma\gamma$. Signatures of such new physics would be visible as an excess in the rate of one or two photons with missing transverse energy. A search is made generally for new physics processes of the type $e^+e^- \to XY$ (single photon) and $e^+e^- \to XX$ (acoplanar photons) where X is neutral and decays radiatively ($X \to Y\gamma$) and Y is stable and weakly interacting.

New-physics contributions to these final states have an irreducible background from the Standard Model process $e^+e^- \to \nu\bar{\nu}\gamma(\gamma)$. However, such contributions can potentially be distinguished from the background by examination of the mass of the system recoiling against the selected photon system which, for the Standard Model process, is dominantly peaked at the mass of the Z^0.

The number of events selected by the single-photon and acoplanar-photons searches are given in Table 4. Also shown, in each case, is the expected number of events from the Standard Model process, from the KORALZ[7] Monte Carlo generator. The recoil mass distributions of events selected by each search are shown in Fig. 3. In general, for both selections, the agreement between the data and the Monte Carlo prediction is good and no evidence is seen for contributions from new physics processes.

Table 4: Numbers of candidate events expected, N_{exp}, and observed, N_{data}, in the search for photons and $\not{E}_T$ at $\sqrt{s} = 189$ GeV, for a total integrated luminosity of 169.3 pb^{-1}. The first error on N_{exp} is statistical, and the second is systematic. N_{bkg} is the number of background events expected from other physics and non-physics background sources.

Selection	N_{data}	N_{exp}	N_{bkg}
$\gamma(\gamma)+\not{E}_T$	543	$573.0 \pm 1.2 \pm 29.0$	5.0 ± 1.2
$\gamma\gamma(\gamma)+\not{E}_T$	20	$26.0 \pm 0.3 \pm 2.1$	0.2 ± 0.1

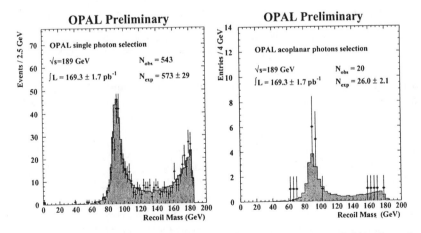

Figure 3: Recoil mass distributions for events passing the single photon (left) and acoplanar photons (right) selections for the $\sqrt{s} = 189$ GeV data sample. In each plot, the points with error bars are the data and the histogram is the expectation from the KORALZ Monte Carlo normalised to the integrated luminosity of the data. The error on the expected number of events is the total error which is dominated by the systematic error.

394

The measured cross-section within the kinematic acceptance of the selection for the single-photon topology is $4.36\pm0.19\pm0.22$ pb (where the first error is statistical and the second is systematic), compared to the KORALZ expectation of 4.64 pb. The cross-section within the kinematic acceptance is shown as a function of energy in Fig. 4 along with the KORALZ expectation.

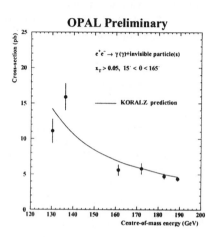

Figure 4: The measured value of $\sigma(e^+e^- \to \gamma(\gamma)$ + invisible particle(s)), within the kinematic acceptance of the OPAL single-photon selection, as a function of $\sqrt{s}$. The data points with error bars are OPAL measurements at centre-of-mass energies of 130, 136, 161, 172, 183 and 189 GeV. The curve is the prediction for the Standard Model process $e^+e^- \to \nu\bar{\nu}\gamma(\gamma)$ from the KORALZ generator. The plotted errors are the statistical and systematic errors added in quadrature. x_T is the photon transverse energy scaled by the beam energy.

To search for new physics processes $e^+e^- \to XY$ and $e^+e^- \to XX$, additional cuts are implemented to reduce the contribution from $e^+e^- \to \nu\bar{\nu}\gamma(\gamma)$. Kinematic consistency with the signal is required, and in the single photon case, cuts are applied to the recoil mass of the event. Based on the efficiencies and numbers of selected events, 95% CL upper limits on $\sigma(e^+e^- \to XY)\cdot BR(X \to Y\gamma)$ are calculated as a function of M_X and M_Y.

The special case of $M_Y \approx 0$ is also studied. This is relevant to processes such as $e^+e^- \to \tilde{\chi}_1^0\tilde{\chi}_1^0$, $\tilde{\chi}_1^0 \to \gamma\tilde{G}$ and $e^+e^- \to \nu^*\nu$, $\nu^* \to \gamma\nu$. In the single photon analysis, no further cuts are applied. For the acoplanar photons analysis, the additional requirements $M_{\mathrm{recoil}} < 80$ GeV and $M_X^{\mathrm{max}} > M_X$ - 5 GeV are implemented, where M_X^{max} is the maximum mass of X consistent with the measured 4-momentum of the 2 photon system.

Fig. 5 shows the 95% CL upper limit on $\sigma(e^+e^- \to XX)\cdot BR^2(X \to Y\gamma)$ for $M_Y \approx 0$ as a function of M_X obtained using the 189 GeV data. The limit ranges from 103 to 53 fb for M_X values from 45 GeV up to the kinematic limit. These limits can be used to set model-dependent limits on the mass of the lightest neutralino in supersymmetric models in which the NLSP is the lightest neutralino and the LSP is a light gravitino ($X = \tilde{\chi}_1^0, Y = \tilde{G}$). Shown in Fig. 5 as a dotted line is the (Born-level) cross-section prediction from a specific light gravitino LSP model [8] in which the neutralino composition is purely gaugino (bino). In this case, the selectron mass and therefore the cross-section depends only on $m_{\tilde{\chi}_1^0}$. Within the framework of this model, $\tilde{\chi}_1^0$ masses below 88.7 GeV are excluded at 95% CL.

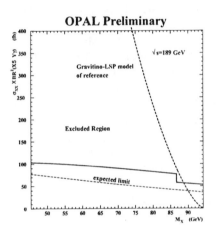

Figure 5: 95% CL upper limit on $\sigma(e^+e^- \to XX) \cdot BR^2(X \to Y\gamma)$ for $M_Y \approx 0$ (solid line). Also shown is the expected limit (dashed line). The dotted line shows the cross-section prediction of a specific light gravitino LSP model. Within that model, $\tilde{\chi}_1^0$ masses below 88.7 GeV are excluded at the 95% CL. These limits assume that particle X decays promptly.

5 Summary

A number of searches for new physics have been performed using total of 265 pb^{-1} of data taken between $\sqrt{s}=2M_W$ and $\sqrt{s}=189$ GeV. A wide range topologies have been studied, each of which are signatures for a variety of new physics processes. No evidence for new particle production has been observed and limits have been set on the production cross-sections and masses of particles within the framework of theoretical models such as the MSSM.

396

References

1. Y. Gol'fand and E. Likhtam, JETP Lett. **13**, 323, (1971);
   D. Volkov and V. Akulov, Phys. Lett. **B46**, 109, (1973);
   J. Wess and B. Zumino, Nucl. Phys. **B70**, 39, (1974).
2. H.P. Nilles, Phys. Rep. **110**, 1, (1984);
   H.E. Haber and G.L. Kane, Phys. Rep. **117**, 74, (1985).
3. OPAL Collaboration, "Search for Acoplanar Lepton Pair Events at $\sqrt{s}$ = 161, 172 and 183 GeV",
   CERN-EP/98-122, submitted to Eur. Phys. J. C.
4. OPAL Collaboration, "Search for Chargino and Neutralino Production at sqrt(s)=189 GeV at LEP",
   OPAL Physics Note PN374, 11th March 1999.
5. OPAL Collaboration, "Search for Anomalous Photonic Events with Missing Energy in e+e- Collisions at sqrt(s)=189 GeV",
   OPAL Physics Note PN386, 29th March 1999.
6. A.G. Frodesen, O. Skjeggestad, and H. Tøfte, *"Probability and Statistics in Particle Physics"*, Universitetsforlaget, 1979, ISBN 82-00-01-01906-3;
   S.L. Meyer, *"Data Analysis for Scientists and Engineers"*, John Wiley and Sons, 1975, ISBN 0-471-59995-6.
7. S. Jadach, B.F.L. Ward and Z. Wąs, Comp. Phys. Comm. **79**, 503, (1994).
8. C.Y. Chang and G.A. Snow, UMD/PP/97-57;
   K.S. Babu, C. Kolda and F. Wilczek, Phys. Rev. Lett. **77**, 3070, (1996).

PROSPECTS FOR STANDARD MODEL HIGGS BOSON SEARCH AT LEP 200

ELIAM GROSS

Department of Particle Physics, The Weizmann Institute of Science, Rehovot 76100, Israel.

E-mail: fhgross@wicc.weizmann.ac.il

The Standard Model Higgs boson search at LEP up to a center-of-mass energy of 189 GeV is reviewed. Prospects for Higgs boson search at LEP up to a center-of-mass energy of 200 GeV are also discussed.

1 Introduction

Since the beginning of LEP four subsequent energy upgrades of the collider contributed to the Standard Model Higgs search at LEP 200. Each experiment collected 10 pb^{-1} of luminosity during the summer of 1996 at a center-of-mass energy of 161 GeV. An additional 10 pb^{-1} were collected the following autumn at a center-of-mass energy of 172 GeV. Late in the summer of 1997 the center-of-mass energy was raised to 183 GeV and LEP provided a luminosity of 55 pb^{-1}. Recently (1998) LEP provided an incredible luminosity of 180pb^{-1} per experiment at a center-of-mass energy of 189 GeV.

This allowed the LEP experiments to probe the difficult mass region of 91 GeV in search for the Higgs boson. Although the Higgs boson has not been discovered yet, the lower bounds put on its mass **are more stringent than all predictions of the LEP 200 workshop held at CERN in 1995** . The experimental lower bounds on the Standard Model Higgs boson will be reviewed here followed by a discussion on future prospects.

2 Is the Higgs Boson around the Corner?

Electroweak measurements can be used to pinpoint the Higgs boson mass in the Standard Model frame. Leaving the Higgs boson mass (or to be more precise $\log m_H$) as a free parameter one arrives at $m_H = 76^{+85}_{-47}$ GeV with an upper bound of 262 GeV on the mass of the Standard Model Higgs boson [1,2]. From Vacuum stability and unitarity arguments one finds out that if the Standard Model is well behaved up to the Planck mass scale the Higgs boson should be heavier than 160 GeV [3,4]. Putting these two results together one concludes that a relatively light Higgs boson is expected, however, if this Higgs boson is discovered at LEP (i.e. its mass is below 110 GeV as will be discussed later) it indicates also the discovery of new physics, maybe supersymmetry. No doubt that these theoretically motivate very strongly the search for Higgs bosons at LEP because they indicate a light Higgs boson, just around the corner.

3   Higgs Boson Search Methodology

At LEP 200 the Standard Model Higgs boson is expected to be produced via the Bjorken mechanism where the Z^0 boson is produced in association with the Higgs boson via the process $e^+e^- \rightarrow H^0Z^0$. If the Z^0 boson is produced on shell it leaves no phase space for a Higgs boson with a mass above $\sqrt{s} - m_{Z^0}$ to be produced. This sets up a natural wall for the maximum Higgs mass one can probe at LEP 200. This wall is 97.8 and 108.8 GeV/c^2 for $\sqrt{s}$= 189 and 200 GeV respectively. In principle there are two ways this wall could be penetrated. The Z^0 boson could be produced off shell in the Bjorken process, or, in the case of electrons or neutrinos in the final state, the Higgs could be produced via the W^+W^- and Z^0Z^0 fusion processes. However, the cross sections for the fusion processes, even though they exceed that of the Bjorken processes for a Higgs mass above the wall, are too small to produce a detectable signal with luminosities less than 1000 pb^{-1}.

The Higgs boson decays predominantly to $b\bar{b}(\sim 86\%)$ while the Z^0 decays $\sim 70\%$ to quarks $(q\bar{q})$, $\sim 20\%$ to neutrinos $(\nu\bar{\nu})$, and $\sim 9\%$ to charged leptons (l^+l^-).

The principal final state topologies, which account for about 84% of all SM Higgs boson final states are:

(i) the four jet channel, $e^+e^- \rightarrow Z^0H^0 \rightarrow q\bar{q}b\bar{b}(\sim 60\%)$;
It is characterized by four energetic hadronic jets, large visible energy and the presence of b-flavored hadron decays. The backgrounds are $Z^0/\gamma^* \rightarrow q\bar{q}$ with and without initial state radiation accompanied by hard gluon emission as well as four-fermion processes, in particular $e^+e^- \rightarrow W^+W^-, Z^0Z^0$. The suppression of these backgrounds relies on the kinematic reconstruction of the Z^0 boson and on the identification of b quarks from the Higgs boson decay.

(ii) the missing energy channel, mainly from $e^+e^- \rightarrow Z^0H^0 \rightarrow \nu\bar{\nu}b\bar{b}$, with a small contribution from the W^+W^- fusion process $e^+e^- \rightarrow \nu\bar{\nu}H^0(\sim 18\%)$;
These events are characterized by large missing momentum and two energetic, acoplanar, hadronic jets containing b-flavored hadrons. The dominant backgrounds are mismeasured $Z^0/\gamma^* \rightarrow q\bar{q}$ events, four-fermion processes with neutrinos in the final state, such as $Z^0Z^0 {}^{(*)} \rightarrow \nu\bar{\nu}\ q\bar{q}$ $W^+W^- \rightarrow \ell^\pm\nu\ q\bar{q}$ $W^\pm\ e^\mp\nu \rightarrow q\bar{q}\ e^\mp\nu$ with the charged lepton escaping detection and, in general, events in which particles go undetected down the beam pipe such as

$e^+e^- \rightarrow Z^0\gamma$ and two-photon events. For most of these backgrounds, the missing momentum vector points close to the beam direction, while signal events tend to have missing momentum in the transverse plane. However, though most of the above mentioned four-fermion background can be largely reduced via b-tagging, the $Z^0Z^{0(*)} \rightarrow \nu\bar{\nu}b\bar{b}$ process is an irreducible background.

(iii) the muon and electron channels, predominantly from $e^+e^- \rightarrow Z^0 H^0 \rightarrow \mu^+\mu^- q\bar{q}$ and $e^+e^- q\bar{q}$, with the latter including a small contribution from the Z^0Z^0 fusion process $e^+e^- \rightarrow e^+e^- H^0 (\sim 6\%)$.

These events yield a clean experimental signature in the form of large visible energy, two energetic, isolated, oppositely-charged leptons of the same species reconstructing to the Z^0 boson mass, and two energetic hadronic jets carrying signs of b-flavored hadrons. The dominant backgrounds are four-fermion processes (Z^0Z^0).

It can be seen that all Higgs boson searches rely on b-tagging and the identification of the candidate Higgs jets via kinematical constrained fits.

A detailed description of the analyses of OPAL and Aleph experiments, which is typical to all LEP Higgs search analyses, can be found in these proceedings [5,6].

4 Mass Limits and the Search for the Higgs boson So Far

In order to clearly announce the observation of the Higgs boson one expects to see an excess of events (over the expected background) with a significance of at least 5 σ. Unfortunately this anticipated event had not occurred yet. In the absence of such excess the experiments derive a lower bound on the Higgs boson mass.

How good is an experiment's sensitivity to discover the Higgs boson? Do we have a way to estimate it? Given the same amount of delivered luminosity, does the lower bound set on the Higgs boson mass by an experiment serve as an estimator for the experiment's sensitivity? The answer is no. The reason is that even in the absence of the Higgs boson signal the background fluctuates and one experiment can be "lucky" and observe no background at all resulting in a very high Higgs mass limit, while another experiment with a more efficient analysis can be "unlucky" in a sense that the background fluctuates and does not allow setting a high mass limit.

In order to quantify the results of a search, a test-statistic X must be defined which discriminates signal [a] from background experimental results. The test-statistic may be as simple as the total number of selected events, or it may be a complex function of both the number of selected events in distinct

---

[a] A "signal experiment" implicitly includes the expected background contribution.

400

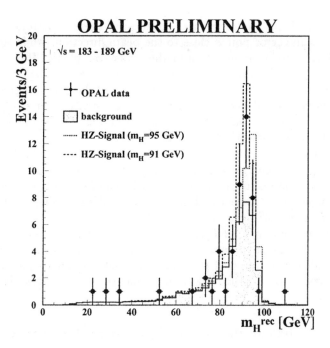

Figure 1: The mass distribution for the selected events in all channels combined (points with error bars), and the expected background. The data and expected backgrounds have been combined for the $\sqrt{s} = 183$ GeV and and $\sqrt{s} = 89$ GeV samples. The expected mass distribution assuming the production of the SM Higgs with a mass of 91 GeV (95 GeV) is added on top of the background and shown with a dashed (dotted) histogram.

search channels and one or more properties of the selected events (such as the invariant mass of the Higgs candidate used in this analysis). The probability distribution of the test-statistic may be known exactly (e.g. Poisson statistics for single-channel counting experiments) or it may be derived by Monte Carlo or other integration techniques for more complex search results.

The confidence in the signal plus background hypothesis CL_{s+b} is defined as the probability to obtain $X < X_{obs}$ in gedanken signal experiments for a specified signal hypothesis, where X_{obs} is the value of the test-statistic observed for the experiment. The confidence in the background hypothesis CL_b is defined as the probability to observe $X < X_{obs}$ in gedanken background experiments. When performing a search with small expected signal rates and non-negligible background rates, the probability of excluding the signal more strongly than if the expected background rate were 0 may be substantial (in such experiments the most probable signal rate may be unphysical). Background fluctuations may also result in such a strong signal exclusion that even infinite Higgs mass (0 signal rate) may be excluded at or above the 95% CL. To prevent a priori such unaesthetic, but nevertheless valid, results from occurring, sometimes the ratio $CL_s = CL_{s+b}/CL_b$ suggested in [7] is used as an approximation to the signal confidence one might have obtained in the absence of background.

The sensitivity of an experiment to probe the Higgs boson is measured by the expected limit. One performs gedanken background experiments and for each one of them measures the $CL \equiv 1 - CL_s$ as a function of m_{H^0}. The expected limit is that value of m_{H^0} for which the average confidence level equals 95%, i.e., a Higgs boson with a mass m_{H^0} is expected to be excluded if $\langle CL(m_{H^0}) \rangle \geq 95\%$.

The observed limit is the Higgs boson mass for which the experiment's measured confidence level equals 95%.

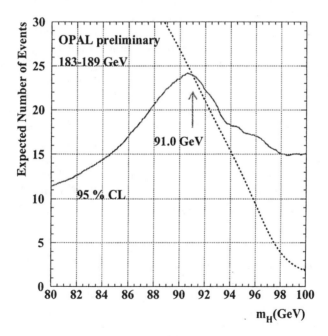

Figure 2: Limit on the production rate for the SM Higgs boson at 95% CL (solid line) and the number of expected signal events (dashed line) as a function of the Higgs mass. The 95% CL lower mass limit on the SM Higgs bosons is set at the point where the solid and dashed curve intersect: $m_{H^0} > 91.0$ GeV.

5 LEP Higgs Mass Limits

The observed and expected lower bounds on the Higgs boson mass are given in Table 1 [8].

Experiment	Observed Limit (GeV)	Expected Limit (GeV)	Probability (%)
ALEPH	90.2	95.7	1%
DELPHI	95.2	94.8	68%
L3	95.2	94.4	55%
OPAL	91.0	94.9	4%

Table 1: The expected and observed mass limits as well as the probability to obtain a mass limit as indicated or less in background-only experiments.

The probability that the experiment will observe that or a lower mass limit in an ensemble of background-only experiments is also given in the Table. Note that ALPEH and OPAL are not able to exceed the mass of the Z boson in their derived observed limit while their expected limit is much higher (around 95 GeV). The reason is that both experiments observe an excess of events around m_{Z^0}. In Fig. 1 [8] one can see that OPAL indeed observe an excess of events which looks compatible with a 91 GeV Higgs boson. However this excess has less than 2 σ significance and if one looks closer at OPAL's limit derivation (Fig. 2) one clearly sees that the expected backgrounds peaks at 91 GeV (as one would expect from the ZZ irreducible background) and had OPAL collected just a bit more luminosity or increased their signal to background ratio their limit could increase to 94 GeV, unless the effect turns out to be real. To put a long story short, only time will tell, if what all experiments observe is a fluctuation (upward or downward) or we finally see the early birds of the Higgs boson.

6 Prospects for Higgs boson search at LEP 200

The two curves shown in Fig. 3 [9] show the minimum luminosities required in order to exclude the SM Higgs boson at the 95% C.L. at $\sqrt{s}$=189 GeV (left curve) and $\sqrt{s}$=200 GeV combined with an integrated luminosity of 150pb^{-1} at $\sqrt{s}$=189 GeV (right curve). The PRELIMINARY results of the experiments as reported in ICHEP98 [10] and afterwards [8] are also shown. The agreement between the expected limits and the prediction is fair. It is interesting to note the downward fluctuation of OPAL with a lower luminosity which drove a

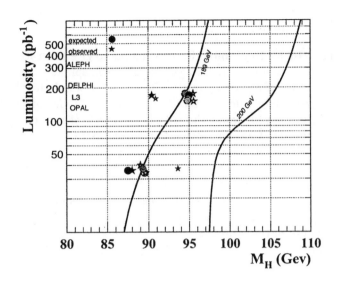

Figure 3: The minimum luminosity required in order to exclude the SM Higgs boson at the 95% Confidence Level at the indicated center-of-mass energies of 189, 200 GeV and combining the luminosity collected at 200 GeV with 150 pb^{-1} collected at 189 GeV. The experimental preliminary results are also shown.

very high observed limit in Vancouver and the flip of this situation when more luminosity was collected. It is to be emphasized again that these results are preliminary and should be taken into consideration with a grain of salt. Based on these curves one can predict that the expected lower limit on the Higgs boson mass combining all LEP experiments at a center-of-mass energy of 189 GeV is around 97 GeV. With 150 pb^{-1} collected per experiment at $\sqrt{s} = 200$ GeV LEP will be able to exclude a Higgs boson close to the kinematical limit of $m_{H^0} \bar{1}09$ GeV.

7 Conclusions

LEP is entering the year 1999 with lots of anticipations from the scientific community. A relatively light Higgs boson is expected from theory and an excess in observed events around 91 GeV from both ALEPH and OPAL left the 1998 run with a big question mark which will hopefully be answered in 1999 when LEP with a center-of-mass energy of up to 200 GeV is expected to be in the reach of a Higgs boson with a mass above 100 GeV.

8 Acknowledgments

I would like to thank Jim Pinfold for inviting me to give this talk. My deepest gratitude is also given to Koichi Nagai, Audrey Schaapman and Bruce Campbell for taking care of me after my head injury in the wonderful slopes of Lake Louise. Unfortunately I do not recall even coming to Canada at that week but I do recall how you, wonderful people took care of me to make sure my recovery is complete. I am fine today, thank you all.

References

1. The LEP electroweak working group. http://www.cern.ch/LEPEWWG/.
2. Jim Pinfold in these proceedings.
3. T. Hambye and K. Riesselmann, hep-ph/9708416; DESY 97-152, D0-TH 97/18.
4. R. Peccei in these proceedings.
5. K. Nagai in these proceedings.
6. Emmerich Kneringer in these proceedings
7. Particle Data Group: R.M. Barnett *et. al.*, *Review of Particle Physics*, Phys. Rev. **D54**, 1, (1996).
8. Results presented by M. Felcini, Moriond EW '99, March 13-20, 1999.
9. E. Gross, A.L. Read and D. Lellouch, Prospects for Higgs boson search at LEP 200. CERN-EP/98-094, June 1998.
10. XXIX International Conference on High Energy Physics, UBC, Vancouver, B.C., Canada July 23-29 1998.

ZZ PRODUCTION AT E_{CMS} = 189 GEV IN DELPHI

R. JACOBSSON

CERN, CH-1211 Geneva 23,
SWITZERLAND
E-mail: Richard.Jacobsson@cern.ch

on behalf of the DELPHI collaboration

During 1998 the LEP accelerator was operated at an energy of 188.6 GeV. This made it possible to study with statistical significance the production of on-shell Z-pairs. This paper briefly outlines the preliminary findings for the different final states considered in the DELPHI experiment. Combining the final states and interpreting them as an on-shell ZZ cross-section gives $\sigma_{ZZ} = 0.58 \pm 0.17$ (stat.) pb as compared to 0.64 pb according to the Standard Model[1].

1 Introduction

The study of doubly resonant production of Z is a relatively new topic as the first instances were observed during 1997 when LEP was operated at an energy of 183 GeV[2]. With the increased cross-section, the run in 1998 at a center-of-mass energy of 188.6 GeV has allowed a measurement of the production.

There are several motivations for studying this channel. Trivially, it is important to verify the production rate in each new energy domain as it should be well predicted by the Standard Model. More importantly, the ZZ production forms an irreducible background to the $H^0 Z$ search when the mass of the Higgs boson is close to that of the Z. In this context, it should also be noted that the analysis techniques used to extract a signal from a Higgs boson with a mass near to that of the Z are very sophisticated. Hence it is interesting to check the behaviour of these methods and improve the understanding of the optimization on an existing similar experimental signature. Particularly important to study is the stability of the methods against systematic errors. In addition, the measurement of the ZZ cross-section also constitutes an intrinsic ingredient in the search for other types of new physics, such as anomalous neutral-current triple gauge couplings (TGC).

At 188.6 GeV the Standard Model predicts a doubly resonant ZZ cross-section of about 0.64 pb. This is to be compared with the order of magnitudes of the main backgrounds from WW and QCD events, which are roughly 25 and 215 times larger, respectively. The predicted cross-section for $H^0 Z$ is 0.32 pb for a 90 GeV/c^2 Higgs boson and 0.14 pb for a 95 GeV/c^2 Higgs boson.

The general signature of a ZZ event is a four-fermion final state. However,

Table 1: Overview of the final states analysed by DELPHI. There is no window cross-section given for the $l^+l^-l^+l^-$ as a different method to extract the cross-section was used in this channel(see section 3).

Channel	Br. ratio	σ_{ZZ}(window)	Exp. ZZ events	Technique
$l^+l^-l^+l^-$	0.94 %	-	1.0	Sequential cuts
$e^+e^-q\bar{q}$	4.70 %	0.0337 pb	5.3	Sequential cuts
$\mu^+\mu^-q\bar{q}$	4.70 %	0.0281 pb	4.4	Sequential cuts
$\nu\bar{\nu}q\bar{q}$	27.97 %	0.1704 pb	26.5	Discriminant
$q\bar{q}b\bar{b}(c\bar{c})$	18.90 %	0.1179 pb	18.6	Probabilistic
Total	57.21 %		55.8	

several additional processes other than the resonant ZZ diagram contribute to the four-fermion event rate (see Fig. 4). In order to take the additional processes and the interferences properly into account, the on-shell signal was defined by a mass-cut at the generator level as being all four-fermion processes fulfilling

$$M_Z - 10\,GeV/c^2 < M_{f\bar{f}} < M_Z + 10\,GeV/c^2 \qquad (1)$$

for both fermion-pairs. In case of pairing ambiguity, the event was taken as signal if by doing all possible pairings one configuration had both fermion-pairs within the mass window. The complementary four-fermion events falling outside the mass window were considered as background in the analyses. In order to select the on-shell events in real data, all analyses also used a similar mass-cut on the reconstructed mass of each fermion-pair. However, due to different mass resolution in different channels, the reconstructed mass window varied.

Converting the results with the signal definition as given above to the cross-section of the true doubly resonant production is described in section 7.

Table 1 shows an overview of the DELPHI analyses which cover about 57 % of the total production. The third and the fourth columns give the expected cross-sections and the expected number of events, respectively, of the signal as defined above.

A detailed description of the DELPHI detector can be found in ref. [3].

2 Data Samples

This paper presents results based on the full real data set recorded by the DELPHI detector during 1998, which corresponds to an integrated luminosity of 158 pb^{-1} at a centre-of-mass energy of 188.6 GeV.

Simulated events were generated with the DELPHI simulation program DELSIM [4] and then passed through the same reconstruction chain as the real data. Processes leading to four-fermion final states ($(Z/\gamma)^*(Z/\gamma)^*$, W^+W^-, $We\nu_e$ and Ze^+e^-) were generated with EXCALIBUR [5] relying on JETSET 7.4 ref. [6] for quark fragmentation. EXCALIBUR includes all tree-level diagrams in a consistent fashion. PYTHIA [6] was used to cross-check EXCALIBUR. The part of the $e\nu q\bar{q}$ phase space in which the finite electron mass is relevant was simulated with the GRC4F generator[7] in which fermion mass effects are included. The background processes $e^+e^- \to f\bar{f}(+n\gamma)$ were generated using PYTHIA[6]. Two-photon interactions were generated using TWOGAM[8] and BDK [9].

3 $l^+l^-l^+l^-$ Channel

These events give rise to a clear signature and the only significant background comes from non-resonant $e^+e^-l^+l^-$ production. The selection required low multiplicity events and imposed simple invariant mass and angle criteria. No particle identification was demanded in the initial selection but if two identified particles were found with invariant mass close to the Z mass, they were required to have the same lepton flavour. The selection is sensitive to all combination of lepton flavours except $\tau^+\tau^-\tau^+\tau^-$.

As an exception, the $l^+l^-l^+l^-$ channel did not use the signal definition described above. Instead, two four-fermion samples were generated: the first containing only events produced via the doubly resonant ZZ diagrams, and the second using all tree-level diagrams excluding the ZZ diagrams. The first was used as signal and the second as background. This procedure neglects interference terms but the effect is negligible as the predicted signal is of the order of one event.

A single event was observed in the 1998 data. Table 2 shows the results.

Table 2: Results of the analysis of the $l^+l^-l^+l^-$ channel.

Channel	Obs. events	Exp. signal	Exp. bkg.	Efficiency
$l^+l^-l^+l^-$	1	0.54	0.30 ± 0.08	57 %

Table 3: Results of the analysis of the $l^+l^-q\bar{q}$ channel.

Channel	Obs. events	Exp. signal	Exp. bkg.	Efficiency
$e^+e^-q\bar{q}$	2	3.69 ± 0.35	0.85 ± 0.16	65.7 ± 6.0 %
$\mu^+\mu^-q\bar{q}$	6	4.02 ± 0.47	0.53 ± 0.15	86.9 ± 5.0 %

4 $l^+l^-q\bar{q}$ Channel

The principal background in this channel originates from non-resonant $l^+l^-q\bar{q}$ events outside the signal definition window and WW events. For the $e^+e^-q\bar{q}$ channel also $q\bar{q}$ events in association with isolated photons which convert are a source of background.

The selection procedure for the $\mu^+\mu^-q\bar{q}$ and $e^+e^-q\bar{q}$ channels was almost the same and differed mainly in the numerical values of the applied cuts. The selection was mainly based on two discriminating variables derived after performing a kinematic fit [11] including four-momentum conservation : the transverse momentum of a lepton candidate with respect to the nearest jet and the χ^2 per degree of freedom of the kinematic fit. The events were required to have at least two lepton candidates of the same flavour and opposite charge, out of which one was requested to pass strong identification criteria and the other one was allowed a less strict identification. A few additional criteria were imposed in the $e^+e^-q\bar{q}$ to suppress background from wrong electron identification and from photon conversions.

The results of the selection are shown in Table 3.

5 $\nu\bar{\nu}q\bar{q}$ Channel

This channel has a conspicuous signature and a sizeable branching ratio but is challenging due to the intractable background. The difficult background comes from $e\nu q\bar{q}$ events, other WW events, and $q\bar{q}$ events with isolated photons escaping detection.

In order to cope with this and perform an efficient selection, the iterated nonlinear discriminant analysis program (IDA) [12] was applied to calculate a second order polynomial from a set of event variables. The polynomial specifies a multidimensional surface which maximizes the separation between signal and background in the event variable space, and its values can be used as weights for signal events. In order to use the combination of variables with the best performance, variables were selected one by one out of a large set using an

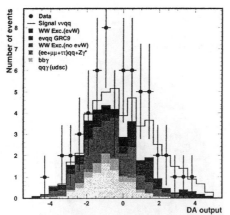

Figure 1: Output of the iterated second order discriminant analysis used in the $\nu\bar{\nu}q\bar{q}$ channel for the signal and the different background types.

automatic procedure until no further improvement in the discrimination was obtained. The automatic procedure computed the second order discriminant for each variable in the set and selected the variable, which together with those already selected, gave the best discrimination power.

In total, nine variables representing the different features in signal and background were selected.

In addition, a special veto algorithm based on the DELPHI hermeticity counters was applied in order to reject events with a $q\bar{q}$ system and a missing or poorly reconstructed photon in regions where the detector has weak coverage.

Fig. 1 shows the final distributions of the discriminant function for the signal and the different background types. The selected data sample at this final level consisted of 59 events. The background and the signal predictions were 34±1 and 15.2±0.3. Imposing a cut to obtain a signal-over-background ratio of 2 gave the results shown in Table 4.

Table 4: Results of the analysis of the $\nu\bar{\nu}q\bar{q}$ channel.

Channel	Obs. events	Exp. signal	Exp. bkg.	Efficiency
$\nu\bar{\nu}q\bar{q}$	10	8.0 ± 0.3	4.1 ± 0.5	30.4 ± 1.1 %

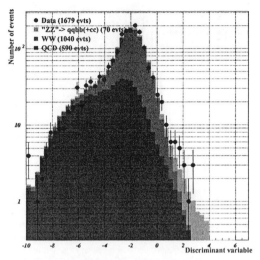

Figure 2: Distribution of the combined discriminant variable for the signal, WW, and QCD events in the $b\bar{b}q\bar{q}$ channel

6 $b\bar{b}q\bar{q}$ Channel

The background in this channel is mainly coming from WW events and QCD processes. As the cross-section of these are orders of magnitude larger than the ZZ cross-section, also in this case powerful signal extraction techniques are needed. In order to achieve an efficient rejection of WW events, this analysis was optimized for final states in which at least one of the on-shell Z bosons decays into $b\bar{b}$ as b quarks are rare in WW events.

The selection made use of a five constraint kinematic fit in which the fifth constraint required that one of the dijet masses be equal to the Z mass. For a set of discriminating variables including kinematics, b-quark identification, and topological variables, the probabilities for signal and background hypotheses were computed. A single discriminating variable was then constructed by combining the ratios of these probabilities. The distribution for signal, WW, and QCD events can be seen in Fig. 2. A cut was imposed on this variable together with a cut on the reconstructed mass of the second Z boson at 80 GeV/c^2. This enabled to obtain a signal-over-background ratio of 3 with the results shown in Table 5. The ZZ signal selected with these cuts was very pure in $b\bar{b}q\bar{q}$ final states. A small contribution from $c\bar{c}q\bar{q}$ (with $q \neq b$) was present, as indicated within the parentheses, and was accounted for in the efficiency.

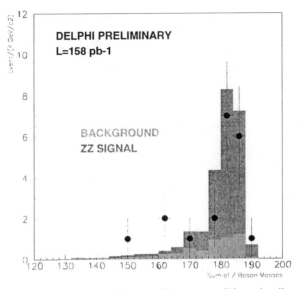

Figure 3: Mass spectrum of the sum of the two Z boson candidates in all events accepted by the five analyses.

7 Results

Summing the results for the different final states leads to an observed total of 25 events in the real data . The expected total is 20.1 signal events and 7.1 background events, and is thus in good agreement. Fig. 3 shows the sum of the two Z boson candidate masses for all the candidates accepted in the five analyses.

As a generator mass window is used for the signal definition, the composition of the observed events is obviously a mix of all four-fermion production mechanisms shown in Fig. 4. In order to convert the result given above to the cross-section of the purely resonant diagrams shown in the box in the figure (referred to as NC02), a scale factor was derived for each channel. The scale factor R was defined as

Table 5: Results of the analysis of the $b\bar{b}q\bar{q}$ channel.

Channel	Obs. events	Exp. signal	Exp. bkg.	Efficiency
$q\bar{q}b\bar{b}(+c\bar{c})$	6	$3.87(+0.05) \pm 0.19$	1.35 ± 0.17	18.2 ± 0.9 %

412

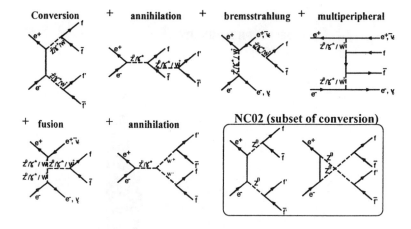

Figure 4: All diagrams contributing to the total rate of four-fermion final states.

$$R = \sigma_{NC02}(total)/\sigma_{4f}(window), \qquad (2)$$

and was computed using EXCALIBUR and WPHACT[13]. They are shown in Table 6. The observed and predicted numbers of events and the scale factors were then used in a maximum likelihood fit to the Poisson probability of observing a given number of events for various values of the NC02 cross-section, constraining the branching ratios of the Z to those in ref.[10]. The value for the NC02 cross-section so obtained is:

$$\sigma_{ZZ} = \sigma_{NC02} = 0.58 \pm 0.17 \text{ pb}, \qquad (3)$$

where the quoted error is statistical only. This is in good agreement with the Standard Model prediction from EXCALIBUR of 0.64 pb.

Table 6: Scale factors to convert from the window cross-sections to the NC02 cross-sections.

Channel	$l^+l^-l^+l^-$	$e^+e^-q\bar{q}$	$\mu^+\mu^-q\bar{q}$	$\nu\bar{\nu}q\bar{q}$	$b\bar{b}q\bar{q}$
Scale factor R	-	0.88	1.05	1.05	1.14

Acknowledgments

We are greatly indebted to our technical collaborators, to the members of the CERN-SL Division for the excellent performance of the LEP collider, and to the funding agencies for their support in building and operating the DELPHI detector.

References

1. Results also submitted to the Rencontre de Moriond 99. DELPHI 99-37 CONF 236.
2. ICHEP'98 Conference, Vancouver, July 22-19. DELPHI 98-104 CONF 172.
3. DELPHI Collaboration, P. Abreu *et al.*, Nucl. Instr. and Meth. **A378**, 57, (1996).
4. DELSIM *Reference Manual*, DELPHI 87-97 PROG-100.
5. F.A. Berends, R. Pittau, R. Kleiss, Comp. Phys. Comm. **85**, 437, (1995).
6. T. Sjöstrand, Comp. Phys. Comm. **39**, 347, (1986); T. Sjöstrand, PYTHIA 5.6 *and* JETSET 7.3, CERN-TH/6488-92.
7. Y. Kurihara *et al.*, Vol. 2 30 2 45 in *Physics at LEP2* G. Altarelli, T. Sjöstrand and F. Zwirner (eds.) CERN 96-01 (1996).
8. S. Nova, A. Olshevski, and T. Todorov, *A Monte Carlo event generator for two photon physics*, DELPHI note 90-35 PROG 152.
9. F.A. Berends, P.H. Daverveldt, R. Kleiss, Comp. Phys. Comm. **40**, 271-284, 285-307, 309, (1986).
10. Particle Data Group, Eur. Phys. **C3**, (1998).
11. see section 5.2 in P. Abreu *et al.* E. Phys. J. **C2**, 581, (1998).
12. T.G.M. Malmgren, Comp. Phys. Comm. **106**, 230, (1997); T.G.M. Malmgren and K.E. Johansson, Nucl. Inst. Meth. **403**, 481, (1998).
13. E. Accomando, A. Ballestrero Comp. Phys. Comm. **99**, 270, (1997).

HIGGS SEARCHES AT LEP2 WITH THE ALEPH DETECTOR

E. KNERINGER

Institut für Experimentalphysik, Universität Innsbruck, 6020 Innsbruck, AUSTRIA
ALEPH Collaboration
E-mail: Emmerich.Kneringer@uibk.ac.at

Data collected with the ALEPH detector at centre-of-mass energies between 130 and 189 GeV are used to search for Higgs bosons of the Standard Model and its supersymmetric extensions. No evidence for a Higgs particle has been found in 256 pb^{-1} of LEP2 data. Mass exclusion limits are set.

1 Introduction

A neutral Higgs boson is required to complete the particle spectrum of the Standard Model. Fits to electroweak data [1] from LEP, SLC, and Tevatron indicate that if the Higgs particle exists it has most probably a mass around 100 GeV/c^2. The situation at LEP, which is summarized in Fig. 1 (left), shows that we are getting more and more sensitive to this mass range.

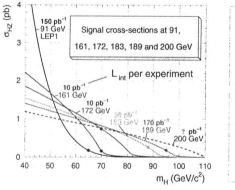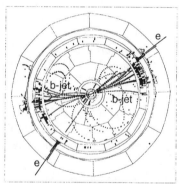

Figure 1: Left: cross section for the Higgs-strahlung process at the various LEP energies. For a given energy the point on the curve indicates the Standard Model Higgs boson mass exclusion limit obtained by a typical LEP experiment with the data taken at that energy. Right: Higgs boson candidate event in the H$\ell\ell$ channel (as seen by the ALEPH detector).

All mass exclusion limits given in this report are at the 95% confidence level. To get an idea of the sensitivity of a specific analysis the expected mass exclusion limit [2] from gedanken experiments usually is also quoted. All the search analyses use cut-based selections and/or neural network methods. We make extensive use of Monte Carlo simulations for the neural net training and the setting of the selection cuts. No details of the preliminary analyses are given; however, the references always point to the corresponding latest published analysis.

2   The Standard Model Higgs Boson

At LEP2, the dominant Higgs production mechanism is the so-called Higgs-strahlung process $e^+e^- \to Z^* \to HZ$. There are also minor contributions from processes where the Higgs boson is generated through the fusion of two electroweak gauge bosons. In the mass range accessible at LEP2, the Higgs boson decays in 92% of the cases into two quark jets (91% of which are $b\bar{b}$) and in about 8% of the cases into $\tau^+\tau^-$. Combining these numbers with the branching fractions of the Z boson we obtain the branching fractions of HZ into the following four characteristic event topologies:

- four jet channel: $q'\bar{q}'q\bar{q}$ (64.6%)

- missing energy channel: $H\nu\bar{\nu}$ (20.0%)

- lepton channel: $H\ell^+\ell^-$, $\ell =$ e or μ (6.7%)

- τ–channel: either H or Z decays to $\tau^+\tau^-$ (8.7%)

Except for the lepton channel, a high b-quark content is required for the quark jets coming from Higgs boson decay. In all channels the invariant mass of the decay products of the Z must be compatible with m_Z. The results of the HZ analyses[3] applied to data taken with the ALEPH detector[4,5] at 189 GeV are summarized in Table 1.

The analyses for the four individual channels are optimized in such a way that the combination of these analyses gives the best global HZ searches analysis. The definition used here for 'best analysis' is: the analysis which gives the highest expected Higgs mass exclusion limit under the hypothesis that there is no signal[2].

No signal is observed in 250 pb^{-1} of data (*cf.* m_h distribution in Fig. 2) which allows to exclude a Standard Model Higgs boson with a mass less than 90.4 GeV/c^2, as shown in Fig. 2 (right). The expected limit is 93.4 GeV/c^2.

3   Neutral Higgs Bosons of the MSSM

The particle spectrum of the Higgs sector of the Minimal Supersymmetric extension of the Standard Model (MSSM) consists of five physical states: two

Table 1: Signal efficiencies ϵ, expected number of signal events N_{95}^{exp} for $m_H = 95$ GeV/c^2, background expectation from Standard Model processes N_{bkg}^{exp}, background from ZZ events N_{ZZ}^{exp} (dominant background source; irreducible), and number of observed events in 170 pb^{-1} of ALEPH data taken at $\sqrt{s} = 189$ GeV for the HZ decay topologies.

Selection	ϵ	N_{95}^{exp}	N_{bkg}^{exp}	N_{ZZ}^{exp}	N_{obs}
$b\bar{b}q\bar{q}$	39.3%	7.7	18.5	10.1	20
$b\bar{b}\nu\bar{\nu}$	24.3%	1.6	3.7	3.1	7
$b\bar{b}\ell^+\ell^-$	72.9%	1.5	13.2	11.8	13
$b\bar{b}\tau^+\tau^-$	20.1%	0.2	0.8	0.6	2
$\tau^+\tau^-q\bar{q}$	18.9%	0.3	1.8	1.2	0
total		11.3	38.0	26.8	42

CP-even neutral bosons h and H (with mixing angle α and masses $m_h < m_H$), one CP-odd neutral boson A, and two charged bosons H$^\pm$. At tree level, the Higgs sector can be parameterized by two independent parameters, $e.g.$ m_h and $\tan\beta = v_2/v_1$, the ratio of the vacuum expectation values of the two Higgs doublets. Only the h and A bosons are within the reach of LEP2. They are expected to be produced in Z decays via the Higgs-strahlung process Z$^* \to$ hZ, with a cross section proportional to $\sin^2(\beta-\alpha)$, and via the associated pair production Z$^* \to$ hA, with a cross section proportional to $\cos^2(\beta - \alpha)$. These two processes are complementary in the sense that if one cross section is maximal the other is minimal and vice versa. Thus both processes must be searched for. For the first process, the results from the hZ searches (Sec. 2) can be used. The second process is searched for in the following two decay channels:

- hA $\to$ b$\bar{\text{b}}$b$\bar{\text{b}}$ (85%)

- hA $\to$ b$\bar{\text{b}}\tau^+\tau^-$, $\tau^+\tau^-$b$\bar{\text{b}}$ (15%)

The performance of the two analyses[6] is reported in Table 2.

Table 2: Performance of the hA analyses at $\sqrt{s} = 189$ GeV: signal efficiencies, expected number of signal events for $m_h = m_A = 85$ GeV/c^2, background expectation, and number of observed events.

Selection	ϵ	N_{85}^{exp}	N_{bkg}^{exp}	N_{obs}
$b\bar{b}b\bar{b}$	58.3%	3.4	9.0	13
$b\bar{b}\tau^+\tau^-$	24.0%	0.2	0.6	0

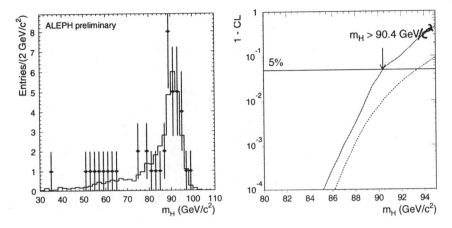

Figure 2: Left: distribution of m_H of the selected events (points) and the expectation from the Standard Model background processes (histogram) in HZ searches at $\sqrt{s} = 189$ GeV. Right: observed (full line) and expected (dashed line) combined confidence levels for the Standard Model Higgs search.

No signal is observed in the data, as can be seen in Fig. 3 (left). This allows to set an upper limit on the cross section for hA production as a function of m_h, which implies an upper limit on $\cos^2(\beta - \alpha)$ as a function of m_h. On the other hand, the hZ searches interpreted in this way result in an upper limit on $\sin^2(\beta - \alpha)$ as a function of m_h. Combining the hZ and hA searches in an optimal way [2] leads to an excluded region in the two-dimensional parameter space $[m_h, \sin^2(\beta - \alpha)]$, which usually is translated to the $[m_h, \tan\beta]$ plane, as shown in Fig. 3 (right). For $\tan\beta \geq 1$ we find that ALEPH data up to centre-of-mass energies of 189 GeV exclude the h and A Higgs bosons of the MSSM with masses less than 80.8 and 81.2 GeV/c^2, respectively, i.e. the neutral Higgs bosons must be heavier than the W boson if $\tan\beta \geq 1$.

4 Charged Higgs Bosons

In the MSSM, charged Higgs bosons are heavier than W bosons. This restriction does not hold for general two doublet models, which are very attractive theoretically because of the absence of flavor changing neutral currents and the relation $m_W = m_Z \cos\theta_W$ holding at tree level. The H$^\pm$ has the same decay modes as the W$^\pm$, but since its coupling is proportional to the charged fermion masses, it dominantly decays into the heaviest energetically allowed fermion pair of the quark and lepton families. Whether H$^\pm$ decay preferentially into

418

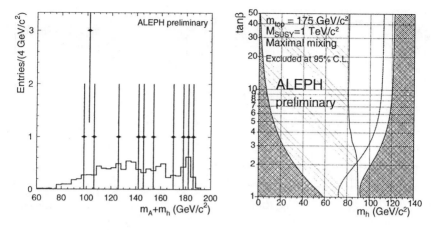

Figure 3: Left: distribution of $m_A + m_h$ of the selected events (points) and the expectation from the Standard Model background processes (histogram) in hA searches at $\sqrt{s} = 189$ GeV. Right: experimentally excluded region in the $[m_h, \tan\beta]$ plane (after combination with the hZ searches). The dark area represents the theoretically forbidden region for the case of 'maximal stop mixing' (dashed-dotted curve for the 'no stop mixing' case).

quarks or leptons depends on other parameters of the model. Therefore, the search for pair-produced H^+H^- is performed in the following three channels:

- leptonic channel: $\tau^+\nu_\tau\tau^-\bar{\nu}_\tau$

- mixed channel: $c\bar{s}\tau^-\bar{\nu}_\tau$

- hadronic or four jet channel: $c\bar{s}s\bar{c}$

The performance of these analyses[7] is summarized in Table 3.

The search for charged Higgs bosons in the three final states $\tau^+\nu_\tau\tau^-\bar{\nu}_\tau$, $c\bar{s}\tau^-\bar{\nu}_\tau$ and $c\bar{s}s\bar{c}$ has been performed using 175 pb^{-1} of ALEPH data collected at $\sqrt{s} = 189$ GeV. No evidence of Higgs boson production was found (Fig. 4 left) and mass limits were set as a function of the branching ratio $\mathcal{B}(H^+ \to \tau^+\nu_\tau)$. The result of the combination of the three analyses is displayed in Fig. 4 (right) where the curves corresponding to expected and observed confidence levels of 95% exclusion are drawn. As can be seen from this figure, charged Higgs bosons with masses below 62.5 GeV/c^2 are excluded at 95% C.L. independently of $\mathcal{B}(H^+ \to \tau^+\nu_\tau)$, where the expected mass exclusion limit is 68.5 GeV/c^2.

Table 3: Efficiencies ϵ, number of Standard Model background events expected N_{bkg}^{exp}, and number of observed candidates N_{obs} for the three charged Higgs boson analyses at a centre-of-mass energy of 189 GeV, as a function of the charged Higgs mass. For the four jet channel, numbers are quoted within a ±3 GeV/c^2 window around the assumed Higgs boson mass.

| $m_{H^\pm}$ | $\tau^+\nu_\tau\tau^-\bar{\nu}_\tau$ | | | $c\bar{s}\tau^-\bar{\nu}_\tau$ | | | $c\bar{s}s\bar{c}$ | | |
(GeV/c^2)	ϵ (%)	N_{bkg}^{exp}	N_{obs}	ϵ (%)	N_{bkg}^{exp}	N_{obs}	ϵ (%)	N_{bkg}^{exp}	N_{obs}
50	33.5	15.5	20	35.6	9.4	11	40.1	15.1	18
55	35.2	15.5	20	37.2	9.4	11	38.4	22.8	23
60	38.2	15.5	20	37.4	9.4	11	37.6	28.0	19
65	35.2	15.5	20	34.8	9.4	11	37.2	30.6	28
70	39.9	15.5	20	28.1	9.4	11	36.4	32.7	35
75	40.8	15.5	20	19.1	9.4	11	34.6	55.1	40

5   Invisible Higgs Boson Decays

Many extensions of the Standard Model allow for the Higgs boson to decay invisibly, e.g. into a pair of lightest neutralinos when the neutralino χ is light enough. Since the Higgs boson is produced through the Higgs-strahlung process hZ, this leads to the following two event topologies:

- a pair of acoplanar leptons, when Z $\to$ e$^+$e$^-$ or Z $\to$ $\mu^+\mu^-$
- a pair of acoplanar jets, when the Z decays hadronically

where the acoplanarity is defined as the azimuthal angle between the two lepton or jet directions. Two analyses [8] are designed for the two channels. When they are applied to ALEPH data taken at $\sqrt{s} = 189$ GeV, 33 candidates are found, in agreement with 33.6 events expected from all background processes (cf. Table 4). The distributions of the reconstructed Higgs boson masses are shown in Fig. 5 (left).

Table 4: Performance of the acoplanar lepton and acoplanar jet pair analyses.

Selection	N_{bkg}^{exp}	N_{obs}
h$\ell^+\ell^-$	4.3	5
hq$\bar{q}$	29.3	28

Quite generally, the production cross section for invisible Higgs boson decay can be parameterized as $\xi^2 \cdot \sigma_{SM}(e^+e^- \to hZ)$, where ξ^2 is a model dependent factor ranging from 0 to 1. One way of presenting the result of the negative searches is to calculate the 95% C.L. level upper limit on the production cross

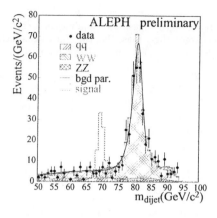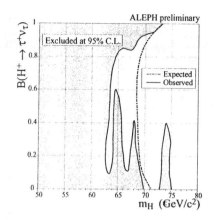

Figure 4: Left: distribution of the dijet mass distribution as obtained after applying all cuts of the c$\bar{s}$s$\bar{c}$ selection; shown are 189 GeV data (dots), background Monte Carlo (histogram), and the polynomial parameterization of the background (full line). Also shown is the Monte Carlo expectation for a signal with $m_{H\pm} = 70$ GeV/c^2 (dashed line) with arbitrary normalization. Right: limit on the mass of charged Higgs bosons as a function of $\mathcal{B}(H^+ \to \tau^+ \nu_\tau)$. Shown are the expected (dash-dotted) and observed (full) exclusion curves for the combination of the three charged Higgs boson analyses. The shaded area is excluded at 95% C.L.

section of the invisibly decaying Higgs boson in units of $\sigma_{SM}(e^+e^- \to hZ)$ as a function of m_h, which is shown in Fig. 5 (right).

6 Summary

In this report, an overview has been given of the present status of Higgs boson searches with the ALEPH detector. The results are mainly based on the analysis of data taken at 189 GeV centre-of-mass energy. When the data taken at lower energies are included in the analyses the 95% confidence level exclusion limits improve slightly.

For the Standard Model Higgs boson we find $m_H > 90.4$ GeV/c^2 at 95% C.L. using 250 pb^{-1} of data taken at $\sqrt{s} = 161, 172, 183$ and 189 GeV. This means that a Higgs boson with a mass equal to the Z mass just cannot be excluded at 95% C.L. with ALEPH data taken up to the year 1998.

Including all lower energy data, for the neutral Higgs bosons of the MSSM we obtain the limits $m_h > 80.8$ GeV/c^2 and $m_A > 81.2$ GeV/c^2 (valid for all values of $\tan \beta \geq 1$).

From the analysis of 175 pb^{-1} of ALEPH data collected in 1998, the charged Higgs bosons of general two doublet models are excluded below 62.5

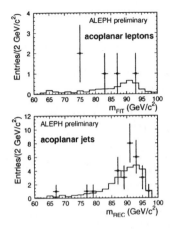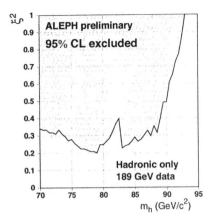

Figure 5: Left: distribution of the reconstructed Higgs boson mass of the selected events compared to the Monte Carlo expectation for the acoplanar lepton and jet pair searches. Right: region in the $[m_h, \xi^2]$ plane excluded at 95% C.L. using only the hadronic channel at 189 GeV.

GeV/c^2 independent of their decay mode.

Finally, using the same data, invisibly decaying Higgs bosons are searched for. For a production cross section equal to that of the Standard Model Higgs boson, *i.e.* $\xi^2 = 1$, masses below 92.8 GeV/c^2 are excluded.

Acknowledgments

I would like to thank all my colleagues from the ALEPH Higgs Task Force for the continuous effort to produce all these results.

References

1. LEP Electroweak Working Group, CERN-EP/99-15.
2. P. Janot and F. Le Diberder, Nucl. Instrum. Methods **A411**, 449, (1998).
3. ALEPH Collaboration, Phys. Lett. **B440**, 403, (1998).
4. ALEPH Collaboration, Nucl. Instrum. Methods **294**, 121, (1990).
5. ALEPH Collaboration, Nucl. Instrum. Methods **360**, 481, (1995).
6. ALEPH Collaboration, Phys. Lett. **B440**, 419, (1998).
7. ALEPH Collaboration, Phys. Lett. **B450**, 467, (1999).
8. ALEPH Collaboration, Phys. Lett. **B450**, 301,(1999).

SOLITONS IN SO(5) SUPERCONDUCTIVITY

R. MacKENZIE

Laboratoire René-J.-A.-Lévesque, Université de Montréal,
Montréal, Québec H3C 3J7
and
Department of Physics and Astronomy, University of British Columbia,
6224 Agriculture Rd, Vancouver BC V6T 1Z4

J.M. CLINE

Department of Physics, McGill University,
3600 University St., Montréal, Québec H3A 2T8

A model unifying superconductivity and antiferromagnetism using an underlying approximate SO(5) symmetry has injected energy into the field of high-temperature superconductivity. This model might lead to a variety of interesting solitons. In this paper, the idea that superconducting vortices may have antiferromagnetic cores is presented, along with the results of some preliminary numerical work. An outlook for future work, including speculations about other possible exotic solitons, is presented.

1 Introduction

The phase diagrams of a variety of exotic superconductors (high-temperature superconductors, heavy fermion superconductors, organic superconductors) have a very rich structure. Although profound differences in these phase diagrams exist, it is surprising that in all of them two features are common: superconductivity and antiferromagnetism. It is extremely enticing to speculate that these materials, despite the incredible range of underlying structures, may have some common underlying reason for the appearance of these two phases.

This idea was formalized by S.C. Zhang [1], who observed that both superconductivity and antiferromagnetism involve spontaneous symmetry breaking. Superconductivity is essentially spontaneous breaking of electromagnetism (it is, in fact, the first example of what we now call dynamical symmetry breaking); antiferromagnetism is spontaneous breaking of spin-rotation symmetry. The first symmetry group is U(1), or equivalently SO(2); the second is SO(3). Zhang's suggestion, borrowing heavily on ideas from particle physics, was that these two symmetries might be unified into a larger symmetry group. He presented a strong case for the group SO(5). His work has given rise to a minor cottage industry of SO(5) phenomenology, not to mention fueling a heated debate (with some of the heavyweights of condensed matter physics appearing on opposite sides) over the merits and possible fundamental flaws of the idea.

In this work (which, admittedly, uses a rather broad interpretation of the theme of this Institute – "Electroweak Physics"), the ABCs of superconductivity and antiferromagnetism will be briefly reviewed, and Zhang's unified description of the two will be outlined. The case for exotic solitons will then be discussed. Superconducting vortices with antiferromagnetic cores will be examined in some detail (though much work remains), and other, even more speculative, possibilities will be discussed.

2 The SO(5) Model

2.1 Superconductivity

Superconductivity is a phenomenon which occurs in a huge number of materials, if a sufficiently low temperature is reached. Superconductors display several striking features, among them zero resistance, the Meissner effect, the existence of a gap in the spectrum of low-energy excitations (this is not always the case, but usually it is so), and a transition to a normal state as the temperature increases.

Most of these phenomena can be described by a phenomenological model, known as a Ginzburg-Landau model. The great success of many workers in the late fifties and early sixties, culminating in the work of Bardeen, Cooper and Schrieffer (who first succeeded in an essentially complete description of all "conventional" superconductors), was the derivation of this GL model from an underlying microscopic model.

For our purposes, the GL model of superconductivity is essentially a nonrelativistic version of what we in particle physics call the Abelian Higgs model. The distinction between nonrelativistic and relativistic models is irrelevant here, so I will discuss only the relativistic case. The Lagrangian is

$$\mathcal{L} = |D_\mu \phi|^2 - \frac{\lambda}{4} \left(|\phi|^2 - v^2 \right)^2 - \frac{1}{4} F_{\mu\nu}{}^2, \tag{1}$$

where $D_\mu \phi = \partial_\mu \phi - i2eA_\mu \phi$ and where $\phi = c_{p,\uparrow}^\dagger c_{-p,\downarrow}^\dagger$ is the Cooper pair creation operator.

When $\langle \phi \rangle \neq 0$ (as is the case in the above Lagrangian), the O(2) symmetry is spontaneously broken, and it is easy to derive such features as the Meissner effect and a gap in the excitation spectrum from this fact.

2.2 Antiferromagnetism

In elementary discussions of spontaneous symmetry breaking, often the first example given is the ferromagnet. At high temperatures a spin system has spins oriented randomly. As the temperature is decreased, it can occur that the spins prefer to be aligned in the same direction as their neighbours. This alignment amounts to the selection of an *a priori* random direction in space which is singled out. Rotations of the resulting configuration about this direction do not alter the system (that symmetry is not broken), yet rotations about any other direction do change the system (those symmetries are broken). Thus the ferromagnet breaks spin rotational symmetry from SO(3) (rotations about any direction when the average magnetism in any region is zero) to SO(2) (rotations about the direction in which the spins are aligned).

Somewhat less familiar, but no more complicated (for our purposes, at least), is the case of antiferromagnetism. There, adjacent spins prefer to be anti-aligned at low energies. Once one spin's direction is chosen, all the others must follow suit (alternating in direction from site to site) in order to minimize the energy. Once again, SO(3) spin rotation symmetry is broken to SO(2).

The order parameter in the case of ferromagnetism is (somewhat loosely) the average value of the spin $\langle \vec{S} \rangle$; in the case of antiferromagnets this averages to zero, but one defines a staggered spin vector $\vec{n} = \langle (-)^n \vec{S} \rangle$, where the extra sign is positive or negative depending whether one is on a site an even or odd number of translations away from some reference site.

2.3 Combined superconductivity and antiferromagnetism

We have seen that superconductivity can be described by a complex field ϕ (or, equivalently, by a real doublet of fields defined by $\phi = (\phi_1 + i\phi_2)/\sqrt{2}$), and that antiferromagnetism can be described by a real triplet field $\vec{n}$. Zhang put forth the idea that, since these two order parameters seem to be relevant to such a wide variety of systems, perhaps there is an underlying (approximate) symmetry which includes these two as subgroups, rather like the central idea of Grand Unified Theories. This theory would then be described by a five-component vector $\vec{N} = (n_1, n_2, n_3, \phi_1, \phi_2)$; if the dynamics of the system is such that the expectation value of $\vec{N}$ lies in the upper three-dimensional subspace the system is antiferromagnetic, while if it lies in the lower subspace the system is superconducting. If the expectation value of $\vec{N}$ is zero the system is neither superconducting nor antiferromagnetic. (And a fourth possibility, seemingly not realized in nature, is that the system could in principle be in a state which breaks both superconductivity and antiferromagnetism.)

Zhang's work suggested that the SO(5) symmetry is explicitly broken by small terms, in a similar way to the explicit breaking of chiral symmetry by small quark masses; as a result some of the Goldstone bosons which would arise due to the breaking of SO(5) would, in fact, be pseudo-Goldstone bosons (low-energy but not quite massless); low-energy excitations seen in high-temperature superconductors were among Zhang's original motivations for introducing this model.

3 Solitons in the SO(5) Model

Whenever a model exhibits spontaneous symmetry breaking, there is a family of equivalent vacua (which are rotated into one another by the broken symmetry generators). The possibility then arises that topological solitons could exist. General topological arguments can be applied to any case to see if, in fact, solitons are realized.

One of the simplest examples of solitons is superconducting vortices. These appear in the appropriate GL model, or equivalently in the Abelian Higgs model (1), in 2+1 dimensions. The potential is the familiar Mexican-hat potential, with a ring of vacua given by $|\phi| = v$. For finiteness of energy, ϕ must go to a vacuum at infinity, but there is no need for this to be the same vacuum along different directions. We can construct a configuration such that the phase of ϕ changes by 2π as we go around a circle at infinity; such a configuration cannot be unwound by continuous deformations (without wandering away from the vacuum at infinity, which would cost an infinite energy). Furthermore, if the field configuration is continuous, it is a topological necessity that somewhere there must be a zero of the field. In the simplest, most rotationally symmetric configuration, this zero will be at the origin; we may write

$$\phi(r, \theta) = v \, f(r) e^{i\theta}, \tag{2}$$

where the function $f(r)$ interpolates from zero at the origin to 1 at infinity.

In the SO(5) theory, the order parameter is considerably more complicated, and interesting and exotic possibilities for solitons might arise, as we will now see [2].

The potential is assumed to be exactly invariant under rotations of the superconducting and antiferromagnetic order parameters, and we may assume that V has the following form, depending only on the magnitudes of these order parameters, $\phi = |\phi|$ and $n = |\vec{n}|$:

$$V(\phi, n) = -\frac{m_1^2}{2}\phi^2 + \frac{\lambda_1}{4}\phi^4 - \frac{m_2^2}{2}n^2 + \frac{\lambda_2}{4}n^4 + \frac{\lambda_3}{2}\phi^2 n^2 + const. \tag{3}$$

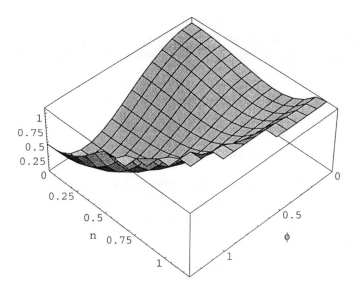

Figure 1: SO(5) potential $V(\phi, n)$.

Here, we have added a constant so that the minimum of the potential is zero, included even terms up to fourth order, and assumed the quadratic terms are such that symmetry breaking in both sectors is favoured. The potential is assumed bounded below at all directions at infinity, which will be true if the quartic couplings satisfy $\lambda_{1,2} > 0$ and $\lambda_3 > -\sqrt{\lambda_1 \lambda_2}$.

Suppose furthermore that the parameters of the potential are such that V is as shown in Fig. 1. There are two important features. First, the global minimum is at a nonzero value of ϕ with $n = 0$; this corresponds to having a superconducting ground state. Second, if we were to force ϕ to be zero, the potential $V(0, n)$ is minimized at a nonzero value of n. These features do occur if the parameters obey the following conditions:

$$\frac{\lambda_3}{\lambda_1} > \frac{m_2{}^2}{m_1{}^2}, \qquad \frac{m_1{}^4}{\lambda_1} > \frac{m_2{}^4}{\lambda_2}. \tag{4}$$

The second feature is no mere mathematical curiosity, since at the core of a superconducting vortex ϕ is indeed zero. Thus, if the potential energy had its way, the superconducting vortex core would surely be antiferromagnetic. In fact, the energetics are somewhat more complicated, and there is a range of parameters where the core is antiferromagnetic; outside this range the core is normal.

To see this, we must solve the coupled equations for ϕ and n. These equations come from the energy functional (we assume planar geometry):

$$E = \int d^2x \left\{ \frac{|\nabla\phi|^2}{2} - \frac{m_1{}^2|\phi|^2}{2} + \frac{\lambda_1|\phi|^4}{4} \right.$$
$$\left. + \frac{|\nabla\vec{n}|^2}{2} - \frac{m_2{}^2|\vec{n}|^2}{2} + \frac{\lambda_2\vec{n}^4}{4} + \frac{\lambda_3|\phi|^2|\vec{n}|^2}{2} \right\} . \tag{5}$$

The equations of motion are straightforward, and can be rewritten in the following form with a rotationally symmetric ansatz $\phi(x) = v\phi(r)e^{i\theta}$, $\vec{n}(x) = \vec{n}_0 n(r)$ (where $\vec{n}_0$ is a constant unit vector), and with appropriate field and coordinate rescalings:

$$\frac{d^2\phi}{du^2} + \frac{1}{u}\frac{d\phi}{du} + \left(1 - \frac{1}{u^2}\right)\phi - \delta n^2\phi - \phi^3 = 0, \tag{6}$$

$$\frac{d^2n}{du^2} + \frac{1}{u}\frac{dn}{du} + \beta n - \alpha\phi^2 n - n^3 = 0, \tag{7}$$

where the constants α, β and δ are $\alpha = \lambda_3/\lambda_1$, $\beta = m_2{}^2/m_1{}^2$ and $\delta = \lambda_3/\lambda_1$.

The equations can be solved subject to four boundary conditions: ϕ must go to zero at the origin and to 1 at infinity, and n must have zero slope at the origin and must go to zero at infinity. The crucial observation, however, is that there is no need for n to be zero at the origin, and we find that for some parameters n goes to a nonzero value, corresponding to an antiferromagnetic core for the vortex.

As an example, Fig. 2 displays the functions $\phi(u)$ and $n(u)$ (u being a scaled radial variable), for specific values of the parameters. Since n is nonzero at $u = 0$, the core of the vortex in this case is superconducting. For other values of the parameters (for example, if β is reduced to a sufficiently low value), one finds that $n(u) = 0$ for all u, indicating a normal core.

Still to be done is a systematic scan of parameter space to see under what conditions vortices do indeed have antiferromagnetic cores. It is also important to make contact between the parameters of the GL model above and experimental parameters of exotic superconductors.

Finally, it is not difficult to see that other types of solitons could in principle have similar exotic structure[3]. One example would occur in the antiferromagnetic phase of these materials, if the antiferromagnetism is of an "easy-plane" variety. (This means that, rather than being truly isotropic, a plane is favoured for antiferromagnetism, due to crystal asymmetry.) In this case, it is easy to

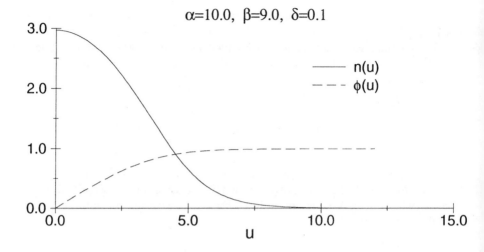

Figure 2: Soliton profile.

imagine antiferromagnetic vortices of a very similar structure to the superconducting ones discussed above, and depending on the parameters it is possible
that such vortices would have superconducting cores.

A second example might actually be observed in certain underdoped high-
temperature superconductors which display striping. In these materials, the
striping can be understood in terms of the formation of antiferromagnetic
domain walls (domain boundaries, really, since the materials are effectively
planar), with superconductivity occuring in the domain wall. This might fit
in nicely with the SO(5) model, since it is fairly straightforward to construct
antiferromagnetic domain boundaries which have a superconducting core.

References

1. S.C. Zhang, Science **275**, 1089, (1997).
2. D.P. Arovas, *et al.*, Phys. Rev. Lett. **79**, 2871, (1997).
3. J. Cline and R. MacKenzie, in preparation.

PERTURBATIVE QCD TESTS IN DIS AT HERA

STEPHEN R. MAGILL

Argonne National Laboratory
9700 South Cass Ave., Argonne, Illinois 60439, USA
E-mail: srm@hep.anl.gov

The investigation of hadronic final states in deep inelastic scattering at HERA has expanded our understanding of perturbative QCD. The Breit frame of reference makes it possible to separate a DIS event into two regions - the current region in which pQCD adequately describes the data; and the target region in which standard properties of pQCD are challenged. This report summarizes some of the studies of hadronic final states performed at HERA, comparing the results obtained in the current and target regions.

1 Introduction

The HERA ep collider, at the Deutsches Elektronen-Synchrotron Laboratory (DESY), has been operating since 1992. Two of the experimental collaborations, ZEUS and H1, have accumulated $\sim 55\ pb^{-1}$ each of ep luminosity. The physics of HERA is rich and diverse, ranging from photoproduction to deep inelastic scattering (DIS), from diffraction to multi-jets, and from electro-weak to strong interactions. In this report, the focus will be to summarize some of the important tests of perturbative QCD which have been performed on the charged particles and jets produced in DIS.

2 DIS Kinematics in the Breit Frame

The following variables characterize DIS kinematics :

$$Q^2 = -q^2 = -(k - k')^2; x = Q^2/(2p \cdot q); y = p \cdot q/(p \cdot k) \qquad (1)$$

where k and k' are the 4-momentum of the incident and scattered electron, p is the 4-momentum of the incident proton, Q^2 is the 4-momentum transfer squared, x represents the momentum fraction of the struck quark, and y is the fractional energy transfer. These variables are related by :

$$Q^2 = sxy \qquad (2)$$

where s is so that only two of the kinematic variables are needed to describe a DIS event. The invariant mass (squared) of the hadronic system is given by :

$$W^2 = (p + q)^2 \simeq \frac{1 - x}{x} Q^2 . \qquad (3)$$

DIS physics results from HERA span the ranges $10^{-6} < x < 1$ and $0.1 < Q^2 < 50000 \, GeV^2$ for $y \sim 0.005$ to $y \sim 1$.

A particularly interesting frame of reference in which to analyze DIS events is the Breit frame - the frame which maximizes the separation of the struck quark (current fragmentation region) and the proton remnant (target fragmentation region). In this frame, the virtual photon is purely spacelike ($q = (0,0,0,Q/2)$) and there is no transverse energy in the (order α_s^0) Born process; transverse energy only results from (higher order α_s) initial or final state gluon radiation (QCDC process) and the Boson-Gluon Fusion (BGF) process. In the Breit frame, a direct comparison of experimental results can be made between the current region of DIS (evolving with energy $Q/2$) and e+e- colliders (evolving with energy $Q/2 = \sqrt{s}/2$ in one hemisphere). However, the target region in DIS evolves longitudinally with energy of $(1-x)/x \times Q/2$, which means that p_L is much larger than in the current region at low x.

3 Inclusive Charged Particle Distributions

Several inclusive charged particle results from HERA will be presented here which consist of measurements made in the current region of DIS and compared to pQCD calculations and to e+e- results. It is expected that since the current region of the Breit frame in DIS is equivalent to a hemisphere in e+e- collisions, pQCD should describe the DIS data in shape and magnitude, including features of pQCD such as Q^2-dependent parton evolution (DGLAP) and coherence effects (angular ordering of gluon emission).

3.1 Event Shapes

The distribution of hadrons in a DIS event, whether isotropic or collimated, is influenced by the underlying parton interactions including gluon radiation effects. Therefore, variables have been defined which can be used to test pQCD at the hadron level, provided hadronization effects are properly modeled. A representative sample of event shape variables includes :

- **Thrust** $T_z = \dfrac{\sum_h p_z}{\sum_h |\mathbf{P_h}|}$: represents the fraction of vector momentum along the z-axis (γ^* direction in the Breit frame).

- **Jet Broadening** $B_c = \dfrac{\sum_h p_\perp}{2 \sum_h |\mathbf{P_h}|}$: represents the fraction of vector momentum perpendicular to the γ^* direction.

- **Jet Mass** $\rho_E = \dfrac{(\sum_h p)^2}{4(\sum_h E_h)^2}$: another measure of degree of collimation in which the mass is compared to the total energy.

As events change from isotropic to collimated, $T_z \to 0$ to 1, $B_c \to 0.5$ to 0, and $\rho_E \to 0.25$ to 0.

In order to compare the pQCD predictions with data, the parton-level calculation must be modified to include hadronization. This is done by including "power corrections" which are calculable in NLO QCD and introduce an additional (expected to be universal) strong coupling constant $\overline{\alpha_0}$ applicable below some scale μ_I. A particular event shape variable can then be calculated to NLO by including a term F^{pow} which depends on $\alpha_s(Q)$, $\alpha_s^2(Q)$, and $\overline{\alpha_0}(\mu_I)$ in addition to the pQCD term. The infrared matching scale, μ_I, is defined by $\Lambda_{QCD} \ll \mu_I \ll \mu_r \equiv Q$.

Fig. 1 shows results from H1 for several event shape variables including the ones defined above, comparing the data to NLO pQCD only (dashed line) and with a fit to the data including NLO pQCD and power corrections (solid line).

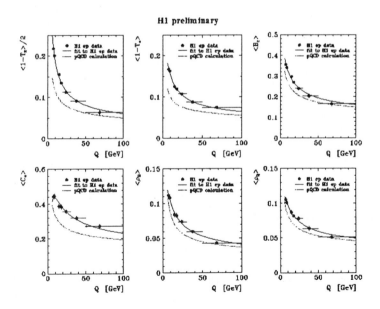

Figure 1: Selected event shape variables from H1 compared to NLO pQCD.

432

The data is well described by the pQCD fit with power corrections to simulate hadronization. In addition, the fits can be used to extract the strong coupling constants α_s and $\overline{\alpha_0}$. The $\alpha_s(M_z)$ values obtained are consistent with the world average and the obtained values of $\overline{\alpha_0}$ are consistent with each other; $\langle \overline{\alpha_0}(\mu_I = 2\ GeV) \rangle = 0.488 \pm 0.050$, where the error is an estimate of the combined statistical and systematic errors which are dominated by the μ_r dependence at NLO. This average value is also consistent with the 'universal' value of $\overline{\alpha_0}(\mu_I = 2\ GeV) \sim 0.5$ obtained in e+e- experiments.

3.2   Multiplicity and Scaled Momentum

Fragmentation functions, $D(x_p, Q^2)$ where x_p is the scaled momentum given by $x_p = p_h/p_{max}$ where p_h is a particular hadron momentum, can also be used to test predictions of pQCD. While the functions themselves can not be calculated perturbatively, their evolution with Q^2 is known as well as the shape and evolution of the x_p distribution. At small x_p, coherence effects (angular ordering of gluon emission such that large p_T occurs at large angles and small p_T occurs at small angles) are expected to influence the rise in parton multiplicity with energy and make the distributions in $\ln(1/x_p)$ appear gaussian in shape.

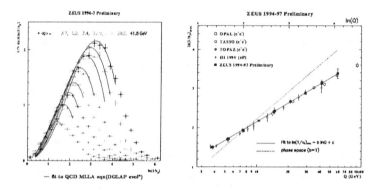

Figure 2:   a) Multiplicity distributions versus $\ln(1/x_P)$ (left) and b) peak positions of $\ln(1/x_P)$ versus $\langle Q \rangle$ (right).

This behavior would indicate that the x_p distributions are described by the Modified Leading Log Approximation (MLLA) combined with Local Parton-Hadron Duality (LPHD). The MLLA incorporates the Double Leading Log Approximation (DLLA) which resums terms of the type $(\ln(1/x)\ln(Q^2))^n$ and incorporates DGLAP parton evolution. At large x_p, scaling violations in $D(x_p, Q^2)$ can be used to determine α_s. Fig. 2a shows multiplicity distributions versus $\ln(1/x_p)$ for various values of $\langle Q \rangle$. The (approximately) gaussian fits are pQCD MLLA fits including DGLAP evolution. In Fig. 2b, the peak positions of the MLLA fits are plotted versus$\langle Q \rangle$. The ZEUS and H1 data are compared to e+e- data and a straight line fit motivated by expected coherence effects (solid line) is seen to describe all of the data, while a fit with no coherence effects (dashed line) fails to describe any of the data.

At high x_p $(x_p > 0.1)$ and for $Q^2 > 100\ GeV^2$, preliminary fits of $1/\sigma_{Tot} d\sigma/dx_p$ versus Q^2 yield a best fit parameter of $\Lambda_4 = 350\ GeV$. Future analyses will extract α_s from higher statistics data.

3.3 Angular Correlations

Complementing global event analyses are studies of n-particle correlations (where $n > 1$) in which the correlation is expressed as a function of a particular choice of variable. At ZEUS, the relative angular distribution of 2 particles in a cone of half-angle Θ around the $\gamma*$ direction (current region) is expressed as a function of the QCD-motivated variable $\varepsilon = \ln(\Theta/\theta)/\ln(P\Theta/\Lambda)$ where θ is the angle between 2 particles, $P(= Q/2)$ is the momentum in the current region, and $\Lambda(= 0.15\ GeV)$ is a scale parameter. If $\varepsilon = 0$, the 2 particles are uncorrelated and if $\varepsilon = 1$, the particles are parallel. If pQCD holds, there should be no dependence on either Q or Θ.

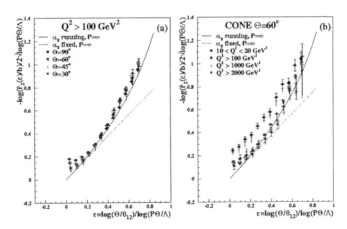

Figure 3: Relative angular distributions a) in various cones versus ε and b) in various Q^2 ranges versus ε (ZEUS Preliminary).

Fig. 3a shows the (normalized) angular distribution for $Q^2 > 100\ GeV^2$

434

in cones from 30° half-angle to 90° (entire current region). Fig. 3b shows the Q^2 variation for a fixed cone of 60°. There is no Θ dependence and no Q^2 dependence for $Q^2 > 100\ GeV^2$. Therefore pQCD is able to describe the data and, furthermore, the observed behavior indicates that Local Parton-Hadron Duality (LPHD) also holds for $P = Q/2 > 5\ GeV$. Some Q^2 dependence is seen in the lowest Q^2 bin - this is expected since in this kinematic region, the BGF process reduces the average number of particles in the current region.

4 Jets in DIS

In order to investigate pQCD in the target region, it is useful to analyze events with a suitable jet algorithm, comparing distribution shapes and cross sections with NLO pQCD calculations. There are presently several NLO QCD calculations available that can predict dijet and forward jet cross sections in DIS.

4.1 Dijets

Dijets in DIS are sensitive to parton distributions, the evolution of partons (QCD dynamics), and the fundamental QCD parameter, α_s. At high x and

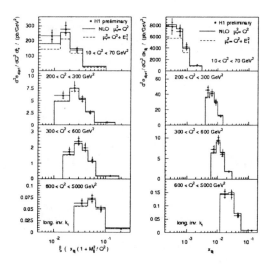

Figure 4: H1 dijets versus ξ and x.

Q^2, it is expected that the QCD Compton (QCDC) process should dominate (more jets in the current region) and that the dijet cross section should be $\propto \alpha_s q(x)$. Since the quark distributions are known in this region, and assuming that DGLAP evolution applies, α_s can be extracted. This has been done by both ZEUS [1] and H1 [2], with $\alpha_s(M_z)$ matching the world average.

Fig. 4 shows dijet data from H1 compared to a NLO QCD calculation (DISENT) for 3 high Q^2 bins and one low Q^2 bin. At high Q^2, the DGLAP-based NLO QCD calculation is able to describe the data. At low Q^2, the description is not as good, and furthermore, the dependence on the choice of renormalization scale μ_r indicates that higher orders are needed in the calculation in order to make a stable comparison to the data.

At low x and Q^2, the Boson Gluon Fusion (BGF) process is expected to dominate the dijet cross section (more target region jets) which is $\propto \alpha_s g(x)$. Assuming a value for α_s consistent with the world average, it should be possible to extract $g(x)$. Fig. 5 shows a comparison of ZEUS dijet data with a NLO QCD calculation (MEPJET) for several variables. DGLAP-based Monte Car-

ZEUS 1995 Preliminary

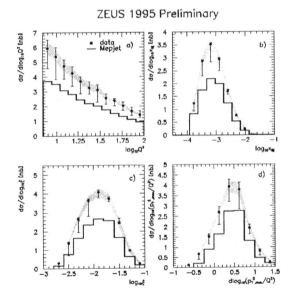

Figure 5: ZEUS dijet cross sections.

los are unable to describe the ZEUS data at low Q^2. In all cases, the measured cross section is much greater than the calculated result. The observed shape

agreement is an indication that the x-dependence of the gluon distribution used in the calculation is appropriate, while the disagreement in magnitude may indicate inadequacy in the (DGLAP-based) parton evolution dynamics.

4.2 Forward Jets

It is expected that in the very low x region at HERA, the contribution to parton evolution by $\ln(1/x)$ terms in the resummation is large, dominating $\ln Q^2$. Therefore, cross section calculations which assume parton evolution based only on resummation of $\ln Q^2$ terms should not be able to describe the data. The calculation of parton evolution which includes leading $\ln(1/x)$ terms is called BFKL dynamics. To investigate QCD dynamics at low x and Q^2, the inclusive forward jet cross section has been measured by both ZEUS[3] and H1[4] in DIS data and compared to Monte Carlo results and NLO QCD calculations.

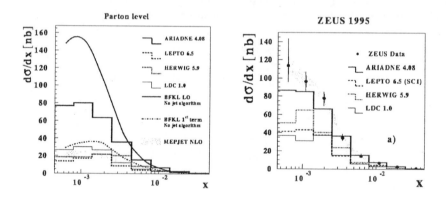

Figure 6: Parton level comparison of forward jets (left), and ZEUS forward jet cross section compared to DIS Monte Carlos (right).

To test the BFKL hypothesis that parton evolution in x occurs, the cross section of events in which a jet is found with momentum fraction $x_{Jet} \gg x$ is measured. The rapidity interval between the selected jet and the current region is given by $\delta\eta = \ln(x_{Jet}/x)$ and is chosen to be as large as possible (high energy jets in low x events), hence the name "Forward Jets". In addition, the transverse energy of the selected jet is required to satisfy $kT_{Jet}^2/Q^2 \sim 1$. These selection criteria ensure that there is a large distance for x evolution to occur and that Q^2 evolution is suppressed.

Since the NLO QCD calculations are done at the parton level, the NLO QCD calculation (MEPJET) will be compared to a leading order BFKL calculation and the parton levels of several Monte Carlo programs. The first plot in Fig. 6 shows that the small, relatively flat cross section expected from the DGLAP-based NLO QCD calculation is reproduced by the LEPTO, Herwig, and LDC Monte Carlos, as well as by the first term in the BFKL calculation. This is the "Born" term of BFKL and is equivalent to (DGLAP) NLO QCD. The Ariadne Monte Carlo ("BFKL-like") compares well in shape to the BFKL calculation and is the only one in which the cross section rises at low x.

The second plot shows the data corrected to the hadron level and compared to the Monte Carlo results. The cross section in the data is much larger than any of the Monte Carlo results, and if the hadron - parton correction factor from Ariadne is applied to the data, the shape of the data is described much better by the BFKL calculation than by NLO QCD. These results indicate that DGLAP evolution is unable to describe the low x HERA data.

5 Conclusions

Separation of DIS into current and target regions of the Breit frame provides unique insights into the dynamics of pQCD. This unique combination of regions with different energy dependencies allows pQCD to be studied in its fine details, including the interplay of parton distributions and evolution dynamics with the QCD parameter - α_s.

In the current region, pQCD at NLO provides a good description of the HERA data. Inclusive charged particle distributions evolve with energy in the same way as in e+e-, parton evolution is dependent only on Q^2 (DGLAP evolution), and there is evidence for angular ordering of gluon radiation (coherence effects). Using these constraints on the partons primarily in the current region (at high x and Q^2), the strong coupling constant α_s can be extracted from DIS dijets by comparing the measured dijet cross section with the calculated NLO pQCD prediction. Each of the results from H1 and ZEUS is consistent with the world average for this parameter.

In the target region, however, cross sections for low Q^2 dijets and forward jets can not be described by DGLAP-motivated pQCD calculations. There is also evidence for the breakdown of angular ordering, indicating that an alternate form of parton evolution is needed in the low x, low Q^2 region. The BFKL parton evolution approach (in which leading $ln(1/x)$ terms are included) is able to reproduce the shape of the forward jet cross section, however (presently without higher order corrections which are presumed to decrease the cross section) it overshoots the data in magnitude.

438

Acknowledgments

I wish to thank the many individuals from the H1 and ZEUS collaborations at the Deutsches Elektronen-Synchrotron Labororatory on whose work this report is based. The author is supported in part by the U.S. Department of Energy. Also, the design, construction, and operation of the HERA accelerator and the experimental detectors would not have been possible without the cooperation of the many funding agencies worldwide.

References

1. Phys. Lett. **B363**, 201-216, (1995).
2. Eur. Phys. Jou. **C5**, 4, 625-639, (1998).
3. Eur. Phys. Jou. **C6**, 239-252, (1999).
4. Nucl. Phys. **B538**, 3-22, (1999).

ELECTROWEAK RESULTS FROM THE CROSS-SECTIONS AND ASYMMETRIES WITH ALEPH AT ENERGIES FROM 130 TO 189 GeV

E. MERLE[a]

[a] *Representing the ALEPH Collaboration*
LAPP, Chemin de Bellevue, BP110,
F-74941, Annecy-le-Vieux, France
E-mail: merle@lapp.in2p3.fr

The cross sections and forward-backward asymmetries of hadronic and leptonic events produced in e^+e^- collisions at center-of-mass energies from 130 to 189 GeV are presented. Results for e^+e^-, $\mu^+\mu^-$, $\tau^+\tau^-$, $q\bar{q}$ and $b\bar{b}$ production show no significant deviation from the Standard Model predictions. This places constraints upon physics beyond the Standard Model such as four-fermion contact interactions and R-parity violating squarks and sneutrinos. Limits on the energy scale Λ of $eeff$ contact interactions are typically in range from 6 to 15 TeV. Limits on R-parity violating sneutrinos reach masses of a few hundred GeV/c^2 for large values of their Yukawa couplings.

1 Introduction

In the period 1995-1998, the center-of-mass energy of the LEP collider increased in six steps from 130 to 189 GeV, opening a new energy scale for electroweak cross section and asymmetry measurements. Such measurements provide a test of the Standard Model and allow to place limits on its possible extensions.

This paper begins by providing the definitions of cross section and asymmetry, as well as a presentation of luminosity measurements. The leptonic and hadronic selections are then described; cross section and asymmetry results are reported for the three lepton species, together with the hadronic cross section and the ratio of the $b\bar{b}$ cross section to the total $q\bar{q}$ cross section. Based on these results, limits are then derived on extensions to the SM involving contact interactions and R-parity violating sneutrinos.

2 Definitions

Cross section results for all fermion species are provided for

1. the *inclusive* process, for events with $\sqrt{s'/s} > 0.1$, including events having hard initial state radiation (ISR).

440

2. the *exclusive* process, for events with $\sqrt{s'/s} > 0.9$, excluding radiative events, such as those in which a return to the Z resonance occurs.

Here, the variable s is the square of the center-of-mass energy. For leptonic final states the variable s' is defined as the square of the mass of the outgoing lepton pair. For hadronic final states s' is defined as the mass squared of the Z/γ^* propagator.

Interference effects between ISR and final state radiative (FSR) photons affect the exclusive cross sections at the level of a few percent. They are not described by Monte Carlo generators and only at the first order by the program Zfitter [2] used to compute the predicted cross sections. They are particularly important when the outgoing fermions make a small angle to the incoming e^+e^- beams. To reduce theoretical uncertainties related to this, the exclusive cross section and asymmetry results presented here include only the polar angle region $|\cos\theta| < 0.95$, where θ is the polar angle of the outgoing fermion.

When selecting events, the variable s'_m is used, which provides a good approximation to s' when only one ISR photon is present:

$$s'_m = \frac{\sin\theta_1 + \sin\theta_2 - |\sin(\theta_1 + \theta_2)|}{\sin\theta_1 + \sin\theta_2 + |\sin(\theta_1 + \theta_2)|} \times s . \tag{1}$$

It is based on the angles of the two outgoing fermions measured with respect to the direction of the incoming e^- beam or with respect to the direction of a photon seen in the apparatus and consistent with ISR.

For dilepton events, the differential cross sections are measured as a function of the angle θ^*, the angle of the outgoing fermion in the center-of-mass frame. The forward-backward asymmetries are determined from the formula

$$A_{FB} = \frac{\sigma_F - \sigma_B}{\sigma_F + \sigma_B} , \tag{2}$$

where σ_F and σ_B are the cross sections to produce events with the negative lepton in the forward ($\theta^* < 90°$) and backward ($\theta^* > 90°$) hemispheres, respectively.

The data used were taken at six center-of-mass energies which are given in Table 1. This table also shows the integrated luminosity recorded at each energy point, together with the statistical and systematic uncertainties.

Bhabha events are generated using BHWIDE v1.01 [3] and muon and tau pairs using KORALZ v4.2 [4]. Simulation of hadronic events is done with either KORALZ or PYTHIA v5.7 [5]. The PYTHIA generator is also used for four-fermion processes such as the Z pair and Ze^+e^- channels. The programs PHOT02 [6], HERWIG v5.9 [7] and PYTHIA are used to generate the two-photon

Table 1: Center-of-mass energies and integrated luminosities of the high energy data samples. The two uncertainties quoted on each integrated luminosity correspond to the statistical and systematic uncertainty respectively.

Energy (GeV)	Luminosity (pb^{-1})
130.2	$6.03 \pm 0.03 \pm 0.05$
136.2	$6.10 \pm 0.03 \pm 0.05$
161.3	$11.08 \pm 0.04 \pm 0.07$
172.1	$10.65 \pm 0.05 \pm 0.06$
182.7	$56.78 \pm 0.11 \pm 0.29$
188.6	$173.59 \pm 0.20 \pm 0.87$

events. Finally, backgrounds from W pair production are studied using the generators KORALW v1.21[8] and EXCALIBUR[9].

3 Event selection

3.1 Lepton pairs

The electron pair production is dominated by the t-channel photon exchange. Events are selected by requiring the presence of two good tracks of opposite charge with a polar angle to the beam axis of $|\cos \theta| < 0.9$. The sum of the momenta of the two tracks must exceed 30% of the center-of-mass energy. The total energy associated in the electromagnetic calorimeter (ECAL) must be at least 40% of the center-of-mass energy. If one of the tracks passes near a crack in the ECAL, the associated energy in the hadronic calorimeter (HCAL) is included if it matches the track extrapolation within 25 mrad. The energy of bremsstrahlung photons is included when found within a 20° cone around the track. The main background is due to ISR and this is reduced to a level of 10-12%, by requiring that the invariant mass of the e$^+$e$^-$ final state exceeds 80 GeV/c^2. The exclusive cross section is determined in two polar angle ranges: $-0.9 < \cos \theta^* < 0.9$ and $-0.9 < \cos \theta^* < 0.7$. The selection efficiencies range between 90 to 95 %.

The muon pair selection requires events to contain two good tracks with opposite charges and with momenta exceeding 6 GeV/c. The sum of the momenta of the two tracks must exceed 60 GeV/c to reject $\gamma\gamma$ background. To limit the background from cosmic ray events, both tracks are required to have at least four associated ITC hits. Both tracks must be muon candidates, where muon identification is based either on the digital hit pattern associated with a track in the HCAL or the muon chambers or on the muon energy

442

deposition in the calorimeters.

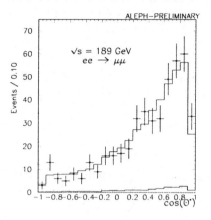

Figure 1: $\cos\theta^*$ distribution of high energy muon pairs at 189 GeV.

In the case of the exclusive selection it is required that the muon pair invariant mass exceeds 110 GeV/c^2, reducing the main background, radiative events, to 8-12 %. The selection efficiencies are around 96 % for the high energy dimuons events. The $\cos\theta^*$ distribution of the exclusive events, used to compute the asymmetries, is represented on Fig. 1.

The tau pair selection begins by clustering events into jets, using the JADE algorithm with a y_{cut} equal to 0.008 . Events with two tau jet candidates are selected, providing that the invariant mass of the two jets exceeds 25 GeV/c^2. This requirement removes $\gamma\gamma$ background. Each tau jet candidate is identified and events are required to have at least one tau jet candidate identified as a decay into a muon or into charged hadrons or charged hadrons plus π^0. To suppress background from $\mu^+\mu^-$, events with two jets identified as muonic tau decay are removed. The main backgrounds, the electron pair production, the WW process and $\gamma\gamma \to \tau^+\tau^-$ events, represent between 10 to 13 % of the high-energy selected events. The corresponding selection efficiencies range between 63 to 68 %.

3.2 Hadronic events

The hadronic event selection require at least seven good charged tracks in the event. All particles are clustered into jets using the JADE algorithm [10] with a y_{cut} of 0.008 . Low multiplicity jets with an electromagnetic energy content of at least 90% and an energy above 10 GeV are considered to be ISR photon candidates.

The visible mass of the event is reconstructed using charged and neutral particles but excluding the ISR candidates and neutral electromagnetic particles within 2° around the beam axis and is required to be more than 50 GeV/c^2. The $\sqrt{s'/s}$ distribution of the selected events is shown in Fig. 2 at 189 GeV. For the exclusive process, two additional cuts are applied. The visible mass is required to exceed 70% of the center-of-mass energy to suppress residual events with a radiative return to the Z. When above the W pair threshold, about 80% of WW background is eliminated by requiring that the thrust of the event exceeds 0.85 . This process is the main remaining background, representing 5.5 % of the high-energy events, for selection efficiencies around 90 %.

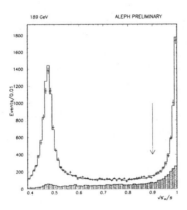

Figure 2: $\sqrt{s'/s}$ distribution of hadronic events at 189 GeV. The background (grey histogram) is dominated by W pair events.

The ratio R_b of the $b\bar{b}$ to $q\bar{q}$ production cross sections is based on the exclusive selection described above. The $b\bar{b}$ events are separated from the hadronic ones using the relatively long lifetime of b hadrons. From the measured impact parameters of the tracks in an event, the confidence level that all these tracks originate from the primary vertex is calculated [12] and required to be less than a certain value, which is chosen to minimize the total uncertainty on $\sigma_{b\bar{b}}$. The largest contribution to the systematic uncertainty arises from the b tagging efficiency. This is evaluated by using the same technique as above to measure the fraction R_b in Z data taken in the same year. The difference between this measurement and the very precisely known world average measurement of R_b at the Z peak is then taken as the systematic uncertainty. The b tagging effi-

444

ciency is almost independent of the center-of-mass energy, because the impact parameters of tracks from a decaying b hadron depend little on its energy. The systematic uncertainties evaluated at the Z peak can therefore be directly translated to higher center-of-mass energies.

3.3 Systematic studies and results

Systematic uncertainties are at 189 GeV of the same order as the statistical ones. They are arising from four sources.

- Tracking effects are estimated by applying the selections with/without the momenta corrections, due to local field distortions.

- The calorimeters energy scale uncertainty is equal to $0.7\% \times E_{ECAL}$ for the electromagnetic calorimeter and $2.0\% \times E_{HCAL}$ for the hadronic one.

- The detector response to jet energy on Z peak hadronic events is compared as a function of the polar angle between Monte-Carlo and data and the difference is taken as a systematic.

- The error on the background estimation due to the uncertainty on the cross-section prediction.

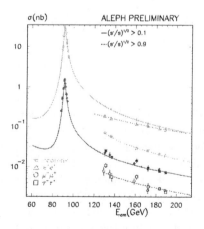

Figure 3: Measured cross sections for fermion pair production. The curves indicate the predictions obtained from BHWIDE for the Bhabha process and from ZFITTER for the other channels.

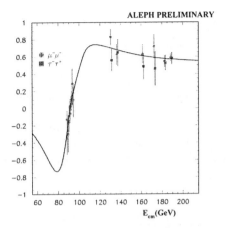

Figure 4: Measured asymmetries for muon and tau pair production. The curves indicate the predictions obtained from ZFITTER.

The measured cross sections and asymmetries are plotted as a function of center-of-mass energy in Figs. 3 and 4, respectively. The results are compared to the Standard Model predictions based on BHWIDE[3] for electron pair production and ZFITTER[2] for all other processes. The ZFITTER predictions are computed from the input values $m_Z = 91.1867$ GeV/c^2, $m_t = 174.1$ GeV/c^2, $m_H = 127.0$ GeV/c^2, $\alpha_{em}(M_Z) = 1/128.896$ and $\alpha_s(M_Z) = 0.120$. The theoretical uncertainties are dominated by the error on the ISF/FSR interference computed only to the first order in Zfitter. They are equal to -1.5 % to -2.0 % for the hadronic process and to 2.4 % to 3.0 % for the dimuon and ditau productions.

4 Interpretations beyond the Standard Model

4.1 Four-fermion contact interactions

A convenient parametrization of possible extensions to the SM is given by the addition of four-fermion contact interactions [13] to the known SM processes. Such contact interactions are characterized by a scale Λ, interpreted as the mass of a new heavy particle exchanged between the incoming and outgoing fermion pairs, and a coupling g giving the strength of the interaction. Following the notation of ref. [14], the effective Lagrangian for the four-fermion contact

446

interaction in the process $e^+e^- \to f\bar{f}$ is given by

$$\mathcal{L}^{CI} = \frac{g^2 \eta_{\text{sign}}}{(1+\delta)\Lambda^2} \sum_{i,j=L,R} \eta_{ij}[\bar{e}_i \gamma^\mu e_i][\bar{f}_j \gamma_\mu f_j] \,, \tag{3}$$

with $\delta = 1$ if $f = e$, or 0 otherwise. The fields $f_{L,R}$ are the left- and right-handed chirality projections of fermion spinors. The coefficients η_{ij} indicate the relative contribution of the different chirality combinations to the Lagrangian. The sign of η_{sign} determines whether the contact interaction interfers constructively or destructively with the SM amplitude. Several different models are considered in this analysis, corresponding to the choices of the η_{sign} and η_{ij} given in Table 2.

Table 2: Four-fermion interaction models.

Model	η_{sign}	η_{LL}	η_{RR}	η_{LR}	η_{RL}
LL$^\pm$	± 1	1	0	0	0
RR$^\pm$	± 1	0	1	0	0
VV$^\pm$	± 1	1	1	1	1
AA$^\pm$	± 1	1	1	-1	-1
LR$^\pm$	± 1	0	0	1	0
RL$^\pm$	± 1	0	0	0	1
LL+RR$^\pm$	± 1	1	1	0	0
LR+RL$^\pm$	± 1	0	0	1	1

In the presence of contact interactions the differential cross section for $e^+e^- \to f\bar{f}$ as a function of the polar angle θ of the outgoing fermion with respect to the e^- beam line can be written as

$$\frac{d\sigma}{d\cos\theta} = F_{\text{SM}}(s,t) \left[1 + \varepsilon \frac{F_{\text{IF}}^{\text{Born}}(s,t)}{F_{\text{SM}}^{\text{Born}}(s,t)} + \varepsilon^2 \frac{F_{\text{CI}}^{\text{Born}}(s,t)}{F_{\text{SM}}^{\text{Born}}(s,t)}\right]$$

with $\varepsilon = g^2 \eta_{\text{sign}}/(4\pi\Lambda^2)$. g is fixed to $\sqrt{4\pi}$.

The predictions of Eq. 4 are fitted to the data using a binned maximum likelihood method. The fit is performed in terms of the parameter ε. Because of the quadratic dependence of the theoretical cross sections upon ε, the likelihood function can have two maxima.

The 95% confidence level limits $\Lambda_{95}^\pm$ are computed and the results for contact interaction affecting leptonic final states and all difermion events are listed in Table 3. These limits obtained on the energy scale Λ are in the range 6 TeV to 15 TeV. Models of $e^+e^-u\bar{u}$ and $e^+e^-d\bar{d}$ contact interactions which violate parity (LL, RR, LR and RL) are already severely constrained by atomic physics

Table 3: Limits on contact interactions at 95% confidence level. The results presented for l^+l^- assume lepton universality and for $\bar{f}f$ assume that the contact interaction couples to all the outgoing fermion types equally.

Model	Λ^- (TeV)	Λ^+ (TeV)
$e^+e^- \to l^+l^-$		
LL	6.3	8.1
RR	6.1	7.8
VV	11.5	13.5
AA	9.4	11.5
LR	6.6	6.8
LL+RR	8.8	11.1
LR+RL	9.4	9.6
$e^+e^- \to \bar{f}f$		
LL	8.0	8.0
RR	6.0	5.5
VV	9.6	9.2
AA	10.5	10.4
LR	3.8	4.1
RL	2.3	5.0
LL+RR	10.0	9.7
LR+RL	3.3	5.6

parity violation experiments, which quote limits of the order of 15 TeV [15]. The LEP limits for the fully leptonic couplings or those involving b quarks are of particular interest since they are inaccessible at $p\bar{p}$ or ep colliders.

4.2   Sneutrinos with R-parity violation

At LEP, dilepton production cross sections could differ from their Standard Model expectations as a result of the exchange of R-parity violating sneutrinos in the s or t channels [16]. Table 4 shows the most interesting possibilities. Those involving s channel sneutrino exchange lead to a resonance. For the results presented here, this resonance is assumed to have a width of 1 GeV/c^2.

Limits on the couplings are obtained by comparing the measured dilepton differential cross sections with respect to the polar angle with the theoretical cross sections in reference [16] using a binned maximum likelihood method. The fit is performed in terms of the parameter λ^2.

In Fig. 5 the limits for processes involving sneutrino exchange in the s channel extracted from the dielectron measurement is shown. Similar limits

Table 4: For each dilepton channel, the coupling amplitude, the sneutrino type exchanged, and the channel s or t of the exchange are indicated.

λ^2	e^+e^-	$\mu^+\mu^-$	$\tau^+\tau^-$
λ^2_{121}	$\tilde{\nu}_\mu$ (s,t)	$\tilde{\nu}_e$ (t)	—
λ^2_{131}	$\tilde{\nu}_\tau$ (s,t)	—	$\tilde{\nu}_e$ (t)
$\lambda_{121}\lambda_{233}$	—	—	$\tilde{\nu}_\mu$ (s)
$\lambda_{131}\lambda_{232}$	—	$\tilde{\nu}_\tau$ (s)	—

are derived from $\mu^+\mu^-$ and $\tau^+\tau^-$ cross sections. Limits on λ_{121} and λ_{131} from t channel exchange of $\tilde{\nu}_e$ in muon and tau pair production, respectively, give much weaker limits. These rise from $|\lambda_{1j1}| < 0.5$ at $\tilde{\nu}_e = 100$ GeV/c^2 to $|\lambda_{1j1}| < 0.9$ at $\tilde{\nu}_e = 300$ GeV/c^2.

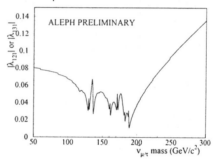

Figure 5: 95% confidence level upper limits, obtained from the Bhabha cross sections on $|\lambda_{121}|$ versus the assumed $\tilde{\nu}_\mu$ or equivalently on $|\lambda_{131}|$ versus the assumed $\tilde{\nu}_\tau$ mass.

5 Conclusions

The difermionic cross sections and asymmetries measured with the ALEPH detector from 130 to 189 GeV are in good agreement with the Standard Model predictions. The errors are dominated by the theoretical uncertainties on the knowledge of the ISR/FSR interference effects. The combination of the four LEP experiments (1995-2000) at the end of LEP2 will allow to study higher orders of such corrections.

Limits on new physics phenomena are derived. On the energy scale Λ of $eeff$ contact interactions they are typically in range from 6 to 15 TeV. Limits on R-parity violating sneutrinos reach masses of a few hundred GeV/c^2 for large values of their Yukawa couplings.

Acknowledgments

Many thanks to Marie-Noelle Minard and the ALEPH-Lapp group for their great help in preparing this talk and to the Beyond-ElectroWeak group of ALEPH for the interesting discussions we exchanged.

We thank our colleagues from the CERN accelerator divisions for the successful operation of LEP at higher energies. We are indebted to the engineers and technicians in all our institutions for their contribution to the continuing good performance of ALEPH.

References

1. L3 Collaboration, M. Acciarri *et. al.*, Phys. Lett. **B371**, 137, (1996).
2. D. Bardin *et al.*, ZFITTER: An analytical program for fermion pair production in e+ e- annihilation, CERN-TH 6443/92 (1992).
3. S. Jadach, W. Placzek and B.F.L. Ward, BHWIDE 1.00: O(alpha) YFS exponentiated Monte Carlo for Bhabha scattering at wide angles for LEP1 / SLC and LEP2, Phys. Lett. **B390**, 298, (1997).
4. S. Jadach, B.F.L. Ward and Z. Was, The Monte Carlo program KORALZ, version 4.0, for lepton or quark pair production at LEP / SLC energies, Comp. Phys. Comm. **79**, 503, (1994).
5. T. Sjostrand, High-energy physics event generation with PYTHIA 5.7 and JETSET 7.4, Comp. Phys. Comm. **82**, 74, (1994).
6. ALEPH Collaboration, An experimental study of $\gamma\gamma \to$ hadrons at LEP, Phys. Lett. **B313**. 509, (1993).
7. G. Marchesini *et al.*, HERWIG: A Monte Carlo event generator for simulating hadron emission reactions with interfering gluons, version 5.1, Comp. Phys. Comm. **67**, 465, (1992).
8. M. Skrzypek *et al.*, Monte Carlo program KORALW 1.02 for W pair production at LEP2 / NLC energies with Yennie Frautschi Suura exponentiation, Comp. Phys. Comm. **94**, 216, (1996).
9. F. Berends, R. Pittau and R. Kleiss, Excalibur: A Monte Carlo program to evaluate all four fermion processes at LEP-200 and beyond, Comp. Phys. Comm. **85**, 437, (1995).
10. JADE Collaboration, Experimental studies on multi-jet production in e^+e^- annihilation at PETRA energies, Z. Phys. **C33**, 23, (1986).
11. D. Bardin, A. Leike and T. Riemann, The process $e^+e^- \to l^+l^-q\bar{q}$ at LEP and NLC, Phys. Lett. **B344**, 383, (1995).
12. ALEPH Collaboration, A measurement of R_b using a lifetime-mass tag, Phys. Lett. **B401**, 150, (1997).

13. E. Eichten, K. Lane and M. Peskin, New tests for quark and lepton substructure, Phys. Rev. Lett. **50**, 811, (1983).
14. H. Kroha, Compositeness limits from e^+e^- annihilation reexamined, Phys. Rev. **D46**, 58, (1992).
15. A. Deandrea, Atomic parity violation in cesium and implications for new physics, Phys. Lett. **B409**, 277, (1997).
16. J. Kalinowski *et al.*, Supersymmetry with R-parity breaking: contact interactions and resonance formation in leptonic processes at LEP2, Phys. Lett. **B406**, 314, (1997).

THE PROPERTIES OF THE CHARGED WEAK BOSON $W^\pm$

PETER MOLNÁR
Humboldt-Universität zu Berlin
Invalidenstr. 110
10115 Berlin
Peter.Molnar@cern.ch

The charged weak boson $W^\pm$ is produced in pairs at LEP since 1996. The large amount of W-pair events that is collected by the L3-collaboration allows to determine the mass of the W-boson and to study with high accuracy its coupling to other electroweak gauge bosons x. These preliminary and final results are compared to results from other experiments and average values are derived. New results obtained from the data taken at 189 GeV which became available only after the conference are added.

1 Introduction

The energy of the Large Electron-Positron collider LEP was increased in steps since 1995, crossing the W-pair production threshold of 161 GeV in 1996. In recent years each of the four LEP experiments collected a luminosity of about 250 pb^{-1} at energy points up to 189 GeV, which corresponds to 3500 W-pair events and about 100 single W events. This large amount of reconstructed Ws allows to determine with high precision [1] the properties of the W-boson, its mass and its couplings to other electroweak bosons. The measurement of the triple gauge boson couplings (TGC) tests for the first time the non-abelian structure of the electroweak Standard Model (SM) [2]. The W-pair production can proceed via s-channel Z^0 or γ exchange or due to t-channel neutrino exchange as shown in Fig. 1. Another diagram which is sensitive to TGCs is the single W production also displayed in Fig. 1.

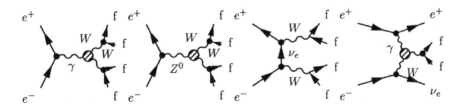

Figure 1: The Feynman graphs for W-pair and single W production.

452

2 The W-mass measurement

The W-mass was measured with high precision at LEP 1 using radiative corrections to the Z^0-fermion couplings. The W-mass was determined to be $M_W = 80.367 \pm 0.029$ GeV [3]. This measurement holds only if one assumes the SM. Hence a direct measurement of the W-mass with the same accuracy will test the SM on this level of precision. Two kinds of four fermion final states can be exploited for the W-mass measurement: the hadronic case where both Ws decay into quarks, which manifest as jets in the detector, and the semileptonic, where one W decays into quarks and the other in a lepton-neutrino pair. Semileptonic events can be exploited since one knows the initial momentum configuration and can apply four-momentum conservation in a kinematic fit. In addition one uses that both Ws have about the same mass.

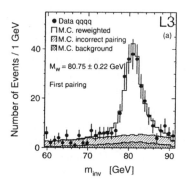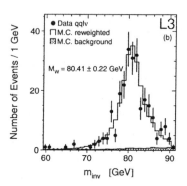

Figure 2: The invariant mass distribution of hadronic (a) and semileptonic (b) W-pair events which were selected at $\sqrt{s} = 183$ GeV.

Therefore one finds two more constraints than free variables and thus can improve the W-mass resolution. Hadronic events have the additional problem of assigning the right jets to the W, the pairing of jets. If the event is clustered in four jets one finds three possible pairings. To resolve this problem one uses also here the kinematic fit, which is five times over-constrained. In about 75 percent of the cases one finds the right pairing of quark jets, if the pairing with the best fit probability is taken. The distribution of the invariant mass of semileptonic and hadronic events are then fitted by comparing fully simulated and reweighted MC samples to the data as shown in Fig. 2. One finds a W mass of 80.41 ± 0.22 GeV for fully hadronic events and 80.75 ± 0.22 GeV for semileptonic events.

One of the main systematic problems in the case of hadronic W-pair decays is the four-momentum exchange between quarks of different Ws. The gluon exchange occurs since the Ws decay within the interaction scale of QCD. This process is called colour reconnection (CR). Computations revealed that perturbative effects are negligible [4]. However, non-perturbative effects can give significant biases [4], but it is complicated to model to which extend the CR will influence the W-mass measurement. One test of CR models is the measurement of the number of charged tracks in hadronic and semileptonic W-pair events. L3 measures $\langle N_{ch}^{qqqq} \rangle - 2\langle N_{ch}^{qql\nu} \rangle = -1.0 \pm 0.9$, which is consistent with zero while CR effects may lead to non-zero values. In addition, a bias on the W-mass in the qqqq channel may be introduced by Bose Einstein correlations among the bosonic hadrons of the jets from different Ws [5].

3 The TGCs

TGCs are extracted not only from the differential but also from the total cross section, as it depends on the couplings too. In addition to the W-pair production already exploited for the W-mass determination one can also use the production of single Ws.

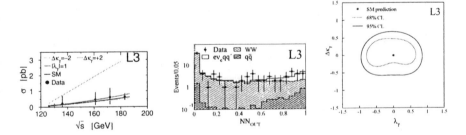

Figure 3: The single W cross section, the neural network output which is used to select single Ws at $\sqrt{s} = 183$ GeV and the two-dimensional contour of the fit result in $\Delta\kappa_\gamma$ and λ_γ.

In the case of single W production the cross section is measured at five different centre-of mass energies. The coupling dependence of the total cross sections is displayed in Fig. 3. In addition the differential distribution of a neural network output, which discriminates signal (at a value of one) and background (at a value of zero), is used for the coupling evaluation. The coupling determination uses the reweighting method as explained for the mass determination. The single W channel is only sensitive to the W-γ coupling constants $\Delta\kappa_\gamma$ and λ_γ, which can be determined simultaneously in a two-dimensional fit, as can be seen in Fig. 3. Both the total cross section as well as the differential prefer the SM prediction of zero for $\Delta\kappa_\gamma$ and λ_γ. In the case of W-pair production one finds that the distributions in five phase space angles have the full sensitivity to TGCs. Since TGCs change the forward-backward asymmetry one chooses $\cos\theta_{W^-}$, the emission angle of the W^- and since TGCs change the polarisation, one chooses the polarisation analysers $\cos\theta_{f_{1,2}}$ and $\phi_{f_{1,2}}$ - the decay angles of both Ws. In semileptonic events one can only determine three phase space angles unambiguously while in hadronic events only $\cos\theta_{W^-}$ can be determined without ambiguity. Thus a fit to the three-dimensional distribution $\cos\theta_{W^-} - \cos\theta_l - \phi_l$ for semileptonic and to $\cos\theta_{W^-}$ for hadronic events as well as to the total W-pair cross section is used to evaluate the TGCs. The fit exploits the reweighting method. The total cross section and the differential cross section of $\cos\theta_{lW}$ are displayed in Fig. 4. W-pair production is sensitive to the coupling of the W to the photon but also to the Z^0. The likelihood curve from the measurement of the coupling Δg_l^Z is displayed in Fig. 4.

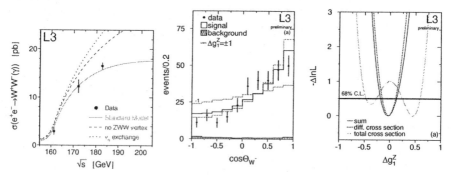

Figure 4: The W-pair production cross section, the W^{-1} emission angle in semileptonic W-pair decays and the log-likelihood for the coupling Δg_l^Z was measured by L3 at $\sqrt{s} = 183$ GeV.

4 Comparison and Summary

The sensitivity to the coupling constants as well as to the mass can be increased by averaging the results of the four LEP experiments. In the case of one-dimensional fits to Wγ couplings also results of the $D\emptyset$ experiment at TEVATRON are included. The comparison of the results for Δg_l^Z and the

W-mass as well as the combined two dimensional contour of $\Delta\kappa_\gamma - \lambda_\gamma$ are displayed in Fig. 5. The combination gives numerically:

$$g_l^Z = 1.00 \pm 0.08 \qquad \kappa_\gamma = 1.13 \pm 0.14 \qquad \lambda_\gamma = -0.03 \pm 0.07$$
$$m_W = 80.36 \pm 0.09 \text{GeV}.$$

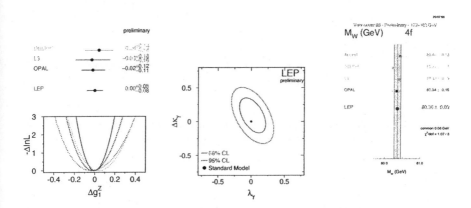

Figure 5: The combination of W-properties as measured by the LEP experiments with the data set collected at $\sqrt{s} = 161 - 183$ GeV.

At the end of LEP 2 each experiment will have collected about 500pb^{-1}, allowing to determine the W-mass with an accuracy of about 30 MeV and $g_l^Z(\kappa_\gamma)$ with an error of less than 0.02 (0.05).

5 After conference update

New much more precise measurements using the data set taken in 1998 at $\sqrt{s} = 189$ GeV became available shortly after the Lake Louise Winter Institute. With these higher statistics the LEP combined W-mass is $80.347 \pm 0.057 \text{GeV}(L3 : 80.404 \pm 0.109 \text{GeV})$. A summary of the new LEP results is displayed in Fig. 6. The error on the TGC results decrease by a factor of two, to be :

$$g_l^Z = 0.99 \pm 0.03 \qquad \kappa_\gamma = 1.05^{+0.09}_{-0.08} \qquad \lambda_\gamma = -0.03 \pm 0.03.$$

The comparison of the one-dimensional fits in the coupling constant $\Delta\kappa_\gamma$ and the contour curve of the combined LEP and $D\emptyset$ measurement in the $\Delta g_l^Z - \lambda_\gamma$ plane are shown in Fig. 6.

456

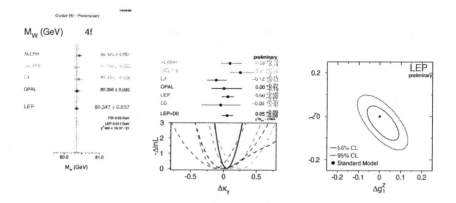

Figure 6: The combination of W-properties as measured by the LEP experiments and $D\varnothing$ including data taken in 1998 at $\sqrt{s} = 189$ GeV.

Acknowledgement

It is a pleasure to thank my colleagues from L3 and the LEP electroweak working group for valuable discussions.

References

1. W. Beenakker et al., Report CERN 96-01 (1996), Vol. 1, p. 79;
   Z. Kunszt et al., Report CERN 96-01 (1996), Vol. 1, p. 141;
   G. Gounaris et al., Report CERN 96-01 (1996), Vol. 1, p. 525.
2. S.L. Glashow, Nucl. Phys. 22, 579, (1961);
   S. Weinberg, Phys. Rev. Lett. 19, 1264, (1967);
   A. Salam, in Elementary Particle Theory, ed. N. Svartholm, Stockholm, Almquist and Wiksell, 367, (1968).
3. LEP Electroweak working group, LEPEWWG/99-01.
4. G. Gustafson et al., Phys. Lett. B209, 90, (1988);
   T. Sjöstrand and V.A. Khoze, Phys. Rev. Lett. 72, 28, (1994);
   E. Accomando et al., Phys. Lett. B362, 141, (1995);
   G. Gustaffson and J. Häkkinen, Z. Phys. C64, 659, (1994);
   L. Lönnblad, Z. Phys. C70, 107, (1996);
   J. Ellis and K. Geiger, Phys. Rev. D54, (1996) 1967 and CERN-TH/97-046.
5. L. Lönnblad and T. Sjöstrand, Phys. Lett. B 351, 293, (1995);
   S. Jadach and K. Zalewski, CERN-TH/97-029.

RECENT RESULTS ON HIGGS SEARCHES AT OPAL

K. NAGAI

EP Division, CERN, CH-1211 Geneva 23, Switzerland
E-mail: Koichi.Nagai@cern.ch

The searches for Higgs bosons in aagreement with the Standard Model and with models beyond the Standard Model are performed on the data collected at $\sqrt{s}$ = 189 GeV with the OPAL detector, which corresponds to integrated luminosities of approximately 150 pb^{-1}. None of these searches has shown an excess over the expected Standard Model background. The lower mass limit for the Standard Model Higgs boson is obtained to be 94.0 GeV at the 95% confidence level. The results for models beyond the Standard Model are also presented.

1 Introduction

The Standard Model (SM) predicts one Higgs boson to produce the masses of particles through spontaneous symmetry breaking, but it does not predict the Higgs mass. The electroweak global fits imply that the SM Higgs mass (m_H) is less than 262 GeV at 95% confidence level (CL)[1]. If there are no SM Higgs boson, the Higgs bosons in models beyond the SM have to be considered. One case of the extensions to the SM are the two Higgs doublet models (2HDM) which predict five physical Higgs bosons including three neutral Higgs bosons (h^0, H^0 and A^0) and two charged Higgs bosons (H^+, H^-). The Minimal Super-symmetric SM (MSSM) is a special case of the 2HDM, which has more constraints and more parameters but can predict the cross section and masses of Higgs bosons precisely under a given set of parameters. The lightest neutral Higgs mass is predicted to be less than Z^0 mass at the tree level. After taking the radiative corrections into account, the mass of the lightest Higgs boson is expected to be less than approximately 130 GeV[2].

For all neutral Higgs bosons, the decays into b-quarks and τ-leptons are considered, and in addition for the Higgs bosons beyond the SM, $h^0 \rightarrow A^0 A^0$, $h^0 \rightarrow \gamma\gamma$ and $h^0 \rightarrow$ invisible particles. For the charged Higgs bosons, the hadronic ($H^\pm \rightarrow cs$) and leptonic ($H^\pm \rightarrow \tau^\pm \nu_\tau$) decays are considered.

The OPAL detector collected approximately 180 pb^{-1} of e^+e^- collision data at $\sqrt{s}$ = 189 GeV in 1998. This high statistics sample has extended the sensitivity of our searches to higher Higgs masses. The results in this report are using data corresponding to an integrated luminosity of approximately 150 pb^{-1} at $\sqrt{s}$ = 189 GeV and/or 54 pb^{-1} at $\sqrt{s}$ = 183 GeV.

2   The Standard Model Higgs boson

The Higgs-strahlung process $(e^+e^- \rightarrow Z^0H_{SM}^0)$ is expected to be the dominant production process of the SM Higgs boson at LEP II, if it is kinematically allowed. Analyses are performed on four search channels: $Z^0H_{SM}^0 \rightarrow q\bar{q}b\bar{b}$, $Z^0H_{SM}^0 \rightarrow \nu\bar{\nu}b\bar{b}$, $Z^0H_{SM}^0 \rightarrow \ell^+\ell^-b\bar{b}$ where $\ell^\pm$ are $e^\pm$ or $\mu^\pm$, and $Z^0H_{SM}^0 \rightarrow \tau^+\tau^-b\bar{b}/b\bar{b}\tau^+\tau^-$. There are contributions on the production cross-section of Higgs boson from the WW $(e^+e^- \rightarrow \nu\bar{\nu}H_{SM}^0)$ and ZZ fusion $(e^+e^- \rightarrow e^+e^-H_{SM}^0)$ processes which become more important near the kinematic limit of the search region for the $Z^0H_{SM}^0 \rightarrow \nu\bar{\nu}b\bar{b}$ and the $Z^0H_{SM}^0 \rightarrow e^+e^-b\bar{b}$, respectively.

The observed and expected numbers of events are shown in Table 1 with the efficiencies for $m_H = 95$ GeV. The Higgs mass distribution of candidates and expected background are shown in Fig. 1(a). The data and expectations are shown, combined with our 183 GeV results [3]. No significant excess is observed. The observed and average expected CL for the signal hypothesis is calculated as a function of the Higgs boson mass as shown in Fig. 1(b). The lower mass limit of the SM Higgs boson is obtained to be 94 GeV at the 95%CL.

Table 1: The numbers of candidates and expected background events in the SM and MSSM Higgs boson search at $\sqrt{s} = 189$ GeV. The efficiencies are given for m_H=95 GeV for all Z^0h channels, for m_h=m_A=80 GeV for Ah-4b and Ah-τ channels, and for m_h=$2m_A$=80 GeV for Ah-6b channel.

Z^0h Channel	Int. lum. (pb^{-1})	data	Background			Efficiency (%)
			Total	$q\bar{q}(\gamma)$	4-fermi.	
Four-Jets	151.0	18	16.9 ± 0.8	4.3	12.6	46.2
Missing-E	150.4	7	6.0 ± 0.4	0.9	5.1	36.1
Tau	149.4	4	4.0 ± 0.6	0.3	3.7	32.7
Electron	151.0	3	1.8 ± 0.3	0.1	1.7	56.6
Muon	142.7	0	1.3 ± 0.3	0.0	1.3	59.8
Ah Channel						
Ah-4b	151.0	12	7.5 ± 0.5	3.2	3.9	51.8
Ah-τ	149.4	5	4.8 ± 0.7	0.9	3.9	41.9
Ah-6b	151.0	8	7.3 ± 0.7	6.0	1.3	59.6

3   MSSM Higgs bosons

In the MSSM, Higgs bosons are produced by pair production of h^0 and A^0 as well as by the Higgs-strahlung process. The analysis for the SM Higgs is reused to search for the Higgs-strahlung process in the MSSM. For the searches

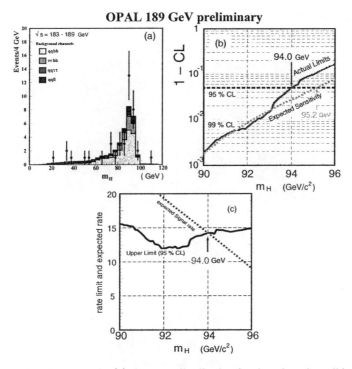

Figure 1: The SM Higgs search: (a) the mass distribution for the selected candidates events (dots with error bars) and the expected background for data taken at $\sqrt{s} = 183$ and 189 GeV. (b) the confidence level for signal-plus-background hypothesis as observed (solid line) and expected on average for background-only experiments (dashed line). (c) Limit on the production rate at the 95%CL (solid line) and the number of expected signal events (dashed line).

for the pair production process, the b-tagging is more important to reduce the background than the case for the SM, because the Z^0 mass constraint cannot be applied. No excess was found and the numbers of candidate and expected backgrounds are shown in Table 1.

The limit is determined in the MSSM parameter space for a constrained MSSM where most of parameters are fixed to "universal" values (see Ref. 4, Scan (A) for the definition of the parameters and their values). Two cases of mixing in the scalar-top sector are investigated: "no mixing" and "maximal mixing". The top quark mass is fixed at 175 GeV. In Fig. 2, the 95% CL of exclusion limits are shown. For $\tan\beta > 1$, the lower mass limits $m_h > 76.0$ GeV and $m_A > 77.0$ GeV are obtained, while the expected limits are

460

78.0 and 78.5 GeV for m_h and m_A, respectively.

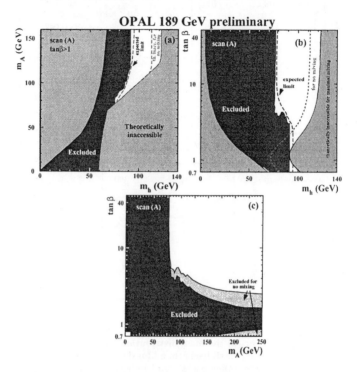

Figure 2: The limit in the MSSM parameter space. Excluded regions are shown for (a) the (m_h, m_A) plane for $\tan\beta > 1$, (b) the $(m_h, \tan\beta)$ plane, and (c) the $(m_A, \tan\beta)$ plane. The black areas are excluded at the 95% CL. The grey areas in (a) and (b) indicate the theoretically inaccessible regions. The expected limits are also shown in the plots. The grey area in (c) corresponds to the case with no mixing in the scalar-top sector.

4   Charged Higgs bosons

The charged Higgs boson is expected to be produced through pair production. The cross section for this process in the 2HDM is completely determined by SM parameters for a given charged Higgs mass. The analysis has searched for three topologies: the leptonic ($H^+H^- \to \tau^+\nu_\tau\tau^-\bar{\nu}_\tau$), semi-leptonic ($H^+H^- \to \tau\nu_\tau q\bar{q}'$) and hadronic ($H^+H^- \to q\bar{q}'q\bar{q}'$) final states. Results are given only for the data at $\sqrt{s} = 183$ GeV [3]. The background for the charged Higgs boson search mainly comes from W^+W^- events. The limits on the production cross section for each of the final states are shown in Fig. 3(a). The lower bound on the mass of charged Higgs boson is presented as a function of $Br(H^+ \to \tau^+ \nu_\tau)$ in Fig. 3(b), combining with the results obtained at lower energies [5]. Charged Higgs bosons are excluded up to a mass of 59.5 GeV at the 95% CL.

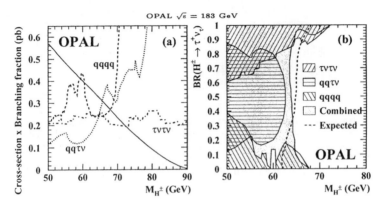

Figure 3: (a) Upper limits at the 95% CL on the production cross-section times branching fraction of the three final states considered. Different centre-of-mass energies are scaled to $\sqrt{s} = 183$ GeVand combined, using the predicted s-dependence of the charged Higgs boson production cross-section. The charged Higgs boson production cross-section at $\sqrt{s} = 183$ GeV is shown as a solid line. (b) Excluded areas at the 95% CL in the $[M_{H\pm}, Br(H^+ \to \tau^+ \nu_\tau)]$ plane. The dashed line shows the expected 95% CL limit from simulated background experiments.

5 Higgs boson decaying to photons

The branching ratio of $h^0 \to \gamma\gamma$ is very small in the SM and the MSSM. However, if the Higgs boson only has couplings to bosons – the "bosonic Higgs Model" [6], the branching ratio is expected to be large. Searches for the process: $e^+e^- \to Z^0h^0 \to q\bar{q}\gamma\gamma$, $e^+e^- \to Z^0h^0 \to ll\gamma\gamma$ and $e^+e^- \to Z^0h^0 \to \nu\bar{\nu}\gamma\gamma$, have been performed. The number of observed (expected background) events are 52 (58.6 ± 2.2), 9 (10.5 ± 1.2) and 5 (7.5 ± 0.2) for the $q\bar{q}\gamma\gamma$, $ll\gamma\gamma$ and $\nu\bar{\nu}\gamma\gamma$ channels, respectively. The di-photon mass distribution is shown in Fig. 4 (a). An upper limit at the 95% CL on $Br(h^0 \to \gamma\gamma)$ is calculated, assuming the production rate of the SM Higgs boson and combining with the results obtained at lower energies [7].

462

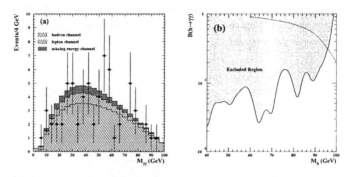

Figure 4: Search for Higgs boson decaying to photons:(a) Distribution of the di-photon invariant mass in the selected events for all search channels combined. The data (expected background) are shown as points with error bars (histograms) (b) Upper limit at the 95% CL on Br($h^0 \to \gamma\gamma$) assuming the production rate of the SM Higgs boson. Results obtained at lower energies are combined with our 189 GeV data. The dotted line is the predicted Br($h^0 \to \gamma\gamma$) in the "bosonic" model.

6   Invisible Higgs decays

The branching ratio of the process $h^0 \to \tilde{\chi}_1^0 \tilde{\chi}_1^0$ in MSSM, where $\tilde{\chi}_1^0$ is the lightest neutralino, taken to be the lightest super-symmetric particle (LSP), may become large, even dominant. The case of "nearly invisible" Higgs boson decays are also considered; $h^0 \to \tilde{\chi}_1^0 \tilde{\chi}_2^0 \to \tilde{\chi}_1^0 \tilde{\chi}_1^0 + \gamma/Z^*$ with ΔM ($\equiv m_{\tilde{\chi}_2^0} - m_{\tilde{\chi}_1^0}$) = 2 (4) GeV, where $\tilde{\chi}_2^0$ is the second-lightest neutralino. The observed (expected) number of events after all selection is 32 (35.8 ± 1.0) where the dominant backgrounds are four-fermion processes. The Higgs mass, m_h, is identified with the missing mass of the event, after scaling the visible mass to the Z^0 mass. The distribution of the missing mass is shown in Fig. 5(a). The upper limit for the production cross section of (nearly) invisible Higgs decays, scaled to the SM Higgs cross section for each decay mode separately assuming 100% branching ratio to each decay mode in term, is derived as shown in Fig. 5(b). If the Higgs boson is produced with the SM Higgs cross section, we set limits on the Higgs mass of 87 (84) GeV for the (nearly) invisible decay mode.

OPAL Preliminary $\sqrt{s} = 189$ GeV

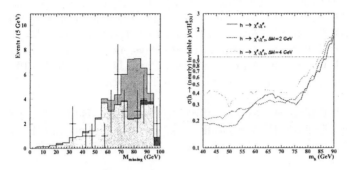

Figure 5: (a) The Higgs mass distribution of selected candidates. Dots with error bars show data and the light (dark) grey histogram shows four-fermion (q$\bar{q}$) processes. The grey histogram on the top of background shows the signal of $m_h = 80$ GeV, normalised to the SM Higgs cross section with 100% branching ratio to the invisible decay. (b) The limits on the production of $h^0 \to$ (nearly) invisible particles relative to the SM Higgs cross section.

7 Summary

The searches for Higgs bosons in agreement with the Standard Model and with models beyond the Standard Model have been performed on the data collected at $\sqrt{s} = 189$ and/or 183 GeV with the OPAL detector. None of the searches has shown an excess over the Standard Model expectation. The limits for the Higgs bosons in the Standard Model and in models beyond Standard Model have been calculated.

Acknowledgements

I would like to thank all of my colleagues from the OPAL collaboration who gave me the chance to give this talk, especially the Higgs search working group in OPAL.

References

1. J. Pinfold, these proceedings.
2. M. Carena, J.R. Espinosa, M. Quirós, and C.E.M. Wagner, Phys. Lett. **B355**, 209, 1995;
   M. Carena, M. Quirós, and C.E.M. Wagner, Nucl. Phys. **B461**, 407,

1996;

H.E. Haber, R. Hempfling, and A.H. Hoang, Z. Phys. **C75**, 539, 1997.

3. OPAL Collab., G. Abbiendi *et al.*, Eur. Phys. Journal **C7**, 407, 1999.
4. OPAL Collab., K. Ackerstaff *et al.*, Eur. Phys. Journal **C5**, 19, 1998.
5. OPAL Collab., K. Ackerstaff *et al.*, Phys. Lett. **B426**, 180, 1998;
   OPAL Collab., K. Alexander *et al.*, Phys. Lett. **B370**, 74, 1996.
6. A. Stange, W. Marciano, and S. Willenbrock, Phys. Rev. **D49**, 1354, 1994.
7. OPAL Collab., K. Ackerstaff *et al.*, Phys. Lett. **B437**, 218, 1998;
   OPAL Collab., K. Ackerstaff *et al.*, Eur. Phys. Journal **C1**, 31, 1998;
   OPAL Collab., K. Alexander *et al.*, Z. Phys. **C71**, 1, 1996.

DELPHI RESULTS ON Z LINESHAPE AND 2 FERMION PHYSICS AT LEP2

M. NIKOLENKO

Joint Institute for Nuclear Research, Dubna, Russia 141980

During the years 1993-1995, about 115 pb^{-1} of data were collected with the DELPHI detector at energies near Z^0 resonance. The lineshape parameters were determined from fits to the measured fermion pair production cross-sections and leptonic forward-backward asymmetries. The data sample collected at higher energies (130-189 GeV) during 1995-1998 amounts to 250 pb^{-1}. The measured cross-sections and asymmetries at this energy range were interpreted in terms of S-matrix fits. Several new physics models have been confronted with the experimental results but no evidence for new physics phenomena was found.

1 Introduction

This article describes the DELPHI analyses and results on lineshape measurements and two fermion physics. It is organized in two parts. The first part presents the results of the measurements and interpretations at energies near Z^0 peak. It mainly concerns the data collected in 1993-1995. The second part describes the results of the analyses of the data collected at higher energies since the fall of 1995 when the LEP collision energy was set to 130 GeV and has been increased from year to year up to 189 GeV in 1998.

The results presented here were already reported in details in other papers[1,2,3]. The readers are also referred to the papers[4] for the description of the DELPHI detector and its performance.

2 Measurements at LEP-1

The two main goals of the lineshape analysis at LEP are to test the validity of the Standard Model and to determine one of its main parameters, M_Z.

LEP is a Z^0 factory which delivered more than 4 millions $e^+e^- \to Z^0$ events for each of four LEP experiments. A high Z^0 production rate makes possible a precise measurement of the fermion-antifermion pair production $e^+e^- \to f\bar{f}$ cross-sections and the leptonic forward-backward asymmetries at different collision energies near the Z^0 resonance. These observables measured at a well determined collision energy can be used to fit the parameters of Z^0 resonance; the so called lineshape parameters M_Z, Γ_Z, etc. M_Z is a fundamental Standard Model parameter. The values of the other parameters are predicted by theory and are sensitive to the masses of the top quark and the Higgs boson through the radiative corrections.

2.1 Cross-section and asymmetry calculation. Event selection.

The cross-sections $e^+e^- \to l\bar{l}$ and forward-backward asymmetries are determined for the three leptonic channels separately. For the hadronic channel the cross-section $e^+e^- \to q\bar{q}$ is determined inclusively for all quark flavors.

The cross-section is calculated as follows:

$$\sigma = \frac{N_s - N_b}{\epsilon \cdot L},$$

where N_s stands for the number of selected events, N_b the number of background events, ϵ the selection efficiency, L the integrated luminosity.

The forward-backward asymmetry A_{FB} is determined from the fit to the lowest order form of the angular distribution:

$$\sigma(\theta) \approx 1 + cos^2\theta + \frac{8}{3}A_{FB}cos\theta,$$

where θ is the angle between the electron beam line and the final state negative lepton direction.

The selection of events is based on the distinctive features of each channel. Hadronic Z^0 decays produce a large number of tracks and thus are well separated from low multiplicity leptonic events. Electron and muon final states are selected by means of the lepton identification. Taus are selected by a kinematical analysis of the event.

Table 1 contains the number of events selected for each channel in 1993-1995. To make a lineshape fit one needs to measure the cross-sections and asymmetries at different collision energies. In 1993 and 1995 data were taken at several energy points: at the peak energy 91.3 GeV and at two "off-peak" energy points: 89.4 GeV and 93.0 GeV. In 1994 all data were collected at the peak energy point.

The statistical errors on the measured asymmetries are of the order of 0.01, and the systematic errors are about 0.002. The statistical errors on the cross-sections are of a few permill. The systematic uncertainty on the cross-section has the following contributions: the error on background subtraction and selection efficiency evaluation is 0.1% for hadrons and 0.3-0.6% for leptons, the experimental error of the luminosity determination is 0.09%, and the theoretical luminosity error is 0.06% Another important systematic uncertainty is the one on the collision energy because it translates directly to the error on M_Z. The LEP energy is measured with an impressive precision of 10^{-5}. The energy measurement is described in detail in ref. [5].

2.2 Fit to the data

In order to fit the hadronic and leptonic cross-sections and the leptonic forward–backward asymmetries, a set of parameters with only small correlations between them was chosen. They are the following: M_Z and Γ_Z (Z^0 mass and width), σ_0 (hadronic pole cross-section), $R_f = \Gamma_{had}/\Gamma_f$ (ratio of hadronic and leptonic partial widths), and A_{FB}^{0f} (leptonic pole asymmetry). If independent couplings for three lepton species are allowed then the set contains 9 parameters. When lepton universality is assumed the number of parameters is reduced to 5. The results of both 9-parameter and 5-parameter fits are given in Table 2. Combining the results of all four LEP experiments one determines M_Z with a precision close to 2 MeV.

Table 1: The number of selected Z^0 decays in 1993-1995

	1993	1994	1995
$e^+e^- \to e^+e^-$	24k	41k	21k
$e^+e^- \to \mu^+\mu^-$	29k	57k	26k
$e^+e^- \to \tau^+\tau^-$	22k	40k	19k
$e^+e^- \to q\bar{q}$	0.7M	1.3M	0.7M

Table 2: The results of the 9-parameter and 5-parameter fits to all DELPHI data on hadronic and leptonic cross-sections and leptonic forward-backward asymmetries.

Parameter	Value
M_Z	91.1864 ± 0.0029
Γ_Z	2.4872 ± 0.0041
σ_0	41.553 ± 0.079
R_e	20.87 ± 0.12
R_μ	20.67 ± 0.08
R_τ	20.78 ± 0.13
R_l	20.728 ± 0.060
$A_{FB}^{0\,e}$	0.0189 ± 0.0048
$A_{FB}^{0\,\mu}$	0.0160 ± 0.0025
$A_{FB}^{0\,\tau}$	0.0243 ± 0.0037
A_{FB}^0	0.0187 ± 0.0019

2.3   Interpretation of the results

Assuming the Minimal Standard Model value for $\Gamma_\nu/\Gamma_l = 1.991 \pm 0.001$ (evaluated for $M_Z = 91.1867\ GeV$, $m_t = 174.1\ GeV$, and $m_H = 150\ GeV$) and using the measured value for $\Gamma_{inv}/\Gamma_l = 5.950 \pm 0.036$, the number of neutrino species can be deduced:

$$N_\nu = 2.988 \pm 0.018.$$

Within the context of the Minimal Standard Model, a fit has been made to the DELPHI data, leaving the values of the top mass m_t and the strong coupling constant $\alpha_s(M_Z^2)$ as free parameters. The results are:

$$m_t = 174^{+15+8}_{-17-5}\ GeV$$
$$\alpha_s(M_Z^2) = 0.110 \pm 0.006 \pm 0.001.$$

The central values were obtained assuming a Higgs boson mass m_H of $150\ GeV$, and the second uncertainty corresponds to the variation of m_H in the range $90 < m_H(GeV) < 300$.

By comparing the measured value of Γ_Z to its Standard Model prediction, one can derive an upper limit on the partial width of Z^0 decay into yet unknown particles. The following 95% confidence level limit was obtained:

$$\Gamma_Z^{new} < 2.9\ \text{MeV}.$$

3   LEP-2 results

Since the end of 1995, LEP runs at high collision energies (so called LEP-2 regime). It started at $130\ GeV$ in 1995 and will reach $200\ GeV$ in 1999-2000. Table 3 shows the LEP energy settings in 1995-1998 and corresponding integrated luminosities.

DELPHI results for the energies 130-172 GeV were published[2]. The analyses of the 183 GeV and 189 GeV data are still under way and the results are preliminary.

At these energies Initial State Radiation (ISR) becomes quite significant. For large part of events it reduces the effective centre-of-mass energy $\sqrt{s'}$; in

Table 3: The collision energy points at LEP-2 and corresponding integrated luminosities.

Year	1995		1996		1997	1998
Energy (GeV)	130	136	161	172	183	189
Luminosity (pb^{-1})	3	3	10	10	55	160

particular down to Z^0 energies. For this reason the cross-sections and asymmetries were computed for two ranges of $\sqrt{s'}$: full range and "non-radiative" range where the energy reduction is less than 15% ($\sqrt{s'}/\sqrt{s} > 0.85$). The second range is interesting for new physics studies.

The cross-sections and asymmetries were determined for the three leptonic channels. For quarks the inclusive cross-section $e^+e^- \to q\bar{q}$ was computed and the flavor exclusive cross-sections and asymmetries were computed as well. The separation of bottom, charm and light quarks final states in the flavor-tagged analysis was based on the quark lifetime.

Fig. 1 shows the measured cross-section and asymmetries and the Standard Model prediction. The measurements were interpreted by performing S-matrix fits. They also were confronted with several new physics models. These results are given below.

3.1 S-matrix fits

In the S-matrix formalism the cross-section $e^+e^- \to f\bar{f}$ is expressed as a sum of three terms: photon exchange, Z^0 exchange and γ-Z interference, represented by S-matrix parameters g, r and j respectively. By making a fit to the measured cross-sections and asymmetries, one can determine the parameters r, j, Γ_Z and M_Z (a more detailed description of this approach can be found in ref. [6]).

The results of the 8-parameter S-matrix fit to the data (Z^0 and high energy combined) are given in Table 4. In this fit lepton universality was assumed.

Table 4: Results of the 8-parameter fits to the combined Z^0 and high energy data. Also shown are the Standard Model predictions for the fit parameters.

Parameter	Value	SM prediction
M_Z [GeV]	91.179 ± 0.005	-
Γ_Z [GeV]	2.489 ± 0.004	2.493
$r_{\text{had}}^{\text{tot}}$	2.955 ± 0.010	2.960
r_{ℓ}^{tot}	0.1423 ± 0.0006	0.1425
$j_{\text{had}}^{\text{tot}}$	0.65 ± 0.25	0.22
j_{ℓ}^{tot}	0.042 ± 0.021	0.004
r_{ℓ}^{fb}	0.0032 ± 0.0004	0.0027
j_{ℓ}^{fb}	0.774 ± 0.024	0.799

3.2 Fits to physics beyond the Standard Model

There is a variety of new physics models which can be tested with the data collected at LEP-2 energies. By comparing the predictions with measurements, one can derive the limits on physics beyond the Standard Model. In this study the following models were considered: contact interactions between fermions, sneutrino and squark exchange, and Z' boson models. The measured lepton and quark cross-sections and asymmetries were used to perform the fits.

The term "contact interactions" is a general description of the interaction between fermion currents [7]. It is valid for the energies much lower than the characteristic energy scale of the process Λ. Fits to the DELPHI lineshape and asymmetry data show no evidence for the presence of these interactions and give a value of $1/\Lambda$ compatible with 0. The following 95% confidence level limits on the energy scale were derived: $\Lambda > 7 - 14\,\mathrm{TeV}$ for leptons and $\Lambda > 1 - 4\,\mathrm{TeV}$ for quarks.

Another set of models consider possible sneutrino and squark exchange in R-parity violating supersymmetry [8]. It is parametrised in terms of coupling constants λ. The experimental results do not hint to these kinds of exchanges. Fits show that a coupling of $\lambda > 0.1$ can be excluded for sneutrino mass in the range 130-190 GeV. The lower limits on the coupling λ for squarks are 0.4-0.8 depending on the quark flavor (assuming squark mass of 200 GeV).

There are several extensions of the Standard Model that predict the existence of additional heavy gauge boson Z'. It is parametrised by the mass, couplings to fermions and mixing angle with a standard Z^0 [9]. Fits were made to the mass $M_{Z'}$ and to the mixing angle $\theta_{ZZ'}$. No indication of the existence of a Z'-boson was found using any of the models. The mixing angle value from the fit is compatible with the Standard Model expectation of $\theta_{ZZ'} = 0$. The 95% confidence level limit on the Z' mass is $M_{Z'} > 250\,\mathrm{GeV}$.

4 Conclusions

Several fundamental parameters have been extracted from the analysis of fermion pair production at LEP. M_Z, the most precisely determined quantity, is now known with a precision close to $2 \cdot 10^{-5}$. The measurements confirm the validity of the Standard Model, up to the level of its fermionic and bosonic electroweak corrections.

All measurements performed at LEP-2, which cover collision energies up to 190 GeV, are in agreement with the Standard Model predictions as well. Several alternate models have been confronted with the data, but no evidence for physics beyond the Standard Model was found.

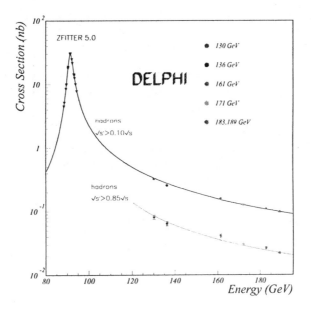

Figure 1: The measured cross-sections together with Standard Model predictions.

References

1. DELPHI Collaboration, P. Abreu *et al*, preprint CERN-EP-99-005.
2. DELPHI Collaboration, P. Abreu *et al*, ICHEP'98-438.
3. A. Behrmann *et al*, ICHEP'98-441 (1998).
4. DELPHI Collaboration, P. Aarnio *et al*, Nucl. Instr. Meth. **A303**, 233, (1991);
   DELPHI Collaboration, P. Abreu *et al*, Nucl. Instr. Meth. **A378**, 57, (1996).
5. The LEP Energy Working Group, R. Assmann *et al*, Eur. Phys. J. **C6**, 187, (1999).
6. A. Leike, T. Riemann and J. Rose, Phys. Lett. **B273**, 513, (1991);
   T. Riemann, Phys. Lett. **B293**, 451, (1992);
   S. Kirsch and T. Riemann, Comp. Phys. Comm. **88**, 89, (1995).
7. E. Eichten, K. Lane, M. Peskin, Phys. Rev. Lett. **50**, 811, (1983).
8. J. Kalinowski *et al*, DESY 97-044 [hep-hp/9703436].
9. A. Leike, S. Riemann, T. Riemann, Munich University preprint LMU-91/6;
   A. Leike, S. Riemann, T. Riemann, preprint CERN-TH 6545/92.

IS CONFORMAL GRAVITY AN ALTERNATIVE TO DARK MATTER?

A. EDERY[1] and M.B. PARANJAPE[2]

[1] *Center for Theoretical Physics, Department of Physics,*
McGill University, 3600 University Street
Montréal, Québec, Canada, H3A 2T8

[2] *Groupe de Physique des Particules, Département de Physique,*
Université de Montréal, C.P. 6128,
succ. centreville, Montréal, Québec, Canada, H3C 3J7

We consider the possibility that conformal gravity could explain the flat rotation curves without the need for dark matter. We find that in a particular choice of the conformal gauge, light and matter are deflected in opposite directions.

Einstein gravity corresponds to the action

$$ S = \frac{1}{16\pi G} \int d^4x \sqrt{-g} R^\alpha_\alpha \,. \tag{1} $$

This theory has been verified by standard tests in the solar system: deflection of light by the sun, radar echo delay, precession of perehelia, etc. Actually, what is really being verified is the Schwarzschild geometry, and the Einsteinian theory only in as much as it predicts this geometry as the solution.

Motion of test particles along geodesics is a consequence of the covariant conservation of the energy momentum tensor, and not a specific prediction of Einsteinian general relativity. Covariant conservation of the energy momentum tensor only requires that the action be a coordinate scalar, and the Bianchi identity. There are many other possibilities to modify the Einstein theory without losing the Schwarzschild solution, especially with small corrections. The Einstein action is however unique if the resulting field equations are demanded to be second order in derivatives of the metric.

Einstein gravity has actually been tested on a very small set of distance scales, from a few centimeters to the scale of the solar system. It is possible that it is not the fundamental theory. The first observational tests of Einstein gravity outside the solar system fail badly. Galactic rotation curves plot the velocity of stars versus their radius. The Keplerian fall off of the velocity is

$$ v \sim \frac{1}{\sqrt{r}} \,. \tag{2} $$

The discrepancy between the observed rotation curves and the predicted fall off is striking.

The solution that is rather readily accepted is to postulate the existence of dark matter, which by its very definition is not visible. We find such an idea quite unsatisfactory although it may well be true. The additional matter will provide the attraction necessary for the orbiting velocities to remain high. Copious quantities of dark matter are required. Some forms of it have actually been discerned. MACHOS, for example, which are massive compact halo objects, have been detected via lensing effects, however not in adequate quantities. Any such dark matter should not of course perturb any cosmological niceties such as inflation or the spectrum of density fluctuations.

Another route is to consider alternative gravitation, invoking invariance principles and higher symmetry. Einstein imposed that the field equations of gravity should be second order in derivatives of the metric, and that the action should be a coordinate scalar. If we relax the former condition, the scalars

$$(R^\alpha_\alpha)^M \; ; (R^{\alpha\beta} R_{\alpha\beta})^N \; (R^{\alpha\beta\gamma\delta} R_{\alpha\beta\gamma\delta})^P \tag{3}$$

are permitted. The unique combination

$$S = -\alpha \int d^4x \sqrt{-g}((R^{\alpha\beta}_{\alpha\beta}) - \frac{1}{3}(R^\alpha_\alpha)^2) \tag{4}$$

contains conformal invariance. Consequently, α is dimensionless. Conformal transformations comprise of the infinite dimensional local rescaling of the metric:

$$g_{\mu\nu}(x) \to \Omega^2(x)g_{\mu\nu}(x) . \tag{5}$$

The action can be equally well written as

$$S = -\alpha \int d^4x \sqrt{-g}(C^{\alpha\beta\gamma\delta} C_{\alpha\beta\gamma\delta}) \tag{6}$$

where $C_{\alpha\beta\gamma\delta}$ is the Weyl tensor, the trace free part of the Riemann tensor. $C_{\alpha\beta\gamma\delta} = 0$ for conformally flat metrics. Conformal symmetry keeps only the causal structure of the space invariant, i.e., it keeps the light cones invariant. The dimensionless coupling constant α implies that the theory is power counting renormalizable as a quantum field theory. The theory has been suggested as an alternative to Einstein gravity as it affords an alternative solution to the missing matter problem.

The spherically symmetric source free solution is given by

$$d\tau^2 = B(r)dt^2 - A(r)dr^2 - r^2 d\Omega^2 \tag{7}$$

with

$$B(r) = \frac{1}{A(r)} = 1 - \frac{2\beta}{r} - 3\alpha\gamma + \gamma r - kr^2 \qquad (8)$$

and α, β, γ, and k are integration constants. If $\beta = GM$ and $\gamma, k \approx 0$, we recover the Schwarzschild metric. β breaks the conformal invariance, but as $r \to \infty$ we can neglect the term $\frac{\beta}{r}$, and the remaining solution is conformal to a Robertson-Walker cosmology with the 3-space curvature

$$K = -k - \frac{\gamma^2}{4}. \qquad (9)$$

Mannheim and Kazanas [1] proposed that the γr term should become important only at galactic scales. At cosmological scales it is the kr^2 term that dominates. They proposed $\gamma r \approx \frac{2\beta}{r} = \frac{2GM}{r}$ for $r \approx 10$ kpc and M around 10^{11} solar masses. Such a choice meant the effective gravitational potential was Newtonian for $r << 10$ kpc, approximately constant for $r \approx 10$ kpc and growing linearly for $r >> 10$ kpc.

With

$$B(r) = 1 + \phi(r) \qquad (10)$$

we get $\phi(r) \approx -\frac{2\beta}{r^2} + \gamma r$ which implies that rotational velocities v satisfy

$$v = \sqrt{\frac{2\beta}{mr^2} + \frac{\gamma r}{m}} \qquad (11)$$

where m is the mass of the orbiting star. This yields a Newtonian behaviour for radii less than 10 kpc and essentially constant for $r \approx 10$ kpc as desired. Hence one reproduces the observed flat rotation curves for stars at the edge of the galaxy. For large r the velocities grow as $\sqrt{r}$, but the number of stars at large r is negligible. Numerically we require that $\gamma \approx 10^{-28}/cm$, which happens to be an inverse Hubble length.

Unfortunately this analysis is flawed. The spherically symmetric line element that is the solution to conformal gravity is determined only up to a spherically symmetric conformal factor.

$$d\tau^2 \to \Omega^2(r,t)d\tau^2 \qquad (12)$$

is equally well a solution of the source free equations. Geodesics of massive test particle (like stars) are sensitive to the conformal factor. Hence to accept the conclusion, one must motivate the choice of conformal gauge.

Null geodesics on the other hand are conformally invariant, they describe trajectories on the light cones, which are left unchanged by conformal transformations. Hence the deflection of light will have an unambiguous, conformal gauge independent prediction in the spherically symmetric conformal geometry. For massless geodesics, the Newtonian potential approximation method is invalid, and we must actually solve for the orbit. The solution[2] gives the orbit equation

$$\varphi(r) = \int dr \frac{A^{\frac{1}{2}}(r)}{r^2 (\frac{1}{J^2 B(r)} - \frac{1}{r^2})^{\frac{1}{2}}} \tag{13}$$

where J is a constant of integration and $\varphi(r)$ is the azimuthal angle in the scattering plane. As $r \to \infty$ the photon simply moves in the (conformal to) Robertson-Walker geometry and it deviates from this (scatters) as it approaches and goes by the source. Taking $\beta\gamma \approx 0$ we find the angular deflection

$$\Delta\varphi \approx \frac{4GM}{r_0} - \gamma r_0 \tag{14}$$

where r_0 is the point of closest approach. Notice k has cancelled altogether reflecting the fact that the deflection of light is insensitive to the cosmology. The first term is just the usual Einstein result, while the γr_0 is negligible for solar system scales. If this second term is to replace dark matter at galactic scales we need $\gamma < 0$. This is contrary to the sign required to explain the flat rotation curves for stars, i.e. massive geodesics.

Observations constrain the value of γ. Solar gravitational deflection of light is measured to $\approx 1\%$, hence does not give any real constraint. Radar echo delay confirms the Einsteinian prediction to $\approx 0.1\%$. This yields $|\gamma| \leq 10^{-23}/cm$, which is still not particularly strong. Finally, looking at galactic lensing data gives a reasonable condition, although it is subject to inherent ambiguities is the large scale structure of the universe via the Hubble constant H_0, $\gamma \approx -7 \times (\frac{H_0}{100}) \times 10^{-28}/cm$. We still harbour the idea that adjusting the conformal factor allows for the explanation of the galactic rotation curves and the deflection of light within the conformal theory of gravitation.

Acknowledgements

We thank NSERC of Canada and FCAR du Québec for financial support.

References

1. D. Kazanas and P.D. Mannheim, ApJ **342**, 635, (1989).
2. A. Edery and M.B. Paranjape, Phys. Rev. **D58**, 024011, (1998).

MEASUREMENT OF THE W MASS AT LEP

H. PRZYSIEZNIAK

CERN, PPE Division, 1211 Geneva, Switzerland

E-mail: helenka.przysiezniak@cern.ch

The mass of the W boson is measured using W-pair events collected with the ALEPH, DELPHI, L3 and OPAL detectors at LEP2. Three methods are used: the cross-section method, the lepton energy spectrum method and the direct reconstruction method, where the latter is described more in detail. For data collected at E_{cm} = 161, 172 and 183 GeV, the following combined preliminary result is obtained: M_W^{LEP} = 80.37 ± 0.08 GeV/c^2.

1   Introduction

The W mass is a key parameter of the Electro-Weak (EW) theory. One can express the Fermi constant G_F as

$$G_F = \frac{\alpha(M_Z^2)\pi}{\sqrt{2}M_W^2(1 - M_W^2/M_Z^2)} \frac{1}{1 - \Delta r}$$

where G_F is determined from muon decay experiments, M_Z is measured at SLD and LEP, $\alpha(M_Z^2)$ is extrapolated from α_0, and the term Δr contains what are known as the radiative corrections, which are a function of m_{top} and $\ln(M_{Higgs})$. The W mass acts as a constraint on the Standard Model (SM) Higgs mass extrapolated in the EW fits, as well as on that of the lightest Minimal SuperSymmetric (MSSM) Higgs.

Data from phase 2 of LEP (LEP2) were taken by the 4 detectors (A: ALEPH, D: DELPHI, L: L3, O: OPAL) situated equidistantly along the 27 km long LEP ring. The centre-of-mass energies of interest are 161 GeV, 172 GeV, 183 GeV, and 189 GeV, for extraction of the W mass via three methods: the direct reconstruction of W pairs, the cross-section measurement, and the lepton energy spectrum measurement for the leptonic channels.

W pairs are produced via three CC03 diagrams: the neutrino exchange t-channel, and the γ and Z exchange s-channel (see Fig. 1). The pairs decay fully hadronically 46 % of the time, when both W bosons decay into a quark-antiquark (4q); 43 % are semileptonic decays, when one of the W bosons decays hadronically, and the other into a neutrino and lepton ($\ell = e$, μ, τ); and 11% are fully leptonic decays, when both W bosons decay leptonically.

The direct reconstruction method is only feasible using the two first channels and for centre-of-mass energies greater or equal to 172 GeV; the cross-

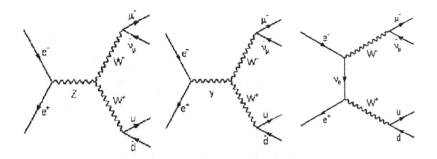

Figure 1: CC03 diagrams for W-pair production.

section method makes use of all three channels and is most sensitive at threshold (161 GeV); the lepton energy spectrum method uses both the semileptonic and fully leptonic channels.

The direct reconstruction analysis is described in detail in Section 2; final 183 GeV results, new results from the lepton energy spectrum, and preliminary 189 GeV results are presented in Section 3; hopes for all data taken up to and including the 1998 run at LEP are presented in Section 4; and in Section 5, the LEP and World Averages for the W mass are given as well as conclusions and outlook.

2 Direct reconstruction analysis

2.1 Event selections

Approximately 10 pb^{-1} at 161 and 172 GeV centre-of-mass energies (E$_{cm}$), 57 pb^{-1} at 183 GeV and 175 pb^{-1} at 189 GeV were accumulated per experiment, which corresponds in total to four samples of about 3000 selected WW pair candidates. The main background processes are e$^+$e$^-$ → q$\bar{q}$γ, e$^+$e$^-$ → Weν, e$^+$e$^-$ → ZZ and e$^+$e$^-$ → Zee.

The four collaborations have developed various techniques which make use of discriminating variables, giving efficient and pure selections. The semileptonic event selections follow a standard scheme. A lepton candidate track (e, μ) or low multiplicity jet associated to a candidate τ are searched for. The candidate lepton is required to be isolated and energetic. Its direction must be opposite to the missing momentum of the event. The tracks not associated to the lepton candidate are then forced into two hadronic jets. An accurate measurement of the missing momentum and lepton energy, and, in the case of the τ channel, the extra undetected neutrinos from the tau decay, make this

Table 1: Semileptonic and 4q selections efficiencies and purities at 183 GeV, for ALEPH (A), DELPHI (D), L3 (L) and OPAL (O).

	A	D	L	O
eμ efficiency (%)	84	71	85	90
eμ purity (%)	93	94	96	94
τ efficiency (%)	69	-	59	75
τ purity (%)	90	-	87	83
4q efficiency (%)	89	85	88	85
4q purity (%)	82	68	85	78

channel delicate.

For the 4q (fully hadronic) channel, neural network or likelihood selections are used. All the tracks in the event are forced into four jets. Jet mixing, higher background, and additional final state interaction systematics (colour reconnnection and Bose Einstein correlations) make this channel difficult. The efficiencies and purities for the eμ channels combined, the τ channel, and the 4q channel at 183 GeV, are given for the 4 experiments in Table 1.

2.2 Invariant mass

Once the events have been selected and the jets reconstructed, the event-by-event invariant mass is calculated. The hadronically decaying W can be used to determine the invariant mass: $M_{q\bar{q}} = \sqrt{2E_{j1}E_{j2}(1 - \cos\theta_{j1j2})}$, where $j1$ and $j2$ denote the two hadronic jets, giving two masses per event for the 4q channel, but only one in the case of the semileptonic channels. Using the energy conservation constraint, the resolution on this mass variable can be improved by rescaling it to the total beam energy: $M_{q\bar{q}} \times E_{\text{beam}}/(E_{j1} + E_{j2})$.

The resolution can still be improved using a kinematic fit. At LEP, the total energy (E_{cm}) and momentum ($\sum \vec{p} = 0$) of an event are very well known. The energy and momentum constraints (1+3 constraints) are used, in addition to an optional constraint of requiring that both W bosons in the event have equal masses (+1 c.). For the semileptonic channels, three constraints are lost due to the undetected neutrino (-3 c.). This gives a five (5C) or four constraint (4C) fit for the 4q channel, and a 2C or 1C fit for the semileptonic channels, in which a χ^2 is minimised, making use of jet and lepton energy and angular

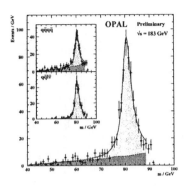

Figure 2: OPAL BW fit to $E_{cm} = 183$ GeV hadronic and semileptonic invariant mass distributions.

resolutions determined using MC (ALO) or data (D).

All experiments use a 2C fit for the $e\mu$ channel. In the case of the τ channel, ALEPH and OPAL use a modified 2C fit, due to the fact that the visible τ jet energy always underestimates the true τ energy. L3 uses the rescaled hadronic mass whereas DELPHI does not analyse this channel. ALEPH and DELPHI use a 4C fit for the 4q channel, whereas L3 and OPAL use a 5C fit.

2.3 Jet pairing

For the 4q channel, three possibilities exist in combining four jets into two W bosons. Only one of those is the correct pairing. The others constitute combinatorial background and contain no M_W information. ALEPH chooses one combination only, L3 and OPAL choose the first and second best combinations, using the χ^2 from the constrained fit to classify them, DELPHI uses all combinations, for four jet (three combinations) and five jet (ten) events.

2.4 Extraction of M_W

The W mass can now be extracted from the invariant mass distribution. Many extraction methods are used, either for the final result or as a cross-check analysis. The simplest and most straightforward method consists of fitting a Breit-Wigner (BW) analytical function to the data invariant mass distribution, e.g. the relativistic BW function $= m_{inv}^2 / [(m_{inv}^2 - M_W^2)^2 - m_{inv}^4 \Gamma_W^2 / M_W^2]$. OPAL has fitted a BW function to their data distribution as a cross-check analysis (see Fig. 2).

480

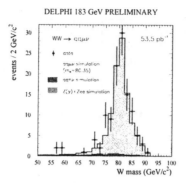

Figure 3: DELPHI convolution fit to the $E_{cm} = 183$ GeV muon invariant mass distribution.

Convolution techniques make maximal use of the available information. The method consists in maximizing a likelihood $\mathcal{L} = \pi_{i=1}^{Nevts}\Phi(m_1, m_2; M_W)$ where Φ is given by: $\Phi = \int de \int dm_1' \int dm_2' \, G \frac{d^3\sigma}{dm_1' dm_2' de}$. The function G describes the detector resolution, and the differential cross section is decomposed into the product of an initial state radiation function, the WW CC03 matrix element, and two Breit-Wigner functions for two masses per event.

A third technique used is the reweighting method, where the MC invariant mass distribution is fitted to the data distribution, as the MC events are reweighted to various W masses. For example, in the ALEPH analysis, the MC event weights are given by:

$$w_i(M_W, \Gamma_W) = \frac{|\mathcal{M}(M_W, \Gamma_W, p_1^i, p_2^i, p_3^i, p_4^i)|^2}{|\mathcal{M}(M_W^{gen}, \Gamma_W^{gen}, p_1^i, p_2^i, p_3^i, p_4^i)|^2}$$

where p_j^i is the four-momentum of the jth outgoing fermion, and $\mathcal{M}$ is the CC03 matrix element for the process $e^+e^- \rightarrow W^+W^- \rightarrow f_1\bar{f}_2f_3\bar{f}_4$, M_W^{gen} is the W mass at which the MC was generated, and M_W is the W mass to which the MC sample is reweighted.

For their final results, ALEPH uses a reweighting technique to extract the W mass, and specific to the 4q channel, performs a 2-dimensional fit using the two masses per event; L3 and OPAL use a 1-d reweighting technique, but also extract the W width in the process of a 2-d fit; DELPHI uses a convolution method and also fits the width. The muon invariant mass distribution at $E_{cm} = 183$ GeV from DELPHI is shown in Fig. 3.

Table 2: Typical systematic errors for the four LEP collaborations.

Uncorrelated	lvqq	4q
Detector effects	40 MeV	30
Fit procedure	25	15
Correlated	lvqq	4q
Hadronisation	25	30
ISR	15	15
LEP Energy	25	25
FSI	–	60

2.5 Systematics

The systematic errors are summarised in Table 2. The numbers quoted are averaged for the 4 LEP experiments. The **Detector effects** systematic consists of calorimeter calibration, MC simulation of the detector, jet energy and angular resolutions. The **Fit procedure** systematic consists of background uncertainties, as well as effects due to limited MC statistics, MC calibration for biased methods, MC diagrams used to determine the weights in the reweighting procedure, fit error reliability, binning for binned likelihoods, and invariant mass window cuts. **Hadronisation and initial state radiation (ISR)** uncertainties are limited by the present MC statistics.

In the future, the most important systematic errors promise to be the **LEP beam energy** uncertainty, and the error coming from final state interactions (FSI) which only affect the 4q channel. The LEP energy is presently being measured using the resonant depolarisation method [1] for beam energies ranging from 45 GeV to 60 GeV. The dominant error on the beam energy comes from the extrapolation to higher beam energies ($\approx$ 80-90 GeV).

Final state interactions which may affect the W mass measurement are colour reconnection effects and Bose-Einstein correlations. The first is a non-perturbative effect occuring during the hadronisation phase of the W bosons decaying hadronically, the perturbative effects being small. Most present day models predict a higher W mass value in the 4q channel than for the semileptonic one. Bose-Einstein correlations occur when same sign soft pions attract each other. Models also predict a higher value of the W mass in the 4q channel.

482

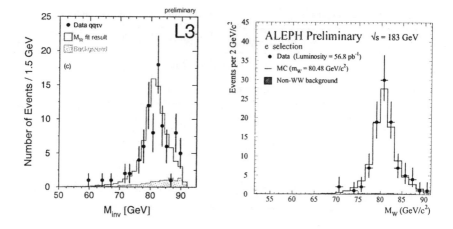

Figure 4: L3 τ and Aleph electron invariant mass distributions at 183 GeV.

2.6 Direct reconstruction results

The combined $E_{cm} = 172$ and 183 GeV results for the W mass extracted using the method of direct reconstruction are, for the four LEP experiments: A : $M_W = 80.44 \pm 0.13$ GeV/c^2, D : $M_W = 80.24 \pm 0.17$ GeV/c^2, L : $M_W = 80.40 \pm 0.18$ GeV/c^2, O : $M_W = 80.34 \pm 0.15$ GeV/c^2, where the error is statistical and systematic. The average of all four results gives:

$$\text{LEP} : M_W = 80.36 \pm 0.09 \text{ GeV/}c^2.$$

The L3 and ALEPH invariant mass distributions for the τ and electron channels respectively, at $E_{cm} = 183$ GeV, are shown in Fig. 4.

3 News since ICHEP98

Since the Vancouver ICHEP98 conference, the four LEP collaborations have published their 183 GeV centre-of-mass energy results [2,3,4,5].

In addition, a new ALEPH analysis at 183 GeV based on extracting the W mass using the lepton energy spectrum has been performed. A reweighting procedure is used, similar to the one described in the direct reconstruction analysis where the data invariant mass variable distribution is fitted. The variables used in this analysis are the lepton energy, the missing energy and the rescaled lepton-neutrino invariant mass for the semileptonic channels; the maximal and minimal lepton energies and the missing energy for the fully leptonic channels. The results are: $M_W(\ell\nu\ell\nu) = 80.34\pm0.53_{stat}\pm0.09_{syst}$ GeV/c^2 and $M_W(\ell\nu q\bar{q}) = 80.09 \pm 0.26_{stat} \pm 0.09_{syst}$ GeV/c^2. When combining the two channels, one obtains: $M_W = 80.14 \pm 0.23_{stat} \pm 0.09_{syst}$ GeV/c^2. Combining this measurement with previous ALEPH direct reconstruction semileptonic measurements at 172 and 183 GeV gives:

$$\text{ALEPH leptonic}: M_W = 80.25 \pm 0.15_{stat} \pm 0.06_{syst} \text{ GeV/c}^2.$$

The preliminary results for the direct reconstruction analysis at 189 GeV are also given by the ALEPH collaboration: $M_W(4q) = 80.56 \pm 0.12_{stat} \pm 0.05_{syst}\pm0.06_{theory}$ GeV/c^2 and $M_W(\ell\nu q\bar{q}) = 80.406\pm0.11_{stat}\pm0.03_{syst}$ GeV/c^2.

4 Hopes for 189 GeV results

As an example of what can be expected for the LEP average W mass measurement to this day, the ALEPH preliminary result is given, for all LEP2 data and for all mass extraction methods used (cross-section at 161 and 172 GeV, direct reconstruction at 172, 183 and 189 GeV, lepton energy spectrum at 183 GeV):

$$\text{ALEPH}: M_W(4q) = 80.42 \pm 0.08 \text{ GeV/c}^2.$$

Taking this result, assuming that the four experiments will obtain similar results, one can expect an error of $\simeq 40 MeV/c^2$ on the LEP average W mass to this day!

5 Conclusions and outlook

The mass of the W boson is measured at LEP using W-pair events collected with the ALEPH, DELPHI, L3 and OPAL detectors at LEP2. For data collected at $E_{cm} = 172$ and 183 GeV, using the direct reconstruction method only, the preliminary average of all four experiments gives:

$$\text{LEP direct reconstruction}: M_W = 80.36 \pm 0.09 \text{ GeV/c}^2.$$

Combining the cross-section mass measurement with the one from direct reconstruction and the ALEPH 183 GeV lepton energy spectrum measurement, the mass is:

$$\text{LEP}: M_W = 80.37 \pm 0.08 \text{ GeV/c}^2.$$

Combining the LEP measurement with the Tevatron and $p\bar{p}$ (UA2) measurements, the World Average (at the time of the Lake Louise 1999 Winter Institute) is given by:

$$W.A. : M_W = 80.39 \pm 0.06 \text{ GeV}/c^2.$$

During the 1998 run at $E_{cm} = 189$ GeV, a total integrated luminosity of approximately 175 pb^{-1} per experiment was collected. Using this data, one can expect to reduce the error of the W mass measurement at LEP to $\simeq 40 MeV/c^2$. Another 150 pb^{-1} per experiment is expected during the upcoming 1999 run, at centre-of-mass energies attaining 200 GeV.

Acknowledgements

It is a pleasure to thank the organisers of the Institute for their efforts in arranging such an inspiring and enjoyable meeting. I would also like to thank the Annecy and Saclay groups, my ALEPH colleagues and specially the WW semileptonic group, my climbing buddies, with whom I've had many fruitful discussions on the W mass topic.

References

1. The LEP Energy Working Group, *Evaluation of the LEP centre-of-mass energy for data taken in 1998*, LEWG 99/01.
2. ALEPH Collaboration, *Measurement of the W mass in* e^+e^- *collisions by the ALEPH detector at LEP*, ICHEP98, Vancouver, 23-29 July, Abstract 899 Session 1.
3. DELPHI Collaboration, *Measurement of the W boson mass and width in* e^+e^- *collisions at* $\sqrt{s} = 183$ *GeV*, ICHEP98, Vancouver, 23-29 July, Abstract 341 Session 1.
4. L3 Collaboration, *Measurement of the Mass and Width of the W Boson in* e^+e^- *Interactions at LEP*, ICHEP98, Vancouver, 23-29 July, Abstract 502 Session 1.
5. OPAL Collaboration, *Measurement of the mass of the W-boson from direct reconstruction*, ICHEP98, Vancouver, 23-29 July, Abstract 392 Session 1.

QUARK MASS MATRICES WITH DIAGONAL ELEMENTS
IDENTIFIED WITH QUARK MASS EIGENSTATES

SUBHASH RAJPOOT

Department of Physics and Astronomy, California State University,
Long Beach, California 90840, USA

In the Standard Model of electroweak interactions, flavour mixing results from quark and lepton mass matrices that in general are neither symmetric nor hermitian. A scheme of quark mass matrices is presented in which the diagonal elements of the quark mass matrices are identified with the eigenvalues of the quark mass matrices. In this scheme, some of the off diagonal elements become complex, thus providing a natural basis for accommodating the phenomenon of CP violation. Implications of the scheme for quark flavour mixing are also discussed.

1 Introduction

At present there is no insight into the origin of quark masses, quark flavour mixing and the phenomenon of observed CP violation. It is a challenge to construct a fundamental theory of quark masses and their mixing that respects all the features of the Standard Model. The Standard Model of electroweak interactions with three families of quarks and leptons has eighteen independent parameters of which ten describe the quark sector. The ten parameters are the six quark masses, the three mixing angles and the phase angle responsible for CP violation. In the absence of a theory, one dabbles into writing down various schemes of mass matrices in the hope that a successful pattern of the mass matrices will pave the way to the correct underlying theory. Schemes with specific form of quark mass matrices with less than ten parameters have by now all have been ruled out by the CDF [1] and DO [2] bounds on the mass of the top quark,

$$m_t^{pole} = 176.0 \pm 6.5 \, \text{GeV} \qquad \text{(CDF Collaboration)}$$
$$m_t^{pole} = 172.1 \pm 7.1 \, \text{GeV} \qquad \text{(DO Collaboration)}. \qquad (1)$$

Evolving m_t to the renormalization point $\mu = 1 \, \text{GeV}$ through the one loop renormalization group equation, the top quark mass is given by

$$m_t(\mu) = m_t(m_t)(1 + \frac{4\pi}{3}\alpha_s(\mu)) \qquad (2)$$

486

where the evolution of the strong coupling constant α_s to the same order is given by

$$\frac{d\alpha_s}{dt} = -\frac{1}{2\pi}(\frac{11}{3}N - \frac{1}{3}N^{fermions} - \frac{1}{6}N^{real\ scalars})\alpha_s{}^2. \tag{3}$$

In the above equation $t = \ln(\mu)$. We take $N = 3$ for $SU(3)$ of colour, $N^{real\ scalars} = 1$, $\alpha_s(\mu = M_Z) \approx 0.12$ and find $\alpha_s(\mu = 1\text{GeV }) \approx 0.17$ when flavour excitations ($N^{fermions}$) are properly taken into account between $\mu = M_Z$ and $\mu = 1\text{GeV }$. The top quark mass equation then gives $m_t(\mu = 1\text{ GeV})$ ~ 300 GeV!

2   New Scheme of Quark Flavour Mixing

In the standard model, quarks and leptons acquire masses through the Yukawa couplings of scalars with the fermion bilinears. The interaction terms in the quark sector which is relevant for our discussion are given by

$$L_{Yukawa} = \sum_{i,j=u,c,t} Y^u_{ij}e^{i\theta^u_{ij}}\bar{\Psi}^i_L\Phi\Psi^j_R + \sum_{i,j=d,s,b} Y^d_{ij}e^{i\theta^d_{ij}}\bar{\Psi}^i_L i\sigma_2\Phi^*\Psi^j_R + (\text{h.c.}). \tag{4}$$

The Yukawa couplings in general are complex. When the scalar field ϕ develops a vacuum expectation value $< \phi >$, one gets for quarks with isospin $+\frac{1}{2}$ i.e., (u,c,t) and quarks with isospin $-\frac{1}{2}$ i.e., (d,s,b) the following mass matrices;

$$M_u = \begin{pmatrix} m_{11u}e^{i\theta_{11u}} & m_{12u}e^{i\theta_{12u}} & m_{13u}e^{i\theta_{13u}} \\ m_{21u}e^{i\theta_{21u}} & m_{22u}e^{i\theta_{22u}} & m_{23u}e^{i\theta_{23u}} \\ m_{31u}e^{i\theta_{31u}} & m_{32u}e^{i\theta_{32u}} & m_{33u}e^{i\theta_{33u}} \end{pmatrix},$$

$$M_d = \begin{pmatrix} m_{11d}e^{i\theta_{11d}} & m_{12d}e^{i\theta_{12d}} & m_{13d}e^{i\theta_{13d}} \\ m_{21d}e^{i\theta_{21d}} & m_{22d}e^{i\theta_{22d}} & m_{23d}e^{i\theta_{23d}} \\ m_{31d}e^{i\theta_{31d}} & m_{32d}e^{i\theta_{32d}} & m_{33d}e^{i\theta_{33d}} \end{pmatrix}, \tag{5}$$

where $m_{iju} = Y^u_{ij} < \Phi >$ and $m_{ijd} = Y^d_{ij} < \Phi >$. As can be noticed, the mass matrices are neither symmetric nor hermitian. One looks for ways to lower the numbers of free parameters so that the matrices lead to some predictability. This is achieved at the expense of imposing restrictions on the mass matrices, i.e., they either be hermitian or symmetric or respect some discrete or continuous symmetries. In some cases, texture in the mass matrices is created by setting some elements of the mass matrices to zero by hand.

The matrices M_u and M_d can be expressed in the generic form $M = P_1 \mathbf{M} P_2$
where

$$\mathbf{M} = \begin{pmatrix} m_{11} & m_{12} & m_{13} \\ m_{21} & m_{22} & m_{23} \\ m_{31} & m_{32} & m_{33} \end{pmatrix}, \quad P_z = \begin{pmatrix} e^{i\alpha_z} & 0 & 0 \\ 0 & e^{i\beta_z} & 0 \\ 0 & 0 & e^{i\gamma_z} \end{pmatrix}, \quad (6)$$

and z=1,2. Thus $\theta_{11} = \alpha_1 + \alpha_2$, $\theta_{12} = \alpha_1 + \beta_2$ and so on and so forth. One can pull out an overall phase factor from the phase matrices P_z (z=1,2) without loss of generality. Also, one can set $P_1 = P_2^\dagger$. This definition is possible since in the Standard Model the phases of the right-handed fermion fields can be transformed independently of the left-handed fermion fields. Further reduction in the number of parameters results if we take the phaseless matrix $\mathbf{M}$ to be symmetric. The matrices become hermitian if one simultaneously takes the phaseless matrix $\mathbf{M}$ to be symmetric and $P_1 = P_2^\dagger$. As a first go round, we will drop all the phases and work with matrix $\mathbf{M}$ that is symmetric. We identify the diagonal elements with the eigenvalues of the symmetric matrix $\mathbf{M}$, i.e., $m_{11} = m_1$, $m_{22} = m_2$, $m_{33} = m_3$. This ansatz is the new ingredient of our scheme. To our knowledge, this possibility has not been entertained in the literature before. For simplicity of notation, we also take $m_{12} = m_{21} = A$, $m_{13} = m_{31} = B$, $m_{23} = m_{32} = C$. The generic matrix takes the simple form

$$\mathbf{M} = \begin{pmatrix} m_1 & A & B \\ A & m_2 & C \\ B & C & m_3 \end{pmatrix}. \quad (7)$$

The eigenvalues (m_1, m_2, m_3) of $\mathbf{M}$ are related to the mass parameters of $\mathbf{M}$ by the following equations,

$$A^2 + B^2 + C^2 = 0, \\ m_1 C^2 + m_2 B^2 + m_3 A^2 = 2ABC. \quad (8)$$

These constraint equations reduce the number of independent parameters to just two. Furthermore, the constraints require that at least one of the parameters A, B, C to be complex. The complexity provides a natural premise for the discussion of CP violation. In what follows, we will work with the generic form of the matrix $\mathbf{M}$ and leave out the phases till the end. Also, we will solve for A and B in terms of C and the eigenmasses m_1, m_2, m_3. Explicitly,

$$A = \pm i \frac{Cf(C)}{\sqrt{1 + f(C)^2}}, \quad B = \pm i \frac{C}{\sqrt{1 + f(C)^2}}, \quad (9)$$

where $f(C)$ is given by

$$f(C) = \frac{C \pm \sqrt{C^2 - (m_3 - m_1)(m_2 - m_1)}}{(m_3 - m_1)}. \tag{10}$$

The matrix $\mathbf{O}$ that diagonalizes $\mathbf{M}$ is determined to be

$$\mathbf{O} =$$

$$\begin{pmatrix} \sqrt{\frac{C^2+(m_3-m_1)(m_1-m_2)}{(m_3-m_1)(m_1-m_2)}} & -\sqrt{\frac{C}{(m_2-m_3)(m_1-m_2)}} & -\sqrt{\frac{C}{(m_3-m_2)(m_1-m_3)}} \\ \sqrt{\frac{B}{(m_3-m_1)(m_1-m_2)}} & \sqrt{\frac{B^2+(m_2-m_3)(m_1-m_2)}{(m_2-m_3)(m_1-m_2)}} & \sqrt{\frac{B}{(m_3-m_2)(m_1-m_3)}} \\ -\sqrt{\frac{A}{(m_3-m_1)(m_1-m_2)}} & \sqrt{\frac{A}{(m_2-m_3)(m_1-m_2)}} & \sqrt{\frac{A^2+(m_3-m_2)(m_1-m_3)}{(m_3-m_2)(m_1-m_3)}} \end{pmatrix} \tag{11}$$

The matrices $\mathbf{O_u}$ and $\mathbf{O_d}$ that diagonalize the up-type and the down-type quark mass matrices M_u and M_d are are gotten by substituting $(m_1 = m_u, m_2 = m_c, m_3 = m_t)$ and $(m_1 = m_d, m_2 = m_s, m_3 = m_b)$ in $\mathbf{O}$. In order to keep matters simple, in analyzing our results, we will only consider values of A, B and C with the positive signature.

3 The Flavour Mixing Matrix

The weak interaction eigenstates (primed) and the mass eigenstates (unprimed) are related via the diagonalizing matrices $\mathbf{O_u}$ and $\mathbf{O_d}$ as follows:

$$\begin{pmatrix} u \\ c \\ t \end{pmatrix} = \mathbf{O_u} \begin{pmatrix} u' \\ c' \\ t' \end{pmatrix}, \qquad \begin{pmatrix} d \\ s \\ b \end{pmatrix} = \mathbf{O_d} \begin{pmatrix} d' \\ s' \\ b' \end{pmatrix}. \tag{12}$$

The charged current weak interaction Lagrangian then translates into the following,

$$L = igW^{\mu}(\bar{u}, \bar{c}, \bar{t})\gamma^{\mu} V_{CKM} \begin{pmatrix} d \\ s \\ b \end{pmatrix} + (\text{h.c}) \tag{13}$$

where the flavour mixing matrix V_{CKM} and the phase matrix P_{ud} are defined as

$$V_{CKM} = \mathbf{O_u} P_{ud} \mathbf{O_d}^{\dagger}, \qquad P_{ud} = P_{1u}^{\dagger} P_{1d} = \begin{pmatrix} 1 & 0 & 0 \\ 0 & e^{i\delta_1} & 0 \\ 0 & 0 & e^{i\delta_2} \end{pmatrix}. \tag{14}$$

Here $\delta_1 = (\beta_{1u} - \beta_{1d}) - (\alpha_{1u} - \alpha_{1d})$ and $\delta_2 = (\gamma_{1u} - \gamma_{1d}) - (\alpha_{1u} - \alpha_{1d})$. The phase combination $P_{ud} = P_{1u}^\dagger P_{1d}$ involves only the relative phase difference. Hence there are only two fundamental observable phases in the scheme. In what follows, we will take α_u, α_d, β_u and β_d to be all zero. In this case the phases δ_1 and δ_2 vanish and the mixing matrix takes the simple form $V_{CKM} = \mathbf{O_u O_d}^\dagger$. This is the matrix we will work with. In this form, we only have two variables to play with, i.e., C_u and C_d.

4 Phenomenological Input and Results

In the Standard Model, the weak interaction and mass eigenstates are related by

$$
\begin{pmatrix} d' \\ s' \\ b' \end{pmatrix} = \begin{pmatrix} V_{ud} & V_{us} & V_{ub} \\ V_{cd} & V_{cs} & V_{cb} \\ V_{td} & V_{ts} & V_{tb} \end{pmatrix} \begin{pmatrix} d \\ s \\ b \end{pmatrix}. \tag{15}
$$

Various phenomenological processes are used to determine the matrix elements V_{ij}. What cannot be determined experimentally, the unitarity of V_{CKM} is employed to get a handle on the range of the left over matrix elements. In particular, V_{ud} is gotten from $0^+ \to 0^+$ nuclear β decay, V_{us} from Kaon ($K \to \pi l \nu_l$) and Hyperon semileptonic decays, V_{cd} from Charm meson semileptonic decays and from dimuon production in ν_μ and $\bar{\nu}_\mu$ interactions, V_{cs} from direct $W^\pm$ decays at LEP II , V_{cb} from B-meson semileptonic decays ($B \to D^* l^- \bar{\nu}_l$) and QCD sum rules using the operator product expansion and methods based on heavy quark effective field theory, V_{ub} from semileptonic decays of B-mesons, V_{td} from CP violation in K^0-$\bar{K}^0$ system and V_{td} from the observed mixing in the B^0-$\bar{B}^0$ system. Limits on the elements V_{tb} and V_{ts} are gotten from unitarity considerations. The absolute values of V_{ij} of the mixing matrix V_{CKM} are given in the Particle Data Group [3]. We require that the matrix elements of V_{CKM} of our model lie within one standard deviation of those determined in the standard model. Agreement is achieved provided

$$
1399\,\text{MeV} \le C^u \le 1415\,\text{MeV}, \qquad -125\,\text{MeV} \le C^d \le -119\,\text{MeV}. \tag{16}
$$

The mixing matrix in our scheme of flavour mixing is determined to be

$$
|V_{CKM}| = \begin{pmatrix} 0.9744 \sim 0.9763 & 0.2198 \sim 0.2289 & 0.0151 \sim 0.0318 \\ 0.2194 \sim 0.2268 & 0.9734 \sim 0.9751 & 0.0313 \sim 0.0326 \\ 0.0151 \sim 0.0322 & 0.0318 \sim 0.0326 & 0.9996 \sim 1.000 \end{pmatrix}. \tag{17}
$$

In working out the flavour mixing matrix, we selected values for the quark masses to lie in the range determined by Gasser and Leutwyler [4]:

$$m_u(1\,\text{GeV}) = 5.1 \pm 1.5\,\text{MeV}, \qquad m_d(1\,\text{GeV}) = 8.9 \pm 2.6\,\text{MeV},$$
$$m_s(1\,\text{GeV}) = 175 \pm 55\,\text{MeV}, \qquad m_c(1\,\text{GeV}) = 1350 \pm 50\,\text{MeV}, \qquad (18)$$
$$m_b(1\,\text{GeV}) = 5300 \pm 100\,\text{MeV},$$

and $m_t(1\,\text{GeV}) = 300 \pm 20\,\text{GeV}$ as determined via the renormalization group equation.

5 Beyond the Present Scheme

The scheme of quark mixing presented here is only an ansatz. It is hoped that by studying the predictions of several ansatz like the one presented here, one may stumble on the correct theory underlying the origin of quark (and lepton) masses. Some interesting facts about fermions of the Standard Model are:

1. They divide themselves into three families.

2. Each family contains exactly fifteen quarks and leptons (no right-handed neutrino).

3. Each family is an exact replica of the other in every quantum number except for the masses of the individual members.

4. Successive families increase in mass, i.e., a family member in the second family is heavier than the corresponding member in the first family.

Since neutrino masses are not known, the lepton mass spectrum is incomplete. However, the quark mass spectrum is complete as given above. Since quarks interact through strong, weak, electromagnetic and gravitational interactions, a mass formula for the quarks may look like

$$m(N) = e^{\cdots f(\texttt{strong})+f(\texttt{weak})+f(\texttt{electromagnetic})+(f\texttt{gravity})+f(N)\cdots}$$
(19)

where dots represent some finer details depending on the weak isospin and yet unknown interactions. Here $f(N)$ represents quantum numbers depending on the family number N. Thus $N = 1$ represents the (e, ν_e, u, d) family, $N = 2$ represents the μ, ν_μ, c, s family and $N = 3$ represents the τ, ν_τ, b, t family. The proposed mass formula can be simplified to

$$m(N, \pm\frac{1}{2}) = m_\pm e^{f_\pm(N)}$$
(20)

where $m_\pm$ represents our present ignorance of the total contribution of the dot dot dot terms and the strong, weak, electromagnetic and gravitational interaction terms. To keep matters simple, we take $f_\pm(N) = \alpha_\pm N$ where $\alpha_\pm$ represent constants. Based on this, we find the following formulae fit the various quark masses,

$$m(\frac{1}{2}, N) = m_+ e^{\alpha_+ N} \text{MeV}, \qquad (21)$$

$$m(-\frac{1}{2}, N) = m_- e^{\alpha_- N} \text{MeV}, \qquad (22)$$

where $m_+ = 0.018$, $m_- = 0.33$, $\alpha_+ = 5,5$ and $\alpha_- = 3.2$. Thus, from Eq. (21), $N = 1$ gives m_u, $N = 2$ gives m_c, $N = 3$ gives m_t and from Eq. (22), $N = 1$ gives m_d, $N = 2$ gives m_s, $N = 3$ gives m_b. All masses fall within the given range in Eq. (18). The next task is to generate the off diagonal elements in the mass matrices within the context of the aforementioned discussion. One possibility is to add inter-family interactions of the form $e^{f_\pm \sqrt{NN'}}$ where $N, N' = 1, 2, 3$ and $N \neq N'$. These and related ideas will be entertained elsewhere.

Acknowledgments

This work was partially supported by grants from California State University, Long Beach. We thank Deepak, Jyoti, Ravi and Urmil for reading the manuscript.

References

1. F. Abe, *et.al.*, CDF Collaboration, Phys. Rev. Lett. **82**, 271, (1999).

2. B. Abbott, *et.al.*, Phys. Rev. Lett. **80**, 2063, (1998); Phys. Rev. **D58**, 052001, (1998); FERMILAB-PUB-98-261-E and hep-ex/9808029.

3. *Review of Particle Physics*, Particle Data Group, C. Caso *et al.*, Euro. Phys. J. **C3**, 1, (1998).

4. J. Gasser and H. Leutweiler, Phys. Rep. **87**, 77, (1992).

RECENT RESULTS FROM THE CHORUS EXPERIMENT

P. RIGHINI[a]

Università di Cagliari and INFN,
Cittadella Universitaria di Monserrato,
P.O. Box 170, 09042 Monserrato (CA), Italy
E-mail: pierpaolo.righini@cern.ch

CHORUS is an experiment designed to look for $\nu_\mu \to \nu_\tau$ oscillation in the CERN SPS Wide Band Neutrino Beam. Data taking with the neutrino beam started in April 1994 and finished in November 1997. A subset of the neutrino interactions collected in all four years by the experiment has been analyzed, searching for ν_τ charged current interactions followed by the τ lepton decay into a muon or into a negative hadron. In a sample of 102,861 events with an identified μ^- and 7,514 events without an identified muon, no ν_τ candidate has been found. As a consequence an upper limit on the oscillation probability of $P(\nu_\mu \to \nu_\tau) \leq 4.6 \cdot 10^{-4}$ at 90% C.L. can be set. The analyses of data taken from emulsion is progressing, improving statistics and efficiency. A sensitivity corresponding to $P(\nu_\mu \to \nu_\tau) \leq 1.0 \cdot 10^{-4}$ is expected in a few years.

1 Introduction

CHORUS (CERN Hybrid Oscillation Research apparatUS) is an experiment searching for $\nu_\mu \to \nu_\tau$ oscillation through the appearance of ν_τ's in a ν_μ beam. The idea is to detect the presence of a τ, produced in a ν_τ charged current interaction, by observing its finite decay length ($1.5\,mm$ on average). As a consequence the target must be active and have very good spatial resolution. This has lead to the use of nuclear emulsions.

The experiment is performed in the CERN Wide Band Neutrino Beam, which contains mainly ν_μ with a contamination from ν_τ well below the level of sensitivity that can be reached in this experiment.

The τ decay channels considered are the one with a muon, the one with a charged hadron and that with three charged hadrons in the final state, although the latter has not been included in the analysis yet.

CHORUS explores the $(sin^2 2\theta_{\mu\tau}, \Delta m^2_{\mu\tau})$ parameter space region corresponding to neutrino masses difference above few eV ($\Delta m^2_{\mu\tau} \gtrsim 1\,eV^2$), a region interesting from the cosmological point of view to propose neutrinos as candidates for the hot component of dark matter[1].

The CHORUS Collaboration has reported [2,3,4] a limit on $\nu_\mu \to \nu_\tau$ oscillation obtained by the analysis of a subsample of neutrino events with an

---

[a] Representing the CHORUS Collaboration.

identified muon (taken in 1994-1996) and without an identified muon (taken in 1994-1995) in the final state. The results presented in this paper are based on a sample with increased statistics, thanks to recent analysis of a significant fraction of the 1997 neutrino interactions with an identified muon.

2 The CHORUS experiment

2.1 The neutrino beam

The CERN Wide Band Neutrino Beam, used in the CHORUS experiment, contains mainly muon neutrinos (with an average energy of $27\,GeV$) with a $\bar{\nu}_\mu$ contamination of 5 % and a ν_e, $\bar{\nu}_e$ contamination at the level of 1 %. These come from the decay in flight of produced π^+ and K^+. The intrinsic fraction of ν_τ's in the beam is estimated [6] to be at the negligible level of $3 \cdot 10^{-6}$ ν_τ charged current interactions per ν_μ charged current interaction.

2.2 The detector

To recognize the short flight path of the τ lepton and its decay topology, CHORUS used a $770\,kg$ nuclear emulsion target for the excellent spatial resolution (of the order of $1\,\mu m$) that emulsions can provide.

The *emulsion target* consists of four stacks of 36 plates each. A plate has two $350\,\mu m$ layers of emulsion on both sides of a $90\,\mu m$ plastic base. Downstream of each emulsion stack there are three sets of *interface emulsion* sheets. They are followed by scintillating fiber trackers to reconstruct the trajectories of the charged particles produced in the neutrino interaction.

In addition to the scintillating fiber trackers, other electronic detectors (an air-core magnet, a lead-scintillator calorimeter and a muon spectrometer), located downstream of the target region, are used for the kinematical reconstruction of the events.

More details about the experimental setup, shown in Fig. 1, can be found in ref. [5].

3 Data collection

In the four years of data taking, from 1994 to 1997, CHORUS collected about 2.31 million triggers corresponding to $5.06 \cdot 10^{19}$ protons on target. Among these events, those characterized by an interaction vertex reconstructed in the emulsion are 458,601 with an identified muon in the final state (the so called 1μ events) and 116,049 without an identified muon (the so called 0μ events).

494

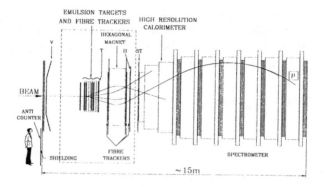

Figure 1: The CHORUS detector.

4 Data analysis

4.1 Analysis steps and event selection

The first step of the analysis is the event reconstruction using the data from the electronic detectors. The event selection for emulsion scanning depends on whether a muon has been found or not. For 1μ events no other muons have to be present at the primary vertex and the muon charge has to be negative. A momentum cut $P_\mu < 30\,GeV/c$ reduces the number of ν_μ induced charged current interactions to be scanned, with a modest loss of ν_τ detection efficiency. In the surviving events the muon track is extrapolated to the emulsions and followed back to primary vertex.

On the other hand, for the 0μ sample, we search in the emulsion for all the negatively charged hadrons with a momentum in the range $1\,GeV/c < P_h < 20\,GeV/c$ (the former cut to reduce the background contamination, the latter one due to poor momentum resolution at higher energy).

The emulsion scanning for the selected tracks is performed by automatic systems. In that way, the emulsion plate where the interaction occurred is located and a search for the decay vertex is carried out. The events not rejected by the automatic scanning are subjected to a computer-assisted *eye scan*. If a candidate is found, the signal/background discrimination is made through complete kinematics reconstruction of the event.

It is worthwhile to notice that the energy measured by the calorimeter has not been used to reject electrons or unidentified muons from the 0μ sample.

Table 1: Status of the analysis in CHORUS.

	1994	1995	1996	1997	Total
$Pot/10^{19}$	0.81	1.20	1.38	1.67	5.06
CHORUS efficiency	0.77	0.88	0.94	0.94	0.90
Emulsion triggers/10^3	422	547	617	719	2305
Expected $CC(\nu_\mu)/10^3$	120	200	230	290	840
1μ to be scanned	66,911	110,916	129,669	151,105	458,601
1μ scanned so far	69%	47%	59%	48%	54%
1μ vertex located	19,581	21,809	30,681	30,790	102,861
0μ to be scanned	17,731	27,841	32,548	37,929	116,049
0μ scanned so far	60%	48%	53%	33%	47%
0μ vertex located	3,491	4,023	5,339	3,837	16,690

Consequently, the contribution offered by the τ leptonic decay modes to this sample is taken into account to evaluate the sensitivity of the experiment.

4.2 Vertex location

The automatic procedure to locate the plate containing the primary vertex has been already described in ref's. [2,3]. It will be briefly recalled here. All the selected tracks (the muon for the 1μ events and all the negatively charged hadrons for the 0μ events) are looked for in the interface emulsion sheets. Once such a track (the so called *scanback track*) has been found, it is followed inside the target, plate by plate, searching for a matching track segment in the $100\,\mu m$ upstream of each plate. This is done until the track disappears in two consecutives plates. The first one is called the *vertex plate*. The scanning results are listed in Table 1. It should be noted that the decay search analysis for the 1996 and 1997 0μ events has not been concluded yet. As a consequence these events have not been included in the statistics considered in this paper.

4.3 Decay search

Once the vertex plate has been determined, the automatic decay search procedure is started. The τ decay signature is given by the presence of a change in direction (*kink*) between the τ track and those of its daughters. Different algorithms (more carefully explained in ref's. [2,3]) have been applied to search for the presence of a kink. In the first procedure, the events are not rejected (and so selected for the manual scanning) either when the scanback track has a large impact parameter with respect to the other tracks or when the change in the scanback track direction, between the vertex plate and the exit from the emulsion stack, corresponds to an apparent transverse momentum P_T larger than $250\,MeV/c$. For the selected events and for those with only one predicted track found in the plate immediately downstream of the vertex one, the so called *video image analysis* is performed. That is digital images of the vertex plate are taken and analysed by means of dedicated algorithms.

In the second procedure (the so called *parent track search*), looking for a compromise between efficiency and scanning speed, we restrict the search to the events whose primary and decay vertices are contained in different plates and for which kink angle is *large*. A kink angle is considered to be *large* when it is greater than the angular tolerance requested to follow back the track, so that for these kind of events the vertex plate contains the decay vertex instead of the primary one. A cone is opened around the direction of the scanback track (with an opening angle given by the inverse of the reconstructed momentum) and we search for a track, in the area defined by the this cone onto the upstream surface of the vertex plate, with a minimal distance in space with respect to the scanback track shorter than $15\,\mu m$. If such a track exists it is considered the *parent track*. Checks are made to be sure that the parent track is not a background track passing by chance near the scanback track. If this is not the case the event is manually scanned.

After the manual scanning, we would have a τ candidate if the presence of the kink is confirmed and the transverse momentum of the particle outgoing from the secondary vertex, with respect to the parent direction, is greater than $250\,MeV/c$.

For the time being no τ candidate has been found.

5 Background

The background for the muonic decay channel is essentially due to charm production in antineutrino charged current events in which the primary lepton (μ^+ or e^+) is not identified. In the present sample, less than 0.1 events are expected from this source of background.

The main potential background to the τ decay in one charged hadron is represented by the elastic hadron scattering without visible recoil of the nucleus in neutral current interactions (called *white kinks*). In the present sample the estimate for this kind of background source is of 0.5 events. As the statistics will increase this background will be kept uder control by means of kinematical cuts.

At the moment, the expected background given by the prompt ν_τ contamination of the beam[6] (common to both the muonic and hadronic decay modes) is much less than 0.1 events.

More details about the background situation are reported in ref's. [2,3,4].

6 Oscillation sensitivity

In a two flavour mixing scheme the $\nu_\mu \to \nu_\tau$ oscillation probability can be expressed as:

$$P(\nu_\mu \to \nu_\tau) = sin^2 2\theta_{\mu\tau} \cdot \int \Psi(E, L) \cdot$$

$$sin^2 \left(\frac{1.27 \cdot \Delta m_{\mu\tau}^2 (eV^2) \cdot L(Km)}{E(GeV)} \right) \cdot dL \cdot dE \qquad (1)$$

where E is the incident neutrino energy, L the neutrino flight length to the emulsion target, $\theta_{\mu\tau}$ the $\nu_\mu - \nu_\tau$ mixing angle, $\Delta m_{\mu\tau}^2 = |m_{\nu_\mu}^2 - m_{\nu_\tau}^2|$ and $\Psi(E, L)$ represents the neutrino flight length distribution at a given energy E.

As said before, the τ decay channels taken into account in this paper to evaluate the CHORUS oscillation sensitivity are: (1) $\tau \to \mu$, (2) $\tau \to h$, (3) $\tau \to e$ and (4) $\tau \to \not\mu$ (muonic τ decay with the μ not identified). As a consequence the maximum number of ν_τ charged current interactions observable *a priori* (that is in case $P(\nu_\mu \to \nu_\tau) = 1$) is given by:

$$N_\tau^{CC}(max) = \frac{\langle \sigma_{\nu_\tau}^{CC} \rangle}{\langle \sigma_{\nu_\mu}^{CC} \rangle} \cdot \frac{\langle A_{\tau \to \mu} \rangle}{\langle A_{\nu_\mu}^{CC} \rangle} \cdot N_{\nu_\mu}^{1\mu} \cdot BR(\mu) \cdot \langle \eta_\mu \rangle \cdot$$

$$\left[1 + \frac{\varepsilon^{0\mu}}{\varepsilon^{1\mu}} \cdot \sum_{i=2}^{4} \frac{BR(i)}{BR(\mu)} \cdot \frac{\langle A_i \rangle}{\langle A_{\tau \to \mu} \rangle} \cdot \frac{\langle \eta_i \rangle}{\langle \eta_\mu \rangle} \right] \qquad (2)$$

where $i = 2$, $i = 3$ and $i = 4$ stand for $\tau \to h$, $\tau \to e$ and $\tau \to \not\mu$ respectively, and

- $\langle \sigma_{\nu_\tau}^{CC} \rangle / \langle \sigma_{\nu_\mu}^{CC} \rangle$ is the neutrino energy weighted cross section ratio. Deep-inelastic reactions, quasi-elastic interactions and resonance production are taken into account;

- $\langle A_{\nu_\mu}^{CC} \rangle$ is the cross section weighted reconstruction and scanning efficiency (up to the primary vertex location) for the ν_μ charged current interactions. $\langle A_{\tau \to \mu} \rangle$ and $\langle A_i \rangle$ ($i = 2, 3, 4$) are the similar quantities for the muonic τ decay channel and for the decay modes contributing to the 0μ sample;

Table 2: 1μ sample: quantities used to estimate the CHORUS sensitivity.

	1994 μ	1995 μ	1996 μ	1997 μ
$N_{\nu_\mu}^{1\mu}$	19,581	21,809	30,681	30,790
$\langle\sigma_{\nu_\tau}^{CC}\rangle/\langle\sigma_{\nu_\mu}^{CC}\rangle$	0.53	0.53	0.53	0.53
$\langle A_{\tau\to\mu}\rangle/\langle A_{\nu_\mu}^{CC}\rangle$	1.07	1.07	1.07	1.07
$BR(\mu)$	0.174	0.174	0.174	0.174
$\langle\eta_\mu\rangle$	0.52	0.35	0.37	0.37
$N_{\nu_\tau}^{max}$	1011	752	1118	1122

- $\langle\eta_\mu\rangle$ and $\langle\eta_i\rangle$ ($i = 2,3,4$) are the corresponding average kink finding efficiencies;

- $BR(\mu)$ and $BR(i)$ are the branching ratios for the considered channels[8];

- $N_{\nu_\mu}^{1\mu}$ is the number of located ν_μ charged current interactions;

- $\varepsilon^{0\mu}/\varepsilon^{1\mu}$ is the fraction of the 0μ statistics scanned so far with respect to the 1μ statistics.

No τ candidate has been found in the considered event sample. This result implies a 90% C. L. upper limit on the oscillation probability of:

$$P(\nu_\mu \to \nu_\tau) \leq \frac{2.38}{N_\tau^{CC}(max)} = 4.6 \cdot 10^{-4}. \tag{3}$$

In the above formula the numerical factor 2.38 takes into account the total systematic error[b] (17%) following the prescriptions given in ref.[7]. The estimated values of the quantities involved in the computation of $N_\tau^{CC}(max)$ (see Eq. 2) are reported in Table 2 and in Table 3 for 1μ and 0μ sample respectively.

The 90% exclusion plot obtained with the present data is shown in Fig. 2. The large $\Delta m_{\mu\tau}^2$ are excluded at 90% C. L. for $sin^2 2\theta_{\mu\tau} > 9.2 \cdot 10^{-4}$; on the other hand maximum mixing is excluded at 90% C. L. if $\Delta m_{\mu\tau}^2 > 0.8\, eV^2$.

7 Experimental check of the τ identification efficiency

The kink finding efficiency η_τ (the efficiency to find the decay vertex once the primary vertex has been located) has been evaluated by Monte Carlo simulation and checked using a dimuon sample and hadron interactions.

---

[b]The systematic uncertainty is mainly due to the reliability of the Monte Carlo simulation of the scanning methods.

Table 3: 0μ sample: quantities used to estimate the CHORUS sensitivity.

	1994 h	1995 h	1994 e	1995 e	1994 $\not{\mu}$	1995 $\not{\mu}$
$N_{\nu_\mu}^{1\mu}$	19,581	21,809	19,581	21,809	19,581	21,809
$\varepsilon^{0\mu}/\varepsilon^{1\mu}$	0.874	0.66	0.874	0.66	0.874	0.66
$\langle\sigma_{\nu_\tau}^{CC}\rangle/\langle\sigma_{\nu_\mu}^{CC}\rangle$	0.53	0.53	0.53	0.53	0.53	0.53
$\langle A_{\tau\to h/e/\mu}\rangle/\langle A_{\nu_\mu}^{CC}\rangle$	0.47	0.47	0.26	0.26	0.073	0.073
$BR(h/e/\mu)$	0.495	0.495	0.178	0.178	0.174	0.174
$\langle\eta_{h/e/\mu}\rangle$	0.26	0.25	0.13	0.13	0.25	0.23
$N_{\nu_\tau}^{max}$	553	445	55	45	29	22

Assuming a charm production rate per charged current interaction of about 5%, in the dimuon sample analysed so far (part of the sample collected in 1995 and 1996) we expect (22.8 ± 3.9) $\nu_\mu N \to \mu^- D^+ X$ interactions with the subsequent muonic decay of the D^+. Some 25 such events have been found, in good agreement with the Monte Carlo simulation.

A sample of about $55\,m$ of hadron tracks has been scanned in emulsion, searching for neutrino interactions with a hadron interaction. 21 events have been detected: a result that is again consistent with the value of (24 ± 2) obtained from Monte Carlo simulation.

These results can be taken as qualitative checks of the simulation of the automatic scanning procedure.

8   Conclusions

We reported the results obtained analysing a subsample of the events collected in all four years (1994-1997) of the CHORUS data taking. No τ decay candidate has been found when searching for ν_τ charged current interactions followed by the τ lepton decay into a muon or into a negative hadron. The corresponding 90% C. L. upper limit on the $\nu_\mu \to \nu_\tau$ oscillation probability is $P(\nu_\mu \to \nu_\tau) \leq 4.6 \cdot 10^{-4}$. Considering the whole statistics by including in the analysis the three prong τ decay channel as well and by improving the global τ identification efficiency (thanks to a second phase of the analysis that will take advantage of higher performance of the automatic emulsion scanning) we expect to reach the proposal sensitivity [9] of the experiment $(P(\nu_\mu \to \nu_\tau) \leq 1.0 \cdot 10^{-4}$, see Fig. 2).

500

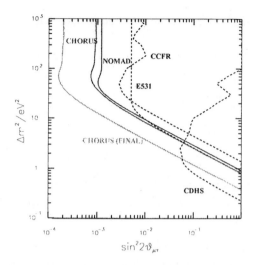

Figure 2: The region of the $(\Delta m_{\mu\tau}^2, sin^2 2\theta_{\mu\tau})$ parameter space excluded by CHORUS at the 90% C.L. compared with the NOMAD result and the previous limits. The final result expected for CHORUS is also shown.

References

1. Ya.B. Zel'dovic, I.D. Novikov, *Relativistic Astrophysics* Nauka, Moscow, 1967; H. Harari, Phys. Lett. **B216**, 413, (1989); J. Ellis, J.L. Lopez, D.V. Nanopoulos, Phys. Lett. **B292**, 189, (1992); H. Fritzsch, D. Holtmannspötter, Phys. Lett. **B338**, 290, (1994).
2. E. Eskut *et al.*, Phys. Lett. **B424**, 202, (1998).
3. E. Eskut *et al.*, Phys. Lett. **B434**, 205, (1998).
4. P. Migliozzi (CHORUS Collaboration), in ICHEP 98, proceedings of the 29th International Conference on High Energy Physics, UBC, Vancouver, B.C., Canada (July 23-29 1998).
5. E. Eskut *et al.*, Nucl. Instum. Methods **A401**, 7, (1997).
6. B. Van der Vyver, Nucl. Instrum. Methods **A385**, 91, (1997); M.C. Gonzalez-Garcia, J.J. Gomez-Cadenas, Phys. Review **D55**, 1297, (1997).
7. R.D. Cousins and V.L. Highland, Nucl. Instrum. Methods **A320**, 331, (1992).
8. C. Caso *et al.*,The European Physical Journal **C3**, 1, (1998).
9. N. Armenise *et al.*, CHORUS Collaboration, CERN-PPE/93-131.

TAU PHYSICS AT OPAL

S.H. ROBERTSON

Department of Physics and Astronomy, University of Victoria,
P.O. Box 3055, Victoria, B.C. Canada V8W 3P6

OPAL Collaboration

The study of the τ lepton can reveal insight into many different aspects of particle physics both within and beyond the Standard Model. In the present work, the results of OPAL τ physics analyses are used to test the universality and the Lorentz structure of the charged-current weak interaction.

1 Introduction

As the third-generation sequential charged lepton, the τ provides a unique laboratory for the study of the weak interaction. Within the Standard Model (SM), the τ is expected to have couplings identical to the electron and muon. However, in contrast to the lighter leptons, the large τ mass permits decays into both leptonic and hadronic final states. The τ decays via the charged-current weak interaction to produce final states composed of neutrinos and charged leptons or mesons composed of u, d and s quarks, as shown in Fig. 1. The SM assumes a pure $V - A$ form for the weak charged-current coupling to the three lepton and quark generations with a universal coupling strength g. The effective charged-current Lagrangian density is given by

$$\mathcal{L}_{cc} = \frac{g}{2\sqrt{2}} W_\mu^\dagger \left\{ \sum_{l=e,\mu,\tau} \bar{\nu}_l \gamma^\mu (1 - \gamma_5) l + \bar{u} \gamma^\mu (1 - \gamma_5) d_\theta \right\} + \ldots \quad . \quad (1)$$

The universality of the strength and the structure of this coupling implies that the ratio, R_τ, of the hadronic and leptonic decay widths is

$$R_\tau \equiv \frac{\Gamma(\tau^- \to \text{hadrons } \nu_\tau)}{\Gamma(\tau^- \to e^- \bar{\nu}_e \nu_\tau)} \simeq n_c = 3 \quad , \quad (2)$$

where n_c is the number of colour degrees of freedom, and furthermore that $B(\tau^- \to e^- \bar{\nu}_e \nu_\tau) \simeq B(\tau^- \to \mu^- \bar{\nu}_\mu \nu_\tau) \approx 1/5$.

The two fully-leptonic τ decay modes are of particular interest since they are both experimentally clean and precisely calculable within the SM. Assuming massless neutrinos, the leptonic partial widths are given by [1]

$$\Gamma(\tau^- \to l^- \bar{\nu}_l \nu_\tau) = \frac{g_\tau^2 g_l^2}{(8m_W^2)^2} \frac{m_\tau^5}{96\pi^3} f\left(\frac{m_l^2}{m_\tau^2}\right) r_{EW} \quad (3)$$

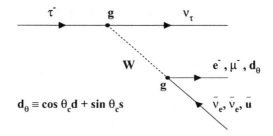

Figure 1: Decay of the τ lepton into leptonic and hadronic final states via the charge-current weak interaction.

where $l = e, \mu$. The factor $r_{EW} = 0.9960$ accounts for electroweak propagator and radiative corrections [2] and $f(x) = 1 - 8x + 8x^3 - x^4 - 12x \ln x$ is a phase-space correction. Universality requires that the coupling constants g_e, g_μ and g_τ are identical. However, even assuming that this picture correctly describes the couplings of W bosons to fermions, "new" physics could introduce corrections to, for example, Eq. 3 and thus could manifest itself as an apparent violation of either the universality of the couplings or of the $V - A$ structure of the charged-current weak interaction. Measurements of the Lorentz structure and couplings can therefore be used to constrain new physics.

2 Selection of $\tau^+\tau^-$ events

During the time that the LEP collider operated near the Z^0 resonance (LEP1), Z^0 bosons decaying at rest in the laboratory frame produced pairs of back-to-back, relativistic τ leptons which subsequently decayed in flight to produce strongly collimated "jets". At LEP energies, multihadronic events ($e^+e^- \to q\bar{q}$) produce significantly higher particle multiplicities than τ-pair events and can be effectively rejected by applying cuts on the number of tracks and EM calorimeter clusters in the event. The unobserved neutrinos produced by the decay of the two τ leptons provide a missing energy signature which distinguishes τ-pair events from Bhabha ($e^+e^- \to e^+e^-$) or dimuon ($e^+e^- \to \mu^+\mu^-$) events, in which the charged particles typically possess the full centre-of-mass energy. Additional kinematic and fiducial requirements [3] are imposed to reject four-fermion, cosmic ray and beam-related backgrounds, resulting in an inclusive sample of τ-pair events with a non-τ background contamination of approximately 1.5%. Different OPAL τ physics analyses impose different requirements on detector performance and acceptance, but typically select $(1\text{-}2) \times 10^5$ $\tau^+\tau^-$ events from the OPAL LEP1 data set.

3   Lepton universality

Tests of the universality of the charged-current couplings to leptons can be obtained using measurements of the τ and μ lifetimes, and of the branching ratios of semi-leptonic and fully-leptonic decays. The ratio of the coupling constants g_μ/g_e can be evaluated by comparing the τ partial decay widths to electrons and to muons using Eq. 3, giving

$$\left(\frac{g_\mu}{g_e}\right)^2 = \frac{\mathrm{B}(\tau^- \to \mu^- \bar{\nu}_\mu \nu_\tau)}{\mathrm{B}(\tau^- \to e^- \bar{\nu}_e \nu_\tau)} \frac{f(m_e^2/m_\tau^2)}{f(m_\mu^2/m_\tau^2)} \quad . \tag{4}$$

Using OPAL measurements of $\mathrm{B}(\tau^- \to e^- \bar{\nu}_e \nu_\tau) = (17.81 \pm 0.09 \pm 0.06)\%$ [3] and $\mathrm{B}(\tau^- \to \mu^- \bar{\nu}_\mu \nu_\tau) = (17.48 \pm 0.12 \pm 0.08)\%$ [4] yields the result $g_\mu/g_e = 1.0046 \pm 0.0051$, thus verifying μ - e universality at the level of 0.5%.

τ - μ universality can be similarly tested by comparing the $\tau^- \to e^- \bar{\nu}_e \nu_\tau$ and $\mu^- \to e^- \bar{\nu}_e \nu_\mu$ partial decay widths. Since the $\mu^- \to e^- \bar{\nu}_e \nu_\mu$ branching ratio is effectively 100%, the ratio of the partial decay widths can be expressed in terms of $\mathrm{B}(\tau^- \to e^- \bar{\nu}_e \nu_\tau)$ and the τ and μ lifetimes:

$$\left(\frac{g_\tau}{g_\mu}\right)^2 = 0.9996 \cdot \frac{\tau_\mu}{\tau_\tau} \left(\frac{m_\mu}{m_\tau}\right)^5 \mathrm{B}(\tau^- \to e^- \bar{\nu}_e \nu_\tau) \quad . \tag{5}$$

Using the OPAL measurement of the τ lifetime [5], $\tau_\tau = 289.2 \pm 1.7 \pm 1.2$ fs, and the world average τ mass [6], $m_\tau = 1777.0 \pm 0.3$ MeV/c^2, gives the ratio $g_\tau/g_\mu = 1.0025 \pm 0.0047$, consistent with τ - μ universality.

A somewhat different test of τ - μ universality can be made by comparing the branching ratios for the semi-leptonic decays $K^- \to \mu \bar{\nu}_\mu$ and $\pi^- \to \mu \bar{\nu}_\mu$ with τ decays into hadronic final states containing a single charged kaon or pion. The decay constants which parameterize the hadronic matrix elements for these decays contain effects due to the strong interaction; however, these constants cancel if one takes the ratio of the semileptonic widths. The ratio of the coupling constants g_τ/g_μ is then obtained from

$$\left(\frac{g_\tau}{g_\mu}\right)^2 = \left\{\frac{2m_\mu^2}{m_\tau^2}\right\} \frac{\mathrm{B}(\tau \to h^- \nu_\tau)}{H_\pi + H_K} \tag{6}$$

where $h = \pi, K$, $\quad H_h = (1 + \delta_h)\left(\frac{\tau_\tau m_\tau}{\tau_h m_h}\right)\left[\frac{1 - (m_h/m_\tau)^2}{1 - (m_\mu/m_h)^2}\right] \mathrm{B}(h^- \to \mu^- \bar{\nu}_\mu)$ and δ_h are radiative corrections. A recent OPAL measurement [7] of the one-prong τ hadronic branching ratio, $\mathrm{B}(\tau^- \to h^- \nu_\tau) = (11.98 \pm 0.13 \pm 0.16)\%$, can be compared with the world-average π^- and K^- muonic branching ratios [6] to

504

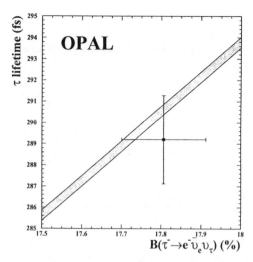

Figure 2: The OPAL τ lifetime is plotted against the $\tau^- \to e^- \bar{\nu}_e \nu_\tau$ branching ratio. The shaded band is the SM prediction under the assumption of lepton universality, and its width represents the uncertainty due to the measured τ mass.

obtain the result $g_\tau/g_\mu = 1.018 \pm 0.010$, which is consistent with universality at the level of $\sim 2\sigma$. If the world average result for $B(\tau^- \to h^- \nu_\tau)$ is used instead of the OPAL result, the agreement is somewhat better (see Fig. 3).

The results of various μ - e and τ - μ universality tests by different experiments are presented in Fig. 4. It is important to recognize that the different tests are complementary, since they are potentially sensitive to different types of new physics. The test using one-prong hadronic τ decays, for example, is only sensitive to the spin-0 part of the charged-current. Eq. 6 can therefore be used to limit possible contributions from the exchange of scalar bosons with Yukawa-like couplings. The OPAL result has been used to place limits on Yukawa couplings in R-parity violating supersymmetric models[7]. In contrast, the two universality tests using leptonic τ decays are also sensitive to transverse W couplings. Combining measurements from CLEO and the four LEP experiments[10] gives a result which is about a factor of two more precise than the individual OPAL measurements: $g_\mu/g_e = 1.0014 \pm 0.0024$ and $g_\tau/g_\mu = 1.0002 \pm 0.0025$. It is worth noting that at this level of precision these tests are sensitive to the electroweak radiative corrections and W boson propagator corrections in Eq. 3, which contribute at the level of $\sim 0.4\%$. A non-zero ν_τ mass would introduce an additional phase-space correction to Eq. 3 which would lead to an apparent violation of universality. The current best limit[6]

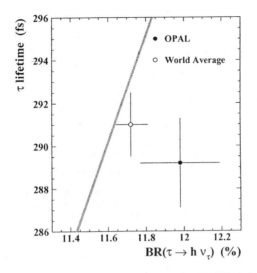

Figure 3: The OPAL τ lifetime is plotted against the OPAL (filled circle) and world average (open circle) B($\tau \to h^-\nu_\tau$) branching ratio. The shaded band is the SM prediction under the assumption of lepton universality.

on the τ neutrino mass ($m_{\nu_\tau} < 18.2$ MeV) is from end-point measurements of the decay spectra for the decays $\tau \to 3\pi^\pm\nu_\tau$ and $\tau^- \to 5\pi^\pm\nu_\tau$. A τ neutrino with this mass would introduce corrections that would contribute at the level of $\sim 0.08\%$, somewhat below the present sensitivity of the universality tests. However, this method does provide a cross-check on the limits obtained from end-point measurements, which are notoriously difficult and sensitive to pathological effects due to background or detector modelling.

Universality constraints from measurements of the W decay leptonic partial widths from W-pair production at LEP2 currently are about a factor of ten worse than those from τ decays[10]. Universality tests based on $\pi^- \to \mu^-\bar{\nu}_\mu$ and $\pi^- \to e^-\bar{\nu}_e$ branching ratios[6] are more precise than those from τ decays, but are only sensitive to longitudinal W couplings.

4 The Strong Coupling Constant

One could in principle test the universality of the W boson couplings to (u, d_θ) compared with (e, ν_e) or (μ, ν_μ). Naïvely, a comparison of the τ hadronic and leptonic partial widths gives the result of Eq. 2; however, QCD corrections to the τ hadronic width introduce a $\sim 20\%$ deviation from this prediction. The

506

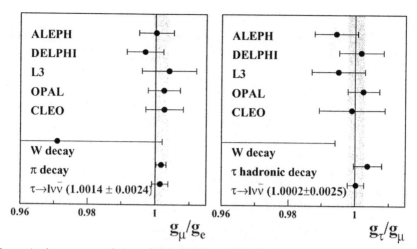

Figure 4: A summary of μ - e (left) and τ - μ (right) universality results. The first five entries in each plot reflect independent measurements by different experiments, while the bottom three represent world-average results. The shaded band represents the uncertainty on the world-average result from leptonic τ decays.

measured discrepancy from the naïve prediction can be used to determine the strong coupling constant α_s at the τ mass scale. The ratio R_τ can be written in terms of the corrections to the naïve prediction:

$$R_\tau = 3\left(|V_{ud}|^2 + |V_{us}|^2\right) S_{EW} \{1 + \delta_{EW} + \delta_{pert} + \delta_{np}\} \qquad (7)$$

where $S_{EW} = 1.0194$ and $\delta_{EW} = 0.0010$ are electroweak corrections. The perturbative QCD correction, δ_{pert}, has been calculated to $\mathcal{O}(\alpha_s{}^3)$ and is partially know to $\mathcal{O}(\alpha_s{}^4)$:

$$\delta_{pert} = \left(\frac{\alpha_s}{\pi}\right) + 5.2023 \left(\frac{\alpha_s}{\pi}\right)^2 + 26.366 \left(\frac{\alpha_s}{\pi}\right)^3 + (78.00 + K_4)\left(\frac{\alpha_s}{\pi}\right)^4 \quad . \quad (8)$$

Non-perturbative and quark mass corrections, δ_{np}, are small and have been estimated to total $\delta_{np} = -0.020 \pm 0.005$. A measurement of R_τ therefore has considerable sensitivity to the value of $\alpha_s(m_\tau^2)$.

Using the completeness of the τ width and assuming μ - e universality, R_τ can be obtained from the OPAL measurement of $B(\tau^- \to e^- \bar{\nu}_e \nu_\tau)$,

$$R_\tau = \frac{1 - B(\tau^- \to e^- \bar{\nu}_e \nu_\tau) \cdot (1.9726)}{B(\tau^- \to e^- \bar{\nu}_e \nu_\tau)} \quad , \qquad (9)$$

yielding a result of $R_\tau = 3.642 \pm 0.033$ (exp) [3]. Extracting $\alpha_s(m_\tau^2)$ from this value gives $\alpha_s(m_\tau^2) = 0.334 \pm 0.010$ (exp) ± 0.016 (theory). This result can then be evolved to the Z^0 mass scale using renomalization group evolution to give $\alpha_s(m_Z^2) = 0.1204 \pm 0.0011$ (exp) ± 0.0019 (theory), which is comparable in precision to the best measurements of $\alpha_s(m_Z^2)$ by any method currently available. It should be noted that fits to the τ hadronic spectral functions yield comparable results[6].

5 Michel parameters

The assumption of a $V - A$ form for the charged-current weak interaction can be tested using the τ decay Michel parameters. If one postulates that the decay $\tau^- \to l^- \bar{\nu}_l \nu_\tau$ proceeds via the most general derivative free, lepton number conserving, four-lepton interaction consistent with locality and Lorentz invariance, the matrix element can be written

$$\mathcal{M} = 4 \frac{G_{\tau l}}{\sqrt{2}} \sum_{\gamma, \epsilon, \omega} g_{\epsilon\omega}^\gamma \langle \bar{l}_\epsilon | \Gamma^\gamma | \nu_l \rangle \langle \bar{\nu}_\tau | \Gamma_\gamma | \tau_\omega \rangle \tag{10}$$

where $\epsilon, \omega = L, R$ label the chiralities of the τ and the lepton l, and $g_{\epsilon\omega}^\gamma$ parameterize the scalar ($\Gamma^S = I$), vector ($\Gamma^V = \gamma^\mu$) and tensor ($\Gamma^T = \sigma^{\mu\nu}/\sqrt{2} \equiv i(\gamma^\mu\gamma^\nu - \gamma^\nu\gamma^\mu)/2\sqrt{2}$) contributions. The coupling g_{LL}^V is defined to be real and positive while the remaining $g_{\epsilon\omega}^\gamma$ are complex with magnitudes limited by a normalization condition imposed on the set of couplings. In the case that $g_{LL}^V = 1$ and all other $g_{\epsilon\omega}^\gamma$ vanish, the $V - A$ form of the SM charged-current interaction is obtained and the constant $G_{\tau l}$ reduces to the Fermi constant G_F.

From an experimental perspective, it is convenient to parameterize the τ decay spectra in the rest frame of the τ:

$$\frac{d^2\Gamma_{\tau \to l}}{d\Omega dx} = \frac{G_{\tau l}^2 \, m_\tau^5}{192 \, \pi^4} \, x^2 \left\{ 3(1-x) + \rho_l \left(\frac{8}{3}x - 2 \right) + 6 \, \eta_l \, \frac{m_l}{m_\tau} \frac{(1-x)}{x} \right.$$

$$\left. -P_\tau \, \xi_l \cos\theta \left[(1-x) + \delta_l \left(\frac{8}{3}x - 2 \right) \right] \right\} \tag{11}$$

where $x \equiv E_l/E_l^{\max}$ is the scaled lepton energy ($E_l^{\max} = \frac{m_\tau^2 + m_l^2}{2m_\tau}$), P_τ is the average τ polarization and θ is the angle between the τ spin direction and the momentum vector of the lepton l. The four Michel parameters ρ, η, ξ and δ are linear combinations of the magnitude-squared of the couplings. In the SM limit of only left-handed vector couplings these parameters have the values $\rho = 3/4, \eta = 0, \xi = 1, \xi\delta = 3/4$. Experimental values of these parameters can be obtained by fitting the $\tau^- \to e^- \bar{\nu}_e \nu_\tau$ and $\tau^- \to \mu^- \bar{\nu}_\mu \nu_\tau$ decay spectra. The left-hand plot of Fig. 5 shows the result of the OPAL fit[8] for $\tau^- \to \mu^- \bar{\nu}_\mu \nu_\tau$

508

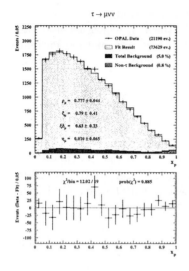
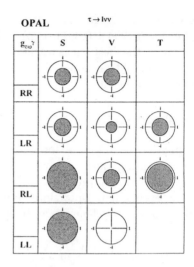

Figure 5: The decay spectra obtained for $\tau^- \to \mu^- \bar{\nu}_\mu \nu_\tau$ decays and the resulting fit values of the Michel parameters (left). 90% confidence limits on the couplings $g^\gamma_{\epsilon\omega}/g^\gamma_{\max}$ ($g^S_{\max} = 2$, $g^V_{\max} = 1$, $g^V_{\max} = 1/\sqrt{3}$) obtained from the combined $\tau^- \to e^- \bar{\nu}_e \nu_\tau$ and $\tau^- \to \mu^- \bar{\nu}_\mu \nu_\tau$ Michel parameter fits (right). The unshaded regions are excluded by this measurement.

decays. A similar fit is obtained for $\tau^- \to e^- \bar{\nu}_e \nu_\tau$ decays, permitting the two results to be combined to give a single set of τ-decay Michel parameters.

Limits on the values of the Michel parameters can then be used to constrain the couplings $g^\gamma_{\epsilon\omega}$ as shown in Fig. 5 (right). Many extensions to the SM predict forms of new physics that introduce additional couplings which would modify the observed τ decay spectra. For example, the charged Higgs bosons predicted by two-Higgs-doublet models and SUSY would introduce a helicity-blind scalar coupling proportional to the lepton masses, $g^S = -m_\tau m_l \left(\tan\beta/m_{H^\pm}\right)^2$, leading to non-SM values of the η and ξ Michel parameters:

$$\eta_l = -\frac{g^S/2}{1 + (g^S/2)^2} \quad , \quad \xi_l = \frac{1 - (g^S/2)^2}{1 + (g^S/2)^2} \quad . \tag{12}$$

The OPAL results can be interpreted as a limit on MSSM charged Higgs bosons of $m_{H^\pm} > (0.97 \text{ GeV}) \times \tan\beta$ (95% C.L.). It is interesting to note that the η parameter introduces a correction to the leptonic decay widths proportional to the ratio of the lepton masses, $\Gamma_l(\eta_l) = \Gamma_l^{(SM)} \left(1 + 4\,\eta_l\,\frac{m_l}{m_\tau} + ...\right)$, thus giving the universality test of Eq. 4 sensitivity to charged Higgs bosons.

As a second example, L - R symmetric models postulate the existence of a heavy right-handed W boson, W_R, which mixes with its left-handed counterpart, W_L, with angle ζ to produce mass eigenstates $W_1 \equiv W^\pm$, and W_2.

These models imply a non-zero g_{RR}^V coupling which causes the ξ and ρ Michel parameters to deviate from their SM values:

$$\xi = \cos^2 \zeta \left(1 - \tan^2 \zeta\right) \frac{1 - (m_{W_1}/m_{W_2})^4}{1 + (m_{W_1}/m_{W_2})^4} \tag{13}$$

$$\rho = \frac{3}{4} \cos^4 \zeta \left(1 + \tan^4 \zeta + \frac{4(m_{W_1}/m_{W_2})^2}{1 + (m_{W_1}/m_{W_2})^4} \tan^2 \zeta\right) . \tag{14}$$

The OPAL Michel parameter results imply the limits $m_{W_2} > 137$ GeV, $|\zeta| < 0.12$ (95% C.L.) on the mass and mixing angle in models of this type. Since the η parameter is not modified by the g_{RR}^V coupling, the universality tests have no sensitivity to right-handed W bosons in these models.

6 Conclusion

The large inclusive samples of τ decays recorded at LEP1 have enabled precise determinations to be made of the τ lifetime, the principal τ branching ratios and the τ decay spectra. These results can be used to test both the universality and the Lorentz structure of the weak charged-current interaction, providing insight into the weak interaction and physics beyond the Standard Model.

References

1. Y.S. Tsai, Phys. Rev. **D4**, 2821, (1971).
2. W.J. Marciano and A. Sirlin, Phys. Rev. Lett. **61**, 1815, (1988).
3. OPAL, G. Abbiendi *et. al.*, A Measurement of the $\tau^- \to e^- \bar{\nu}_e \nu_\tau$ Branching Ratio, CERN-EP/98-175, (submitted to Phys. Lett. B.) and references therein.
4. J.M. Roney, Recent Tau Decay Results from OPAL, Proceedings of the International Europhysics Conference on High Energy Physics, Jerusalem, Israel 19-26 August 1997, to be published.
5. OPAL, G. Alexander *et. al.*, Phys. Lett. **B374**, 341, (1996).
6. Particle Data Group, C. Caso *et. al.*, Eur. Phys. J. **C3**, 1, (1998).
7. OPAL, K. Ackerstaff *et. al.*, Eur. Phys. J. **C4**, 193, (1998).
8. OPAL, K. Ackerstaff *et. al.*, Measurement of the Michel Parameters in Leptonic Tau Decays, CERN-EP/98-104, (submitted to Eur. Phys. J.).
9. OPAL, K. Ackerstaff *et. al.*, Eur. Phys. J. **C5**, 229, (1998).
10. B. Stugu, Summary on Tau Leptonic Branching Ratios and Universality, Proceedings of the Fifth International Workshop on Tau Lepton Physics (TAU98) Santander, Spain, Sept 14 - 17 1998 (to appear in Nucl. Phys. B Proc. Suppl.s) and references therein.

EXPERIMENT E614 AT TRIUMF:
A CLOSE LOOK AT MUON DECAY

N.L. RODNING,[4] P. AMAUDRUZ,[3] W. ANDERSSON,[3] M. COMYN,[3]
Yu. DAVYDOV,[1,3] P. DEPOMMIER,[6] J. DOORNBOS,[3] W. FASZER,[3]
C.A. GAGLIARDI,[2] D.R. GILL,[3] P. GREEN,[4] P. GUMPLINGER,[3]
J.C. HARDY,[2] M. HASINOFF,[5] R. HELMER,[3] R. HENDERSON,[3]
A. KHRUCHINSKY,[1] P. KITCHING,[4] D.D. KOETKE,[10] E. KORKMAZ,[7]
Y. LACHIN,[1] D. MAAS,[3] J.A. MACDONALD,[3] R. MANWEILER,[10]
T. MATHIE,[8] J.R. MUSSER,[2] P. NORD,[10] A. OLIN,[3] D. OTTEWELL,[3]
R. OPENSHAW,[3] L. PIILONEN,[11] T. PORCELLI,[7] J-M. POUTISSOU,[3]
R. POUTISSOU,[3] M.A. QURAAN,[4] J. SCHAAPMAN,[4] V. SELIVANOV,[1]
G. SHEFFER,[3] B. SHIN,[8] F. SOBRATEE,[4] J. SOUKUP,[4]
T.D.S. STANISLAUS,[10] G. STINSON,[4] R. TACIK,[8] V. TOROKHOV,[1]
R.E. TRIBBLE,[2] M.A. VASILIEVv,[2] H-C. WALTER,[3] D. WRIGHT[3]

1) RRC "Kurchatov Institute", Moscow, Russia.

2) Texas A&M University, College Station, Texas, USA.

3) TRIUMF, Vancouver, British Columbia, Canada.

4) University of Alberta, Edmonton, Alberta, Canada.

5) University of British Columbia, Vancouver, BC, Canada.

6) University of Montreal, Montreal, Quebec, Canada.

7) University of Northern British Columbia, Prince George, BC, Canada.

8) University of Regina, Regina, Saskatchewan, Canada.

9) University of Saskatchewan, Saskatoon, Saskatchewan, Canada.

10) Valparaiso University, Valparaiso, Indiana, USA.

11) Virginia Polytechnic Institute and State University,
Blacksburg, Virginia, USA.

Muon decay, a process which involves only the weak interaction, is precisely described within the Standard Model. Therefore, a precision measurement of the decay distribution will provide an unambiguous test of physics beyond the Standard Model. Experiment E614, an experiment under preparation at TRIUMF, will determine the energy and angular distributions of positrons from μ^+ decay to a precision of one part in 10^4. After discussing the general framework for muon decay, the experiment and its sensitivity to new physics are described.

1 Muon Decay - Formalism and Motivation

A general expression for the muon decay distribution can be written in terms of the weak interaction couplings. In the case of the Standard Model, only the vector coupling of left-handed to left-handed leptons is allowed. A more general expression, including scalar, vector, and tensor interactions can be written as

$$\frac{d^2\Gamma}{dx\,d(cos\theta)} = |\sum_{\substack{i=left,right \\ j=left,right \\ \gamma=Tensor,Vector,Scalar}} g_{ij}^{\gamma} < \overline{\psi_{e_i}}|\Gamma^{\gamma}|\psi_{\nu_e} >< \overline{\psi_{\nu_\mu}}|\Gamma_{\gamma}|\psi_{\mu_j} > |^2 ,$$

(1)

where the chiralities of the neutrinos are determined by selection rules on i, j, and γ. This expression allows for couplings of arbitrary combinations of left-handed and right-handed leptons. This expression can be greatly simplified by defining the Michel parameters [1] in terms of the coupling constants. The Michel parameters ρ, δ, ξ, and η are defined as follows:

$$\rho = \frac{3}{4} - \frac{3}{4}[|g_{LR}^V|^2 + |g_{RL}^V|^2 + 2|g_{LR}^T|^2 + 2|g_{RL}^T|^2$$

(2)

$$+ Re(g_{RL}^S g_{RL}^{T*} + g_{LR}^S g_{LR}^{T*})] ,$$

$$\xi\delta = \frac{3}{4} - \frac{3}{4}[|g_{LR}^V|^2 + |g_{RL}^V|^2 + 4|g_{LR}^T|^2 + 2|g_{RL}^T|^2 + 2|g_{RR}^V|^2$$

(3)

$$+ \frac{1}{2}|g_{RR}^S|^2 + \frac{1}{2}|g_{LR}^S|^2 + Re(g_{RL}^S g_{RL}^{T*} - g_{LR}^S g_{LR}^{T*})] ,$$

$$\xi = 1 - [\frac{1}{2}|g_{RR}^S|^2 + \frac{1}{2}|g_{LR}^S|^2 + 2|g_{RR}^V|^2 + 4|g_{RL}^V|^2 - 2|g_{LR}^V|^2$$

(4)

$$- 2|g_{LR}^T|^2 + 8|g_{RL}^T|^2 + 4Re(g_{RL}^S g_{RL}^{T*} - g_{LR}^S g_{LR}^{T*})] ,$$

$$\eta = \frac{1}{2}Re[g_{LL}^V g_{RR}^{S*} + g_{RL}^V(g_{LR}^{S*} + 6g_{LR}^{T*})$$

(5)

$$+ g_{LR}^V(g_{RL}^{S*} + 6g_{RL}^{T*}) + g_{RR}^V g_{LL}^{S*}] .$$

The notation is that of Fetscher and Gerber [2]. The subscripts RL refer to the coupling of a right-handed electron current to a left-handed muon current, LR to the coupling of a left-handed electron current to a right-handed muon current, and so on. In the limit where the electron mass and radiative corrections are neglected, the differential distribution for positrons from μ^+ decay may then be written in terms of these parameters,

$$\frac{d^2\Gamma}{x^2 dx\ d(cos\theta)} \propto 3(1-x) + \frac{2}{3}\rho(4x-3) \pm P_\mu \xi \cos\theta[(1-x) + \frac{2}{3}\delta(4x-3)]\ , \quad (6)$$

where θ is the angle between the muon polarization and the outgoing positron direction, $x = E_e/E_{max}$, and P_μ is the muon polarization. A fourth Michel parameter, η, contributes to the energy spectrum when the electron mass is included in the analysis.[†]

Muon Decay Distribution

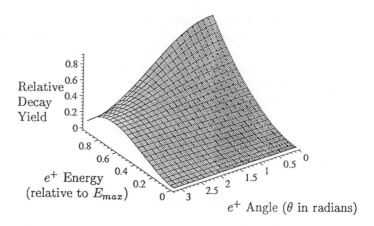

Figure 1: The distribution of e^+ from μ^+ decay as given by equation 6 .

In the Standard Model, only g_{LL}^V is non-zero. This is because the other couplings have been set to zero by fiat. In fact, the experimental case for setting these couplings to zero is not strong, as is shown in Table 1.

The measured values for the Michel parameters [2] are shown in Table 2 along with their Standard Model values. Measurements to date support the Standard Model to within the quoted uncertainties.

The decay distribution predicted by the Standard Model - as given by Eq. 6 - is shown graphically in Fig. 1. Deviations from the Standard Model values

---

[†]In the actual experiment, the data will be compared with a decay distribution which includes the effects of radiative corrections and finite electron mass.

	Standard Model	Current Limits	E614(A)	(B)	(C)	(D)
$\lvert g_{RR}^S \rvert$	0	<0.066	—	—	0.034	0.045
$\lvert g_{RR}^V \rvert$	0	<0.033	0.012	0.017	0.017	0.022
$\lvert g_{RR}^T \rvert$	0	$\equiv 0$				
$\lvert g_{LR}^S \rvert$	0	<0.125	—	—	0.034	0.046
$\lvert g_{LR}^V \rvert$	0	<0.060	0.012	0.015	0.015	0.018
$\lvert g_{LR}^T \rvert$	0	<0.036	—	0.010	—	0.013
$\lvert g_{RL}^S \rvert$	0	<0.424	—	—	—	—
$\lvert g_{RL}^V \rvert$	0	<0.110	0.012	0.012	0.012	—
$\lvert g_{RL}^T \rvert$	0	<0.122	—	0.008	—	—
$\lvert g_{LL}^S \rvert$	0	<0.55	—	—	—	—
$\lvert g_{LL}^V \rvert$	1	>0.96	>0.99977	>0.99942	—	—
$\lvert g_{LL}^T \rvert$	0	$\equiv 0$				

Table 1: Standard Model values and experimental upper limits (90% CL) for weak coupling constants with current limits taken from Fetscher and Gerber. Improved limits set by E614 based on our measurements of ρ, ξ, and δ assume: (A) V, A couplings only, (B) V, A and T couplings, (C) V, A and S couplings or (D) most general (V, A, S, and T) derivative-free couplings. For case (C), our measurement of η will also set a more stringent limit than above on the real (*CP*-conserving) part of g_{RR}^S.

for the Michel parameters would be manifest in deviations from this decay distribution.

The goal of E614 at TRIUMF is to determine $d^2\Gamma/dx\ dcos(\theta)$ with improved precision, allowing the determination of the Michel parameters ρ, δ, and ξ to within a few parts in 10^4 (η will also be determined, but with a precision of roughly a part in 10^3). This will test for physics beyond the Standard Model. For example, these extra couplings may give evidence for weak gauge bosons which have not been included in the Standard Model.

514

	Current Experimental Value	Standard Model Value
ρ	0.7518 ± 0.0026	$\frac{3}{4}$
δ	$0.7486 \pm 0.0026 \pm 0.0028$	$\frac{3}{4}$
$P_\mu\xi$	$1.0027 \pm 0.0079 \pm 0.0030$	1
η	-0.007 ± 0.0013	0

Table 2: The accepted values of the Michel parameters [2], along with the Standard Model values.

2 The Experiment

E614 at TRIUMF will be the first experiment in which the Michel parameters will be extracted from a simultaneous measurement of the entire muon decay distribution. This will be accomplished by stopping polarized μ^+ at the center of a highly symmetric detector, sitting in a nearly uniform 2T solenoidal magnetic field. Data will be taken at a high rate, allowing the study of 10^9 μ^+ decays in approximately two weeks.

2.1 The μ^+ Beam

The muons will be selected from a sample of pions decaying at rest, in which case the muons have a well defined helicity within the Standard Model. Beamline M13 [3] at TRIUMF will be used to select muons which originate from pions which stop very near the surface (within roughly 15 microns of the surface for a channel resolution $\frac{\Delta p}{p} \approx \frac{1}{2}\%$) of the pion production target. In this case, the muons lose little momentum in emerging from the target. Similarly, these "surface muons" undergo limited multiple scattering, and consequently lose very little polarization. The selected muons are expected to have a polarization of 0.9999.

The beam at TRIUMF is pulsed with a period of 43 ns. Most of the beam-related background is correlated in time with this accelerator time structure. The surface muons, however, are delayed in time by the π decay distribution (mean lifetime of 26 ns). A timing gate is used to select surface muons while discriminating against much of the beam-related background. The relatively long period of the TRIUMF beam is ideal for this experiment.

2.2 The Spectrometer

The spectrometer is required to obtain a precise measurement of the energy and angle of the decay positron. This is challenging, in part due to the limited energy (4.1 MeV), and consequently the limited range (≈ 1 mm in CH_2), of the incident muons. This constrains the amount of wire chamber and trigger material that can be placed in the path of the incoming muons but which are required for the tracking of the positrons.

The spectrometer, shown in Fig. 2, is highly symmetric. The wire positions must be accurately determined in three dimensions. The wire positions with respect to the two coordinates transverse to the beam can be accurately verified by fitting straight tracks in the absence of a magnetic field. The coordinate along the beam is difficult to verify. For this reason, precise assembly techniques involving optical interferometry are used to verify the position of the wire planes in the dimension along the beam.

3 Sensitivity to new physics

The goal of the experiment is to test the values of the Michel parameters for consistency with the Standard Model at a level of 1 part in 10^4. For example, the present limits on δ are approximately 50 parts in 10^4 and are shown in Fig. 3 as the extreme curves on either side of the central curve corresponding to the Standard Model value. The E614 sensitivity to variations in δ would lie within the width of the central line. Therefore, if E614 were to measure a distribution falling between the existing limits and the E614 sensitivity, this would be indicative of a failure of the Standard Model. There is no evidence either for or against such a signal at this time.

More fundamentally, the experiment will provide constraints on possible non-zero values for weak couplings other than g_{LL}^V. In this case, the sensitivity to new physics is somewhat model dependent since specific models - such as models exhibiting manifest symmetry between the left-handed and right-handed CKM matrices - are chosen to reduce the number of free parameters. Expected limits on the coupling constants under various assumptions are given in Table 1.

To obtain new upper limits on the couplings implied by the precision in the Michel parameters expected from E614, the magnitudes of eight coupling constants were varied independently from zero up to the current limit for each

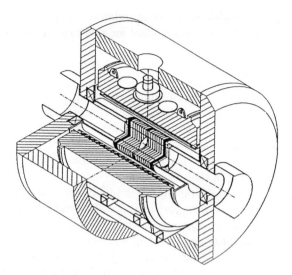

Figure 2: The E614 Spectrometer. The precision circular planar wire chambers with an active area ∼ 40 cm in diameter are shown at the center of the 2T superconducting solenoid. The scale of the drawing is set by the bore of the solenoid, which is ∼ 1m in diameter. Surrounding the magnet is an early concept drawing for the return yoke.

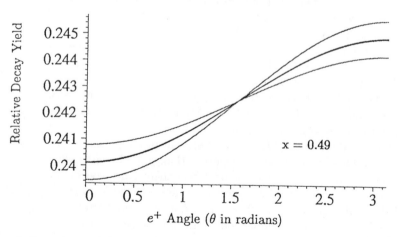

Figure 3: Dependance of the decay yield upon δ. The outer bands indicate the uncertainty in the decay yield due to the current experimental limits on δ (see equation 6 and Table 2). The distribution given by the Standard Model is at the center of this band, bounded by the sensitivity of E614. These limits are narrow enough to be difficult to discern on this plot, so that the three curves merge into a thick band. The distribution is evaluated at x = 0.49.

coupling, and the two relative phases between tensor and scalar couplings were varied from 0 to π. For each combination of couplings and phases which satisfied the constraints imposed by the sensitivity of E614, maximum values of the couplings were saved. These values are shown in column D of Table 1. The second column shows the current limits as quoted in ref. 2.

In the most general case D), limits on all of the RR and LR couplings are improved over current values but the RL and LL limits are unchanged. However, since very few extensions of the Standard Model include both scalar and tensor couplings, three additional cases were studied: A) only vector and axial vector couplings are allowed, B) all but scalar couplings are allowed, and C) all but tensor couplings are allowed. In all three cases, the analysis is greatly simplified because there are fewer couplings to vary and because the interference terms between scalar and tensor couplings in ρ, ξ, and δ disappear, allowing one to take advantage of the positive-definite nature of many of the terms.

Case A is a minimal extension of the Standard Model, in which only three independent vector coupling constants are needed. In this case the limit on the total deviation from V-A is quite stringent, as seen by the deduced lower limit on g_{LL}^V. This leads to significant new limits on the existence of right-handed currents in left-right symmetric theories.

Assuming that the right-handed neutrinos are light, non-zero g_{RR}^V is dependent upon the mass of the right-handed vector boson W_R, while non-zero g_{RL}^V and g_{LR}^V indicate mixing between the intermediate vector bosons W_L and W_R. A measurement of the parameters ξ and ρ will therefore set limits on both the left-right mixing angle ζ and the right-handed boson mass M_R. E614 will actually measure the quantity $P_\mu \xi$, and not ξ and P_μ separately. The value of P_μ depends on assumptions regarding the structure of the right-handed CKM matrix in left-right symmetric theories. Therefore, to be conservative one may take the limit for ξ to be the measured limit for $P_\mu \xi$. In that case ξ and ρ define contours in mixing angle - mass space given by Herczeg [4]:

$$\frac{1 - \xi}{2} = \zeta^2 + \frac{M_L^4}{M_R^4} \, , \tag{7}$$

$$\rho = \frac{3}{4}(1 - 2\zeta^2) \, . \tag{8}$$

These two expressions assume that $g_R = g_L$. More general expressions that remove this assumption may be found in Herczeg [4]. Note that left-right symmetric theories require $g_R > 0.55 g_L$ to be consistent with existing experimental results [5].

The experimentally allowed region at 95% CL is shown in Fig. 4 under the assumption that the left- and right-handed CKM matrices are similar, indicating that E614 is sensitive to new particles with masses up to 819 GeV/c^2 and W_L–W_R mixing angles as small as 0.01. Fig. 4 also shows the existing limits based on the muon decay experiments of Peoples [6,7] and Strovink [8], and the limit expected when the analysis of a new ρ measurement at Los Alamos is completed [9], as well as the limit from the direct search from D0 [10]. If no assumptions are made about the form of the right-handed CKM matrix, E614 is sensitive to new particles with masses up to 740 GeV/c^2 and mixing angles as small as 0.012. In this case, the limits from D0 and Strovink are much weaker than shown in Fig. 4.

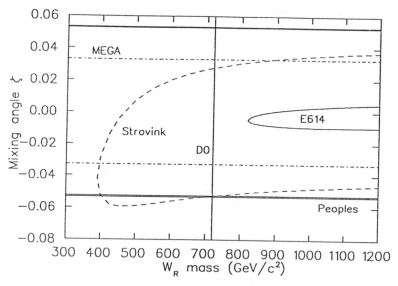

Figure 4: Current and proposed 95% CL limits on W_R mass vs. mixing angle. Solid vertical line: W_R mass limit from D0[10] assuming $g_R = g_L$ and $V^R = V^L$. Solid curve: allowed region from E614 under the same assumptions. Dashed curve: allowed region from Strovink et al.[8] Dot-dashed lines: expected limits on mixing angle for M_R infinite, from MEGA.[9] Double solid lines: mixing angle limits from Peoples et al.[6,7]

These limits can be put into context by comparing them to other experimental results. With the exception of muon decay studies, essentially all experimental tests of left-right symmetric theories are sensitive to the form assumed for the right-handed CKM matrix. For example, the well-known limit that $M_R > 1.6$ TeV, based on the K_L–K_S mass difference [11], only holds for *mani-*

fest left-right symmetry, in which the left- and right-handed CKM matrices are assumed to be identical. By contrast, Langacker and Sankar [12] critically reviewed all the existing experimental limits as of 1989. They found that classes of models with either $V_{ud}^R \approx 1$ or $V_{us}^R \approx 1$ are consistent with M_R as light as 300 GeV, without "fine-tuning" the parameters. They also determined that the current muon decay results provided the most stringent limits on M_R and ζ for a broad class of left-right symmetric models, especially those in which $V_{us}^R \approx 1$. E614 will significantly extend the limits in these cases.

4 Time-line for the Experiment

E614 was approved with high priority by the TRIUMF Experiments Evaluation Committee in July 1996, and was given final funding approval in April 1998. The pre-production prototype of the precision wire chamber package is under construction (Spring of 1999), with test beam scheduled for the summer of 1999.

Final assembly is expected to begin in the fall of 1999, with the experiment "on the floor" in January, 2001.

1. L. Michel, Proc. Phys. Soc. **A63**, 514, 1371, (1950).
2. W. Fetscher and H.-J. Gerber, Phys. Rev. **D54**, 251, (1996).
3. C.J. Oram *et al.*, NIM **179**, 95, (1981).
4. P. Herczeg, Phys. Rev. **D34**, 3449, (1986).
5. M. Cvetic, P. Langacker and B. Kayser, Phys. Rev. Lett. **68**, 2871, (1992).
6. J. Peoples, Columbia University Report No. NEVIS-147 (1966), unpublished.
7. S.E. Derenzo, Phys. Rev. **181**, 1854, (1969).
8. A. Jodidio *et al.*, Phys. Rev. **D37**, 237, (1988); Phys. Rev. **D34**, 1967, (1986).
9. MEGA collaboration, LAMPF Experiment 1240, M.D. Cooper, R.E. Mischke and L.E. Piilonen co-spokesmen.
10. S. Abachi *et al.*, (D0 collaboration), Phys. Rev. Lett. **76**, 3271, (1996).
11. G. Beall, M. Bander and A. Soni, Phys. Rev. Lett. **48**, 848 (1982).
12. P. Langacker and S.U. Sankar, Phys. Rev. **D40**, 1569, (1989).

DOUBLE AND SINGLE W BOSON PRODUCTION

SASCHA SCHMIDT-KÄRST

III. Physikalisches Institut, RWTH Aachen
Physikzentrum, 52056 Aachen, Germany
E-mail: S.Schmidt-Kaerst@cern.ch

W boson production is studied with the L3 detector at the e^+e^- collider LEP. W-pair events, $e^+e^- \to W^+W^- \to ffff$, are selected as four-fermion events with pairs of hadronic jets or leptons with high invariant masses. Branching fractions of W decays into fermion-antifermion pairs are determined. Combining all final states, the total cross section for W-pair production is measured as a function of the centre-of-mass energy from the threshold at 161 GeV up to 189 GeV. In the production of single W bosons, $e^+e^- \to e\nu_e W$, the electron is scattered preferentially into the very forward direction and is therefore not visible in the detector. The signal consists of the decay products of the W boson, either a single energetic lepton or two hadronic jets. The cross section is measured from 130 GeV to 189 GeV centre-of-mass energies.

1 W pair production

1.1 Introduction

Since 1996, the e^+e^--collider LEP is operated at centre-of-mass energies, $\sqrt{s}$, from 161 GeV up to 189 GeV. These energies are above the threshold of W^+W^- production which at the Born level is described by three Feynman diagrams: the s-channel γ/Z exchange and the t-channel ν_e exchange[1]. Depending on the final state of the decaying W^+W^- pair, many more diagrams may contribute to the same 4-fermion final state, in particular in the decay mode $e\nu_e f\bar{f}'$. Therefore, the measured cross section is corrected by a factor determined from Monte Carlo comparisons to correspond to the W^+W^- production cross section. In the following section, the analysis of the three different W^+W^- decay channels, are reported: the leptonic channel $\ell\nu\ell\nu$, (branching fraction 10.6%), the semileptonic channel $qq\ell\nu$, (43.8%), and the fully hadronic channel $qqqq$, (45.6%), at energies above 161 GeV[2] including preliminary results at 189 GeV[3].

1.2 Analysis

The leptonic final state, $\ell\nu\ell\nu$, includes up to 112 Feynman diagrams. The signal is characterized by two acoplanar leptons, either electron, muon or hadronic τ-jet, in any combination. Missing transverse momentum due to the neutrinos from the W decay is required. In total, selection efficiencies between

Table 1: Preliminary results from 189 GeV on the cross sections for the different decay channels of W^+W^- production. The first error is statistical and the second systematic.

Channel	σ [pb]	σ_{SM} [pb]
$e^+e^- \to \ell\nu\ell\nu$	$1.56 \pm 0.13 \pm 0.04$	1.77 ± 0.02
$e^+e^- \to qqe\nu$	$2.31 \pm 0.13 \pm 0.05$	2.43 ± 0.02
$e^+e^- \to qq\mu\nu$	$2.32 \pm 0.14 \pm 0.05$	2.43 ± 0.02
$e^+e^- \to qq\tau\nu$	$2.59 \pm 0.21 \pm 0.09$	2.43 ± 0.02
$e^+e^- \to qqqq$	$7.46 \pm 0.25 \pm 0.22$	7.59 ± 0.08
$e^+e^- \to W^+W^-$	$16.20 \pm 0.37 \pm 0.27$	16.64 ± 0.17

20% ($\tau\nu\tau\nu$) and 60% ($e\nu e\nu$) are reached while the background, mainly from 2-fermion production, Bhabha scattering and $\gamma\gamma$-interactions, is reduced to approximately 15%. At 189 GeV, 186 events of this class have been selected.

The experimental signature of the semileptonic decay channel, $qq\ell\nu$, consists of two hadronic jets accompanied by an isolated lepton (e, μ or hadronic τ-jet) and missing transverse momentum carried by the neutrino. The invariant masses of the two W decay products, M_{qq} and $M_{\ell\nu}$, are required to be in a window around the W mass (Fig. 1a). The signal is selected with efficiencies ranging between 50% ($qq\tau\nu$) and 80% ($qqe\nu$); the background from hadronic processes only accounts for 3% ($qq\mu\nu$) to 13% ($qq\tau\nu$). A total of 1035 semileptonic events are selected at 189 GeV.

A neural network approach is chosen to distinguish between the signal of hadronic W^+W^- decay and the QCD background from $e^+e^- \to q\bar{q}(\gamma)$ after preselecting high-multiplicity events with 4-jet topology and no missing energy. Eight input nodes are fed into the neural net including discriminating variables such as the jet resolution parameter Y_{34} and the minimal jet-jet angle. The cross section for the hadronic decay mode of W^+W^- production is then determined by a maximum likelihood fit to the neural network response.

The preliminary results from 189 GeV are summarized in Table 1. The cross sections are determined simultaneously in one maximum likelihood fit taking into account cross feed between the different channels. The total cross sections measured at all energies between 161 GeV and 189 GeV are shown in Fig. 1b compared to the theorectical expectation calculated with GENTLE[4]. Good agreement with the Standard Model including the ZWW vertex and the t-channel neutrino exchange is observed. The hadronic branching fraction of the W boson is determined to be $68.69 \pm 0.68 \pm 0.39\%$. This parameter depends on the element of the CKM mixing matrix V_{CKM}[5] not including the top quark, $BR(W \to q\bar{q})/[1 - BR(W \to q\bar{q})] = (1 + \alpha_s/\pi) \cdot V^2$. Thus, the presented measurements correspond to $V^2 = \sum_{i,j} |V_{ij}|^2 = 2.113 \pm 0.067 \pm 0.038$. Taking into account the current world-average values for all the matrix elements but

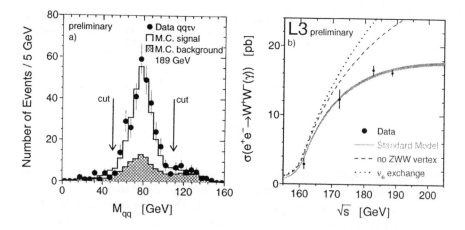

Figure 1: a) Invariant mass distribution of the two jets, M_{qq}, for the $qq\tau\nu$ channel. The mass is required to be in a window around the W mass. b) The cross section, σ_{WW}, of W-pair production as a function of the centre-of-mass energy, $\sqrt{s}$. The measurements including the preliminary result from 189 GeV are compared to the Standard Model expectation taking into account a 1% theoretical error.

V_{cs}, this element is determined to $|V_{cs}| = 1.032 \pm 0.037 \pm 0.018$.

2 Single W boson production

2.1 Introduction

The production of single W bosons, $e^+e^- \to e\nu_e W$, is dominated by contributions from the two diagrams shown in Fig. 2. This process provides one of the best experimental grounds for precision measurements of the electromagnetic couplings of the W boson since its cross section depends only on the gauge couplings κ_γ and λ_γ parametrizing the γWW vertex [6].

The characteristic feature of the single W production is the very low polar angle of the electron (positron) in the final state due to its small momentum transfer. Therefore, the electron remains inside the beam pipe and is not detected. The associated neutrino may carry a large transverse momentum, thus the experimental signature consists of large missing transverse energy and the decay products of the W boson, either a single energetic lepton or two hadronic jets. Due to this signature, the single W boson production constitutes a background to missing energy searches for new physics beyond the Standard Model.

The results from the L3 experiment from the measurement of single W

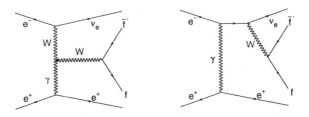

Figure 2: Feynman graphs for single W production, $e^+e^- \to e^+\nu_e W^-$.

2.2 Signal definition

boson production at energies from 130 GeV to 183 GeV [7] as well as preliminary results from the data collected at 189 GeV [8] in 1998 are reported.

In order to separate the single W production from other processes contributing to the same $e^+\nu_e f\bar{f}'$ final state, the signal is defined as

$$|\cos\theta_{e^+}| > 0.997$$
$$\min(E_f, E_{\bar{f}'}) > 15 \text{ GeV}$$
$$|\cos\theta_{e^-}| < 0.75 \quad \text{for } e^+\nu_e e^-\bar{\nu}_e \text{ only,}$$

where θ_{e^+} (θ_{e^-}) is the polar angle of the final state positron (electron), and E_f and $E_{\bar{f}'}$ are the fermion energies. The restriction of the polar angle θ_{e^+} separates the signal from the double resonant W^+W^- production. Requiring a minimum energy of the fermions supposed to be the W decay products removes non-resonant contributions from the signal. In the case of $e^+\nu_e e^-\bar{\nu}_e$, the additional cut on the polar angle θ_{e^-} is imposed in order to separate the signal from contributions where the $\nu_e\bar{\nu}_e$-pair originates from a Z-decay in $e^+e^- \to e^+e^-Z$. Among the events satisfying those requirements, the production of single W bosons dominates. Its fraction accounts for approximately 90% whereas the rest mainly comes from non-resonant contributions.

The Monte Carlo generators GRC4F [9] and EXCALIBUR [10] are used to simulate the production of single W bosons. The Standard Model expectation for this process within the phase space cuts defined above accounts for 0.53 pb calculated with GRC4F whereas EXCALIBUR predicts 0.60 pb for $\sqrt{s} = 189$ GeV.

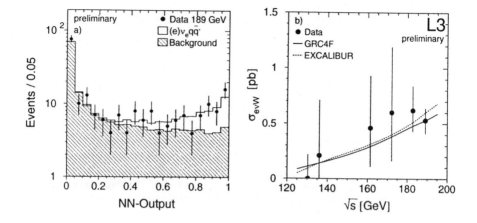

Figure 3: a) Output of the neural net for the hadronic decays of single W boson production, $e\nu_e q\bar{q}'$. Signal-like events have values close to 1. b) Measured total cross sections of the process $e^+e^- \to e\nu_e W$ as a function of the center-of-mass energy, $\sqrt{s}$. The solid and the dashed line show the predicitions of the two generators GRC4F and EXCALIBUR, respectively.

2.3 Analysis

The signal for the process of hadronically decaying single W bosons, $e^+e^- \to e\nu_e W \to e\nu_e q\bar{q}'$, has to be extracted from a large background of hadronic processes with significantly higher cross sections. The experimental signature consists of two acoplanar jets, no identified lepton and missing transverse momentum.

The main background source is the semileptonic decay of W^+W^- production, $e^+e^- \to \ell\nu q\bar{q}'$, if the lepton cannot be identified. Thus the most significant contribution comes from the channel $\tau\nu q\bar{q}'$ since the τ-jet might be difficult to isolate within the hadronic system of the two quarks.

The process $e^+e^- \to q\bar{q}(\gamma)$ constitutes a source of background, especially in the case of hard initial-state radiation when the emitted photon remains undetected within the beampipe.

At $\sqrt{s} = 189$ GeV centre-of-mass energy, the kinematic threshold for the production of two on-shell Z bosons is exceeded. Therefore, the process $e^+e^- \to ZZ \to \nu\bar{\nu}q\bar{q}$ is a considerable source of a nearly undistinguishable background. Due to the ZZ production near to the kinematic threshold, the Z bosons have low momenta. Thus, their contribution can be significantly reduced requiring a high velocity, β, of the hadronic system $\beta = P_{mis}/E_{vis} > 0.35$. In total, the signal is selected with an efficiency of 51% while improving the signal to background ratio to approximately 1:5. Then, a neural net is trained to separate the single W boson production from the most significant background from $e^+e^- \to W^+W^-$. This approach combines nine of the most discriminant variables which are fed into the neural net: global event shape variables as the velocity, β, and the visible mass M_{vis}; variables from the 2-jet topology of the event as the distance scale parameter Y_{23} for the transition from two to three jets; and variables calculated after forcing the event into three jets such as the stereo angle Ω defined by the direction of the jets. The output of the neural net is shown in Fig. 3a. The signal in the selected sample is significantly increased for values close to unity. The cross section for the process $e^+e^- \to e\nu_e W \to e\nu_e q\bar{q}'$ is determined by a binned maximum-likelihood to the neural network output, fixing the background to its Standard Model expectation.

The leptonic decays of singly produced W bosons, $e^+e^- \to e\nu\ell\nu$, is characterized by only one high-energy lepton within the detector, either electron, muon or τ-jet. Excluding other significant activity in the detector, the signal is selected with a high purity since background contributions with such a signature are rather small. The main background is coming from two fermion production, $e^+e^- \to \ell^+\ell^-(\gamma)$, in case the second lepton is not detected. Typically, the purity ranges between 50% ($W \to \tau\nu$) and 70% ($W \to \mu\nu$). Selection efficiencies from 25% in the single τ channel up to 70% for single electrons are reached. The cross section for the leptonic decay mode of single W production is determined by a binned maximum-likelihood fit to the energy spectrum of the single lepton candidates. Combining the results for the hadronic and leptonic channels yields the total cross section for single W boson production, shown in Fig. 3b as a function of the centre-of-mass energy, $\sqrt{s}$. The results are summarized in Table 2. The measurements are compared with the Stan-

Table 2: Measured total cross sections of the process $e^+e^- \to e\nu_e W$ for the different center-of-mass energies, $\sqrt{s}$. In the case of $\sqrt{s} = 133 - 136$ GeV, a zero signal cross section cannot be excluded at 68% C. L. Therefore, the negative error is omitted. The result from 189 GeV is preliminary.

$\sqrt{s}$ [GeV]	$\sigma_{e\nu W}$ [pb]	$\Delta\sigma_{stat}$ [pb]		$\Delta\sigma_{syst}$ [pb]
130	0.00	+0.22	–	± 0.02
136	0.21	+0.50	–	± 0.02
161	0.46	+0.47	−0.35	± 0.04
172	0.60	+0.59	−0.44	± 0.04
183	0.62	+0.22	−0.19	± 0.04
189	0.60	+0.12	−0.11	± 0.04

526

dard Model predictions calculated with the Monte Carlo generators GRC4F and EXCALIBUR, respectively. The disagreement between the theoretical predictions is the main source of the systematic error of approximately 5%.

3 Conclusion

Double and single W boson production is studied with the data collected by the L3 experiment at energies between 130 GeV and 189 GeV. Good agreement with the Standard Model expectations is observed. Until the year 2000, LEP will be operated at energies around 200 GeV. Expecting a total integrated luminosity of $500\,\mathrm{pb}^{-1}$ data per experiment at higher energies, W boson production will be measured with increased precision.

Acknowledgments

This work has been supported by the German Bundesministerium für Bildung und Forschung.

References

1. D. Bardin *et al.*, Nucl. Phys. **B37**, 148, (1994); F.A. Berends *et al.*, Nucl. Phys. **B37**, 163, (1994); W. Beenakker *et al.*, in *Physics at LEP 2*, Report CERN 96-01 (1996), eds G. Altarelli, T. Sjöstrand, F. Zwirner, Vol. 1, p.79; D. Bardin *et al.*, in *Physics at LEP 2*, Report CERN 96-01 (1996), Vol. 2, p. 3.
2. M. Acciari *et al.*, Phys. Lett. **B398**, 223, (1997); Phys. Lett. **B407**, 419, (1997); Phys. Lett. **B436**, 437, (1998).
3. L3 internal note 2376, contribution to the 1999 winter conferences.
4. GENTLE version 2.0 is used. D. Bardin *et al.*, CPC1041611997.
5. N. Cabibbo, Phys. Rev. Lett. **10**, 531, (1963); M. Kobayashi and T. Maskawa, Prog. Theor. Phys. **49**, 652, (1973).
6. T. Tsukamoto and Y. Kurihara, Phys. Lett. **B389**, 162, (1996).
7. M. Acciari *et al.*, Phys. Lett. **B403**, 168, (1997); Phys. Lett. **B436**, 417, (1997).
8. L3 internal note 2367, contribution to the 1999 winter conferences.
9. J. Fujimoto *et al.*, Comp. Phys. Comm. **100**, 128, (1997).
10. F.A. Berends, R. Pittau and R. Kleiss, Comp. Phys. Comm. **85**, 437, (1995).

BaBaR: STATUS AND PROSPECTS

A. SOFFER

Colorado State University, Ft.Collins, CO 80523, USA

BaBar Collaboration

1 Introduction

BaBar is an international collaboration of 652 physicists and engineers from 73 institutions in 9 countries. The purpose of the collaboration is to run a high statistics experiment at the $\Upsilon(4S)$ resonance. The BaBar detector will collect data produced in e^+e^- collisions at the PEP-II asymemtric B factory at the Stanford Linear Accelerator Center. At the design PEP-II luminosity, BaBar should accumulate 30 fb^{-1} per year, corresponding to 32×10^6 $\Upsilon(4S)$ decays.

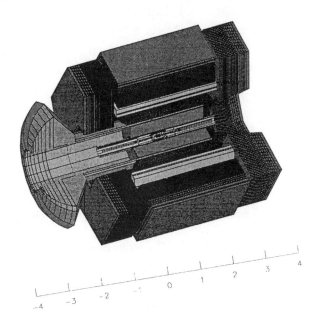

Figure 1: The BaBar detector.

528

2   BaBar Physics

The analysis which has the greatest impact on the parameters of the experiment is the measurement of $\sin(2\beta)$, where β is the phase of the CKM [1] element V_{td}. I will explain this measurement focusing on the "golden" mode $B^0 \to J/\psi K_S$. In this case, $\sin(2\beta)$ is obtained with negligible hadronic uncertainty from the time-dependent decay rate asymmetry [2]

$$a(t) = \frac{\Gamma(B^0(t) \to \psi K_S) - \Gamma(\bar{B}^0(t) \to \psi K_S)}{\Gamma(B^0(t) \to \psi K_S) + \Gamma(\bar{B}^0(t) \to \psi K_S)} = -\sin(2\beta)\sin\Delta mt. \quad (1)$$

Here $t = 0$ is the proper time at which the B meson was identified as a B^0 ($\bar{B}^0$), $B^0(t)$ ($\bar{B}^0(t)$) is the state of this meson at its proper decay time t, Δm is the mass difference between the two neutral B mass eigenstate, and Γ stands for a partial decay width. To obtain $a(t)$, one must thus determine the flavor of the B meson at some point in time, and the proper time that elapses from then until the particle's decay.

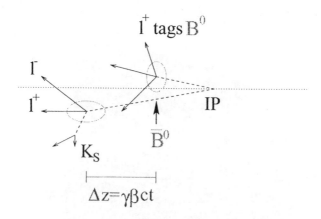

Figure 2: Schematic configuration of a $B^0 \to J/\psi K_S$ event.

In the center of mass (CM) frame, B mesons move at about 6% of the speed of light. Being slow, they decay on average within 30 μm of their production point. This is comparable to or smaller than the spatial resolution obtained with vertex detectors, making accurate time measurements in the CM frame very difficult. The asymmetric B factory is a collider that addresses this problem. PEP-II's electron beam has an energy of 9 GeV, while the positron beam's energy is 3.109 GeV. The asymmetric configuration boosts the B mesons along the z axis, increasing the average decay length to about 270 μm. This reduces

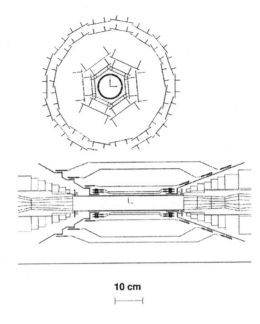

10 cm
|⸺⸺⸺⸺|

Figure 3: The silicon vertex detector.

the amount of data required to conduct the $\sin(2\beta)$ measurement by an estimated factor of five compared to a symmetric B factory[3].

The boost results in the event topology shown in Fig. 2. The two B mesons are producd at the interaction point (IP) and fly mostly along the z axis (the figure is not to scale). One of the mesons decays semileptonically, and the charge of the lepton tags its flavor, B^0 in this case. Since the mesons are created in a correlated state, the other meson must have the opposite flavor at that very point in time (ie., at approximately that z). The distance along z between the decay vertices of the two mesons is $\Delta z = \gamma\beta ct$. Hence, knowing the boost parameter $\gamma\beta$, one obtains the time t.

With the 30 fb^{-1} collected during one year of full luminosity running, BaBar should obtain an error of ~ 0.1 on $\sin(2\beta)$. This estimate was obtained by studying CP-eigenstates final states with a J/ψ, ψ', ϕ or η' and a neutral kaon (K_S, K_L, K^{*0}), and the states[2] $D^{(*)+}D^{(*)-}$.

High statistics will enable BaBar to do a lot of physics in addition to the measurement of $\sin(2\beta)$. BaBar will contribute, for example, to measurements of Δm and the B lifetimes. The magnitudes of CKM matrix elements are as important as the CKM phases for testing the standard model, and BaBar

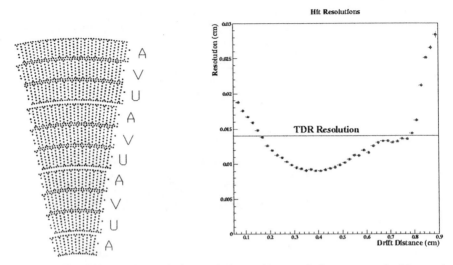

Figure 4: The drift chamber end-plate and the tracking resolution, measured with cosmic rays.

will conduct high statistics measurements of V_{cb} and V_{ub} in $B \to D^{(*)}l^+\nu$ and $B \to \pi^{(*)}l^+\nu$. Rare penguin B decays or decays involving the interference between penguin, electroweak penguin and tree amplitudes are likely to yield interesting physics, with possible limits on or an actual measurement of the CKM phase γ. In addition, $\sin(2\beta + \gamma)$ may be measured in $B^0 \to D^{*-}\pi^+$. Decays such as $B \to \pi\pi$ can in principle be used to measure $\sin(2\alpha)$, although CLEO results [4] suggest that prohibitively large luminosities will be needed to conduct this measurement. In addition, a wide selection of charm, τ and, $\gamma\gamma$ physics topics will be accessible at BaBar.

3   The BaBar Detector

A cutaway section of the BaBar detector [2] is shown in Fig. 1. From the inside out, the detector components, along with their acronyms, are the microstrip silicon vertex detector (SVT), the drift chamber (DCH), the Čerenkov particle identification system (DIRC), and the electromagnetic calorimeter (EMC). Outside the calorimeter is a superconducting coil which provides a magnetic field of 1.5 T. The magnet's instrumented flux return (IFR) region is the muon/neutral hadron detector.

The SVT, shown in Fig. 3, provides accurate spatial measurements of charged tracks close to the IP. This is necessary for the measurement of Δz and particle lifetimes, and helps separate signal from background in many

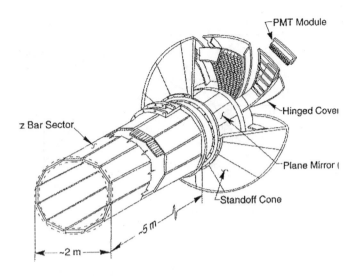

Figure 5: The DIRC.

channels. The SVT has five active layers with radii between 3.2 and 14.2 cm. The design intrinsic resolution in the inner three layers is $10 - 12$ μm. A resolution of $26 - 35$ μm was observed with cosmic ray data before alignment of the detector.

The DCH has forty layers arranged in ten axial and stereo super-layers, and occupies the volume between the radii 22.5 cm and 80 cm. It has been collecting cosmic data for several months, and has already obtained the 140 μm resolution claimed in the technical design report, as shown in Fig. 4.

The DIRC (Detection of Internally Reflected Čerenkov Light) consists of 144 quartz bars layed out around the barrel (Fig. 5). When a charged track passes through a bar (Fig. 6), it emits Čerenkov light which undergoes total internal reflection until it reaches the end of the bar. There the light enters an expansion volume (standoff box) filled with water, to match the quartz index of refraction. 11,000 photomultiplier tubes at the back of the standoff box record the positions of the photons, and the radius of the observed Čerenkov light ring is a measure of the velocity of the charged particle. Fig. 7 shows signal from a cosmic ray going through the bars.

A cutaway section through the upper half of the EMC is shown in Fig. 8. The barrel contains 5760 CsI crystals, and its inner (outer) radius is 90 cm (135.6 cm). An endcap section with 820 crystals occupies the forward region of the detector. The EMC will be calibrated at high and medium energies with

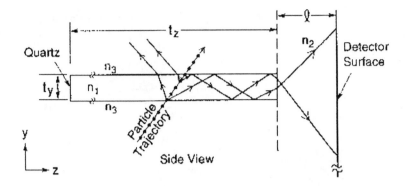

Figure 6: DIRC operation principle.

physics processes such as radiative Bhabhas. Low energy calibration and single crystal monitoring is done by circulating a liquid radioactive source along the inner face of the crystals. The source is activated by a neutron accelerator, and emits 6.13 MeV photons in the ^{16}N β-γ cascade, with a half life of 7 sec.

The IFR, shown in Fig. 9, consists of eighteen iron plates of varying thickness, instrumented with resistive plate counters (RPCs). The barrel section has 21 layers: one layer outside the EMC, one between the magnet cryostat and the innermost iron plate, seventeen between the iron plate gaps, and one outside the outermost plate. There are 18 layers in the endcaps. This fine longitudinal segmentation is designed to maximize the detection efficiency and the identification of muons and neutral hadrons. The total active area of the detector is greater than 1000 m^2.

4   Current Status and Near Future Plans

PEP-II commissioning milestones were met successfully and on schedule. The machine has so far achieved a peak instantaneous luminosity of 0.5×10^{33} cm^{-2} s^{-1}, about 17% of the design luminosity, 3×10^{33}cm^{-2}s^{-1}. Running conditions are good, but machine-produced background is higher than expected. We are attacking this problem by working to reduce the background using masks and machine operating parameters, and at the same time learning to reconstruct our events in the presence of high background.

Most of BaBar was assembled off the beampipe so as not to interfere with PEP-II commissioning. In March the assembled detector, excluding the SVT

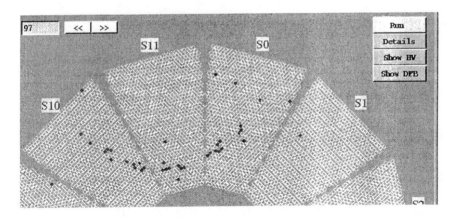

Figure 7: A Čerenkov ring left by a cosmic ray going through the DIRC.

and the DIRC quartz bars, was moved to its position in PEP-II. The SVT has since been installed in BaBar, and its cables are now being connected. 48 of the 144 DIRC quartz bars are due to arrive in time for installation before PEP-II turns on in May. Since quartz bar production has been slow, the rest of the bars are scheduled to be installed in September, so as not to delay the first run. During this May through August run, we expect to put the detector in good operational shape, and to learn to cope with machine-related background. Our hope is to start taking physics-quality data in October.

References

1. For a review, see the contribution by Andrzej Buras in these proceedings.
2. For details, see BaBar Collaboration (P.F. Harrison, H.E. Quinn, ed.), The BaBar Physics Book, 1999.
3. See the contribution by Andy Foland. in these proceedings.
4. CLEO Collaboration (R. Godang *et al*), Phys. Rev. Lett. **80**, 3456, 1998.

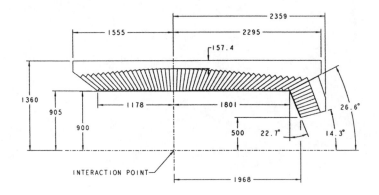

Figure 8: The electromagnetic calorimeter.

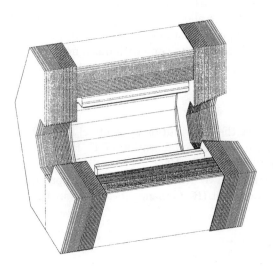

Figure 9: The instrumented flux return.

PHYSICS BEYOND THE STANDARD MODEL AT HERA

ARND E. SPECKA

Laboratoire de Physique Nuclaire des Hautes Energies
Ecole Polytechnique – IN2P3 – CNRS
91128 Palaiseau CEDEX, France
E-mail: specka@poly.in2p3.fr

Direct searches at HERA for leptoquarks, squarks in R-parity violating SUSY, lepton flavor violation, MSSM squarks and selectrons for masses up to 270 GeV are presented. No significant deviation from the Standard Model is observed. Rejection limits on the masses and couplings of new bosons are derived and compared to other experiments. Existing limits are significantly improved for low branching ratios. Indirect searches for new physics via contact interactions are discussed and show no significant deviation from the SM either.

1 Introduction

HERA is the first and only existing collider for electrons or positrons and protons. In its eight years of existence HERA has been running mainly with positrons at 27.5 GeV and protons at 820 GeV energy corresponding to a center-of-mass energy $\sqrt{s}$ of 300 GeV.

In its four years of running with positrons the two collider experiments, H1 and ZEUS, have accumulated data for an integrated luminosity of 37.0pb^{-1} and 46.6pb^{-1} respectively which is the data sample used in the analyses discussed in this paper.

Besides being a powerful "lepton-microscope" of the proton with a resolution of 10^{-18} m, HERA also offers the unique possibility to search for new bosons coupling to leptons and quarks. Theoretical motivations beyond the Standard Model for the existence of such bosons are numerous including Grand Unified Theories (GUTs), Supersymmetry (SUSY), and composite models.

The unique feature of HERA is the possibility of resonant s-channel production of such bosons with masses up to the kinematical limit of 300 GeV.

Searches for physics beyond the Standard Model at HERA have attracted increased attention in the past years when both H1 and ZEUS observed a significant deviation from the Standard Model expectation in the data collected from 94 to 96[1]. This observation has spawned many theoretical interpretations including leptoquarks and squarks (cf. in ref. [3]). However, the 97 data have not confirmed this excess, and thus the overall significance of the excess has decreased.

536

More recently, the observation of events with isolated leptons with high transverse momentum has renewed the interest in searches for new physics at HERA (see ref. [2] and Claude Vallée's paper in these proceedings).

2 Leptoquarks

Leptoquark (LQ) color triplet bosons coupling to quarks and leptons appear naturally in various unifying theories such as GUTs, Superstring inspired E_6 models, and in some compositeness and technicolor models.

At HERA, leptoquarks coupling to the first generation would be singly produced in the s-channel via the fusion of a positron and a quark (or antiquark), and subsequently decay into either a positron and a quark or a neutrino and a quark. Therefore, leptoquark events are individually indistinguishable from the dominant SM background, neutral current (NC) and charged current (CC) deep inelastic scattering (DIS) from t-channel virtual photon/Z^0 and W^+ exchange.

The kinematics of inclusive DIS are determined by two Lorentz invariants usually chosen among the square of the four-momentum transfer Q^2, and Bjorken variables x (momentum fraction of the incoming quark) and y (inelasticity). The mass M of the leptoquark and its positron decay angle in the center-of-mass system θ^* are related to the kinematic variables x and y by $M = \sqrt{sx}$ and $\cos\theta^* = 1 - 2y$.

Hence, leptoquarks are expected to show up as narrow resonances in x above the DIS background. Furthermore, scalar leptoquarks decay isotropically in their rest frame leading to a flat $d\sigma/dy$ spectrum[a], which is markedly different from the $d\sigma/dy \propto y^{-2}$ distribution of DIS.

Neutral current events (i.e. where a leptoquark decays into positron and quark) are selected at H1 by requiring an isolated positron with a transverse energy E_T of more than 15 GeV and a good balance in transverse momentum p_T. The kinematic variables are reconstructed from the positron and the DIS background is reduced further by a mass dependent y-cut: $y > y_c(M)$. The observed number of 312 events is in good agreement with the SM expectation of 306±23 events [3].

Alternately, charged current events (i.e. where a leptoquark decays into a neutrino and a quark) are observed requiring the absence of a positron with $E_T > 5$ GeV and a missing transverse energy in the detector bigger than 30 GeV. The kinematic variables are reconstructed from the hadronic final state. 210 CC events are observed which is in good agreement with the SM expectation of 194±29 events [3].

---

[a] Vector leptoquarks show an angular distribution that varies as $(1 - y)^2$.

With a different selection (E_T(detector)>60 GeV, positron with $E_T > 25$ GeV, jet with $p_T >10$ GeV) the ZEUS collaboration observes 68 events with an invariant mass of the positron and the jet above 200 GeV, in agreement with the SM expectation [4] of 43^{+14}_{-12}.

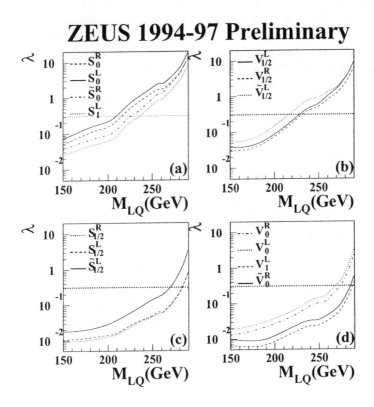

Figure 1: Exclusion limits for scalar (left) and vector (right) leptoquarks that couple to $e^+\bar{q}$ (upper) and to e^+q (lower) in the BRW model. The dotted horizontal line indicates a LQ coupling strength equal to the electro-magnetic coupling.

Since no significant deviation from the SM expectation is observed, one derives exclusion limits for leptoquark production as a function of M and λ the leptoquark coupling constant. Fig. 1 shows the 95% confidence level (CL) limits obtained by ZEUS for leptoquarks described in the phenomenological model of Büchmller-Rückl-Wyler (BRW) [5] which fixes the branching ratios β of the leptoquark decay modes.

Exclusion limits are stronger for leptoquarks with zero fermionic number

F than for those with $F = -2$ since the former couple mainly to the valence quarks whereas the latter couple to the anti-quarks of the sea only. For a coupling strength λ equal to the electro-magnetic coupling $\sqrt{4\pi\alpha_{\rm EM}}$, leptoquark masses bigger than 200 GeV (255 GeV) are excluded for $F = -2$ ($F = 0$) leptoquarks. Comparable limits have been determined by H1[3].

Moving away from the BRW model, H1 has also derived exclusion limits on generic leptoquark production for fixed values of λ as a function of M and the branching ratio β, as shown in Fig. 2 for leptoquarks coupling to positrons and the up quark[3]. The excluded domain extends beyond the rejection limits derived by the TeVatron experiment D0[6], especially at low branching ratios. For λ as small as 0.05 and $\beta = 0.1$, leptoquarks that couple to e^+u (e^+ d) are excluded up to masses of of 190 GeV (160 GeV).

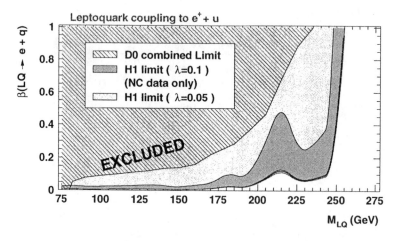

Figure 2: Exclusion limits for generic leptoquarks (arbitrary branching ratio β) for different values of the coupling λ (H1 preliminary data). The limits derived by D0 are given in comparison.

3 R-Parity Violating Supersymmetry

The Minimal Supersymmetric Standard Model (MSSM) assumes a discrete symmetry called R-parity with a conserved multiplicative quantum number $R_p \equiv (-1)^{3B+L+2S}$ which is equal to $+1$ for particles and -1 for their superpartners. If however R-parity is broken ($\not{R}_p$) the following lepton- and baryon-number violating Yukawa couplings are present in the superpotential[7]:

$$W_{\not{R}_p} = \lambda_{ijk}[L_iL_j\overline{E}_k] + \lambda'_{ijk}[L_iQ_j\overline{D}_k] + \lambda''_{ijk}[\overline{U}_i\overline{D}_j\overline{D}_k]. \qquad (1)$$

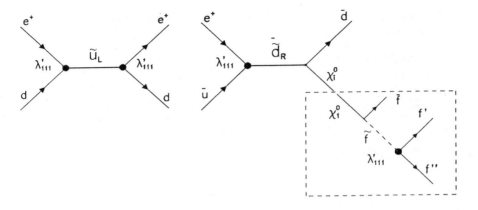

Figure 3: Production of with subsequent $R_p v$ decay (left) and gaugino decay (right).

This implies that the lightest supersymmetric particle (LSP) is unstable, and, provided that λ'_{i1k} is non-vanishing, that squarks can be singly produced at HERA via the reactions $e^+d \to \tilde{q}_L$ ($\tilde{q} = \tilde{u}, \tilde{c}, \tilde{t}$) and $e^+\bar{u} \to \tilde{q}_R^*$ ($\tilde{q} = \tilde{d}, \tilde{s}, \tilde{b}$).

The squark then undergoes either a (direct) $\not{R}_p$ decay, $\tilde{q} \to eq'$ (Fig. 3), in which case the event topology is the same as for leptoquarks, or a gauge-decay cascade, e.g. $\tilde{q} \to q\tilde{\chi}_1^0 \to q\bar{q}'q''e$ (Fig. 3b) and $\tilde{u}_L \to d\tilde{\chi}_1^+ \to dW\tilde{\chi}_1^0 \to d q\bar{q}q'\bar{q}''e$ leading to multi-jet topologies or even to "wrong-sign" leptons.

A search for $\not{R}_p$ squark production has been carried out by the H1 collaboration imposing the same selection criteria for $\not{R}_p$ decays as for leptoquarks (cf. Sec. 2), and by additionally selecting multi-jet topologies with "right-sign" and "wrong-sign" leptons where at least two jets with a transverse momentum bigger than 15 GeV and 7 GeV have to be present [8]. Further conditions are imposed on the jets to improve the signal-to-background ratio, and on the "wrong-sign" lepton track to ensure good charge identification.

The observed numbers of "right-sign" and "wrong-sign" lepton multi-jet events, 289 and 1 respectively, are in good agreement with the SM expectation of 286 ± 28 and 0.49 ± 0.2.

Fig. 4 shows the 95% confidence level rejection limits as a function of $m_{\tilde{q}}$ and λ'_{1j1} derived by H1 under the assumption that the LSP is a neutralino, that the gluino is heavier than the squark, and that the coupling λ'_{1j1} dominates. The dependence of the limits on the MSSM parameters (μ, M_2, $\tan\beta$) has been studied by choosing extreme cases where the lightest neutralino is either zino-like or photino-like. Far more stringent limits on the up squark (λ'_{111}) are imposed by results on neutrino-less double beta decay [9]. The limits on the charm and top squarks derived by H1 are competitive with those obtained from atomic parity violation (APV) measurements [10], for squark masses below 200 GeV and in MSSM scenarios where the gaugino is zino-like.

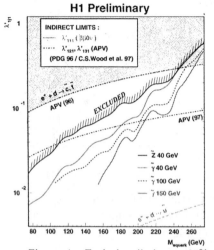

Figure 4: Exclusion limits at 95% confidence level for the Yukawa coupling strength λ'_{1j1}. Also represented are the most stringent indirect limits on λ'_{111} (dotted) and $\lambda'_{1j1}, j = 2, 3$ (dash-dotted).

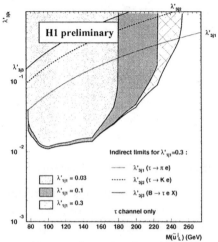

Figure 5: Exclusion limits at 95% confidence level for the third generation Yukawa coupling strength λ'_{3jk} as function of squark mass for several values of the first generation coupling strength λ'_{1j1}

4 Lepton Flavor Violation

Leptoquarks or squarks in $\not{R}_p$ SUSY coupling to two or three generations will result in lepton flavor violation (LFV) which will manifest itself at HERA by the presence of a μ or a τ in the final state instead of a positron.

H1 has searched for leptoquarks and squark decaying into τq with subsequent hadronic decay of the τ. The events were required to have no positron with $E_T > 5$ GeV but two back-to-back jets of which one must be "pencil-like" (low multiplicity, high p_T leading track) and a missing p_T bigger than 10 GeV from the escaping $\bar{\nu}_\tau$ [3,8]. Further cuts are imposed to distinguish the τ jet from NC DIS events with a misidentified positron and photo-production events with low multiplicity jets. No event satisfies all these conditions in agreement with the SM expectation of 0.8 ± 0.3 events.

For the derivation of rejection limits it is assumed that the leptoquark or squark couples only to the first and third generations ($\beta_e + \beta_\tau = 1$). At low values of β_e the rejection limits on λ_{11}, the strength of coupling of the leptoquark to the first generation, are improved by about a factor of three over a mass range from 80 to 270 GeV when the τ decay channel for the leptoquark is included. For $\lambda_{11} = \sqrt{4\pi\alpha_{EM}}$ and for $\beta_e < 0.1$, leptoquark masses below 255 GeV (237 GeV) are excluded for leptoquarks coupling $e^+u(e^+d)$, which is a substantially stronger bound than derived at the TeVatron (100 GeV).

Alternately, for fixed values of λ_{11} (in the case of leptoquarks) or λ'_{1j1} (in the case of $\not{R}_p$ SUSY) strong limits on λ_{3j} or λ'_{3jk} can be derived, as shown in Fig. 5. The excluded domain in the $M(\tilde{u}^j_L)$-λ'_{3jk} plane extends for beyond the relevant indirect limits from rare τ and B-meson decays [11]. For coupling strengths λ'_{1j1} and λ'_{3jk} as small as 0.03 (10% of EM coupling), $\tilde{u}^j_L$ squarks lighter than 165 GeV are excluded.

5 R-Parity Conserving Supersymmetry

In the framework of the minimal supersymmetric standard model (MSSM) sparticles must be produced in pairs given R-parity conservation. ZEUS has searched for associated squark and spositron production via neutralino exchange, and subsequent sparticle decay by emission of the lightest neutralino, assumed to be the LSP, stable in the MSSM: $e^+q \to \tilde{e}\tilde{q} \to e^+q\tilde{\chi}^0_1\tilde{\chi}^0_1$ [12].

Events were required to have one positron with E_T bigger than 4 to 10 GeV depending on its angle, a hadronic final state with p_T bigger than 4 GeV, and missing transverse momentum of at least 14 GeV from the neutralinos escaping detection. Further cuts on the longitudinal and transverse missing momentum were applied to maximize the ratio of the signal to the background of various nature. Only one event satisfies these cuts in good agreement with the value of $1.99^{+0.57}_{-0.84}$ expected from the SM.

Since the cross-section depends mainly on the sum of the selectron and squark masses rejection limits on the sum $m_{\tilde{e}} + m_{\tilde{q}}$ alone can be derived. The MSSM parameter $|\mu|$ is chosen large because the cross-section is high only when the $\tilde{\chi}^0_1$ is gaugino-like. Assuming that the squarks (except the $\tilde{t}$) and also the $\tilde{e}_L$ and the $\tilde{e}_R$ are degenerate and fixing the branching ratios by choosing $M_1 = \frac{5}{3}\tan^2\theta_W M_2$, values of $m_{\tilde{e}} + m_{\tilde{q}}$ below 77 GeV are excluded at 95% confidence level for $m_{\tilde{\chi}^0_1} = 40$ GeV.

6 Contact Interactions

In addition to direct searches up to the kinematical limit, HERA is also sensitive to new vector bosons or currents at a mass scale Λ of the order of the TeV predicted by many unified theories or compositeness models. At energies small compared to Λ, these new interactions can be parameterized by adding four-fermion contact terms to the SM Lagrangian in analogy to the Fermi theory of beta-decay. By interference, these contact interactions (CI) then lead to a deviation of the differential DIS cross-section from the SM value.

Since strong constraints have already been imposed on the scalar and tensor contact terms by other measurements, ZEUS restricted their analysis on the vector terms of the effective Lagrangian [13]:

$$\mathcal{L}_{\text{eff}}^V = \frac{g^2}{\Lambda^2} \Big(\quad \eta_{LL}^q (e_L \gamma^\mu e_L)(\bar{q}_L \gamma_\mu q_L) + \eta_{LR}^q (e_L \gamma^\mu e_L)(\bar{q}_R \gamma_\mu q_R)$$

$$+ \eta_{RL}^q (e_R \gamma^\mu e_R)(\bar{q}_L \gamma_\mu q_L) + \eta_{RR}^q (e_R \gamma^\mu e_R)(\bar{q}_R \gamma_\mu q_R) \Big)$$

where η_{ab}^q. The convention $g^2 = 4\pi$ is used. APV measurements impose strong limits on Λ which can be avoided only if $\eta_{LL}^q + \eta_{LR}^q - \eta_{RL}^q - \eta_{RR}^q = 0$. This implies in particular that couplings of purely chiral type are excluded. Furthermore, SU(2) invariance requires that the coupling to left-handed quarks be independent of the quark flavor ($\eta_{aL}^u = \eta_{aL}^d$).

These constraints limit the number of possible scenarios for $(\eta_{LL}^u, \eta_{LR}^u, \eta_{RL}^u, \eta_{RR}^u)$ to 18. In addition, six SU(2) violating scenarios for the coupling to the u quarks are also considered in the ZEUS analysis.

The effect of each contact interaction scenario on the expected event distribution is estimated by reweighting SM Monte Carlo samples appropriately. The most likely value of the CI mass scale Λ_0 and Poissonian limits on Λ are determined by extracting a log-likelihood function from the comparison of measured and simulated event distributions.

The values of Λ_0 range from about 2 TeV to infinity. No significant deviation from the SM prediction is found. The lower 95% confidence level limits range from 1.5 TeV to almost 5 TeV. Limits on seven scenarios (VA, X1, X2, X5, X6, U1 and U5 in the conventional nomenclature) are reported by ZEUS for the first time.

Table 1 shows the limits obtained for some scenarios in comparison with results from LEP and from the TeVatron when available. Note that CI searches in $e^+ p$ collisions are complementary to those in $e^+ e^-$ and $p\bar{p}$ collisions, because the sign of the SM-CI interference for a given scenario, and therefore experimental sensitivity, is opposite in the two cases.

7 Summary and Conclusions

This brief overview of searches for new physics at HERA based on the analysis of the data taken by H1 and ZEUS from 94 to 97 has shown that HERA

	$(\eta^u_{LL},\eta^u_{LR},\eta^u_{RL},\eta^u_{RR})$	ZEUS	CDF	OPAL	ALEPH	L3
VV	(+1, +1, +1, +1)	**4.9**	3.5	3.3	4.0	3.2
VV	(-1, -1, -1, -1)	**4.6**	5.2	4.3	5.2	3.9
X4	(0, +1, +1, 0)	**4.5**	—	2.5	3.0	2.4
X4	(0, -1, -1, 0)	**4.1**	—	4.1	4.9	3.7
U4	(0, +1, +1, 0)	**4.6**	—	2.0	2.1	1.8
U4	(0, -1, -1, 0)	**4.4**	—	4.3	2.6	2.2
U5	(0, +1, 0, +1)	**4.2**	—	—	—	—
U5	(0, -1, 0, -1)	**3.6**	—	—	—	—

Table 1: Upper limits on the coupling strength Λ in TeV for 8 of the 24 scenarios of CI considered in the ZEUS analysis. VV and X4 respect SU(2) invariance ($\eta^u_{ab} = \eta^d_{ab}$), U4 and U5 violate it ($\eta^d_{ab}=0$). The limits obtained at e^+e^- and $p\bar{p}$ colliders are shown for comparison.

offers the unique possibility to search directly for exotic particles produced by lepton-quark fusion. In particular, HERA has a high discovery potential for leptoquarks, squarks from SUSY violating R-parity, and for lepton flavor violation. At present no signal of new physics has been detected. The rejection limits on leptoquarks and squarks are complementary to or, at low branching ratios, better than those derived from TeVatron data.

Indirect searches of new physics via contact interactions cover a wider range of scenarios than previous measurements by other experiments, and limits on the interaction strength are complementary to those obtained at other colliders.

From 1998 on, HERA has been running with electrons. This will allow to test other types of couplings, e.g. λ'_{11k} in $\not{R}_p$ SUSY. In 2000/2001 HERA will undergo a major upgrade aiming for a fourfold luminosity increase, a longitudinal electron/positron polarization of 70%, and a 10% increase in center-of-mass energy. Therefore, HERA can be expected to improve its discovery potential for new physics.

References

1. H1 Collaboration, C. Adloff *et al.*, Z. Phys. **C74**, 191, (1997);
   ZEUS Collaboration, J. Breitweg *et al.*, Z. Phys. **C74**, 207, (1997).
2. H1 Collaboration, C. Adloff *et al.*, Eur. Phys. J. **C5**, 575, (1998).
3. H1 Collaboration, *A Search for Leptoquark Bosons in DIS at High Q^2 at HERA*, proceedings of the XXIX Int. Conf. on High Energy Physics 1998, Vancouver, Canada.
4. ZEUS Collaboration, em Search for Narrow High Mass States in Positron-

544

Proton Scattering at HERA, proceedings of the XXIX Int. Conf. on High
Energy Physics 1998, Vancouver, Canada.

5. W. Buchmüller, R. Rückl and D. Wyler, Phys. Lett. **B191**, 173, (1987).
6. DØ collaboration, B. Abbot *et al.*, Phys. Rev. Lett. **79**, 4321, (1997);
   DØ collaboration, B. Abbot *et al.*, Phys. Rev. Lett. **80**, 20511, (1998).
7. J. Butterworth and H. Dreiner, Nucl. Phys. **B397**, 3, (1993), and references therein.
8. H1 Collaboration, *A Search for Squarks of R-Parity Violating SUSY at HERA*, proceedings of the XXIX Int. Conf. on High Energy Physics 1998, Vancouver, Canada.
9. R. Mohapatra, Phys. Rev. **D34**, 3457, (1986);
   J.D. Vergados, Phys. Lett. **B184**, 55, (1987);
   M. Hirsch, H.V. Klapdor-Kleingrothaus, S.G. Kovalenko, Phys. Lett. **B352**, 1 (1995), Phys. Rev. Lett. **75**, 17, (1995), Phys. Rev. **D53**, 1239, (1996).
10. P. Langacker, Phys. Lett. **B256**, 277, (1991);
    J.E. Kim and P. Ko, Phys. Rev. **D57**, 489, (1998); and references in [8].
11. S. Davidson, D. Baily and B. Campbell, Z. Phys. **C61**, 613, (1994).
12. ZEUS Collaboration, J. Breitweg *et al.*, Phys. Letters **B434** (1998).
13. ZEUS Collaboration, J. Breitweg *et al.*, submitted to the European Physical Journal;
    ZEUS Collaboration, *Investigation of eeqq Contact Interactions in Deep Inelastic e⁺p Scattering at HERA*, proceedings of the XXIX Int. Conf. on High Energy Physics 1998, Vancouver, Canada.

LATEST RESULTS FROM THE LSND EXPERIMENT

REX TAYLOE

Los Alamos National Laboratory, Los Alamos, New Mexico 87545, USA
E-mail: rex@lanl.gov

representing the LSND collaboration[1]

The LSND experiment at Los Alamos has searched for $\bar{\nu}_\mu \to \bar{\nu}_e$ oscillations using $\bar{\nu}_\mu$ from μ^+ decay at rest and for $\nu_\mu \to \nu_e$ oscillations using ν_μ from π^+ decay in flight. An excess of events attributable to neutrino oscillations has been observed in both of these channels in data collected in 1993-1995. A recent preliminary analysis of the decay at rest $\bar{\nu}_\mu \to \bar{\nu}_e$ data collected in 1996-1998 with a different ν source configuration is consistent with the earlier data. The BooNE experiment that is planned to run at FNAL will further test these results.

1 Introduction

The phenomenon of neutrino oscillations, where a neutrino of one type (e.g. $\bar{\nu}_\mu$) spontaneously transforms into a neutrino of another type (e.g. $\bar{\nu}_e$), has important and far-reaching consequences for particle physics and cosmology. For this phenomenon to occur, at least one neutrino must be massive and the so far observed lepton flavor conservation law must be violated.

In 1995, the LSND experiment reported the observation of candidate events in a search for $\bar{\nu}_\mu \to \bar{\nu}_e$ oscillations [2]. The evidence for oscillations has grown with additional data from the $\bar{\nu}_\mu \to \bar{\nu}_e$ channel [3] and with new results on $\nu_\mu \to \nu_e$ oscillations [4].

2 Experiment

The Liquid Scintillator Neutrino Detector (LSND) experiment [5] at Los Alamos was designed to search with high sensitivity for $\bar{\nu}_\mu \to \bar{\nu}_e$ oscillations from μ^+ decay at rest. The main concept of the experiment is to create a large flux of $\bar{\nu}_\mu$ with little $\bar{\nu}_e$ background then to look for $\bar{\nu}_e$ interactions in the detector and compare the rate with that expected from conventional (non-oscillation) sources. An excess may be attributed to neutrino oscillations.

2.1 Neutrino Source

The LANSCE accelerator at Los Alamos National Lab (LANL) is an intense source of low energy neutrinos due to its 1 mA proton intensity and 800 MeV energy. The neutrino source is well understood because almost all neutrinos arise from π^+ or μ^+ decay — the π^- and μ^- are readily captured in the Fe of the shielding and Cu of the beam stop[6]. This π^+/π^- and μ^+/μ^- asymmetry in the beam stop results in a low relative production rate of $\bar{\nu}_e$ as compared to $\bar{\nu}_\mu$. The $\bar{\nu}_e$ rate is calculated to be 4×10^{-4} times that of the $\bar{\nu}_\mu$ in the $36 < E_\nu < 52.8$ MeV energy range. The $\nu_\mu \to \nu_e$ search utilizes ν_μ from π^+ decay in flight. There are few ν_e in the beam above 60 MeV because few μ^+ decay in flight and the π^+ decay to a ν_e only rarely (with a branching ratio of $\sim 10^{-4}$).

The energy spectra of the main components of the neutrino flux at the LSND detector are shown in Fig. 1. An important verification of the accuracy of these calculated fluxes are the cross section measurements of two well-understood reactions: $\nu_e C \to e^- N_{g.s.}$[7] and $\nu_\mu C \to \mu^- N_{g.s.}$[8].

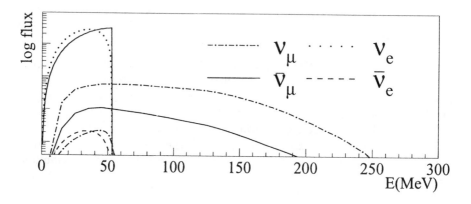

Figure 1: Simplified figure showing the main components of the neutrino flux with an arbitrary scale. Note the log scale.

The neutrino source was reconfigured after the 1995 run to study tritium production. A comparison of the data collected after this change to that collected before allows another important check of the flux calculations and the neutrino oscillation hypothesis.

2.2 The Detector

The LSND detector[5] consists of an approximately cylindrical tank 8.3 m long by 5.7 m in diameter and is located 30 m from the neutrino source. The detector is well-shielded by 9 m steel-equivalent between detector and neutrino source, an 8 m water plug downstream of the detector, and 2 kg/cm^2 overburden. A veto shield [9] with active and passive shielding surrounds the detector tank and tags incoming (or outgoing) charged particles.

The inside surface of the tank is lined with 1220 8-inch Hamamatsu phototubes providing 25% photocathode coverage. The tank is filled with 167 metric tons of liquid scintillator consisting of mineral oil and 0.031 g/l of b-PBD. This low scintillator concentration allows the detection of both Čerenkov light and scintillation light and yields a relatively long attenuation length of more than 20 m for wavelengths greater than 400 nm.[10] A typical 45 MeV electron created in the detector produces a total of $\sim$ 1500 photoelectrons, of which $\sim$ 280 photoelectrons are in the Čerenkov cone.

3 $\bar{\nu}_\mu \to \bar{\nu}_e$ Oscillation Search from μ^+ DAR

The signature for a $\bar{\nu}_e$ interaction in the detector is the detection of the e^+ from the reaction $\bar{\nu}_e p \to e^+ n$ followed by the detection of the γ from the neutron-capture reaction ($np \to d\gamma(2.2 \text{ MeV})$). The e^+ signature is an event with phototube signals that fit to a non-cosmic e^+ hypothesis (small spread in space and time, a good Čerenkov cone, and no sign of cosmic ray correlation). The e^+ ID parameter distribution is shown in Fig. 2 for positrons and compared to that for neutrons.

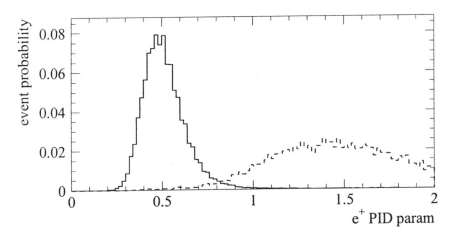

Figure 2: The e^+ ID parameter for a sample of positrons (solid histogram) and for neutrons (dashed).

The neutron-capture γ is identified via a correlated to accidental likelihood ratio, R, that is formed from these distributions: primary to γ time,

PMT multiplicity, and distance from primary to γ. The combination of these quantities serves to separate a correlated γ from an accidental with high efficiency and low background as may be seen by the difference between accidental and correlated γ R distributions in Fig. 3.

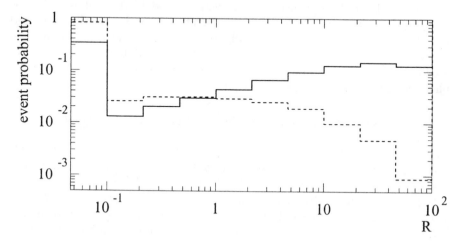

Figure 3: Measured distribution of the γ likelihood ratio, R, for correlated (solid) and accidental (dashed) γs.

The energy distribution for events passing the e^+ and stringent γ cuts ($R > 30$) from the entire LSND data set (1993–1998) are shown in Fig. 4. From this sample, with positron energies from 36 to 60 MeV, a preliminary analysis yields 33 oscillation events with 9.5 ± 0.9 background events expected. The favored regions obtained from a likelihood analysis of this data set are shown in Fig. 6.

4 $\nu_\mu \to \nu_e$ Search from π^+ DIF

The signature for $\nu_\mu \to \nu_e$ oscillations is an electron from the reaction $\nu_e C \to e^- X$ in the energy range $60 < E_e < 200$ MeV. Using two different analyses,[4] a total of 40 beam-related events and 175 beam-unrelated events are observed, corresponding to a beam excess of 27.7 ± 6.4 events. The neutrino-induced backgrounds are dominated by $\mu^+ \to e^+ \bar{\nu}_\mu \nu_e$ and $\pi^+ \to e^+ \nu_e$ decays-in-flight in the beam-stop and are estimated to be 9.6 ± 1.9 events. Therefore, a total excess of $18.1 \pm 6.6 \pm 3.5$ events is observed above the background. The beam excess energy distribution is shown in Fig. 5. If interpreted as $\nu_\mu \to \nu_e$ oscillations, this data yields an oscillation probability of $(0.26 \pm 0.10 \pm 0.05)\%$.

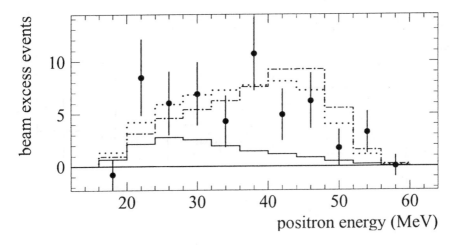

Figure 4: The energy distribution for e^+ candidates that also have a high γ-likelihood ($R > 30$) from the 1993-1998 data set. The points show the beam excess data. Also shown are the estimated neutrino background (solid) and the estimated distributions for oscillations at small (dotted) and large (dot-dash) Δm^2 plus neutrino background (solid).

The 95% confidence region obtained from this data set is shown along with the $\bar{\nu}_\mu \to \bar{\nu}_e$ results in Fig. 6.

5 The BooNE Experiment

It is clear that these results from LSND, while convincing evidence for neutrino oscillations, should be verified with a followup experiment that is capable of detecting hundreds of events for neutrino oscillations in the parameter space suggested by LSND. The BooNE (Booster Neutrino Experiment) experiment at Fermi National Accelerator Laboratory (Fermilab) is designed for just this purpose. BooNE will search for $\nu_\mu \to \nu_e$ oscillations in the region of the LSND excess. Fig. 7 shows the expected sensitivities for $\nu_\mu \to \nu_e$ appearance after one calendar year of operation.

The BooNE detector will consist of a spherical tank, $\sim$ 12 m in diameter and covered on the inside by 1280 8-inch phototubes ($\sim$10% coverage), filled with 800 t of mineral oil, resulting in a $\sim$450 t fiducial volume. The volume outside of the phototubes would serve as a veto shield for identifying particles both entering and leaving the detector. The detector will be located 550 m from the neutrino source.

The neutrino beam will be fed by the 8 GeV proton Booster at Fermilab.

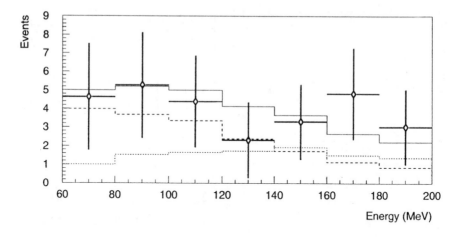

Figure 5: The energy distribution for the $\nu_\mu \to \nu_e$ oscillation sample. The points show the beam excess data. Also shown are expectation for backgrounds (dotted histogram), the oscillation signal for large values of Δm^2 (dashed), and the sum of the two (solid).

The neutrino beam line would consist of a target followed by a focusing system and a ∼50 m long pion decay volume. The low energy, high intensity, and 1 μs time structure of the Booster neutrino beam will be ideal for this experiment.

The BooNE experiment provides an opportunity to resolve the neutrino oscillation question on a short-time scale. Within the upcoming five years, no existing or approved experiments will be able to test conclusively the LSND signal. Thus BooNE represents an important and unique addition to the Fermilab program. BooNE has received Stage I approval, design is underway, and construction of the detector will begin in the latter part of 1999.

6 Conclusion

The LSND experiment observes excesses of events in both the $\bar{\nu}_\mu \to \bar{\nu}_e$ and $\nu_\mu \to \nu_e$ oscillation searches, corresponding to oscillation probabilities of $(0.31 \pm 0.12 \pm 0.05)\%$ and $(0.26 \pm 0.10 \pm 0.05)\%$, respectively. These two searches have different backgrounds and systematics and together provide strong evidence for neutrino oscillations in the range $0.2 < \Delta m^2 < 2.0$ eV^2. New data taken in 1996–1998 with a different ν source configuration strengthens this. In the near future, the BooNE experiment at Fermilab will definitively confirm or refute these observations.

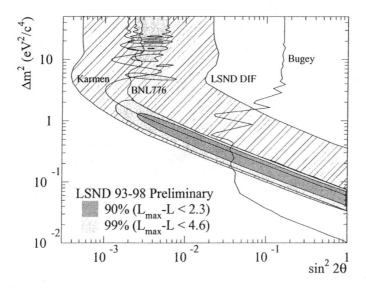

Figure 6: The LSND likelihood regions (shaded) obtained from the 1993–1998 (preliminary) $\bar{\nu}_\mu \to \bar{\nu}_e$ data set along with the 95% confidence region (lightly hatched) for 1993–1995 $\nu_\mu \to \nu_e$ data set. The 90% confidence limits from KARMEN[11], BNL776[12], and Bugey[13] are also shown.

References

1. The LSND Collaboration consists of the following people and institutions: E.D. Church, K. McIlhany, I. Stancu, W. Strossman, G.J. Van-Dalen (Univ. of California, Riverside); W. Vernon (Univ. of California, San Diego); D.O. Caldwell, M. Gray, S. Yellin (Univ. of California, Santa Barbara); D. Smith, J. Waltz (Embry-Riddle Aeronautical Univ.); I. Cohen (Linfield College); R.L. Burman, J.B. Donahue, F.J. Federspiel, G.T. Garvey, W.C. Louis, G.B. Mills, V. Sandberg, R. Tayloe, D.H. White (Los Alamos National Laboratory); R.M. Gunasingha, R. Imlay, H.J. Kim, W. Metcalf, N. Wadia (Louisiana State Univ.): K. Johnston (Louisiana Tech Univ.); R.A. Reeder (Univ. of New Mexico); A. Fazely (Southern Univ); C. Athnassopoulos, L.B. Auerbach, R. Majkic, D. Works, Y. Xiao (Temple Univ.).

2. C. Athanassopoulos *et. al.*, Phys. Rev. Lett. **75**, 2650, (1995).

3. C. Athanassopoulos *et. al.*, Phys. Rev. C **54**, 2685, (1996); C. Athanassopoulos *et. al.*, Phys. Rev. Lett. **77**, 3082, (1996).

4. C. Athanassopoulos *et. al.*, Phys. Rev. C **58**, 2489, (1998); C. Athanas-

552

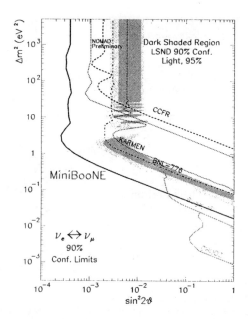

Figure 7: Expected sensitivity for MiniBooNE for $\nu_\mu \to \nu_e$ appearance after one year of running, including systematic and statistical errors, if LSND signal is not observed (solid line). Results from past experiments through December, 1997, also are shown.

sopoulos *et. al.*, Phys. Rev. Lett. **81**, 1774, (1998).

5. C. Athanassopoulos *et. al.*, Nucl. Instrum. Methods A **388**, 149, (1997).

6. R.L. Burman, M.E. Potter, and E.S. Smith, Nucl. Instrum. Methods A **291**, 621, (1990); R.L. Burman, A.C. Dodd, and P. Plischke, Nucl. Instrum. Methods in Phys. Res. A **368**, 416, (1996).

7. C. Athanassopoulos *et. al.*, Phys. Rev. C **55**, 2087, (1997).

8. C. Athanassopoulos *et. al.*, Phys. Rev. C **56**, 2806, (1997).

9. J.J. Napolitano *et. al.*, Nucl. Instrum. Methods A **274**, 152, (1989).

10. R.A. Reeder *et. al.*, Nucl. Instrum. Methods A **334**, 353, (1993).

11. B. Bodmann *et. al.*, Phys. Lett. B **267**, 321, (1991); B. Bodmann *et. al.*, Phys. Lett. B **280**, 198, (1992); B. Zeitnitz *et. al.*, Prog. Part. Nucl. Phys. **32**, 351, (1994).

12. L. Borodovsky *et. al.*, Phys. Rev. Lett. **68**, 274, (1992).

13. B. Achkar *et. al.*, Nucl. Phys. B **434**, 503, (1995).

14. E. Church *et. al.*, FNAL Proposal P898, (1997).

ELECTROWEAK RESULTS FROM THE Z RESONANCE CROSS-SECTIONS AND LEPTONIC FORWARD-BACKWARD ASYMMETRIES WITH THE ALEPH DETECTOR

E. TOURNEFIER

CERN, Switzerland

E-mail: Edwige.Tournefier@cern.ch

The measurement of the Z resonance parameters and lepton forward-backward asymmetries are presented. These are determined from a sample of 4.5 million Z decays accumulated with the ALEPH detector at LEP I.

1 Introduction

From 1990 to 1995 the LEP e^+e^- storage ring was operated at centre of mass energies close to the Z mass, in the range $|\sqrt{s} - M_Z| < 3$ GeV. Most of the data have been recorded at the maximum of the resonance (120 pb^{-1} per experiment) and about 2 GeV below and above (40 pb^{-1} per experiment).

The measurement of the hadronic and leptonic cross sections as well as the leptonic forward backward asymmetries performed with the ALEPH detector at these energies are presented here. The large statistics allow a precise measurement of these quantities which are then used to determine the Z lineshape parameters: the Z mass M_Z, the Z width Γ_Z, the total hadronic cross section at the pole σ^{0had} and the ratio of hadronic to leptonic pole cross sections $R_l = \sigma^{0had}/\sigma_l^0$.

Here we will give some details of the ALEPH experimental measurement of these quantities. A review of the whole LEP electroweak measurements and a discussion of the results as a test of the Standard Model can be found in another talk of this conference[1].

2 Cross sections and leptonic Forward-Backward asymmetries measurement

The cross section and asymmetries are determined for the s-channel process $e^+e^-R_I \to Z, \gamma R_I ghtarrow f\bar{f}$. The cross section is derived from the number of selected events N_{sel} with

$$\sigma_{f\bar{f}} = \frac{N_{sel}(1 - f_{bkg})}{\epsilon} \frac{1}{\mathcal{L}} \tag{1}$$

where f_{bkg} is the fraction of background, ϵ is the selection efficiency and $\mathcal{L}$ is the integrated luminosity. Note that in the case of $e^+e^-e^+e^-$ final state the irreducible background originating from the exchange of γ (Z) in the t-channel is subtracted from N_{sel} to obtain the s-channel cross section.

The leptonic forward-backward asymmetry (A_{FB}) is derived from a fit to the angular distribution

$$\frac{d\sigma_{f\bar{f}}}{d\cos\theta^*} \propto 1 + \cos^2\theta^* + \frac{8}{3}A_{FB}\cos\theta^* \qquad (2)$$

where θ^* is the centre of mass scattering angle between the incoming e^- and the out-going negative lepton.

To achieve a good precision on the cross section a high efficiency and low background is necessary while the asymmetry is insensitive to the overall efficiency and to a background with the same asymmetry as the signal. This justifies the use of different leptonic selections for cross section and asymmetry measurement.

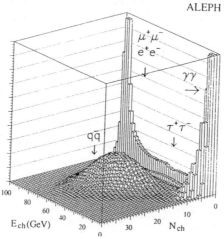

Figure 1. Distribution of charged track multiplicity (N_{ch}) versus charged track energy (E_{ch}).

2.1 Hadronic cross sections

Four million of $e^+e^- \rightarrow q\bar{q}$ events have been recorded at the Z peak, leading to a statistical uncertainty of 0.05%. The systematic uncertainty has been reduced to the same level.

The hadronic cross section measurement is based on two independent selections. The first selection is based on charged track properties while the second is based on calorimetric energy. Details of these selections can be found in previous publications [2]. These two measurements are in good agreement and are combined to obtain the final result. The systematic uncertainties of these selections are almost uncorrelated because they are mainly based on uncorrelated quantities therefore the combination of the 2 selections allows to reduce the systematic uncertainty.

Fig. 1 shows the distribution of charged multiplicity versus the charged track energy for signal and background. These variables are used to separate $e^+e^- \rightarrow q\bar{q}$ events from background in the charged track selection. The most dangerous background in both selections is $\gamma\gamma$ events because Monte Carlo prediction is not fully reliable. Therefore a method to determine this background from the data has been developed. This is achieved by exploiting the different $\sqrt{s}$ dependence of the resonant (signal) and non-resonant (background) contributions. The resultant systematic error (0.04%) reflects the statistics of the data.

Table 1: Efficiency, background and systematic errors for the two hadronic selections at the peak point.

	Charged tracks	Calorimeter
Efficiency (%)	97.48	99.07
Background (%)		
$\tau^+\tau^-$	0.32	0.44
$\gamma\gamma$	0.26	0.16
e^+e^-	-	0.08
Systematic (%)		
Detector simulation	0.02	0.09
Hadronisation modelling	0.06	0.03
$\tau^+\tau^-$ bkg	0.03	0.05
$\gamma\gamma$ bkg	0.04	0.03
e^+e^- bkg	-	0.03
Total syst.	0.087	0.116
Combined	0.071	

The dominant systematic error in the calorimetric selection comes from the calibration of calorimeters (0.09%) and, in the charged track selection from the determination of the acceptance (0.06%). Table 1 gives a breakdown of the efficiency, the background and the systematic uncertainties of both selections.

2.2 Leptonic cross sections

The statistical uncertainty in the leptonic channel is of the order of 0.15%, The aim of the analysis is to reduce the systematic to less than 0.1%. Two analyses were developed for the measurement of the leptonic cross sections. The first one, referred to as *exclusive* is based on three independent selections each aimed at isolating one lepton flavour and still follows the general philosophy of the analysis procedures described earlier [2]. The second one is new and has been optimised for the measurement of R_l. It is refered to as *global* di-lepton selection. The results of these selections agree within the uncorrelated statistical error and have been combined for the final result. These selections are not independent since they make use of similar variables therefore their combination does not reduce the systematic uncertainty.

We concentrate here on the *global* analysis. This selection takes advantage of the excellent particle identification capabilities (dE/dx, shower develope-ment in the calorimeters and muon chamber information) and the high granu-larity of the ALEPH detector. First, di-leptons are selected within the detector acceptance with an efficiency of 99.2% and the background arising from $\gamma\gamma$, $q\bar{q}$ and cosmic events is reduced to the level of 0.2%. Then the lepton flavour separation is performed inside the di-lepton sample so that the systematic un-certainties are anti-correlated between 2 lepton species and that no additional uncertainty is introduced on R_l. Table 2 gives a breakdown of the systematic errors obtained with 1994 data. The background from $q\bar{q}$ and $\gamma\gamma$ events affects mainly the $\tau^+\tau^-$ channel, therefore this channel is affected by bigger selection systematics than e^+e^- and $\mu^+\mu^-$.

As an example we consider the systematic errors related to the $\tau^+\tau^-$ selection efficiency. This efficiency is measured on the data: $\tau^+\tau^-$ events are selected using tight selection criteria to flag τ-like hemispheres; with the sample of opposite hemispheres, *artificial* $\tau^+\tau^-$ events are constructed by associating two back-to-back such hemispheres and the selection cuts are applied. In order to assess the validity of the method and to correct for possible bias of this method, two different Monte Carlo reference samples are used. On the first sample the same procedure of artificial $\tau^+\tau^-$ events is applied. On the second one the selection cuts are applied directly. The uncertainty on the efficiency measured with this method is dominated by the statistics of the artificial events used in the data. This method is applied in order to measure the inefficiency arising from $q\bar{q}$ cuts and in the flavour separation.

The dominant systematic in e^+e^- channel arises from t-channel subtrac-tion and is given by the theoretical uncertainty [3] on the t-channel contribution to the cross section.

Table 2: Systematic uncertainties in percent of dilepton cross sections for peak 1994 data. Correlations between lepton flavours are taken into account in the l^+l^- column.

	e^+e^-	$\mu^+\mu^-$	$\tau^+\tau^-$	l^+l^-
Global selection				
Tracking efficiency	0.05	0.03	0.03	0.04
Angles measurement (*)	0.02	0.01	0.01	0.02
ISR and FSR simulation (*)	0.03	0.03	0.03	0.03
$\gamma\gamma$ cuts (*)	0.02	-	0.05	0.02
$q\bar{q}$ cuts	-	-	0.11	0.04
$\gamma\gamma$ background (*)	-	-	0.02	-
$q\bar{q}$ background(*)	-	-	0.04	0.01
Flavour separation				
$\mu^+\mu^-/\tau^+\tau^-$	-	0.03	0.03	-
$e^+e^-/\tau^+\tau^-$ $cos\theta^* < 0.7$	0.08	-	0.07	0.01
$cos\theta^* \geq 0.7$	-	-	0.06	0.02
t-channel subtraction				
(*)	0.11	-	-	0.04
Monte Carlo statistics				
	0.05	0.06	0.07	0.04
Total	0.16	0.08	0.19	0.09

(*)uncertainties completely correlated among all energy points.

This leptonic cross section measurement contributes to a systematic uncertainty on R_I of 0.08%.

2.3 Leptonic Forward-Backward asymmetries

The measurement of the asymmetries is dominated by the statistical uncertainty equal to 0.0015. Special muon and tau selections have been designed for the asymmetry measurement while the e^+e^- *exclusive* selection is used for the e^+e^- asymmetry measurement. The e^+e^- angular distribution needs to be corrected for efficiency before subtracting the t-channel and therefore relies on Monte Carlo while $\mu^+\mu^-$ and $\tau^+\tau^-$ angular distributions do not need to be corrected with Monte Carlo since the selections are designed such that the efficiency is symmetric.

Because of the $\gamma - Z$ interference, the asymmetry varies rapidly with the centre of mass energy $\sqrt{s'}$ around the Z mass. Cuts on energy induce a dependence of the efficiency with $\sqrt{s'}$ and therefore could introduce a bias in the measurement since the efficiency would no longer be symmetric. To minimise these effects the selections are mainly based on particle identification instead of kinematic variables.

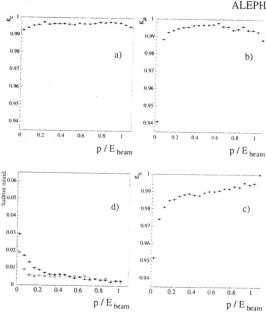

Figure 2. Identification efficiencies for a) electrons, b) muons and c) pions as a function of the momentum to the beam energy ratio. In d) the probability for a pion to be misidentified as an electron (squares) or a muon (circles) is shown.

Fig. 2 shows the particle identification efficiency as a function of the ratio of the momentum to the beam energy.

The asymmetry is extracted by performing a maximum likelihood fit to the differential cross section. The dominant systematic uncertainty arises from t-channel subtraction in the Bhabha channel (0.0013 on $A_{FB}^{0,e}$) and other systematic errors are smaller than 0.0005.

3 Results

Fig's. 3 and 4 show the measured cross sections and asymmetries. The Z line-shape parameters are fitted to these measurements with the latest version of ZFITTER[4]. The error matrix used in the χ^2 fit includes the experimental sta-

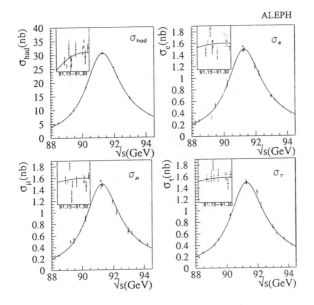

Figure 3. Measurement of cross sections. The inserts show enlarged views of the peak region.

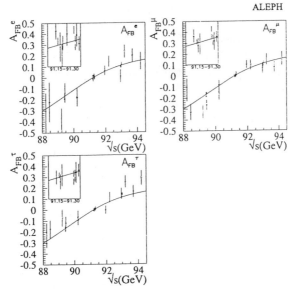

Figure 4. Measurement of the asymmetries. The inserts shows enlarged views of the peak region.

Table 3: Result of the fit to experimental measurement of cross sections and leptonic forward-backward asymmetries. The total error is split into statistical and systematic errors, theoretical error on luminosity and on t-channel and LEP beam energy measurement error.

	value	stat	exp	lumi	t-ch	E_{beam}
M_Z	91.1883±0.0031	0.0024	0.0002			0.0017
Γ_Z	2.4953±0.0043	0.0038	0.0009			0.0013
σ^{0had}	41.557±0.058	0.030	0.026	0.025		0.011
$R_e l$	20.677±0.075	0.062	0.033		0.025	0.013
R_μ	20.802±0.056	0.053	0.021			0.006
R_τ	20.710±0.062	0.054	0.033			0.006
$A_{FB}^{0,e}$	0.0189±0.0034	0.0031	0.0006		0.0013	0.0002
$A_{FB}^{0,\mu}$	0.0171±0.0024	0.0024	0.0005			0.0002
$A_{FB}^{0,\tau}$	0.0169±0.0028	0.0026	0.0011			0.0002
R_l	20.728±0.039	0.033	0.020		0.005	0.002
$A_{FB}^{0,l}$	0.0173±0.0016	0.0015	0.0004		0.0002	0.0001

tistical and systematic uncertainties, the LEP beam energy measurement uncertainty and the theoretical uncertainties arising from the small angle Bhabha cross section in the luminosity determination and the t-channel contribution to wide angle Bhabha events. The results are shown in Table 3.

The value of the Z couplings to charged leptons $|g_V|$ and $|g_A|$ can be derived from these parameters. The experimental measurement is shown in Fig. 5 with the Standard Model prediction. The data favor a light Higgs.

The value of α_s can also be extracted from R_l, Γ_Z and σ^{0had}:

$$\alpha_s = 0.115 \pm 0.004_{exp} \pm 0.002_{QCD} \qquad (3)$$

where the first error is experimental and the second reflects uncertainties on the QCD part of the theoretical prediction [6]. Here the Higgs mass has been fixed to 150 GeV and the dependence of α_s with M_H can be approximately parametrised by $\alpha_s(M_H) = \alpha_s(M_H = 150 GeV) \times (1 + 0.02 \times ln(M_H/150))$ where M_H is expressed in GeV.

4 Conclusion

The high statistic accumulated by ALEPH during LEP 1 running allow to measure the hadronic and leptonic cross sections and the leptonic forward-backward asymmetries with statistical and systematic precision of the order of

1 permil. These measurements are turned into precise determinations of the Z boson properties and constraints on the Standard Model parameters.

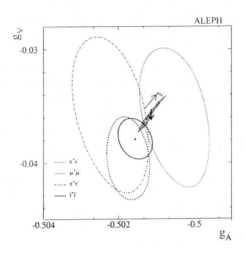

Figure 5. Effective lepton couplings. Shown are the one-σ contours. The shaded area indicates the Standard Model expectation for $M_t = 174 \pm 5$ GeV/c^2 and $90 < M_H GeV/c^2 < 1000$; the vertical fat arrow shows the change if the electromagnetic coupling constant is varied within its error. The sign of the couplings were assigned in agreement with τ polarisation measurements and neutrino data scattering.

Acknowledgments

I would like to thank D. Schlatter for his help in preparing this talk. I also would like to thank the organisers of the Lake Louise Winter Institute conference for the interesting program and the nice atmosphere at the conference.

References

1. Talk given by J. Pinfold at this conference, to appear in these proceedings.
2. ALEPH Collaboration, Z. Phys. **C48**, 365, (1990);
   ALEPH Collaboration, Z. Phys. **C53**, 1, (1992);
   ALEPH Collaboration, Z. Phys. **C60**, 71, (1993);
   ALEPH Collaboration, Z. Phys. **C62**, 539, (1994).

3. W. Beenakker, F.A. Berends and S.C. van der Marck, "Large Angle Bhabha Scattering", Nucl. Phys. **B349**, 223, (1991);
W. Beenakker, G. Passarino, "Large Angle Bhabha Scattering at LEP1", Phys. Lett. **B425**, 199-207, (1998), 199-207.

4. D. Bardin et al., Z. Phys. **C44**, 493, (1989); Comp. Phys. Comm. **59**, 303, (1990); Nucl. Phys. **B351**, 1, (1991); Phys. Lett. **B255**, 290, (1991) and CERN-TH 6443/92 (May 1992)
newline recently updated with results from [5], from A. Czarnecki and J.H. Kühn, "Nonfactorizable QCD and Electroweak Corrections to the Hadronic Z Boson Decay Rate", hep-ph/9608366 preprint, (Aug. 1996), from G. Montagna et al. "On the QED radiator at order α^3", preprint hep-ph/9611463, and from G. Degrassi, P. Gambino and A. Vicini, Phys. Lett. **B383**, 219, (1996) and G. Degrassi, P. Gambino and A. Sirlin, Phys. Lett. **B394**, 188, (1997).

5. *Reports of the working group on precision calculations for the Z resonance*, eds. D. Bardin, W. Hollik and G. Passarino, CERN Yellow Report 95-03, Geneva, 31 March 1995. and references therein.

6. T. Hebbeker, M. Martinez, G. Passarino and G. Quast, Phys. Lett. **B331**, 165, (1994), and references therein.

THE W CROSS SECTION
AND LEPTON PRODUCTION WITH MISSING P_T
AT HERA

C. VALLEE

Centre de Physique des Particules de Marseille,
163 Avenue de Luminy, case 907, F-13288 Marseille cedex 9, France
E-mail: vallee@cppm.in2p3.fr

The H1 and ZEUS Collaborations have performed searches for events with high energy leptons and missing transverse momentum in the positron-proton collisions collected at $\sqrt{s} = 300\ GeV$ between 1994 and 1997. Altogether 9 candidates were found, of which 4 $e^{\pm}$ and 2 $\mu^{\pm}$ events behave as expected from exclusive W-boson production, whereas 3 $\mu^{\pm}$ events show kinematic properties atypical of all Standard Model processes considered. These results are summarized and discussed within the Standard Model and its possible extensions.

1 Analysis Strategies

The analysis procedures used by ZEUS and H1 are described in detail in ref's.[1,2]. At HERA, the main Standard Model (SM) processes expected to yield isolated leptons with high transverse momentum (P_T^l) are Neutral Current interactions (NC), photon-photon interactions ($\gamma\gamma$) and exclusive W-boson production with subsequent leptonic decay. The corresponding cross sections are o(100 pb), o(1 pb) and o(0.1 pb) respectively.

High-P_T leptons ($P_T^e \geq 10\ GeV$, $P_T^\mu \geq 10\ GeV$ (H1) or $5\ GeV$ (ZEUS)) are identified from the match of a high-P_T track with a calorimetric electromagnetic shower (e) or minimum ionizing particle pattern (μ). H1 μ candidates must in addition have a significant signal in the muon detector. Lepton isolation is quantified in the pseudo rapidity-azimuth (η, ϕ) plane by the distance of the lepton from the closest track (≥ 0.5 radian) and either the distance of the lepton from the closest hadronic jet (≥ 1 radian, H1) or the calorimetric energy measured in the vicinity of the lepton ($\leq 4\ GeV$, ZEUS). Once an isolated lepton is identified, a hypothetical transverse neutrino is reconstructed from the calorimetric missing transverse momentum (e-channel) or the transverse momentum imbalance between the μ track and complementary calorimetric deposits (μ-channel).

The kinematic cuts are designed to suppress NC and $\gamma\gamma$ contributions compared to W production. The calorimetric missing transverse momentum P_T^{calo} must exceed $25\ GeV$ (H1), $20\ GeV$ (e-channel, ZEUS) or $15\ GeV$ (μ-channel, ZEUS). Note that P_T^{calo} is equal to the reconstructed neutrino transverse mo-

Table 1: Comparison of the observed event yields with SM expectations.

	e-channel		μ-channel	
	ZEUS	H1	ZEUS	H1
DATA	3	1	0	5
W	2.0	1.7 ± 0.5	0.6	0.5 ± 0.1
OTHERS	1.0 ± 0.5	0.7 ± 0.1	0.6 ± 0.2	0.3 ± 0.2

mentum P_T^ν in the e-channel and approximately equal to the recoil hadronic transverse momentum P_T^X in the μ-channel. In the e-channel the NC contribution is further reduced by requiring an azimuthal acoplanarity between the identified e and the hadronic system ($\Delta\phi_{e-X}$) larger than 5° (H1) or 17° (ZEUS). In the μ-channel, ZEUS compensate their lower P_T^{calo} threshold by an additional cut at 15 GeV in P_T^ν.

2 Results

The above selection applied to the ZEUS (integrated luminosity L = 47 pb^{-1}) and the H1 (L = 37 pb^{-1}) samples yields 3 e^+ events for ZEUS and 1 e^-, 2 μ^+, 2 μ^- plus 1 μ of undetermined sign for H1. The distribution of the high-P_T track isolation variables made by H1 [1] prior to lepton identification shows that the candidates are well separated from the bulk of events with calorimetric missing transverse momentum, mostly due to Charged Current interactions. Table 1 compares the observed yields to the SM predictions. The most important SM process is W production, followed by NC interactions in the e-channel and $\gamma\gamma$ processes in the μ-channel. Contributions from lepton mis-identification or hadron punch-through are small compared to genuine lepton production. The detailed kinematic parameters of the observed events are given in ref's. [1,2]. Some of them are compared to the expectations from dominant SM processes in Figs. 1 and 2.

In the e-channel all candidates have a small or moderate P_T^X, and a lepton-neutrino transverse mass $M_T^{e\nu}$ close to the W mass as expected from W production. Given this agreement of their e^+ candidates properties with SM expectations, ZEUS derive [2] a total cross section for exclusive W production:

$$\sigma(e^+p \rightarrow e^+ W^\pm X) = 1.0^{+1.0}_{-0.7} \text{ (stat)} \pm 0.3 \text{ (syst) pb}$$

compatible with the SM estimation of 0.94 pb. The corresponding upper limit on abnormal W boson production can be translated into the following limits on anomalous $WW\gamma$ couplings: -7.6 $\leq \Delta\kappa \leq$ 3.9 (95% CL, $\lambda = 0$).

In the μ-channel, it is remarkable that despite the fact that no explicit cut is made on these variables in the H1 selection scheme, the μ-events show

ZEUS 1994−97 PRELIMINARY

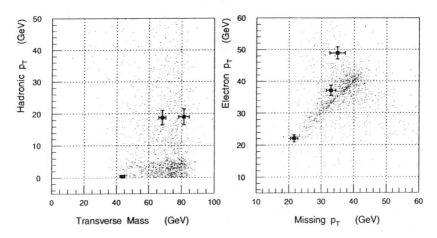

Figure 1: Kinematics of the 3 ZEUS e^+ candidates compared to the distributions expected from W production, for an integrated luminosity 1000 times higher than in the data.

both a non-zero P_T^ν and (except event $\mu 1$) a significant acoplanarity $\Delta\phi_{\mu-X}$ as additional evidence for undetected high-P_T particles. Events $\mu 3$ and $\mu 5$ lie at moderate P_T^X and show $M_T^{\mu\nu}$ values compatible (within 2 standard deviations for event $\mu 5$) with the W mass. Thanks to the observation of the scattered positron, event $\mu 3$ can have its full μ-ν mass reconstructed and found to be equal to $82^{+19}_{-12}\ GeV$. Event $\mu 5$ however shows the unexpected features of a very stiff μ track and a neutral hadronic jet [1]. The other 3 μ-events show either a lower $M_T^{\mu\nu}$ or a higher P_T^X than expected from most W events or $\gamma\gamma$ interactions. The indication of a rate excess in the μ channel therefore corresponds to events with properties not typical of SM processes.

3 Discussion of H1 atypical Muon Events

3.1 H1-ZEUS Compatibility

For a more direct comparison of their data with H1, ZEUS have redone their analysis using similar kinematic cuts as H1 [2,3]. They still find no event with a high-P_T μ associated to a hadronic system of $P_T^X \geq 30\ GeV$. Assuming the same acceptance as for W-production they set an upper limit of 0.17 pb (95 % CL) for the production cross section in this kinematic domain. Taking into account the H1 integrated luminosity and selection efficiency, this translates

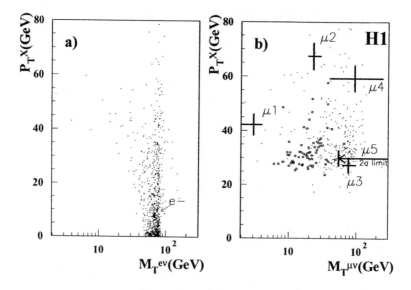

Figure 2: Kinematics of the H1 e^- (a) and μ (b) candidates compared to the distributions expected from W production (dots) and $\gamma\gamma$ interactions (circles), for an integrated luminosity 500 times higher than in the data.

into an upper limit of 3.2 events to be compared to the 4 events measured by H1 at $P_T^X \geq 30\ GeV$. The difference between the μ-event yields observed by H1 and ZEUS is therefore at the level of 2 standard deviations only.

Similarly the apparent asymmetry between the e and μ results at high P_T^X has a statistical significance of order 2 standard deviations.

3.2 W Interpretation

The interpretation of all muon events as SM W production can be further questionned when considering the measured kinematic properties of the candidates in addition to the indication of a rate excess. For example, reconstructing the neutrino using the W mass constraint yields for event $\mu 2$ a very high $x_{Bjorken}$ value, which together with the high P_T^X value makes this event particularly unlikely as a SM W.

However when dealing with W-production it must be remembered that the predictions mentioned here were computed at leading order only[4]. Higher order QCD terms might enhance W-boson production at high P_T^X. NLO calculations were recently made available[5] for W production sub-processes induced by resolved photon interactions, but the experimental conditions make the μ-channel measurements only sensitive to the high-P_T^X part of the cross section, which is dominated by direct photon interactions. In this domain no NLO calculation is available yet. Full NLO computations[6] made for the similar (apart for the W mass) process of prompt photon production predict an increase of the cross section of order 10 % at high P_T^X.

3.3 Interpretations beyond the Standard Model

Considering the low probability of the observed signal within the SM, it is worthwhile investigating which processes beyond the SM could explain the measured events. These can range from anomalous 3-boson couplings to decays of new heavy resonances of various types.

It has been argued[7] that anomalous $WW\gamma$ couplings could enhance W-production in ep collisions preferentially at high P_T^X. Combining the ZEUS and H1 results in the e and μ channels, the observed excess of events at $P_T^X \geq 25\ GeV$ is about a factor 2. In this interpretation this would correspond to an anomalous coupling $\Delta\kappa$ which has almost been excluded by the Tevatron experiments[8].

A lepto-quark produced by positron-quark fusion and decaying into a muon-quark system can be discarded from the measured event kinematics, which is incompatible with 2-body decays[9].

In the hypothesis of a resonance decaying into a quark and a W, the escaping neutrino can be fully reconstructed using the W mass constraint in addition to transverse momentum conservation. The resulting μ-ν-X masses of events $\mu2$ and $\mu4$ are found (within measurement errors of $\approx 15\ GeV$) to be compatible with each other and with the top mass. Exclusive top quark production is expected to be completely negligible at HERA in the SM, but could be significantly enhanced in dynamic models for fermion mass generation[10], provided a significant amount of intrinsic charm is present in the proton.

Another possible source of events with high-P_T jets, leptons and missing transverse momentum is the production of squarks in Supersymmetry models with R-parity violation. A favoured process is the fusion of the incident positron and a d quark into a $\tilde{t}$ squark, decaying into either a $\tilde{b}$ squark and a W[11], or a b quark and a chargino[3]. In the latter case the $\tilde{t}$ and $\tilde{\chi}$ masses can be reconstructed from the hadron jet parameters using 2 independent events only. Taking $\mu2$ and $\mu4$ the mass ranges required to accommodate this interpretation are found to be $\approx 200 - 300\ GeV$ and $\approx 100 - 200\ GeV$ for the $\tilde{t}$

568

and $\tilde{\chi}$ respectively.

4 Outlook

Since 1998, the HERA Collider has been operating in e^-p mode and should produce by the end 1999 an integrated luminosity comparable to that delivered in the e^+p runs. These new data will provide complementary information on the processes discussed here. The question of the origin of the e^+p muon events will also be addressed with higher precision once a significantly larger e^+p data sample has been accumulated, which is foreseen after the HERA luminosity upgrade starting in the year 2000.

Acknowledgments

I wish to thank particularly all my ZEUS and H1 Colleagues who contributed to the results discussed here.

References

1. C. Adloff *et al.* (H1 Collaboration) Eur. Phys. J. **5**, 575, (1998).
2. ZEUS Collaboration, contributed paper 756 to ICHEP98, Vancouver, Canada, July 1998.
3. C. Diaconu *et al.*, hep-ph/9901335, to appear in the proceedings of the 3rd UK Phenomenology Workshop on HERA Physics, Durham, UK, September 1998.
4. U. Baur *et al.*, Nucl. Phys. **B375**, 3, (1992).
5. P. Nason *et al.*, hep-ph/9902296, contribution to the 3rd UK Phenomenology Workshop on HERA Physics, Durham, UK, September 1998.
6. L.E. Gordon and J.K. Storrow, Z. Phys. **C63**, 581, (1994).
7. M.N. Dubimin and H.S. Song, hep-ph/9708259.
8. S. Abachi *et al.* (D0 Collaboration), Phys. Rev. Lett. **78**, 3634, (1997).
9. H1 Collaboration, contributed paper 579 to ICHEP98, Vancouver, Canada, July 1998.
10. H. Fritzsch and D. Holtmannspötter, hep-ph/9901411.
11. T. Kon *et al.*, Phys. Lett. **B376**, 227, (1996), and hep-ph/9707355.

ELECTROWEAK RESULTS FROM THE SLD EXPERIMENT

M. WOODS

Representing the SLD Collaboration

Stanford Linear Accelerator Center
Stanford, CA 94025, USA
E-mail: mwoods@slac.stanford.edu

We present an overview of the electroweak physics program of the SLD experiment at the Stanford Linear Accelerator Center (SLAC). A data sample of 550K Z^0 decays has been collected. This experiment utilizes a highly polarized electron beam, a small interaction volume, and a very precise pixel vertex detector. It is the first experiment at a linear electron collider. We present a preliminary result for the weak mixing angle, $\sin^2(\theta_W^{eff}) = 0.23110 \pm 0.00029$. We also present a preliminary result for the parity violating parameter, $A_b = 0.898 \pm 0.029$. These measurements are used to test for physics beyond the Standard Model.

1 Electroweak Physics program of the SLD Experiment

The SLD Experiment began its physics program at the SLAC Linear Collider (SLC) in 1992, and has accumulated a total data sample of approximately 550K hadronic Z^0 decays between 1992 and 1998. This data sample is a factor 30 smaller than the Z^0 sample available from the combined data of the 4 LEP experiments, ALEPH, DELPHI, L3 and OPAL. Yet the SLD physics results in many areas are competitive with the combined LEP result, and for some measurements SLD has the world's most precise results.

There are 3 features that distinguish the SLD experiment at the SLC: a small, stable interaction volume; a precision vertex detector; and a highly polarized electron beam. SLD is the first experiment at an electron linear collider. The collision volume is small and stable, measuring 1.5 microns by 0.7 microns in the transverse dimensions by 700 microns longitudinally.

These key features for the SLD experiment result in the world's best measurement of the weak mixing angle, a precise direct measurement of parity violation at the $Zb\bar{b}$ vertex, A_b, and a good measurement of the $Zb\bar{b}$ coupling strength, R_b. The weak mixing angle measurement provides an excellent means to search for new physics that may enter through oblique (or loop) corrections, while the A_b and R_b measurements are excellent means to search for new physics that may enter through a correction at the $Zb\bar{b}$ vertex.

In its near (analysis) future, SLD is also exploiting its capabilities to search for B_s mixing. The analysis for this is evolving to take full advantage of the precise vertexing information, and by the time of the summer 1999 conferences SLD should have a measurement of B_s mixing comparable in sensitivity with the combined LEP result. SLD estimates it should have a reach for Δm_s of $12 - 15 ps^{-1}$, in the region where it is predicted in the SM.

2 Z^0 Coupling Parameters

At the $Z f \overline{f}$ vertex, the SM gives the vector and axial vector couplings to be $v_f = I_f^3 - 2Q_f \sin^2(\theta_W^{eff})$, and $a_f = I_f^3$, where I_f is the fermion isospin and Q_f is the fermion charge. Radiative corrections are significant and are treated as follows. First, vacuum polarization and vertex corrections are included in the coupling constants, and an effective weak mixing angle is defined to be $\sin^2(\theta_W^{eff}) \equiv \frac{1}{4}(1 - v_e/a_e)$. Second, experimental measurements need to be corrected for initial state radiation and for $Z - \gamma$ interference to extract the Z-pole contribution.

One can define a parity-violating fermion asymmetry parameter, $A_f = \frac{2v_f a_f}{v_f^2 + a_f^2}$. The cross-section for $e^+ e^- \rightarrow Z^0 \rightarrow f \overline{f}$ can be expressed by

$$\frac{d\sigma^f}{d\Omega} \propto [v_f^2 + a_f^2] \left\{ \begin{array}{l} (1 + \cos^2 \theta)(1 + PA_e) + \\ 2\cos\theta A_f(P + A_e) \end{array} \right\} \tag{1}$$

where θ is the angle of the outgoing fermion with respect to the incident electron, and P is the polarization of the electron beam (the positron beam is assumed to be unpolarized). We can then define *forward, backward,* and *left, right* cross-sections as follows: $\sigma_F = \int_0^1 \frac{d\sigma}{d\Omega} d(\cos\theta)$; $\sigma_B = \int_{-1}^0 \frac{d\sigma}{d\Omega} d(\cos\theta)$; $\sigma_L = \int_{-1}^1 \frac{d\sigma_L}{d\Omega} d(\cos\theta)$; $\sigma_R = \int_{-1}^1 \frac{d\sigma_R}{d\Omega} d(\cos\theta)$. Here, σ_L (σ_R) is the cross-section for left (right) polarized electrons colliding with unpolarized positrons.

At the SLC, the availability of a highly polarized electron beam allows for direct determinations of the A_f parameters via measurements of the *left-right forward-backward asymmetry,* A_{LR}^{FB}, defined by

$$A_{LR}^{FB} = \frac{(\sigma_F^L - \sigma_F^R) - (\sigma_B^L - \sigma_B^R)}{\sigma_F^L + \sigma_F^R + \sigma_B^L + \sigma_B^R} = \frac{3}{4} P_e A_f$$

Additionally, a very precise determination of A_e is achieved from the measurement of the *left-right asymmetry,* A_{LR}, which is defined as

$$A_{LR} = \frac{1}{P_e} \cdot \frac{\sigma_L - \sigma_R}{\sigma_L + \sigma_R} = A_e$$

All Z decay modes can be used, and this allows for a simple analysis with good statistical power for a precise determination of $\sin^2(\theta_W^{eff})$.

3   The SLAC Linear Collider

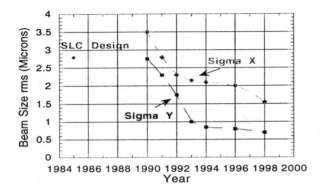

Figure 1: Spotsizes at the SLC.

LEP200 is the last of the large electron storage rings, and a new technology is needed to push to higher center-of-mass energies. Electron linear collider technology provides a means to achieve this, and the SLC is a successful prototype for this. It has reached a peak luminosity of $3 \cdot 10^{30} cm^{-2} s^{-1}$, which is within a factor two of the design luminosity [1]. The spotsizes at the Interaction Point (IP) are actually significantly smaller than design, and Fig. 1 indicates how the spotsizes have improved with time. With the small spotsizes, there is an additional luminosity enhancement from the "pinch effect" the two beams have on each other. At the higher luminosities achieved in the last SLC run, the pinch effect enhanced the luminosity by a factor of two.

The luminosity is limited due to the maximum charge achievable in single bunches because of instabilities in the Damping Rings.

4   SLD's Vertex Detector

SLD's vertex detector [2] consists of 3 layers, with the inner layer located at a radius of 2.7cm. Angular coverage extends out to $\cos(\theta) = 0.90$. It has 307 million pixels, with a single hit resolution of 4.5 microns. There are 0.4% radiation lengths per layer. The capability of SLD's vertex detector is illustrated in Fig. 2, which is a histogram of the reconstructed jet mass. With a mass cut of $2.0 GeV/c^2$, SLD can identify b jets with 50% efficiency and 98% purity.

Pt Corrected Mass

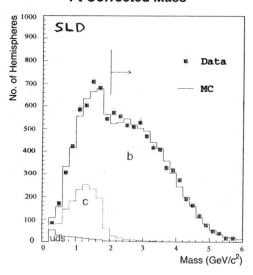

Figure 2: Vertex mass.

5   SLD's Compton Polarimeter

This polarimeter[3], shown in Fig. 3, detects both Compton-scattered electrons and Compton-scattered gammas from the collision of the longitudinally polarized 45.6 GeV electron beam[4] with a circularly polarized photon beam. The photon beam is produced from a pulsed Nd:YAG laser with a wavelength of 532 nm. After the Compton Interaction Point (CIP), the electrons and backscattered gammas pass through a dipole spectrometer. A nine-channel threshold Cherenkov detector (CKV) measures electrons in the range 17 to 30 GeV.[5] Two detectors, a single-channel Polarized Gamma Counter (PGC)[6] and a multi-channel Quartz Fiber Calorimeter (QFC),[7] measure the counting rates of Compton-scattered gammas.

Due to beamstrahlung backgrounds produced during luminosity running, only the CKV detector can make polarization measurements during beam collisions. Hence it is the primary detector and the most carefully analyzed. Its systematic error is estimated to be 0.7%. Dedicated electron-only runs are used to compare electron polarization measurements between the CKV, PGC and QFC detectors. The PGC and QFC results are consistent with the CKV result at the level of 0.5%. Typical beam polarizations for the SLD experiment

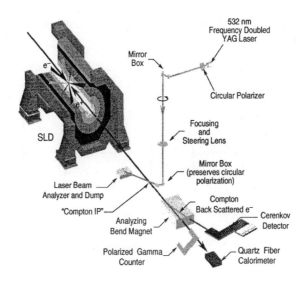

Figure 3: SLD and the Compton Polarimeter.

have been in the range $73 - 78\%$.

6   Measurements of $\sin^2(\theta_W^{eff})$, and testing oblique corrections

For the A_{LR} analysis, all Z decay modes can be used, though in practice the leptonic modes are excluded. They are analyzed separately in the measurements of A_{LR}^{FB} described below. The A_{LR} event selection requires at least 4 charged tracks originating from the IP and greater than 22 GeV energy deposition in the calorimeter. Energy flow in the event is required to be balanced by requiring the normalized energy vector sum be less than 0.6. These criteria have an efficiency of 92% for hadronic events, with a residual background of 0.1%.

SLD's 1998 running yielded 225K hadronic Z decays, with $N_L = 124,404$ produced from the left-polarized beam and $N_R = 100,558$ produced from the right-polarized beam. For the measured beam polarization of 73.1%, this yielded $A_{LR}^{meas} = 0.1450 \pm 0.0030(stat)$. Correcting for initial state radiation and $Z-\gamma$ interference effects, gives $A_{LR}^0 = 0.1487 \pm 0.0031(stat) \pm 0.0017(syst)$. The systematic error includes a contribution of 0.0015 from uncertainties in the polarization scale and 0.0007 from uncertainties in the energy scale. This result determines the weak mixing angle to be $\sin^2(\theta_W^{eff}) = 0.23130 \pm 0.00039 \pm$

0.00022. Combining all of SLD's A_{LR} results from 1992-98, gives $\sin^2(\theta_W^{eff}) = 0.23101 \pm 0.00031$.

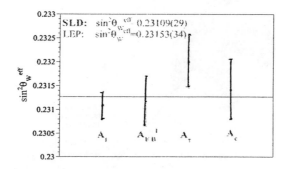

Figure 4: Weak Mixing angle measurements.

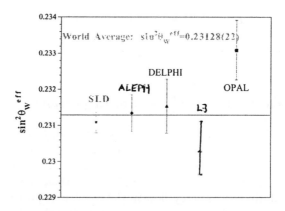

Figure 5: Weak Mixing angle measurements.

For the A_{LR}^{FB} analysis, leptonic Z decay events are selected as follows. The number of charged tracks must be between 2 and 8. One hemisphere must have a charge of -1 and the other hemisphere a charge of +1. The polar angle is required to have $\cos(\theta) < 0.8$. For ee final state events, the only additional requirement is a deposition of greater than 45 GeV in the calorimeter. The $\mu\mu$ final state events must reconstruct with a large invariant mass and have less than 10 GeV per track deposited in the calorimeter. The $\tau\tau$ final state events must reconstruct with an invariant mass less than 70 GeV, and deposit less than 27.5 GeV per track in the calorimeter. One stiff track is required (> 3 GeV), the acollinearity angle must be greater than $160°$ and the invariant mass in each hemisphere must be less than 1.8 GeV. Event selection efficiencies are 87.3% for ee, 85.5% for $\mu\mu$, and 78.1% for $\tau\tau$. Backgrounds are estimated to be 1.2% for ee (predominantly $\tau\tau$), 0.2% for $\mu\mu$ (predominantly $\tau\tau$), and 5.2% for $\tau\tau$ (predominantly $\mu\mu$ and 2γ).

We use Eq. 1 in a maximum likelihood analysis (which also allows for photon exchange and for $Z - \gamma$ interference) to determine A_e, A_μ and A_τ. The results are $A_e = 0.1504 \pm 0.0072$, $A_\mu = 0.120 \pm 0.019$, and $A_\tau = 0.142 \pm 0.019$. These results are consistent with universality and can be combined, giving $A_{e,\mu,\tau} = 0.1459 \pm 0.0063$. This determines the weak mixing angle to be $\sin^2(\theta_W^{eff}) = 0.2317 \pm 0.0008$.

Combining the A_{LR} measurements that use hadronic final states and the A_{LR}^{FB} measurements that use leptonic final states, we determine the weak mixing angle to be $\sin^2(\theta_W^{eff}) = 0.23110 \pm 0.00029$. This is a preliminary result.

A comparison of SLD's result with leptonic asymmetry measurements at LEP [8] is given in Fig. 4. These results are compared by technique, where A_l is SLD's combined result from A_{LR} and $A_{LR}^{FB}(leptons)$ described above; A_{FB}^l is the LEP result using the *forward-backward* asymmetry with leptonic final states; A_τ and A_e are the LEP results from analyzing the τ polarization for the $\tau\tau$ final state. We do not include in this comparison the LEP results using hadronic final states. These results are discussed below, when we examine SLD's A_b measurement and tests of b vertex corrections. The SLD and LEP data in Fig. 4 are replotted in Fig. 5 by experiment rather than by technique. The data are consistent and can be combined to give a world average $\sin^2(\theta_W^{eff}) = 0.23128 \pm 0.00022$.

A convenient framework for analyzing the consistency of the $\sin^2(\theta_W^{eff})$ measurement with the SM and with other electroweak measurements is given by the Peskin-Takeuchi parametrization [9] for probing extensions to the SM. This parametrization assumes that vacuum polarization effects dominate and expresses new physics in terms of the parameters S and T, which are defined in terms of the self-energies of the gauge bosons. In S-T space, a measurement of an electroweak observable corresponds to a band with a given slope. Fig. 6 shows the S-T plot for measurements of the weak mixing angle ($\sin^2(\theta_W^{eff})$), the Z width (Γ_Z), [8] and the W mass (M_W), [8]. The experimental bands shown correspond to one sigma contours. The elliptical contours are the error ellipses (68% confidence and 95% confidence) for a combined fit to the data. The SM allowed region is the small parallelogram, with arrows indicating the dependence on m_t and m_H. The Higgs mass is allowed to vary from 100 GeV to 1000 GeV and from 165 GeV to 185 GeV. The measurements are in reasonable

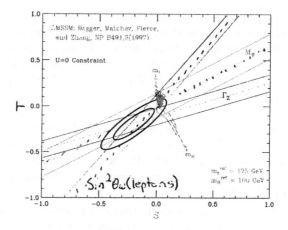

Figure 6: Testing oblique corrections.

agreement with the SM and favour a light Higgs mass. A comparison is also given to a prediction for the parameter space of the Minimal Supersymmetric Model [10] (region of dots in Fig. 6). The combined SLD and LEP measurement for the weak mixing angle gives the narrowest band in S-T space, and provides the best test of the SM for oblique corrections. Improved measurements of M_W from LEP and FNAL are eagerly awaited to further constrain and test the SM in this regard.

7 Measurements of A_b, and testing vertex corrections

The measurement technique for determining A_b is similar to that for determining A_e, A_μ and A_τ. For this analysis, the capabilities of SLD's vertex detector is critical and good use is also made of SLD's particle identification system to identify kaons with the Cherenkov Ring Imaging Detector (CRID). Three different analyses are employed with different techniques for determining the b-quark charge. The **Jet Charge** analysis uses a momentum-weighted jet charge to identify the b-quark charge, and it requires a secondary vertex mass greater than 2.0 GeV. The **Kaon Tag** analysis uses the kaon sign in the cascade decay ($b \to c \to s$) to identify the b-quark charge, and it requires a secondary vertex mass greater than 1.8 GeV. The **Lepton Tag** analysis uses the lepton charge in semileptonic decays to identify the b-quark charge; it has no secondary vertex mass requirement.

The three analyses yield the following results: A_b (Jet Charge) $= 0.882 \pm$

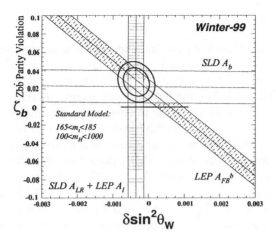

Figure 7: Testing vertex corrections.

0.020 ± 0.029; A_b (Kaon Tag) $= 0.855 \pm 0.088 \pm 0.102$; and A_b (Lepton Tag) $= 0.924 \pm 0.032 \pm 0.026$. These results can be combined, giving $A_b = 0.898 \pm 0.029$. This is a preliminary result.

Similar to the S-T analysis for testing oblique corrections, one can utilize an extended parameter space for testing vertex corrections. This is done in Fig. 7, where we plot the deviation in A_b from the SM prediction versus the deviation in $\sin^2(\theta_W^{eff})$ from the SM prediction. The three bands plotted are SLD's A_b measurement, the combined leptonic $\sin^2(\theta_W^{eff})$ measurement from SLD and LEP, and LEP's *forward-backward b* asymmetry. The elliptical contours are the error ellipses (68% confidence and 95% confidence) for a combined fit to the data. The horizontal line is the SM prediction. We note that the data are in excellent agreement, but differ from the SM prediction by 2.6σ. Unfortunately, there will be no new data to indicate whether this deviation results from a statistical fluctuation, a problem in the b physics analysis, or new physics. We also note the discrepancy of where the SLD-LEP $\sin^2(\theta_W^{eff})$ and the LEP A_{FB}^b measurement bands intersect the SM line. This reflects their 2.2σ discrepancy in determining $\sin^2(\theta_W^{eff})$ within the SM framework.

8 Conclusions

The SLD experiment has been the first experiment at an electron linear collider. The viability of a linear collider has been demonstrated and this technology is now being proposed for future e^+e^- colliders with center-of-mass energies up to 1 TeV. The SLD has made many important contributions to precision electroweak physics. SLD has made the best measurement of the weak mixing angle, $\sin^2(\theta_W^{eff}) = 0.23110 \pm 0.00029$ *(preliminary)*. This provides a stringent test of oblique corrections; our measurement is consistent with SM predictions and favours a light Higgs mass. SLD makes the only direct measurement of A_b, which we determine to be $A_b = 0.898 \pm 0.029$ *(preliminary)*. This measurement, together with measurements by SLD and LEP of $\sin^2(\theta_W^{eff})$ and LEP's measurement of A_{FB}^b, can be used to test b vertex corrections. The data are consistent, but indicate a 2.6σ discrepancy with the SM prediction.

Acknowledgments

This work is supported in part by Department of Energy Contract DE-AC03-76SF00515 (SLAC).

References

1. P. Raimondi *et al.*, SLAC-PUB-7955 (1998).
2. K. Abe *et al.*, Nucl. Instrum. Methods **A400**, 287, (1997).
3. M. Woods in *SPIN96 Proceedings*, ed. C.W. de Jager *et al.* (World Scientific, Singapore, 1997), p.843.
4. M. Woods in *SPIN96 Proceedings*, ed. C.W. de Jager *et al.* (World Scientific, Singapore, 1997), p.623.
5. M. Fero *et al.*, SLD-Physics-Note-50 (1996).
6. R.C. Field *et al.*, IEEE Trans. Nucl. Sci. **45**, 670, (1998).
7. S.C. Berridge *et al.*, in *Calorimetry in High Energy Physics Proceedings*, ed. E. Cheu *et al.* (World Scientific, Singapore, 1998), p. 170.
8. D. Abbaneo *et al.*, CERN-EP-99-015 (1999).
9. M.E. Peskin and T. Takeuchi, Phys. Rev. **D46**, 381, (1992).
10. J. Bagger *et al.*, Nucl. Phys. **491**, 3, (1997).
11. T. Takeuchi *et al.*, published in *DPF 94 Proceedings*, ed. S. Seidel (World Scientific, Singapore, 1995), p. 1231.

Lake Louise Winter Institute - 1999

LIST OF PARTICIPANTS

F. Al-Shamali	University of Alberta
A. Anastassov	The Ohio State University
D. Asner	University of California, Santa Barbara
G. Azuelos	Universite de Montreal
S. Bhadra	York University
S. Bright	The University of Chicago
A. Buras	Technische Universität München
J. Burke	Lawrence Berkeley Laboratory
B. Campbell	University of Alberta
R. Carnegie	Carelton University
J. De Jong	University of Alberta
L. Di Lella	CERN
J. Dilling	Darmstadt, Germany
M. Dobbs	University of Victoria
L. Duong	University of Minnesota
A. Duperrin	Université Claude Bernard de Lyon, France
Y. Efremenko	Oak Ridge National Laboratory
K. Eitel	Los Alamos National Laboratory
E. El Aaoud	University of Alberta
M. Falagan	CIEMAT, Madrid, Spain
A. Foland	Cornell University
S. Freedman	University of California at Berkeley
D. Futyan	The University of Manchester
G. Greeniaus	University of Alberta
E. Gross	Weizmann Institute of Science, Israel
J. Hardy	Carleton University, Ottawa
R. Hossain	University of Alberta
P. Jackson	TRIUMF
R. Jacobsson	CERN
K. Kaminsky	University of Alberta

E. Kneringer	Universität Innsbruck, Austria
M. Lefebvre	University of Victoria
G. Lotz	Augustana University College, Camrose
R. MacKenzie	Université de Montréal
S. Magill	Argonne National Laboratory
J. Martin	University of Toronto
J. McDonald	University of Alberta
E. Merle	LAPP, France
R. Migneron	University of Western Ontario
P. Molnar	Humboldt- Universität zu Berlin
K. Nagai	CERN
M. Nikolenko	Joint Institute for Nuclear Research, Dubna
T. Noble	SNO, Lively, Ontario
D. O'Neil	University of Victoria
M. Paranjape	Universite de Montreal
R. Peccei	UCLA, Los Angeles
M. Picariello	Universita degli studi di Pisa
J. Pinfold	University of Alberta
S. Polenz	DESY
H. Przysienzniak	CERN
S. Rajpoot	California State University
P. Righini	Università di Cagliari/INFN
S. Robertson	University of Victoria
N. Rodning	University of Alberta
S. Schaller	Max-Planck-Institut fur Physik
S. Schmidt-Kaerst	III. Physikalisches Institut, Aachrn, Germany
D. Shaw	University of Alberta
F. Sobratree	University of Alberta
A. Soffer	Cornell University
A. Specka	Ecole Polytechnique, France
R. Tayloe	Los Alamos National Laboratory
E. Tounefier	CERN
C. Vallee	Centre de Physique des Particles de Marseille
M. Vincter	University of Alberta
M. Woods	Stanford Linear Accelerator Center